Greek Epic, Lyric, and Tragedy

Greek Epic, Lyric, and Tragedy

THE ACADEMIC PAPERS OF SIR HUGH LLOYD-JONES

CLARENDON PRESS · OXFORD
1990

Oxford University Press, Walton Street, Oxford OX2 6DP
Oxford New York Toronto
Delhi Bombay Calcutta Madras Karachi
Petaling Jaya Singapore Hong Kong Tokyo
Nairobi Dar es Salaam Cape Town
Melbourne Auckland
and associated companies in
Berlin Ibadan

Oxford is a trade mark of Oxford University Press

Published in the United States
by Oxford University Press, New York

British Library Cataloguing in Publication Data
Greek epic, lyric, and tragedy: the academic papers of
Sir Hugh Lloyd-Jones
1. Latin literature history and criticism
I. Title
880.9001
ISBN 0–19–814680–9

Library of Congress Cataloging in Publication Data
Greek epic, lyric, and tragedy : the academic papers of
Sir Hugh Lloyd-Jones.
Includes bibliographical references.
1. Greek literature—History and criticism. 2. Greek drama
(Tragedy)—History and criticism. I. Title.
PA3133.L5 1990 880.9'001—dc20 89–77424
ISBN 0–19–814680–9

Set by Latimer Trend & Company Ltd, Plymouth
Printed in Great Britain by
Biddles Ltd, Guildford & King's Lynn

DIS MANIBVS
MATRIS OPTIMAE

Preface

PEOPLE whose shorter writings are reprinted often make trouble by wanting to rewrite them. They wish to correct mistakes, to correct references to texts so that they refer to the most up-to-date editions, and to take account of work done upon the problems they have discussed since the appearance of their own discussion. Readers, on the other hand, would often prefer to have the articles or reviews just as they originally were.

I have compromised by restricting myself to correcting errors, to updating references to texts, and to adding references to relevant literature that has appeared since my own works were published; I do not necessarily agree with what that literature contains. Although some matter I now think erroneous has been excised, I have in general chosen not to print extracts from articles or reviews, which means that some things that might by themselves have seemed worth reprinting have been left out. But anyone who wants to look up anything that has been omitted will find his task made easier by the Bibliography. (It is to this that parenthetical numbers in the chapter-headings refer.)

It remains to thank heartily four scholars who have helped me. Professor Rudolf Kassel has supplied many valuable corrections. Dr Leofranc Holford-Strevens, whom I had the great good fortune to have as copy-editor, has done far more to improve the book than could possibly have been expected of him; I have persuaded him to add some notes followed by his own initials. Professor Mary Lefkowitz by compiling the Bibliography and the Table of Contents has made the production of the book possible. Professor F. J. Williams and Mr Nigel Wilson have generously helped with the correction of the proofs.

HUGH LLOYD-JONES

Christ Church, Oxford
29 April 1989

Acknowledgements

Oxford University Press is grateful for permission to reprint papers from the following (numbers in brackets refer to chapters):

History and Imagination: Essays Presented to H. R. Trevor-Roper, ed. H. Lloyd-Jones, V. Pearl, and B. Worden (London: Duckworth, 1981) (1).
Atti del XVII Congresso internazionale di papirologia (Naples, 1984) (2).
Classical Review (Oxford University Press) (3, 15, 16, 22, 29, 32, 34, 36, 38, 39).
Greek, Roman and Byzantine Studies (Duke University, North Carolina) (4).
Zeitschrift für Papyrologie und Epigraphik (Bonn: Dr Rudolf Habelt GmbH) (5, 28).
Classical Philology (University of Chicago Press) (6).
The British Academy (London) (7).
Entretiens Hardt (Geneva: Fondation Hardt) (8).
Journal of Hellenic Studies (London: Society for the Promotion of Hellenic Studies) (9, 18, 20, 35).
Harvard Studies in Classical Philology (Harvard College) (10, 24).
Maia (Bologna: Nuova Casa Editrice Licinio Cappelli SpA) (11).
Gnomon (Munich: C. H. Beck'sche Verlagsbuchhandlung) (12, 14).
Review of English Studies (Oxford University Press) (13).
Fundación Pastor de Estudios Clásicos (Madrid) (17).
L'antiquité classique (Louvain) (19).
Classical Quarterly (Oxford University Press) (21, 26, 30, 31).
Rheinisches Museum für Philologie (Frankfurt: J. D. Sauerlander's Verlag) (23).
Dionysiaca: Festschrift for Denys Page (Cambridge University Press) (25).
American Society of Papyrologists (Wayne State University, Detroit, Michigan) (27).
Yale Classical Studies, collected in *Studies in Fifth Century Thought and Literature*, ed. A. M. Parry (Cambridge University Press, 1972) (33).
Würzburger Jahrbücher für die Altertumswissenschaft (37).

Contents

I. EPIC

II. LYRIC

III. TRAGEDY

I. EPIC

1(75)

Remarks on the Homeric Question

ABOUT the turn of the seventeenth century the great debate between the champions of the ancients and the champions of the moderns that is known as the Battle of the Books marks the moment when the French decided that they could now manage without the ancient writers, and turned their backs upon the study of antiquity. The Abbé d'Aubignac[1] undertook the task of exploding the reputation of Homer; Homer, he claimed, was an inferior poet, whose theology was deplorable, and whose poems were in any case nothing but a collection of old rhapsodies patched together and handed down under his name. Having failed to be elected to the Academy, d'Aubignac had to found one of his own; his dissertation on the *Iliad* was not published until 1715, nearly forty years after his death. Still, it expressed an attitude by no means uncommon at that place and time, and the theory of a collection of rhapsodies pointed in a direction soon to be followed by a person eminently fitted to discuss Greek poetry.

In the reply to the deist Anthony Collins which he published in 1713, Richard Bentley[2] took exception to the trite and empty assertion that 'Homer designed his poems for eternity, to please and instruct mankind.' 'Take my word for it,' he wrote, 'poor Homer, in those circumstances and early times, had never such aspiring thoughts. He wrote a sequel of songs and rhapsodies, to be sung by himself for small earnings and good cheer, at festivals and other days of merriment; the *Ilias* he made for the men, and the *Odysseïs* for the other sex. These loose songs were not connected together in the form of an epic poem till Pisistratus' time, above 500 years after.' Unlike d'Aubignac, Bentley did not venture to dispute the very existence of Homer; but to his rigorous eighteenth-century logic the poems seemed deficient in unity and organization. Even in the great epic of Milton, composed during the preceding century, Bentley found an absence of these qualities; his

[1] François Hédelin, abbé d'Aubignac et de Meimec, *Conjectures Académiques, ou Dissertation sur l'Iliade*, ed. V. Magnien (1925). For 18th-c. views of Homer, see most recently K. Simonsuuri, *Homer's Original Genius* (1979).

[2] *Remarks upon a Late Discourse of Freethinking*, in *Works*, ed. A. Dyce, iii (1838, repr. 1966), 304.

H. Lloyd-Jones, V. Pearl, and B. Worden (eds.), *History and Imagination: Essays in Honour of H.R. Trevor-Roper* (London, 1981), 15–29.

new edition || was based on the hypothesis that the blind poet had dictated the poem to an amanuensis who had taken advantage of its author's blindness to abuse his trust. Bentley dated Homer as early as the eleventh century BC; only some five hundred years later, he thought, following an assertion of Cicero,[3] did the Athenian tyrant Pisistratus arrange the poems as we now have them.

Giambattista Vico in his essay on universal law of 1722[4] argued, much as Bentley had done, that the Homeric poems were not a conscious effort of profound philosophy, but the mirror of a simple age. In the second *Scienza nuova*, which appeared in 1730, he followed d'Aubignac in taking 'Homer' to be simply a collective name and Bentley in arguing that the Homeric poems came together only when collected by Pisistratus; but his judgement of the poems was more sympathetic, and in claiming that they were a collective work of the Greek people he anticipated a notion which later in the century was to be made highly fashionable by the work of Herder. Thomas Blackwell in his *Enquiry into the Life and Writings of Homer* (1735) represented Homer as a primitive bard, whose gifts uniquely fitted the society into which he had been born. Among his pupils at Marischal College, Aberdeen, was James Macpherson; the belief that early epic poems emanate from the people certainly contributed to the vast popularity of Macpherson's *Ossian*.

Not long after the appearance of that fraudulent best-seller, Robert Wood privately circulated the first edition of his *Essay on the Original Genius of Homer* (1767).[5] While accepting the existence of Homer, whom he placed at the same early date as Bentley, Wood explained the phenomena which had led Bentley to suppose that the epics consisted of various rhapsodies which were put together by Pisistratus by means of a different theory. He suggested that they had been composed orally, and were written down only centuries afterwards by Pisistratus or Solon or the Spartan legislator Lycurgus. The guess that Homer was an oral poet is as old as Josephus, who threw it out in the course of an attempt to prove the superiority of Hebrew literature over Greek; but Wood was the first to make a serious case on its behalf.

[3] Cicero, *De oratore*, 3. 137: 'Quis doctior eisdem temporibus illis aut cuius eloquentia litteris instructior fuisse traditur quam Pisistrati? qui primus Homeri libros confusos antea sic disposuisse dicitur ut nunc habemus?' For a view not unlike Bentley's, see R. Merkelbach, *RhM* 90 (1952), 23 ff., and D. L. Page, *The Homeric Odyssey* (1955), 143 ff.; for a sceptical reaction, see J. A. Davison, *TAPA* 86 (1955), 1 ff.; (or in A. J. B. Wace and F. H. Stubbings (eds.), *A Companion to Homer* (1962), 219 ff.); cf. A. Lesky, *Homeros* (Sonderausgabe der Paulyschen Real-Encyclopädie der classischen Altertumswissenschaft, 1967), 146.

[4] *Il diritto universale*, in *Opere*, ed. F. Niccolini (1931–40); see J. L. Myres, *Homer and his Critics* (1958), 57.

[5] The work was not published until 1775, four years after its author's death. Josephus: see *Contra Apionem*, 1. 12.

During the eighties of the eighteenth century Herder made the educated public familiar with the distinctive characteristics of ancient folk poetry, as it was represented by Homer, early Hebrew poetry, and German folksong. In 1788 the French scholar Villoison published for the first time the detailed commentary on the *Iliad* contained in the tenth-century manuscript of Homer known as Venetus A, so throwing much light upon the treatment of || the text and its problems by the scholars of the Hellenistic age, and upon the difficulty of tracing its history back before that time.[6] This publication led directly to the working out of a systematic theory of oral composition and transmission by one of the best-equipped scholars of the time; Friedrich August Wolf published his *Prolegomena ad Homerum* in 1795.

Wolf's theory had obvious weaknesses. He placed the introduction of the Greek alphabet far too late; he was unaware, though he could have learned it from Wood, that a long poem can be transmitted orally; in taking up Josephus' suggestion that Homer was an oral poet he failed to bear in mind that no such idea had occurred to any of the Hellenistic scholars whose opinions were recorded in the scholia. But his theory seemed to many to give the best explanation of the facts, and it was congenial to the spirit of the time; Goethe, then working at *Hermann und Dorothea*, accepted it with enthusiasm, though thirty years later he was to change his view. Wolf had posed a problem that was endlessly debated, with great learning and assiduity, throughout the great age of German scholarship that followed.

Gottfried Hermann[7] in 1832 argued for an original nucleus by a single poet, which had later undergone successive additions and alterations. For the remainder of the nineteenth century this continued to be the most popular kind of view; it was made familiar to the English public by Grote in the second volume of his *History of Greece*, published in 1846. Wilamowitz in his book *Die Ilias und Homer*, published in 1916, placed Homer not at the beginning but in the middle of the Homeric tradition; Homer, he thought, had made use of various early poems, and later poets in their turn had added to or changed his work. This became the most popular type of view, especially in Germany, and continued to be so until recent times. But in England and America certain scholars argued for a unitarian view, sometimes under the influence of anti-German prejudice, but not

[6] For the impact of the scholia upon Wolf, see H. Erbse, 'F. A. Wolf e gli scoli all'Iliade', *ASNP*, 3rd ser., 9/1 (1979), 39 ff. Wolf's *Prolegomena ad Homerum, sive de operum Homericorum prisca et genuina forma variisque mutationibus et probabili ratione emendandi* appeared in 1795 (3rd edn. by R. Peppmüller, Halle, 1884 [English translation by A. Grafton, G. W. Most, and J. E. G. Zetzel, Princeton, 1985]).

[7] *Opuscula*, v (1834), 52 ff.

without putting forward arguments that deserved serious consideration.

This was the situation in 1928, when Milman Parry published his two major contributions to Homeric scholarship.[8] Using and carrying further work on Homer's language and dialect by other scholars which indicated that they were artificial constructions called into being by the requirements of the epic form itself, he was able to give substance to the notion of oral composition by showing in detail that many features of the style, language, and dialect of the poems can only be explained by the assumption that they were dictated by its needs. Parry went on to strengthen the proof by examining other traditions of oral poetry, thus creating a whole new science of the comparative study of this kind of composition. But for many years German and other Continental scholars clung to their established methods of analysis,‖ taking no notice of this American innovation.

Parry certainly proved that the tradition to which the Homeric poems belonged must have been for many centuries an oral one. But did he prove, as many scholars have assumed, that they themselves were composed without the aid of writing?[9]

During the last years before his death at the age of thirty-three, Parry was occupied in studying and recording the last phases of the Serbo-Croatian poetic tradition, which was the last in Europe to retain its oral character. After his death this work was continued by his assistant A. B. Lord, now Professor of Slavic and Comparative Literature at Harvard. Lord in his book *The Singer of Tales* (1960) and in many other writings has contended that the Homeric epics must have been composed orally. He has argued that the techniques of written and of oral composition exclude each other; he has mantained, at least until recently, that the coming of literacy brings immediate death to any oral tradition. Like Parry and most other modern scholars, Lord puts Homer at the end of the tradition, in that eighth century to which our earliest specimens of Greek alphabetic writing belong. He thinks Homer must have dictated his poems to someone else who wrote them down.

The late Sir Denys Page was an enthusiastic advocate of the belief that Homer's epics were composed orally. He placed their authors, whom he held to be two different people, in the late ninth or early

[8] *L'Épithète traditionelle dans Homère* and *Les Formules et la métrique d'Homère* were both published at Paris in 1928; for English versions, see *The Making of Homeric Verse*, ed. Adam Parry (Oxford, 1971).

[9] This has been alleged by many scholars, including M. P. Nilsson, *Homer and Mycenae* (1933), 179, and E. R. Dodds, 'Homer as Oral Poetry', in M. Platnauer (ed.), *Fifty Years of Classical Scholarship* (1950), (2nd edn. as *Fifty Years (and Twelve) of Classical Scholarship* (1962), 13 ff.). See *The Making of Homeric Verse*, pp. li ff.

eighth century. Homer, he thought, adapting the already well-established nucleus theory, 'developed, expanded, transmuted' earlier poems; the structure of his epics was 'loosely knit' and 'the connexions between the chief structural features are usually rather loose and sometimes very weak'.[10] Page accounts for these features, which had been the subject of complaint as early as the time of d'Aubignac, by arguing that the poems were not written down before the sixth century. That is not very different from the view of Wolf. G. S. Kirk in *The Songs of Homer* (1962) took a similar view. At that time he accepted from Lord the proposition, which both have now abandoned, that literacy is at once fatal to any oral tradition; but since unlike Lord he held that a poem could be transmitted orally over a long period without substantial alteration, he believed that two generations had elapsed between the composition of the poems and the time when they were written down. It was during this period, he argued, that they had added to them certain parts commonly held to be later than the time of Homer, such as the lay of Dolon in the tenth book of the *Iliad*, the visit of Odysseus to the underworld in the eleventh book of the *Odyssey*, and the conclusion of that poem.

Although these writers have had great influence, especially in England and America, their view that the epics were composed without the aid of writing is by no means universally accepted, and is now more than ever under attack. In *Heroic Poetry* (1952) Sir Maurice Bowra offered the results of a study not ‖ merely of Serbo-Croatian epic but of material supplied by many different epic traditions; he argued that Homer had indeed many centuries of oral tradition behind him, but that he himself composed his epics with the aid of writing. In the same year H. T. Wade-Gery brought out a small, learned, and ingenious book called *The Poet of the Iliad*. He suggested that the alphabet which the Greeks adapted from that of the Phoenicians was invented for the express purpose of recording the great epics, which he believed were intended from the first for continuous recitation by relays of performers. Such continuous performance is known to have taken place at the Panathenaic festival in Athens, founded during the sixth century, the occasion for which Peisistratus is alleged to have arranged the works of Homer. Wade-Gery conjectured that the poet might have composed his poems for continuous recitation at the common festival of the Ionian Greeks known as the Panionion, which took place at Mount Mycale in Asia Minor, just opposite the island of Samos.

Bowra's book seems to have led directly to the writing of an

[10] See *History and the Homeric Iliad* (1959), 259 ff., and *The Homeric Odyssey* (1955), 142.

important article, in which for the first time a leading German-speaking scholar took account of the importance of Parry's work; this was Albin Lesky in his 'Mündlichkeit und Schriftlichkeit im homerischen Epos' (1954).[11] Like Bowra, Lesky held that Homer had composed his epics with the aid of writing; and he developed this view in the successive editions of his history of Greek literature (from 1957) and in his article *Homeros* in Pauly–Wissowa's *Real-Encyclopädie* (1967). In 1966[12] Adam Parry, the son of Milman Parry, strongly argued that Homer had used writing; his article drew a reply from Kirk in 1970.[13] In 1971[14] Adam Parry brought out a collection of his father's writings, including English versions of the two important works that had originally appeared in French. He prefaced to the book what seems to me still the best general account of the current state of the Homeric Question. Like Bowra and Lesky, Adam Parry argued that though the Homeric epics bore many signs of belonging to a poetic tradition that had for many centuries been oral they could not have been composed without the aid of writing.[14a]

Kirk in replying to Adam Parry[15] abandons the proposition, which he had formerly taken over from Lord, that literacy must at once be fatal to any oral poetic tradition; even Lord now seems to have changed his mind about this matter. Since then such material that makes against this opinion has been accumulated; much of it is conveniently collected in Ruth Finnegan's *Oral Poetry* (1977). First Miss Finnegan points out that several early English poems which contain repeated formulas and other features associated with an oral technique are known to have been composed with the aid of writing; some have even turned out to have been translated from the Latin. Next she mentions living Bantu poets who write their poems in a formulaic style ‖ elaborated from their originally oral poetic tradition.[16] 'In practice,' writes Miss Finnegan, 'interaction between oral and written forms is extremely common, and the idea that the use of writing automatically deals a death-blow to oral literary forms has

[11] *Festschrift für Dietrich Kralik* (1954), 1 ff. (or in *Gesammelte Schriften* (1966), 63 ff.).

[12] 'Have we Homer's *Iliad*?' *YCS* 20 (1966), 175 ff. [= *The Language of Achilles* (1989), 104 f.]

[13] 'Homer's *Iliad* and Ours', *PCPS* 196 (1970), 48 ff.

[14] See n. 8, above.

[14a] [See A. Heubeck in *Omero, Odissea*, i: *Libri I–IV*, ed. A. Heubeck and S. West (1981), p. xxi = A. Heubeck, S. West, and J. B. Hainsworth, *A Commentary on Homer's* Odyssey, i: *Introduction and Books I–VIII* (1988), 12].

[15] See n. 13, above.

[16] Miss Finnegan cites L. D. Benson, 'The Literary Character of Anglo-Saxon Formulaic Poetry', *Publications of the Modern Language Association* (1966), 81, and J. Opland, 'Praise Poems as Historical Sources', in C. Saunders and R. Derricourt (eds.), *Beyond the Cape Frontier: Studies in the History of the Transkei and the Ciskei* (1974). The third, fourth, and fifth chapters of her book deserve the close attention of Homeric scholars.

nothing to support it.' Miss Finnegan can have no axe to grind, for her book shows no sign of acquaintance with the most recent developments in Homeric scholarship; she never mentions Adam Parry. That makes it all the more significant that her study of a large body of material derived from various epic traditions makes strongly against a dogma that was once one of the main props of belief in the oral composition of the Homeric epics.[17]

Adam Parry affirmed against Kirk Lord's other central proposition, that a poem must continue to suffer 'fundamental change' so long as it continues to be transmitted orally. Kirk in his reply to him claims that if as he thinks the poems were not written down until two generations after their author's death it is easy to understand how those parts of them commonly thought to have been late additions came to be added to the poem. But one can hardly be certain that the existence of written copies must have protected the poems against change and interpolation; written copies would soon have multiplied, and the text would have been subject to alteration.

But if scholars cling to the opinion that the epics were composed orally they do so above all because they hold that, as Kirk puts it, 'a large number of crystallised formulas are employed with an astonishing economy and lack of unnecessary variation' (*Homer and the Oral Tradition*, 115–16). He holds that this 'suggests strongly, and indeed imperatively, that the oral technique was used in full and undiminished degree for the final main act of composition of each of the two monumental poems'. But each of the scholars who have examined the 'formula' in all possible senses of that elusive term[18] has ended by stressing, more strongly than his predecessors, its extreme flexibility and adaptability. The extensive and economical system of formulas revealed by Milman Parry has proved, after close testing, a good deal less tight than people had supposed.‖

Lord's interdiction of what he has called 'the subjective interpretation and appreciation of the Homeric poems' would seem to deny to

[17] See also Miss Finnegan's paper in *Oral Literature and the Formula*, ed. B. A. Stolz and R. S. Shannon (1977), 127 ff. In replying to it, Lord (op. cit. 175) claims that he did not say 'in exactly those terms' that once a singer learned to write he lost the ability to compose orally, while acknowledging that this is how his meaning has been interpreted. He makes it clear that he does not now maintain this view.

[18] See e.g. A. Hoekstra, *Homeric Modifications of Formulaic Prototypes* (1965); J. B. Hainsworth, *The Flexibility of the Homeric Formula* (1968); M. Nagler, *Spontaneity and Tradition* (1975). J. A. Russo in his valuable contribution to Stolz and Shannon (eds.), *Oral Literature and the Formula*, 31 ff., argues convincingly that a formulaic style is not necessarily oral. In reply Lord (p. 67) can only reiterate his conviction that without 'really knowing' an 'oral model' like that furnished by the Serbo-Croatian poems to which he has devoted so much labour 'one is in a rather shaky position for saying that Homer is or is not' an oral poet. It also helps to know some Greek, and to know something about poetry.

Homer any power to adapt his style and language to the requirements of an individual context. Not surprisingly it has incurred severe criticism, notably from the late Anne Amory Parry.[19] Lord is not a professional Hellenist; but the paradoxical consequences for literary appreciation that result from belief in such theories as his may be amply illustrated from the work of Greek scholars of great eminence. 'Subtlety of soul, complexity of character, true portrayal of personality,' wrote Page (*The Homeric Odyssey*, 142–3), 'for these we must wait until the practice of the art of writing affords the poet the necessary leisure and the necessary means for reflexion, for planning the future in some detail, and for correcting the past.' 'Intricacy of design and subtleties of soul', he added later, 'wholly alien to the oral technique of composition have been sought (and found) in him.'

Bentley, whom Sir Denys Page in many ways resembled, was led by the irregularities of the Homeric epics to embrace the theory of a Homer whose loose songs were not put together to form continuous wholes before the age of Peisistratus. Page's words take us back to the perennial question: do the many occurrences in Homer of what by the standards of much later poetry seem to be inconsistencies, inconcinnities, and irrelevances constitute a proof of multiple authorship, or at least of large-scale revision and interpolation? The old controversy between unitarians and analysts returns in a new form.

English-speaking followers of Parry justifiably complain of his neglect by continental scholars; but they themselves have unjustifiably neglected some important continental writings about Homer. Wolfgang Schadewaldt in his *Iliasstudien* of 1938 drew attention to the presence of elaborate links between different sections of the *Iliad*—the working-out of pervasive motives, and the presence of an unmistakable unity despite all diversity. Karl Reinhardt[20] in his posthumous book *Die Ilias und ihr Dichter* carried this kind of study further. Hartmut Erbse in his *Beiträge zum Verständnis der Odyssee* (1972) has effectively criticized arguments supposed to prove multiple authorship which had long remained unquestioned; he has even defended the authenticity of the conclusion of the *Odyssey* in a fashion which commands respect. I cannot here deal adequately with the complex problems which the mention of this topic brings to mind; but I will offer a few examples of cases that seem to me to reveal weaknesses in the arguments of those who believe in multiple authorship.

Let me take first Page's treatment of the episode from the nineteenth book of the *Iliad*. 'The start is unpromising,' he writes (*History and the*

[19] 'Homer as Artist', *CQ*, NS 21 (1971), 1 ff.

[20] For an account of Karl Reinhardt, see my introduction to the translation of his *Sophocles* by Hazel and David Harvey, reprinted in my *Blood for the Ghosts* (1982), ch. 20.

Homeric Iliad, 313). 'Achilles has lost his friend; he is distraught with grief and rage; Hephaestus has made him a suit of Olympian armour; his mother Thetis brings it to him. Now surely he will rush into battle intent upon the killing of Hector? And so he might; only his mother unexpectedly and very abruptly tells him that he must first make friends with Agamemnon—though she does ‖ not explain (and we cannot guess) why he must do so at this time or indeed at any time.' The reason is that since Agamemnon is the commander of the army, and Achilles now wants to fight, they have to fight together. Achilles now has ample assurance that Agamemnon will speak and act so as to give him full satisfaction, and so he does. Agamemnon in his speech in the assembly (78 ff.) saves face by blaming Zeus and his portion and the Erinys for his rash behaviour in provoking Achilles; but in Homer human beings always have to take responsibility for their actions, even though a god has suggested them, and Agamemnon never thinks of withholding the vast indemnity that he has already offered to Achilles.

Achilles replies briefly to Agamemnon's speech, saying that he is eager to go into battle. 'Now, at last,' writes Page (p. 314), 'let us hope, the action will begin. But hope is to be long deferred, and the heart to be made moderately sick. Odysseus has a word to say, indeed he has a great many words to say, about food. You cannot expect the army to fight before breakfast; no doubt Achilles is in a hurry, but the men must have their meal.' Meanwhile the indemnity is paid over. 'More than 180 lines have passed', Page complains, 'since luncheon stole the limelight, and nothing has been achieved.' The gifts are taken to Achilles' tent, Patroclus is again lamented, and in the end Zeus sends Athene to drip nectar and ambrosia into Achilles so that his strength shall not fail. Page remarks that the story 'is none the better for being padded as it is with lecture after lecture about food'.

To me the discussion about food seems to be wholly to the point.[21] Achilles in his grief for his friend can ignore the ordinary processes of life; but Odysseus, always a realist, acts wholly in character in politely reminding Achilles that an army fights better on a full stomach. Agamemnon accepts his suggestion, and also agrees to send at once for the indemnity and to swear that he has laid no finger on Briseis. Achilles protests that he cannot think of eating at such a time; but in a second speech (216 ff.) Odysseus tells him that though Achilles is the greater warrior he himself is older and has more experience. 'So,' says he, 'let your heart bear with what I say. Men soon grow weary of battle; in battle the bronze strews many a stalk upon the ground, but the harvest reaped is scanty, when Zeus the master of war inclines his

[21] The same point has now been made by Jasper Griffin, *Homer on Life and Death* (1980), 15 ff.

scales. The Achaeans cannot mourn for the dead man with their bellies. All day men are falling thick and fast; if we did so, when could we have respite from labour? Those who die we must bury, keeping in check our pity, shedding the tears the day demands; and those who survive hateful war must think of food and drink so that we may the better fight unremittingly against our enemies . . .'. It is important to remember that the technique of epic is not the same as that of tragedy; we must learn the technique of epic by patient study of the texts, pausing to ask ourselves whether difficulties may be explained by other means before putting them down to assumed multiple authorship. In a tragedy food would not be likely to find mention at such a moment; but epic presents a broader canvas, on which men of different natures are portrayed, and many actions, arousing different feelings in different persons, are ‖ presented. Achilles may grieve passionately for Patroclus and live only for revenge on Hector. But the army must still eat; ordinary life has to go on. The epic has room for Achilles and also for Odysseus; nowhere is the contrast between their characters more clearly drawn. Provided the reader is willing and able to read the text in accordance with the proper canons of the epic poet's art, the value of this episode as a proof of multiple authorship amounts to nothing.

In the fifteenth book of the *Odyssey* Telemachus, who has set out on a journey to ask Nestor and Menelaus if they have news of his absent father, is about to embark on the ship that will carry him back from Sparta to Ithaca when he is accosted by a stranger (223). This man has killed an enemy in Argos and is on the run, with powerful pursuers on his track. He is most ceremoniously introduced, for thirty lines are devoted to his genealogy; his name is Theoclymenus, and he comes of one of the most distinguished families of prophets in the epic world. 'It is natural to presume', writes Page (*The Homeric Odyssey*, 84), 'that a person whose introduction is so long and loudly trumpeted will do or say something of importance; but the truth turns out to be the reverse. His part is very small, and wonderfully unimportant.' Let us see what part he plays and consider whether we agree.

Theoclymenus accosts Telemachus, imploring him to rescue him from certain death by taking him aboard his ship. His words, Page complains, are 'offensive to the custom of the Greek Epic; they are offensive because ancient custom would frown upon a suppliant who began his prayer by asking the name and address of his protector; it is for Telemachus to ask Theoclymenus, not *vice versa*.' It seems to me doubtful whether a man in peril of his life would worry about such niceties before accosting a potential saviour; and I should expect an epic poet to be cognizant of this.

Theoclymenus is taken on board, and when they reach Ithaca

(502 f.) and are about to land, he asks Telemachus where he can find a lodging. Telemachus replies that ordinarily he would be glad to accommodate him himself, but that as things are he cannot allow him to enter the house without being present himself to ensure that his guest comes to no harm. Since Telemachus plans not to return home before visiting Eumaeus, he will not be able to accompany his guest; and as he points out Penelope seldom enters the hall while the suitors are present, so that she will be unable to protect him. Eurymachus, Telemachus says, is a man greatly respected, and he advises Theoclymenus to go to his house and ask for his protection. Eurymachus is of course one of the most formidable and one of the most offensive of the suitors, being second only to Antinous in those respects.

At that moment a hawk flies by holding a pigeon in its claws and stripping off its feathers. Theoclymenus beckons Telemachus to come aside out of earshot of the rest, takes his hand, and tells him that this is an omen sent by a god; it means that the house of Telemachus and his family is the one royal house of Ithaca, and that its members will always rule. Telemachus thanks him, and instructs Peiraeus, one of his closest friends, to take him to his house and entertain him.‖

'There is no need to dwell on this,' says Page (p. 85); 'one moment Telemachus proposes that his friend shall stay with his arch-enemy Eurymachus; the next moment, without another word about that fantastic proposal, he commits him to the hospitality of his most loyal companion, Peiraeus. What the poet has in mind is beyond our comprehension.' It seems to me obvious that the prediction, though Page dismisses it as 'a drab *non sequitur*', has shown Telemachus that the stranger is indeed a genuine prophet, acquainted by supernatural means with the situation in Ithaca. That is surely enough to explain why he revokes his decision to send him to Eurymachus, and instead sends him to a house where he can count on suitable entertainment.

In the seventeenth book Telemachus returns to the palace of Odysseus and is greeted by Penelope, who is immensely relieved to see him, knowing as she does that the suitors have been plotting against his life. She asks him to tell her what people he has seen (44), but Telemachus defers complying with her request, because he must go to the market-place to pick up Theoclymenus. When he finally comes back and tells his mother what has happened on the journey, he can tell her no more than that Menelaus has been told by Proteus in Egypt that Odysseus is with Calypso. But Theoclymenus insists that he himself has more important information than that, and swears a solemn oath that he is not mistaken (155 f.). Odysseus, he declares, is already in Ithaca; he knows about the conduct of the suitors, and is planning their destruction. All this, Theoclymenus explains, he has learned from the portent he observed during the conversation with Telemachus described earlier. Penelope replies to him with courtesy, wishing that his prophecy may be fulfilled and promising him rich presents if it is. But it is clear that she does not allow herself to place much hope in its fulfilment, and we know that she has previously been approached by many strangers claiming to have news of her husband whose claims have afterwards come to nothing. So when she hears a true prophecy from a prophet of impeccable credentials, she pays no attention to it, although it is vastly more significant than anything Telemachus has brought back from his long and perilous journey. The irony lends the episode an extra point, which Page has not observed.

In the twentieth book the suitors sit feasting in the hall for the last time, just before the trial of the bow with its terrible consequences for all of them. As they eat and drink, their insolence rises; Ctesippus hurls the hoof of a cow at the disguised Odysseus. Telemachus remonstrates with indignation; in response, Agelaus tells his comrades to restrain themselves, but insists that Odysseus must now be presumed dead and that Penelope must put an end to the uncertainty by choosing a new husband. Telemachus replies by saying, as he has said earlier, that he would be glad if his mother did remarry, but that he cannot turn her out; the situation could hardly be explained more clearly.

At this climactic moment, the suitors break into mad laughter; Athene has turned aside their wits. They are laughing 'with other men's jaws'; that means 'involuntarily', and indicates the mirthlessness of hysterical laughter. The meat they are eating is flecked with blood, and their eyes fill with tears. Theoclymenus now addresses the company. 'Wretches!' he cries out (20.|| 351). 'What is the matter with you? Your heads and faces and limbs are shrouded in darkness; a groan is heard; your cheeks are wet with tears; the walls and roof beams are splashed with blood; the porch and courtyard are full of ghosts hurrying to Erebus in the dark; the sun has vanished from the sky, and an evil darkness covers it.' The suitors burst into laughter, and Eurymachus exclaims that the stranger had better leave, since he is talking nonsense. Theoclymenus walks out, not without prophesying that disaster will presently come upon all the suitors. The suitors remark to Telemachus that he is unfortunate in his choice of guests. He sits waiting for his father to make a move; and Athene prompts Penelope to fetch the great bow, so that the suitors can try to string it.

Even Page concedes (p. 86) that the speech of Theoclymenus among the suitors is 'brief but memorable'. It is indeed memorable; it brings the whole preceding narrative to the point of climax. In early Greek literature we often encounter the figure of the warner, the man who plainly advises the person or persons who are bent on a disastrous course of the dangers which they are incurring; they pay no attention, since delusion sent by the gods has robbed them of their wits. Often the warner is a prophet; early Greek literature is pervaded by the belief in prophecy, and many of its legends feature members of the great families in which the gift was hereditary, as it was in certain families in historical times. Theoclymenus belongs to such a family. I find that the part he plays in the *Odyssey* fully justifies his ceremonious introduction, and see no need for Page's conjecture that there must have been an earlier version of the story in which Theoclymenus turned out to be the disguised Odysseus.[22]

Now let us consider a problem that affects the central plot of the *Iliad*. In the ninth book Agamemnon sends an embassy to Achilles, offering him a huge compensation if he will return to battle; after a long scene in which the ambassadors implore him to relent, Achilles refuses to yield to their solicitations. Page complains that the rest of the *Iliad* reveals no awareness that this celebrated episode ever occurred, and holds that after the embassy Achilles seems twice to deny that any approach has been made to him by Agamemnon.

The fatal decision of Achilles to compromise by allowing his friend Patroclus to put on his armour and enter the battle in his place is led up to in a series of passages that are subtly linked together in a way I find hard to reconcile with the idea that the *Iliad* is solely a product of oral composition. In the eleventh book Nestor takes out of the battle in his chariot the doctor Machaon, who has been wounded in the shoulder by an arrow. Achilles, standing on his ship where it lies on the beach, notices this, but cannot identify the wounded man. 'Now,' he exclaims to Patroclus (609), 'I think the Achaeans will soon stand about my knees and implore me; for a need that is no longer bearable has come upon them.' Then he sends Patroclus to Nestor's tent to learn the identity of the wounded hero.

'It seems very obvious,' Page writes (*History and the Homeric Iliad*, 305), 'that these words were not spoken by Achilles about whose knees the || Achaeans were in fact standing in supplication on the previous evening; an Achilles who had rejected their prayers, who had made it clear that he would never accept apology or compensation, but would wait till Hector was killing the Greeks in their tents and burning their ships.' But do Achilles' words really imply that no attempt to persuade him to return to the battle has yet been made? The ambassadors of the ninth book do not actually clasp his knees in supplication; neither do

[22] Page's treatment of the Theoclymenus episode has already been refuted by Erbse, *Beiträge zum Verständnis der Odyssee* (1972), 42 ff. [Cf. B. Fenik, *Studies in the Odyssey* (*Hermes* Einzelschriften 30, 1974), 233–44.]

they frankly admit the high-handedness of Agamemnon's behaviour and signify his willingness to express regret.

Nestor seizes the opportunity furnished by Patroclus' visit to talk seriously with him about the situation, using all his considerable powers of persuasion to try to ensure that Patroclus will put the case for reconsidering his attitude to his friend Achilles. On the way back Patroclus meets another wounded Achaean hero, Eurypylus, takes him back to his tent, and tends his wound.

From Eurypylus' tent, Patroclus hears the noise of the Trojan attack on the ships and sets off on his way to Achilles (15. 390). When he arrives he is in tears, and Achilles inquires the reason for his distress. In reply, Patroclus enumerates the Achaean heroes who are now incapacitated by wounds. 'You are impossible, Achilles,' he says (16. 29), 'may I never be overcome by such wrath as that which you guard! You are no son of Peleus and Thetis; your parents are the grey sea and the steep rocks, so unyielding is your purpose. If you are afraid of some prophecy from Zeus which your mother has warned you of, send me!' Here Patroclus is carrying out the instructions given him by Nestor; at least scholars are now less quick to assume that repetition is necessarily the result of interpolation.

Replying, Achilles denies that he fears any prophecy; he feels bitter pain, he says, at the thought that a man should wish to deprive his peer of his prize; he repeats these words with intense emotion. In the ninth book the third of the ambassadors, Ajax, has made the briefest speech, but the one that has come nearest to moving Achilles; he has reproached him with his inflexibility, and his indifference to the plight of his fellow-warriors. 'Ajax,' says Achilles in reply (9. 644), 'I think all you say is right; but my heart swells with anger, when I remember how Agamemnon treated me injuriously, as though I were an immigrant without honour.' In his speech in the sixteenth book (59), Achilles again says that Agamemnon treated him 'like an immigrant without honour'; and it is hard to resist the impression that he is echoing his speech to Ajax here. In that speech, he has said that he will not fight until Hector threatens his own ships (9. 650 f.); now he alludes to that statement, but so far modifies it as to permit Patroclus to take the field. The view that whoever wrote the speech of Achilles in the sixteenth book takes no account of the ninth book seems to me hardly tenable.

The Trojans, Achilles continues, are now filled with confidence; 'they would soon be filling the ditches with their dead, if Agamemnon were to feel kindly towards me' (71 f.). Page complains that Agamemnon must have felt kindly when he sent the embassy; but the offer sent through the ambassadors was accompanied by no adequate apology, nor has any been offered since. Achilles tells Patroclus to go, so that the

Greeks may bring back the girl, and || noble gifts in addition. This, Page complains, is what has been offered by the embassy, only to be rejected. But now the situation is different; the Greek predicament has become critical, and if Patroclus succeeds Achilles will get Briseis back, and the gifts into the bargain, without having had to fight himself. The supposed proof that the poet of the sixteenth book had no knowledge of the embassy in the ninth has not a leg to stand on.

The discussion of these problems has revealed some of the links between the episodes of the poem which seem to me to indicate that it can hardly have been composed without the aid of writing. 'Delicate and subtle preparations for what will follow, in five hundred lines' time,' writes Page (*The Homeric Odyssey*, 142), 'veiled and indirect allusion now to what happened five hundred lines ago—such artifice lies beyond his power, even supposing that it lay within the bounds of his imagination.' References back and forward (over more than a short space) will be more or less explicit, and limited to the broad outlines of the story. I have space only for these few examples; but I leave it to anyone who has leisure to work through the works of Schadewaldt and Reinhardt to determine whether this is the right way to deal with Homer. If one insists on applying modern standards of strict consistency, at the same time refusing to learn from the patient study of the poems the canons which the poets have observed in their practice, then it is easy to accumulate evidence of multiple authorship. But the underlying unity of both great epics can be discerned, as it was discerned by Aristotle. The human memory is capable of amazing feats, and in theory the poems, with all the links between their various parts, could have been composed without the aid of writing. Only so far no example of anything comparable having been achieved has been adduced; and no one able and willing to perceive the unity of the epics is likely to believe it possible.

If the Homeric poems were indeed composed with the aid of writing as early as the eighth century, the fact is only a little less astonishing than if they were composed orally. Our earliest specimens of Greek alphabetic writing date from the third quarter of that century.[23] More than half consist, like the Homeric poems, of hexameter verse, so that the connection between such verse and writing at that time is not a matter of conjecture. A word for writing material found in Homer is derived from the name of the Syrian town of Byblos; Egypt was open to the Greeks as early as the seventh century, so that papyrus may have been available; but we know that in early times the Greeks used some

[23] They are conveniently collected by L. H. Jeffery, *The Local Scripts of Archaic Greece* (1961, 1990); on 'Nestor's Cup', see R. Meiggs and D. M. Lewis, *A Selection of Greek Historical Inscriptions* (1969, 1988²), 1 ff., and M. L. West, *ZPE* 6 (1970), 171 ff.

form of parchment as writing material. The writing of the great book will have been a most laborious business; but in view of the immense value it will have had for the author and his dependants, the effort will have been worth while. Wade-Gery's suggestion that the alphabet was actually introduced in order for the epics to be written down seems to me unlikely; more probably it was invented for practical reasons by merchants trading with those Phoenicians from whose alphabet it was adapted.[24] Kirk ‖ writes (*Homer and the Oral Tradition*, 130) that 'the agents of traditional culture are normally the last, not the first, to make use of radically new techniques'. That may be true in our day, but it was not necessarily true in Homer's. The early Greeks were nothing if not enterprising, and the value of writing to professional poets will have been immediately apparent.

Wade-Gery believed that the great epics were designed for continuous recitation at a festival, as they were recited at the Athenian festival of the Panathenaea from the sixth century on; he thought of the joint festival of the Ionian Greeks at Mycale. Poetry was in those days designed for recitation, epic poetry being accompanied on the lyre; and though extracts from the great epics were no doubt recited in early, as they were in later, times, their authors surely envisaged at least continuous performances. The tradition must have started centuries earlier, doubtless with the composition of lays as inferior in quality as most of the Yugoslav material[25] which Professor Lord thinks so important for the understanding of Homer; it must have culminated with the creation of the two extensive masterpieces, designed to be appreciated as wholes. For a long time the text will have enjoyed no protection, and addition and interpolation will have occurred easily. It may well be that large sections of the epics, such as the lay of Dolon, the visit of Odysseus to the underworld, and perhaps the final episode of the Odyssey entered the poems in this fashion.

Our lack of information about Homer as a person is not necessarily an indication that he lived at a very early date. The ancients on the whole thought that he and Hesiod were contemporary, or very near each other in time; rather more of them made Hesiod the older than made Homer.[26] We know a fair amount about Hesiod as a person, as we know a fair amount about the rather later Archilochus, because he himself has told us something of himself. But poets whose work belonged to a genre which did not allow them to write about themselves left no materials for biography; of Stesichorus and Tyr-

[24] See A. Heubeck, 'Schrift' in F. Matz and H.-G. Buchholz (eds.), *Archaeologia Homerica*, iii, ch. 10 (1979), 151 ff.

[25] Specimens are given by F. Dirlmeier, *SB Heidelberg*, ph.-hist. Klasse (1971), Abh. 1.

[26] See M. L. West, *Hesiod, Theogony* (1966), 40 ff.

taeus, who wrote later still, we know virtually nothing.[27] Certain authors allege that there were people living on Chios and calling themselves Homeridae, 'sons of Homer', who gave recitations and were regarded as authorities on the poems; but we must be wary of Wade-Gery's suggestion that they may have preserved an ancient tradition, for Detlev Fehling has lately argued that the whole story depends on a misunderstanding by Hellanicus (*FGrH* 4 F 20) of the opening words of Pindar's second Nemean ode, and he may well be right.[28]

We have seen that in the great age of analysis one group of scholars followed Hermann in placing Homer at the beginning of his tradition, and another followed Wilamowitz in placing him at the middle.[29] The modern || scholars whose views I have discussed all place him at the end. Mysterious as Homer is and is likely to remain, I feel a good deal more confident that he existed and that we know certain things about him than I did thirty years ago.

Without a detailed re-examination of the text of the two great poems, summary treatments of the complicated problems of Homeric scholarship are of very limited value. Such a study, based on the results of the most recent scholarship, is an urgent need; and it will offer a release from the rigours of the period of Homeric investigation which I like to think is now drawing to an end. For far too long Homeric scholars have seemed to be bogged down in a kind of Passchendaele, from which the generals have seen no way out but a minute analysis of formulas, an excessive preoccupation with material objects, an exaggerated confidence in positive historical conclusions based upon the poems.[30] That work has not been futile, but it has gone on long enough, and it is time we took advantage of its results in returning to a detailed study of the poems. We should do so not in a spirit of positivist reductionism, but in that of scholars who while remembering that no labour is too arduous if it will throw light on Homer realize that they are dealing not simply with documents, but with documents that contain great poetry.[31]

[27] See M. R. Lefkowitz, 'The Poet as Hero: Fifth-Century Autobiography and Subsequent Biographical Fiction', *CQ*, NS 28 (1978), 459 ff.; also *The Lives of the Greek Poets* (1981).

[28] See Wade-Gery, *The Poet of the Iliad*, 19 ff.; T. W. Allen, *Homer: Origins and Transmission* (1924), 42 ff.; D. Fehling, *RhM* 122 (1979), 193 ff.

[29] See Dodds, op. cit. (n. 9).

[30] As to this, see M. I. Finley, 'Homer and Mycenae: Property and Tenure', *Historia*, 6 (1957), 133 ff., or in G. S. Kirk (ed.), *Language and Background of Homer*, 191 ff.; 'The Trojan War', *JHS* 84 (1964), 1 ff. (where I think Finley has the best of his exchange with G. S. Kirk and J. L. Caskey); 'Schliemann's Troy: One Hundred Years After', *PBA* 60 (1974), 393 ff.

[31] The modern literature about the topic is vast: for intelligent surveys of modern Homeric criticism with a bibliography, see James Holoka, 'Homer and Modern Literary Critical Discourse' (Diss. Michigan, 1974); 'Homeric Originality: A Survey', *Classical World*, 66 (1973), 257 ff.; 'Homer Studies, 1971–7', *Classical World*, 73 (1979), no. 2, and A. Heubeck, *Die Homerische Frage*

(1974). See L. E. Rossi, 'I poemi omerici come testimonianza di poesia orale', in R. Bianchi Bandinelli (ed.), *Storia e civiltà dei greci*, i (1978), 73 ff. for the view that the Homeric poems are partly oral and partly literary compositions, a view developed earlier by A. Dihle, *Homerprobleme* (1970).

I am grateful to J. Griffin, C. W. Macleod, and O. P. Taplin for discussing these questions with me; and particularly to Professor G. S. Kirk, who with characteristic kindness has done so much to improve the statement of a point of view decidedly different from his own. 'Many learned readers of Homer', he wrote in 1970 (*Homer and the Oral Tradition*, 129), 'evidently have little patience for inevitably arid surveys of the formular language and schematic structure of the great poems; they will gratefully accept the simple picture of Homeric composition that Adam Parry has unfashionably and even daringly revived.' Like Adam Parry, whose position I feel special pleasure in defending, I admit that I have little patience left for the dryness of much modern Homeric scholarship. Griffin, in the book cited in n. 21 and in *Homer* (in the OUP's *Past Masters* series), and Macleod, in *Homer, Iliad Book XXIV* (1982), write from a standpoint with which I feel much sympathy. [See now David Shive, *Naming Achilles* (1987), for a criticism of Parry's 'theory of economy'.]

2(80)

The *Meropis* (*SH* 903 A)

Ed.: L. Koenen et R. Merkelbach, P. Coll. Youtie (= *Collectanea Papyrologica: Texts Published in Honour of H. C. Youtie*, ed. A. E. Hanson, Bonn, 1976) I 1; denuo (ed. B. Kramer) P. Köln III 126. pap. s. a. C. i.

Disc.: Henrichs, *Cron. erc.* 5 (1975), 23; 7 (1977), 124. Führer, *ZPE* 24 (1977), 42. Henrichs, *ZPE* 27 (1977), 69. S. M. Sherwin-White, *Ancient Cos* (1978), 48 n. 96.

Tab.: P. Coll. Youtie I t. i.

ἔνθ᾽ ὁ μὲν αλ |[46][. . .]. . εν Μερόπων κατὰ νη|[47][. . .] φῦλα
νευρῆι ἔπι ψάλλων

— — — —

ἀλλ᾽ οὐκ Ἄστερον ἰὸς ἐ|[49]δάμνατο· τρὶς γὰρ ἐπ᾽ [αὐτῶι]|[50]
ἧκε διὰ ννευρῆς· τ̣[ῶι δ᾽ οὐ]|[51] χρόα τῆνος ἵκανεν
. [ς]|[52]κληρῆς πέτρης ἐξᾶλτ[ο χαμᾶ]|[53]ζε
δριμὺ βέλος. πικρ[ὸν δ᾽ ἄ]|[54]χος ἔσχεθεν Ἡρακλε̣[ῖα
ὡς]|[55] ἴδεν̣

— — — —

κ̣[αί νύ]|[56] κεν Ἡρακλεῖα κατέκτ[ανεν,]|[57] εἰ μὴ Ἀθήνη
λάβρον [ἐπεβρόν]|[58]τησε διὲγ νεφέων κα[ταβᾶ]|[59]σα
πληξαμένηιθε . [ ]|[60] δ᾽ ἁπαλὸν χρόα πρόσθ[ε]|[61]θη
Ἡρακλεῖος ἄνακτ[ος· ὁ δ᾽ εἴ]|[62]σιδεν ἄσθματι θυίω̣[ν
γνῶ τε]|[63] θεόν

— — — —

ἔνθ᾽ ὁ μὲν ε[ἰς πλη]|[68]θὺν Μερόπων κίεν. ἡ [δὲ δια]|[69]πρὸ
αἰχμῆι στῆθος [ἔλασσεν.]|[70] ὃ δ̣᾽ ἐ̣ξέχυτ᾽· οὐ γὰρ [ὁμοῖαι
ἀ]|[71]θάναται θνηταῖσι βολ[αὶ κατὰ]|[72] γαῖαν ἔασιν.
πρη⟦μ⟧νη̣[ ]|[73]τησε, μέλας δ᾽ ἐπερ{ι}ε⟨ί⟩δ[ετο εὐ]|[74]ρὼς
ὀφθαλμοῖς, Ἀϊδός[δε δ᾽ ἀπή]|[75]λυθε θυμὸς ἀναιδής.

— — — — ‖

καὶ τοῦ [μὲν βού]|[77]λευσε πε̣ρ̣ὶ χρόα ῥ̣ι̣ν̣ὸ̣ν̣ [Ἀθήνη]|[78]
ἕσσασθαι

— — — —

Atti del XVII Congresso internazionale di papirologia (Naples, 1984), i. 141–52.

ἔγδε[ιρεν δ' ἄρ']|[79] ἅπαν cκύλος ἄλκιμον· [αἶψα δ' ἔ]|[80]πειτα
αὐάνθη. τοῦ μὲν .[ ]|[81] κατέχευε φέρουcα
.[ ]|[82] περὶ cῶμα δ' ἑλιξα[ἕρ]|[83]κοc
αὐταῖc cὺν χ[είρεccι καὶ]|[84] εὐρήεccι πεδίλοιc
]|[85]μηθεῖcα . . . . [

P. Köln III 126 (Apollodorus Atheniensis, *Περὶ θεῶν*: vid. ed. pr.), de Athena et de Aegide . . . (17) *περιεπέcομεν δὲ ποιήμαcιν,* | *ἐφ' ὧν ἦν ἐπιγραφὴ Μεροπίc,* | *οὐ δηλοῦcα τὸν ποηcο*[. . .] (voluit *τὸν ποήcαντα*? ed. pr.), *ἢ* | (20) *πρὸc τοῖc ἄλλοιc Μέροψι [τ]οῖc* | *ἐν τῆι Κώιαι καταριθμουμέ* | *νοιc* ⟦. . .⟧ *καὶ Ἄcτερον ὑπάρ|χειν διεcάφει. τοῦτον δ' εἶναι* | *μάχιμόν τινα καὶ δυνατόν, ἔτι* | (25) *δὲ ἄτρωτον. ἐπὶ τοῦ οὖν Ἡρα|κλῆν εἰc τὴν Κώιαν παραγε|νέcθαι καὶ πόλεμ[ον] τοῖc |[Κώ]ιοιc cυcτήcαcθαι, θλιβο|-[μέ]νων δὲ αὐτῶν ὑπὸ τοῦ Ἀc|*(30)*[τέ]ρου τὴν Ἀθηνᾶν ἐπιβοη|[θῆ]cαι εἰδυῖαν ὅτι ὑπὸ μὲν Ἡρα|[κλέ]ουc οὐθὲν ἂν πάθοι, μᾶλ|[λο]ν δ' ἂν ὑπ' αὐτοῦ Ἡρακλῆc ἀ|[ναι]ρεθείη{ι}. τέλοc δὲ ἀπολο|*(35)*μένου τούτου ὑπ' αὐτῆc cυν|θεωρῆcαι χρήcιμον αὐτοῦ* | *τὸ{υ} δέροc ἐcόμενον πρὸc τοὺc* | *[ἄλ]λουc κινδύνουc· διόπερ ἐ|[ξη]λύτρωcεν αὐτὸ cύν τε τοῖc* | (40) *[πο]cὶν καὶ ταῖ⟨c⟩ χερcὶ περιθέc|[θα]ι* (claudicat syntaxis; de coniecturis vid. P. Köln). *ἐδόκει δέ μοι τὰ ποήμα|[τα] νεωτέρου τινὸc εἶναι· διὰ* | *[δὲ] τὸ ἰδίωμα τῆc ἱcτορίαc* | *[ἐξε]λάβομεν αὐτό. εἶχεν δὲ|*(45) *[τὰ ἔ]πη οὕτω· ἔνθ' ὁ μὲν — ψάλλων* (vv. 1–2). (48) *[εἶτε]ν· ἀλλ' οὐκ Ἄcτερον – ἴδεν* (vv. 3–7). *μετὰ δὲ ὀλίγα· κ[αί νύ] κεν Ἡρακλεῖα — θεόν* (vv. 8–12). *εἶτεν μετά τιν[α ἡ Ἀθηνᾶ]* | *κελεύει τὸν μὲν Ἡρα[κλῆν πε]|*(65)*ρὶ τοὺc ἄλλουc γίνεc[θαι Μέρο]|παc, αὐτὴ{ι]} δὲ τῶι Ἀcτ[έρωι ἀν]|θίcταται· ἔνθ' ὁ μὲν — ἀναιδήc* (vv. 13–17). *[εἶτεν αὐ]|τὸν ἐγδείρει· καὶ τοῦ — ἕccαcθαι* (vv. 18–19). *εἶτεν· ἔγδε[ιρεν —]μηθεῖcα* · · · · · · · · · [*c.* 10] | (86) *Ἀcτέρου δορὰν ἡ Ἀθ[ηνᾶ δέδυκεν]* | *ὁμολογεῖ, ἀλλ' οὐ τ.* [*c.* 8 *Μέ*]|*ροποc· τούτου μὲν* [

carmen inter cetera tamquam prosam orationem exscripsit librarius (versus carminis 'v.' notavimus, prosae lineas 'lin.'). unde fit ut lacunas in mediis versibus (id est, ad initia vel ad fines linearum) satis incerte metiaris.

carmen anonymum novit Apollodorus, novit ex Apollodoro Philodemus (*De piet.* ap. Henrichs, *Cron. erc.* 7 (1977), 124): ergo secundo a.C. saeculo antiquius. ceteroqui de aetate ambigitur. (*a*) sexto saeculo assignavit ed. pr.: anonyma enim huius modi epica, quae certae regionis ‖ fabulas tractent, *Minyadem*, *Phocaidem* etc., ante quintum saeculum omnia extitisse: hoc ex eo ipso apparere, quod nulla auctorum nomina tradita essent. (*b*) saeculo tertio tribuit Sherwin-White, quia (i) *νεωτέρου τινόc* crediderit Apollodorus (pap. lin. 42); (ii) Coam materiem item tetigerint Herodas (2. 95) et Theocritus (7. 5). neque illud satis certum neque hoc (de *νεωτέρου* vid. 944. 6 adn.) sed tertio (vel quarto) suffragatur stilus (notate Homerum exquisitius variatum, ex. gr. v. 16); suffragatur etiam illud, quod demonstravit Henrichs (*ZPE* 27 (1977), 69), poetam nostrum ex eodem fonte haurire atque idyllii 'Theocritei' 25 poetam (hunc autem fontem antiquius aliquod *Herculem Leontophonum* fuisse; hinc absurdum illud de soleis, v. 23, cum pellem giganteam ad exemplum pellis leoninae inscite describeret).

fabulam narrat inauditam (nam quod alibi de Asteri pelle legerat Apollodorus, pap. lin. 86, plane incertum; vid. ad v. 24), fort. ab ipso poeta consutam. alibi (*a*) Hercules in Coum egreditur (*Il.* 14. 255); insulam vastat, regnante Chalcone filio Eurypyli (Hes. fr. 43a. 60 M.–W.; *aliter* Eurypylum occidit, sed a Chalcodonte vulneratus a Iove eripitur, Apollod., *Bibl.* 2. 7. 1); in Meropas tamquam fulmen incidit (Pind. fr. 33a S.–M.). (*b*) Hercules leonem Nemeae invulnerabilem occidit, pellem induit; astat Athena (hoc soli Bacchyl. 13 (?), Callim. [*SH*] 264. 4). (*c*) Athena gigantas occidit, in eisque Astera (Aristot. fr. 637 Rose); gigantis pellem induit, quae est aegis (Epicharm., in hoc pap. lin. 8–16 = C. Austin, *Comicorum Graecorum fragmenta in papyris reperta* fr. 85a; etc.). in *Meropide* leonem cum gigante, pugnam Coam cum pugna Phlegraea commixtam reperimus.

ars metrica regulis Callimacheis ut vid. nusquam repugnat, nisi 11 (in nomine proprio).

in papyro nulli accentus, spiritus, apostrophi; interpunctio nulla, nisi spatia vacua (his versuum initia et fines plerumque indicantur), paragraphi et coronides (haec prosam orationem dividunt). iota adscriptum (falso lin. 34. ?59 (= v. 10), 66), nusquam omissum. elisio: hic illic scriptio plena (lin. 34, 55; 44 ad finem sententiae, 80 (= v. 20) metro convenienter). de lacunarum mensura vid. supra.

supplementa, quae silentio recepimus, ex ed. pr. desumpta 1 *αλ*|[. . .]. . *εν*: *ἄλ*|[*λ'*] *ἧ̣κ̣εν* ed. pr., cf. 4, 'alia (sc. *βέλη*) emisit Hercules': spatio brevius, nisi *αλ*|[*λα*]*ηκ̣εν* plene scribas (fort. contra librarii normam, vid. supra). *νη*|[. . .]: *νή*|[*ριτα*] ed. pr., cf. Hes. *Op.* 511, Livrea ad A.R. 4. 158: *Il.* 2. 840 *φῦλα Πελαcγῶν* /, 17. 220 *μυρία φῦλα . . . ἐπικούρων*. spatio fort. convenientius *νη*|[*λέα*]: *Od.* 7. 206 *ἄγρια φῦλα Γιγάντων* /   2 *ἔπι* ‖ *ψάλλων* (*Il.* 4. 118 *ἐπὶ νευρῆι κατεκόcμει πικρὸν ὀϊcτόν*) potius quam *ἐπιψάλλων*? ed. pr.   3 *Il.* 5. 106 *τὸν δ' οὐ βέλοc ὠκὺ δάμαccε*   4 *Il.* 4. 118 *ἐπὶ νευρῆι*, 11. 476 *ἀπὸ νευρῆc*: de orthographia vid. S. West, *Ptolemaic Papyri of Homer* (1967), 113   *τ̣*[*ῶι δ' οὐ*]: primo loco *π* vel *τ* vel fort. *η*, ut vid. (*ἀ̣*[*λλ' οὐ*] ed. pr.): *Il.* 11. 352 *οὐδ' ἵκετο χρόα καλόν*, 23. 819 *νύξ', οὐδὲ χρό' ἵκανεν*   *τῆνοc*: notate formam Doricam (Coam?)   5 . [: hastae verticalis pars superior, ut vid. *ἠ̣*[*ὖτ' ἀπὸ* ed. pr.; nos coniunctionem desideramus: *ἠ̣*[*ὕτε δ' ἐκ*? *Il.* 13. 137 *ὀλοοίτροχοc ὣc ἀπὸ πέτρηc*, 3. 29 *ἐξ ὀχέων . . . ἆλτο χαμᾶζε*, 4. 125 *ἆλτο δ' ὀϊcτὸc* / *ὀξυβελήc*   6 *Ἡρακλ̣ε*[*ῖα*: cf. 8, 11; de orthographia vid. S. West, *Ptolemaic Papyri of Homer*,

116 s. 8 ss. *Il.* 5. 311 *καί νύ κεν ἔνθ' ἀπόλοιτο ἄναξ ἀνδρῶν Αἰνείας, / εἰ μὴ ἄρ' ὀξὺ νόηςε Διὸς θυγάτηρ Ἀφροδίτη / . . . ἀμφὶ δ' ἑὸν φίλον υἱὸν ἐχεύατο πήχεε λευκώ,* 8. 131 *καί νύ κε cήκαcθεν . . ., / εἰ μὴ ἄρ' ὀξὺ νόηςε πατὴρ ἀνδρῶν τε θεῶν τε· / βροντήcαc δ' ἄρα δεινὸν ἀφῆκ' ἀργῆτα κεραυνόν, / κὰδ δὲ πρόcθ' ἵππων Διομήδεοc ἧκε χαμᾶζε* 10 *θε*̣[: fort. *ν.* tum *πληξαμένη{ι} θέṿ[αρι]| δ'* (vel *θέṿ[ε τόν]|δ'*) *ἁπαλὸν χρόα πρόcθ[ε φαάν]|θη* (vel *πρόcθ[ε δὲ φάν]|θη*) ed. pr.: metro convenientius *θέṿ[αροc]* (*Il.* 5. 339 *θέναροc πρύμνον*)? sed etiam de sensu haeremus; cuius carnem pulsavit Athena? Herculis (ita ed. pr.)? Asteri, ut ictum averteret? ipsius, quo dolorem exprimeret (*Il.* 15. 397 *ὤιμωξεν δ' ἄρ' ἔπειτα καὶ ὣ πεπλήγετο μηρὼ / χερcὶ καταπρηνέcc',* 16. 125, *HHCer.* 245)? 11 Vel *ὁ δ' ὡ]|c ἴδεν . . . [γνῶ ῥα]| θεόν]*: *ὁ δ' ἐν ἄcθματι θυμ[οῦ] / γνῶ ῥα]| θεόν* ed. pr., contra vestigia. *Il.* 21. 234 *οἴδματι θύων,* 22. 272 *ἔγχεϊ θύων.* de orthographia (*θυι / θυ*) cf. Allen–Halliday ad *HHMerc.* 560 13 *Il.* 7. 307 *ὁ δ' ἐc Τρώων ὅμαδον κίε* *Il.* 5. 66 *ἡ δὲ διαπρό,* 13. 388 *διαπρὸ δὲ χαλκὸν ἔλαccεν,* 5. 41 *διὰ δὲ cτήθεcφιν ἔλαccεν* 14 *ẹχξεχυτ* (vel fort. *ẹḳξεχυτ*? Mayser, *Gramm. d. griech. Pap. aus d. Ptolemäerzeit,* i/1² (1970), 184) pap. 14 s. *Il.* 5. 441 *ἐπεὶ οὔποτε φῦλον ὁμοῖον / ἀθανάτων τε θεῶν χαμαὶ ἐρχομένων τ' ἀνθρώπων* 16 *Il.* 5. 57 *διὰ δὲ cτήθεcφιν ἔλαccεν, / ἤριπε δὲ πρηνήc, ἀράβηcε δὲ τεύχε' ἐπ' αὐτῶι,* 15. 578 *δούπηcεν δὲ πεcών, τὸν δὲ cκότοc ὄccε κάλυψεν* *πρημ̇νη̣[* (*μ* puncto deletum) pap.]|*τηcε*: subiectum aut Asterus aut Athena; si illud, *πρηνή[c* expectes, tum verbum aut cadendi (*ἤριπε* *Il.* 5. 58, *ἐλιάcθη* 15. 543; *πέcε* 12. 395, *κάππεcε* 16. 311, *κεῖτο* 21. 119) aut sonandi (*δούπηcεν* *Il.* 15. 578); si hoc, *πρηνῆ* et verbum sternendi (*τανύccαc* *Il.* 23. 25). ad rem *κρό]|τηcε,* sed obiectum desiderat; *πρηνῆ [δ' ἄψ οὔ]|τηcε, πρηνὴ[c δ' ἠΰ]|τηcε* possis, nec tamen expectes. *δ' ἐπερ{ι}ε⟨ί⟩δ[ετο* Führer: *δεπεριεδ̣[*pap. (ex *δεπερῖδ[* ortum? Führer): *δὲ περί{ε}δ[ραμεν* Koenen, haud recte (pontem Hermannianum infringit). hoc aut 'gravabat' accipias (Eur. *Alc.* 385 *ὄμμα βαρύνειν,* Ov. *Met.* 4. 145) aut 'tangebat' (Callim. *hy.* 4. 234 *ἐπὶ πτερὸν ὕπνοc ἐρείcει*: sim. apud ‖ alios poetas Alexandrinos, Gow ad Theoc. 5. 24) *εὐ]|ρώc* Merkelbach (vix *ἴδ]|ρώc* Koenen, contra et scribarum normam, qui *ι|δρωc* dividunt, et heroum, qui pugnantes sed non morientes sudant): id est *cκότοc*: Hesych. *ε* 7159 *εὐρώεντα . . .* (Cyrill. codd. ASn) *cκοτεινά* (ad *Il.* 20. 65), 7160 *εὐρώεντοc· cκοτεινοῦ. ἀπὸ τοῦ εὐρῶτοc* 17 *Od.* 10. 560 *ψυχὴ δ' Ἄϊδόcδε κατῆλθεν,* *Il.* 7. 131 *θυμὸν . . . δῦναι δόμον Ἄϊδοc εἴcω* 20 *cκύλοc . . .* (22) *ἕρ]|κοc*: cf. Callim. [*SH*] 268 B *cκύλοc . . . ἔρυμα,* [Theocr.] 25. 142 *cκύλοc αὖον . . .* 279 *ἕρκοc . . . ἰωχμοῖο* (Henrichs) [*αἶψα δ' ἔ]|πειτα* (Hes. *Scut.* 465) vel [*αὐτὰρ ἔ]|πειτα* (Hes. *Theog.* 116, 132, *Op.* 607, A.R. 1. 598, 1058): [*αἶψα δὲ*] | *πέπλ⟨οc⟩* ed. pr., contra vestigia 21 s. cf. (*a*) *Il.* 4. 445–6 *ἀμβροcίην ὑπὸ ῥῖνα ἑκάcτωι θῆκε φέρουcα / ἡδὺ μάλα πνείουcαν,*

ὄλεccε δὲ κήτεοc ὀδμήν, (*b*) *Il.* 19. 38–9 *Πατρόκλωι δ' αὖτ' ἀμβροcίην καὶ νέκταρ ἐρυθρὸν / cτάξε κατὰ ῥινῶν, ἵνα οἱ χρὼc ἔμπεδοc εἴη* (sim. 16. 670, 23. 186, Hes. fr. 23a. 22 M.–W.). unde *τοῦ μέν ν̣[υ θεὸc] | κατέχευε φέρουcα / ἀ̣[μβροcίην]|* vel *τοῦ μὲν ν̣[έκταρ] | κατέχευε φέρουcα ; ἀ̣[θανάτη]|* ed. pr.   22 *ἐλιξα[μένη* (ed. pr.) *φέρεν* (Burkert), vel *ἐλίξα[το* (ed. pr.), *Ἄρεοc* (West). *ἕρ]|κοc* Hübner: *Il.* 4. 137 *μίτρηc θ', ἣν ἐφόρει ἔρυμα χροόc, ἕρκοc ἀκόντων*: vid. ad 20   23 *πεδίλοιc*: [*πο*] *cίν* interpretatur Apollodorus (pap. lin. 40), haud recte   24 . . . . ^οδε[].^ [*c.* 10] | *Ἀcτέρου δορὰν ἡ Ἀθ[ηνᾶ δέδυκεν] | ὁμολογεῖ κτλ.* pap.: ad init. *ὧc ἄρα κοc]|μηθεῖcα* (Burkert) *τε̣τ̣ά̣[ρπετ]ο[* legit, litteras suprascriptas ad orationem quae sequitur prosam rettulit ed. pr. (e.g. *ὁ δὲ Ὀ[ρφεὺc*?? — *ὅτι τὴν]| Ἀcτέρου δορὰν* — |*ὁμολογεῖ*). omnia satis incerta.

THIS is the text, as it appears in H. Lloyd-Jones and P. Parsons, *Supplementum Hellenisticum* (1983) of the fragments of a local Coan epic called the *Meropis* which are quoted in what must be a fragment of Apollodorus, *Περὶ θεῶν*, contained in a Cologne papyrus of the first century BC.

The first editors, and also Frau Kramer when she republished it in P. Köln III, assigned the poem to the second half of the sixth century, to which a number of local epics are commonly assigned, such as the Argive *Phoronis*, the Orchomenian Minyas, the *Atthis* of Hegesinous, the *Phocais*, the *Thesprotis*, the *Corinthiaca*, and perhaps the *Naupactia*. But S. M. Sherwin-White in her comprehensive work on *Ancient Cos* (Hypomnemata 51 (1978), 48 n. 96) tentatively suggests a Hellenistic date. 'The ancient commentator', she writes, 'considers the poem to be recent.' That argument is worthless, for the *Περὶ θεῶν* is essentially a work of Homeric scholarship (see Pfeiffer, *History of Classical Scholarship*, i (1968), 261), and in the language of ancient Homeric scholars *οἱ* || *νεώτεροι* means 'poets later than Homer'. 'The choice of subject-matter', she adds, 'fits well the early Hellenistic period', since Theocr. 7. 3 and Herodas 2. 95 f. mention Coan myths; does that mean that no one did so earlier? 'The pervasive Homeric imitation in the poem', she adds, 'could be Hellenistic imitation.' Is it in fact more like the imitation of Homer that we find in Hellenistic poetry than it is like the Homeric imitation in the brief and scanty fragments of post-Homeric epic of the early period that we possess?

During the Hellenistic period epic poems about the early history and mythology of a particular locality were by no means uncommon (see K. Ziegler, *Das hellenistische Epos* (2nd edn. 1966), 18 f.). Of a number of these we can get little notion; the *Lacedaemonica* and *Macedonica* of Phaestus (*CA*, p. 28; *SH*, p. 316), the *Arcadica* of Ariaethus (see *SH*, p. 5, on Agathyllus), the *Aitia Romana* of Butas (*SH*, p. 88), Hegemon's poem on the war of Leuctra (*SH*, p. 236), the *Thebaica* of Menelaus (*SH*, p. 271), the *Sicelica* of Polycritus (*SH*, p. 337), the poem on Smyrna of Theodorus (*SH*, p. 359). A second category consists of poems by accomplished poets, by no means always natives of the cities or countries about which they wrote, and works of a very different order from the *Meropis*; here we may place the *Ktiseis* of Alexandria, Caunus, Cnidus, Naucratis, and Lesbos by Apollonius of Rhodes, the *Messeniaca* and *Thessaliaca* of Rhianus, the *Mopsopia* of Euphorion, the *Aetolica*, *Colophoniaca*, *Oetaica*, *Thebaica*, *Sicelica* of Nicander; the *Lyrcus* of Nicaenetus (*CA*, p. 1) also seems to have belonged in this group. But there is a third category of poems to which it would not be unfair to compare the *Meropis*, one consisting of poems by local poets writing about their own localities. Unfortunately we know very little of their character. The *Cymaica* of Hyperochus (*SH*, p. 249) and the *Rhodiaca* of Idaeus (*SH*, p. 250) are mere names to us. Of the *Bithyniaca*, in not less than ten books, by Demosthenes of Bithynia (*CA*, p. 25), we have one substantial fragment, fr. 4; the elegant style, correct poetic diction and Callimachean metre show it to be the work of a highly skilful writer, very different from our author. A passage about the antiquities of Cos itself is quoted from one Phylarchus (*SH*, p. 336), whose name it seemed to the editors of *SH* hazardous to emend; the fragment is corrupt and the nature of the poem is obscure. The Jewish epics of Philo Senior on Jerusalem (*SH*, p. 328) and Theodotus (*SH*, p. 360) belong in a different subdivision of this category; but if their text is properly edited they are seen to be competent versifiers of the second order, not indeed comparable with Demosthenes of Bithynia, but far || superior to the author of the *Meropis*. None of these poems show Homeric imitation anything like as crude as his; but if we turn to the fragments of post-Homeric early epic, it is a different story. Such relatively long fragments as we have, like those of the *Cypria*, best read in Erich Bethe's *Der troische Epenkreis* (reprinted in 1966 from *Homer, Dichtung und Sage* (1929), ii/2. 149–297)[1] reveal a kind of sub-Homeric style that is a good deal more like that of the Coan poet; and the fragments of the *Phoronis* (Davies, *Epicorum Graecorum Fragmenta*, 153 f.) which are the largest we have from early epics about localities, are epic verse of the same order, though the difference of the subject-matter makes detailed comparison seem unprofitable.

The epic passages relevant to the text are indicated in the notes; but let us consider the nature of the imitation.

[1] [But now in Malcolm Davies' *Epicorum Graecorum Fragmenta* (1988), 13 f.]

3 ἰὸϲ will mean the arrow, not the poison; cf. Theocr. 25. 238 *ἀλλ' οὐδ' ὣϲ ὑπὸ βύρϲαν ἔδυ πολυώδυνοϲ ἰόϲ.*

4 The flatness of the pronoun, standing where we might have expected another word for 'arrow', is no less striking than the incursion of a dialect form into the epic. In Coan we should expect *κῆνοϲ*, a word whose vocalization F. Bechtel, *Die griechischen Dialekte*, i (1923), 585 thought might have been influenced by that of *τῆνοϲ*.

5 Henrichs compares Theocr. 25. 230 *οὐ γάρ τι βέλοϲ διὰ ϲαρκὸϲ ὄλιϲθεν ὀκριόεν, χλώρηι δὲ παλίϲϲυτον ἔμπεϲε ποίηι.*

11 *οἴδματι θυίων* is used of the sea *Il.* 23. 230 (cf. Anacreon fr. 2 = *PMG* 347 fr. 1. 17) and of a river (*Il.* 23. 234), but 'raging with his breath' strikes one as clumsy.

14 *ἐξέχυτ'* in the sense of 'crumpled up', 'collapsed', is unusual; the nearest parallel I can find is in Dioscorides, *AP* 5. 55. 8 = Gow–Page, *HE* 1490, where Gow and Page cite *ἔκχυτοϲ ὕπνωι* from Paulus Silentiarius, *AP* 5. 275. 'Mortal casts are not like immortal casts' seems to be a feeble adaptation of phrases like that found at *Il.* 5. 441.

16 Führer's *ἐπερ{ι}ε⟨ί⟩δετο* seems likely to be right. In classical poetry *ἐρείδω* usually means 'to prop' or 'to lean', but in Hellenistic poetry it means little more than *τίθημι* or *κατατίθημι* (see Gow on Theocr. 5. 24). But this use is not only Hellenistic; cf. *Il.* 7. 145 *οὔδει ἐρείϲθη*, Aesch. *Agam.* 64 *γόνατοϲ κονίαιϲιν ἐρειδομένου*, Eur. fr. 541 *ἡμεῖϲ δὲ Πολύβου παῖδ' ἐρείϲαντεϲ πέδωι*, *HF* 1362 *νεκροὺϲ πρὸϲ ϲτέρν' ἐρείϲαϲ μητρί.* The notion of darkness as something that can be put over or taken away from a man's eyes is common in Homer; cf. e.g. *Il.* 5. 127 *ἀχλὺν . . . ἀπ' ὀφθαλμῶν ἕλον, ἣ πρὶν ἐπῆεν.* The verb usually denoting its being put on is *καταχέω*: note also *Il.* 15. 668 *τοῖϲι δ' ἀπ' ὀφθαλμῶν νέφοϲ ἀχλύοϲ ὦϲεν Ἀθήνη.* ||

More surprising in a Hellenistic author would be the hiatus in the fifth foot of the hexameter. Hermann, *Orphica* (1805), 741 observes that hiatus is relatively frequent after the third person of a passive verb; cf. W. Christ, *Metrik der Griechen und Römer* (1874), 212. If the author is Hellenistic, this is a deliberate Homericism.

18 The first editors observed that only at *Il.* 7. 131 does Homer speak of the *θυμόϲ* as leaving the body on death; believers in a certain kind of *Geistesgeschichte* no doubt take that line to be 'late'. This is an abnormality that seems to be more likely to occur in post-Homeric early epic than in Hellenistic verse.

19 For this use of *βουλεύω* to mean 'to plan to' do something, cf. *Il.* 9. 458 and *Od.* 9. 299. Note Theocr. 25. 272 *καὶ τότε δὴ βούλευον ὅπωϲ λαϲιαύχενα βύρϲαν / θηρὸϲ τεθνειῶτοϲ ἀπὸ μελέων ἐρυϲαίμην.*

20 *ἔγδ[ειρεν*: cf. *Il.* 10. 19 *ἐκδείραϲ*, Theocr. 25. 278 *ἀπέδειρα. ϲκύλοϲ* is not found before the Hellenistic period (Callim. fr. 677 Pf. = *SH*

268 B, Theocr. 25. 142, Nicander, *Alex.* 270, Leonidas of Tarentum, *AP* 6. 35. 2 = Gow–Page, *HE* 2256, Phalaecus (?), *AP* 6. 165. 2; but who shall say that it did not exist in earlier poetry?

21 Note *αὐανθέν* at the beginning of the line in *Od.* 9. 421.

22 The *middle* of *ἑλίϲϲω* seems not to be used in the sense of 'winding round' by Homer or any other author. Henrichs compares Theocr. 25. 278–9 *ἀμφεθέμην μελέεϲϲιν / ἕρκοϲ ἐνυαλίου ταμεϲίχροοϲ ἰωχμοῖο*, and Callim. fr. 677 (= *SH* 268 B) *τὸ δὲ ϲκύλοϲ ἀνδρὶ καλύπτρη / γιγνόμενον, νιφετοῦ καὶ βελέων ἔρυμα.*

23 Editors have supplemented *χ[είρεϲϲι* and have taken the sense to be 'with front paws and large back paws'. But *πεδίλοιϲ* means not 'feet' but 'boots', so that symmetry would seem to demand *χ[ειρῖϲι*. Not only footguards but armguards, like those worn by Laertes to protect himself against thorns (*Od.* 24. 230) were a feature of early armour; see A. M. Snodgrass, *Early Greek Armour and Weapons* (1964), 88, with n. 54 on p. 240. Why should Athene not have made herself armguards and footguards out of the hide? Her footguards will have been wide, like the *κόθορνοι* worn by Alcmaeon for his famous visit to the treasure-house of Croesus (Hdt. 6. 125).

εὐρήεϲϲι involves an arbitrary lengthening made for the sake of the metre, a phenomenon not uncommon in epic verse; see the variant *εὐρῆοϲ* at Hes. fr. 13. 2, and see Merkelbach and West ad loc. and W. Schulze, *Quaestiones epicae* (1892), 67 n. 4.

Almost any of the peculiarities of style and diction found in these fragments might conceivably have occurred in Hellenistic verse, and though no surviving fragment of that verse seems to me like this, there || is always the possibility that an obscure provincial poet might have written thus during the Hellenistic period. But as Francis Vian has written to me, the fragments contain nothing that is specifically Hellenistic, and taken together they surely suggest rather a late follower of the poets of the archaic period. The lumbering and pedestrian narrative well accords with the flatness and dullness commonly attributed by ancient critics to these writers and exemplified in their remains; the poet repeats Homeric commonplaces in a form suggestive of but different from that given them by Homer, and awkwardness results.

The central motif is of an embarrassing simplicity; the story of how Heracles put to his own use the impenetrable hide of the Nemean lion is crudely adapted to provide an *aition* of the aegis, like that known from Epicharmus and another known from Diodorus (3. 70). Henrichs has drawn attention to similarities in phrasing between this poem and the narratives of Heracles' killing of the lion in Theocr. 25 and in the fragments of the third book of Callimachus' *Aitia*; and although in different narratives of the same event we must expect certain similarities of wording, these seem to be significant. The *Meropis* can hardly have been a source of these poets, and as Henrichs remarks its author must have been familiar with early epics about Heracles. Panyassis fr. 1 *δέρμα τε θήρειον Βεμβινήταο λέοντος* and fr. 2 *καὶ Βεμβινήταο πελώρου δέρμα λέοντος* must be noted; Henrichs points out that *δέρμα . . . θηρός* occurs in Theocr. 25. 175; fr. 2 recalls *τοῦδε πελώρου* in l. 195 of that poem; and ll. 20–3 mention Bembina as the locality most afflicted by the lion. This slender link might be held to indicate a possibility that the epic of Panyassis may have influenced all three poets. If so, the *Meropis* can hardly have been earlier than the fifth century, for that seems to have been when Panyassis lived (see V. J. Matthews, *Panyassis of Halikarnassos* (1974), 12 f.). But stories about Heracles were told in verse long before Panyassis; and the *Suda* says that some date the Heracles epic of Peisandros of Cameiros as early as 648 BC, while others make him contemporary with Eumolpus or with Hesiod. According to Eratosthenes, *Catast.* 12, he wrote of how Heracles took the lion's skin (fr. 2a Davies); see Gow, *Theocritus*, ii. 546 on the epigram Theocritus wrote for a bronze statue of this poet (22), calling Heracles *τὸν λεοντομάχαν, τὸν ὀξύχειρα*. Strabo 15. 688 says that the club and lionskin of Heracles were *πλάσμα τῶν τὴν Ἡράκλειαν ποιησάντων, εἴτε Πείσανδρος ἦν εἴτ' ἄλλος τις*. The epigram of Theocritus proves that there was interest in Peisandros at this time, and it seems possible that his narrative lies behind the three other poems.||

Cos during the Hellenistic age was hardly an intellectual backwater; the home of Philitas, birthplace of Ptolemy Philadelphus, and scene of the *Thalysia* would hardly have tolerated a local poet so distressingly out of date as the author of the *Meropis*. I think it highly probable that he belongs to the archaic period, most likely the second half of the sixth century, as the first editors suggested.

[F. Vian, 'Nouvelles Réflexions sur la *Gigantomachie*', in *Studi in onore di Adelmo Barigazzi* (1984), 259 n. 34 writes: 'C'est peut-être l'œuvre de Pisandre qui a influencé la *Méropide* (connue depuis peu grâce à P. Colon. 5604: cf. H. Lloyd-Jones et P. Parsons, *SH* (1983) fr. 903 A. En tout cas, cette épopée locale (fin du VIme siècle?) racontait la guerre d'Héraclès contre les Méropes de Cos en utilisant des thèmes manifestement empruntés à la *Gigantomachie*.']

II. LYRIC

3(R50)

Review of J. Pouilloux *et al.*, *Archiloque* (*EH* 10)

J. Pouilloux, N. M. Kontoleon, Anton Scherer, K. J. Dover, Denys Page, Winfried Bühler, Erik Wistrand, Bruno Snell, Olivier Reverdin, Max Treu, *Archiloque*. (Entretiens sur l'antiquité classique 10.) Vandœuvres (Geneva): Fondation Hardt (Cambridge: Heffer), 1964. Pp. 307, 4 pls.

ARCHILOCHUS was a good subject for a discussion at the Fondation Hardt; a distinguished company was assembled; and the resulting book is of great interest.

Pouilloux ('Archiloque et Thasos: histoire et poésie') starts off with a somewhat highly coloured picture of the poet's life on Thasos built up from fragments, but goes on to give a useful description of the discoveries of the French ‖ excavators during the last ten years; the tomb of Glaucus, son of Leptines, the seventh-century objects from the sanctuary on the hill that dominates the city, the shrine of Artemis with its important deposit of archaic articles and the still more ancient relics found in the residential quarter and described in *BCH* 88 (1964). He compares what the excavations teach us about the cults of Paros in Archilochus' time with the references to gods in Archilochus' poems. Pouilloux's attempt to establish a similarity between what we know of seventh-century Thasian art and the poetry of Archilochus is not especially enlightening; it is hard to see a special affinity between the art of Archilochus and that of the dish with Bellerophon and the Chimera first published by Salviat and Weill in *BCH* 84 (1960) and again reproduced in this volume (opposite p. 16). It is more interesting to learn that the Parian colonists on Thasos in the middle of the seventh century seem to have been a comparatively rich community, and so far as we can guess from the physical remains, their relations with the Thracians seem to have been good. Page rightly observes (p. 80) that we need not be alarmed at an apparent inconsistency with the evidence of Archilochus' poems.

Kontoleon, with his unrivalled knowledge of the Cyclades, first sets out the chief facts known about the honours done to Archilochus in antiquity. Instead of repeating what he has already written about the

Classical Review, NS 15 (1965), 263–7.

Monumentum Archilochi and the Elitas inscriptions (see n. 1 on p. 44), he gives most of his space to the grave monument of Archilochus built about the middle of the fourth century and discovered in 1962 by Orlandos, and to the Boston pyxis, formerly supposed to represent the encounter with the Muses of Hesiod but, since his own important discoveries, seen by Rumpf to show that of Archilochus. Kontoleon gives solid support to Rumpf's arguments; the case for his own suggestion that the Archilocheion was a gymnasium seems to me to be much strengthened by the analogy of the Mimnermeion at Colophon.

Kontoleon next summarizes what we know of Paros in the time of Archilochus. He emphasizes that Paros belonged to the group of Cycladic islands whose closest connections were with Ionia and the east, notably Miletus, in contrast with the 'westernizing' group of islands whose chief member was Naxos. Pointing out that almost half the seventh-century vases found on Delos are of the type known as 'Melian', Kontoleon makes what seems a strong case for thinking that they were really made in Paros; the so-called Siphnian vases he thinks may also have been Parian, but of earlier date. He has less to say about the bronzes and the sculpture; the well-known Kriophoros must be later than Archilochus.

Scherer ('Die Sprache des Archilochos') gives a competent account of the poet's language. The language of his elegiac poems he finds close to that of epic; that of his iambi he finds to be not epic with a strong Ionic admixture, but colloquial Ionic with selected epic elements. Most interestingly, he points out that Archilochus on the whole avoided the Aeolic elements of epic diction.

Page ('Archilochus and the Oral Tradition') inquires whether Archilochus was or was not an oral composer. With the aid of Miss L. H. Jeffery's *Local Scripts of Archaic Greece*, he shows 'how few and how small are the scraps of evidence for the use of the alphabet in the generation preceding the birth of Archilochus'; it is most useful to have these scraps set out (on pp. 121 f.). He then proceeds to a detailed demonstration that not only the elegiac but also || the epodic poems are dominated by the phraseology of the epic. 'The formula-element', he concludes (p. 161), 'comes almost exclusively from the epic, and the new style is formed by more or less extensive adaptation of traditional phrases combined with components, generally in moderate measure, of pre-meditated word-selection. In brief, the non-dactylic compositions of Archilochus reveal the transition from oral to written verse.'

The diction of the epodic poems is not easy to analyse; the lack of material combines with our ignorance of colloquial Ionic in the seventh century to obscure its nature. The few Ionic inscriptions of archaic date often contain words known to us only from epic or early Ionic poetry (so Dover, p. 213): is this because the vernacular contained many such words, or because the language of inscriptions, and particularly of archaic inscriptions, tends to differ from the vernacular? None the less, Page shows that even the Ionic poems contain a very large proportion of epic elements; it is relevant to note that in the following paper Dover argues with much plausibility that the epodic poems may not have been separated from the elegiac poems at the time of their origin, at least, by any definite difference of genre. Does this really indicate that Archilochus wrote at a transitional stage between oral and written composition? Certainly his elegiac poems are written in the language of the epic, and Page has shown that even his epodic poems contain more of this language than is usually supposed. But even a poet who used writing might well have been powerfully influenced by the epic tradition. By no means all the instances of epic language in Archilochus can usefully be classified as 'formulaic', a concept that needs careful definition. The discussion that followed Page's paper shows that we cannot rule out the possibility that several different kinds of writing material may have been available to Archilochus. The suggestion that the poems of Archilochus may have been transmitted orally cannot be proved wrong, and that is itself a fact of great interest; but it remains a possibility and not, so far, a certainty.[1]

Dover begins a paper of great interest and importance with the argument, already mentioned, that no fundamental difference of genre separated the iambic and epodic from the elegiac poems. True, both an elegy and an epode consist of distichs, a longer followed by a shorter; but the difference between 'double short' and 'single short' metres is fundamental, and the iambic poems contain certain colloquialisms and obscenities not imaginable in elegiacs. Both Scherer and Page allow that the iambic poems are less strongly coloured by the epic than the elegies.

Dover next argues that elegiacs were imported from Ionia to the mainland and not from the mainland to Ionia, resting his case on a thoroughly convincing demonstration that the language of Tyrtaeus, whom he rightly regards as a Spartan, shows Ionic elements not easily explained on any other hypothesis; in particular, he points to several instances of non-observance of the digamma.

Coming now to an assessment of the originality of Archilochus, Dover points out that it is usual to assume that 'Homer, Hesiod and Archilochus represent successive stages in the spiritual development of the Greek people'. He suggests that we may profitably consider an alternative hypothesis, one that 'would regard Hesiod and Archilochus

[1] [I should now (1989) hardly consider it a possibility.]

as two different personalities through whom, shortly after the introduction of writing, poetic genres of long standing found expression at a very high artistic level' (p. 198). Archilochus, he reminds us,‖ wrote songs, and songs often contain words put into the mouths of imaginary characters in imaginary situations, a fact which he abundantly illustrates from primitive examples, most of them taken from Bowra's valuable book *Primitive Song*. This leads him to deliver a most salutary warning against the all too prevalent habit of assuming that sentiments expressed in isolated quotations are the poet's own. Dover rightly points out that we cannot even be sure that Critias (fr. 44 D.–K.) did not fall into this error. 'We must go', he writes, echoing Plato, 'where the wind of reason blows, however barren the shore upon which it casts us.' These sensible observations deserve to be taken notice of by those who try to extract information about the poet's biography from even the slightest scraps of evidence. Dover might have gone further and warned against the distortions of those who try to squeeze as much *Geistesgeschichte* as possible out of each fragment, grossly exaggerating the differences of outlook between Archilochus and Homer. Page rightly protests (pp. 159 and 214) against the often repeated allegation that Archilochus' portrait of the ideal general (fr. 114 W.) testifies to a rejection of the heroic ideal, pointing out its resemblance to Homer's account of Tydeus (*Il.* 5. 801, which is relevant to Pind. *Isth.* 4. 57: compare also the account of Odysseus at *Il.* 3. 216 f.). Archilochus' satisfaction in having saved his life by abandoning his shield (fr. 5 W.) is quite in the spirit of Homer's Odysseus; in any case, Reverdin is right to say (p. 285) that the concept of the ῥίψαςπις derives from hoplite warfare and is scarcely relevant to the fighting in which Archilochus took part. Nor is the sexual phraseology of epic always more 'circumscribed and decorous' than that of Archilochus (p. 197); with Archilochus frr. 191 and 194 W. we may compare *Il.* 3. 441 f.; 14. 294 f., 313 f. Is it so surprising that Archilochus should call himself a servant of Enyalios who knows the gifts of the Muses (fr. 1 W.) when one remembers that Achilles in his tent played the lyre and sang of the famous deeds of men (*Il.* 9. 185–9)? Every imaginable difference between Homer and Archilochus has been magnified to the utmost extent by those who insist on seeing early Greek literature mainly in terms of the now overworked concept of 'paideia'; we have reached a point where it may prove more interesting to notice the similarities between them.

Bühler ('Archilochos und Kallimachos') gives a highly competent account of the use made of Archilochus and Hipponax by Callimachus in his *Iambi*, skilfully describing the difference of tone between them in a way reminiscent of Leo's well-known paper on Horace and Archilochus (*Ausgewählte kleine Schriften*, ii. 139 f.). In the observation that the fifth Iambus (fr. 195 Pf.) imitates Hipponax fr. 118 Masson, he has been anticipated by Rudolf Kassel (*RhM* 101 (1958), 235 f.). His suggestion (p. 241) that the Diegetes is probably wrong in saying that the expostulation with the boy Euthydemus was only a subsidiary topic of the third Iambus (fr. 193) is likely to be right.

Wistrand ('Archilochus and Horace') refers us for a comprehensive treatment of the topic to Leo's paper, cited above, and to Eduard Fraenkel's book on Horace and devotes his space to points of detail. His suggestion, that when Horace writes (*epist.* 2. 2. 50–1) *paupertas impulit audax ut uersus facerem* the word *audax* has the semi-political connotation of 'audacious, reckless, rebellious', seems to me to lend the passage a heaviness altogether foreign to its author. The line makes use of the commonplace first exemplified by Telemachus' remark (*Od.* 12. 347) that αἰδώϲ is no useful quality for a man in need; it also implies with a light irony that the act of writing verse, at least on the || author's part, implies a certain ἀναίδεια or θραϲύτηϲ. Wistrand's contention that in this *audacia* Horace saw a quality he had in common with Archilochus is not convincing. Wistrand next reopens the vexed question of the correct interpretation of *epist.* i. 19. 23–32, on which I will not now embark.

P. 32: 'Faut-il admettre', says Pouilloux, discussing fr. 21 W., 'que le sommet chauve émerge au-dessus des forêts qui couvrent les pentes?' I suggest that ἐπιϲτεφήϲ is *active* in sense, so that the treeless ridge is said to crown the forests. Wilamowitz may have exaggerated when he said that in early Greek *all* verbal adjectives might be active as well as passive (*Griechische Verskunst*, 182 n. 1), but we are often inclined to forget that a given adjective may be either. Pp. 85–6: a problem posed by the account of the Muses' appearance to Archilochus in the inscription published in 1952 that greatly puzzled the symposiasts has been solved by M. L. West (*CR*, NS 14 (1964), 141–2).

The book contains many valuable observations on points of detail, to which its excellent indexes can quickly guide the reader.

4(41)

The Cologne Fragment of Alcaeus [= Page, *SLG* S 262]*

Reinhold Merkelbach has lately published a new fragment of Alcaeus which is of great interest.[1] Following the civilized practice of Girolamo Vitelli and his Florentine successors, Merkelbach has shown the piece to many scholars before its publication, and the first edition contains suggestions by many experts, including the two leading authorities on Lesbian poetry, Edgar Lobel and Denys Page. The first publication is admirably done and is accompanied by a good photograph of the manuscript, which probably dates from the first century of our era; Merkelbach compares the writing of P. Oxy. 2295. I have little to contribute in the way of new readings, and owe much of the material here presented to Merkelbach's publication and to the commentary on a fragment of the same poem given by Page in *Sappho and Alcaeus*;[2] readers will find it useful to have Merkelbach's publication in hand while studying this article. Still, I hope a rather fuller commentary on this important piece may be found useful.

Text

πολεμος

]υτι

μηνδικ^{α} τα μη ε[νδικα τ]α μη εν δικ^{η} κε[ι]μεν^{α}

[ ]cαντας αἰcχυν[. . .]τατα μὴνδικα

[ ]ην δὲ περβάλον[τ' ἀν]άγκα — την αγχονην / ουτως ειρη / κε δεcμον

λε[υ]cμω

1 δρά]cαντας Merkelbach: αἰcχύν[νον]τα τὰ Lobel: αἰcχύν[θεν]τα τὰ Page 2 utrum ἀν]άγκαι (Merkelbach) an ἀν]άγκα (Page) incertum

* [For a bibliography of this fragment, see A. M. von Erp Taalman Kip in *Some Recently Found Greek Poems* (*Mnemosyne*, Suppl. 99 (1987), 95–128.]

[1] P. Colon. inv. no. 2021, in *ZPE* 1 (1967), 81 f. (cited hereafter as Merkelbach).

[2] P. Oxy. 2303 fr. 1(*a*) and (*b*) = fr. 298 L.–P.; for commentary, see D. L. Page, *Sappho and Alcaeus* (Oxford, 1955), 283 ff. (cited hereafter as Page, *S&A*). The poem is mentioned in the ancient book on the early lyric poets published in Part XXIX of the *Oxyrhynchus Papyri*; see p. 51 below. The piece appears as Alcaeus fr. 138 in D. L. Page's *Lyrica Graeca Selecta* (Oxford, 1968), 75.

Greek, Roman and Byzantine Studies, 9 (1968), 125–39.

[ ]χ̣ε̣νι λ̣α̣β̣ολίωι π. [. .]αν· την αναγ̄κ ‖
[ ] Ἀχαίοιϲ' ἦϲ πόλυ βέλτερον ην
[ ]. . η̣ν̣τ̣α̣ κατέκτανον θεοϲυλη(ν)οντα
[π]α̣ρ̣π̣λ̣έοντεϲ Αἴγαιϲ
[ ]ϲ ἔτυχον θαλάϲϲαϲ
[ ]ε̣ν ναύωι Πριάμω πάϊϲ
[Ἀ]θανάαϲ πολυλάϊδοϲ
[ ] ἐπαππένα γενήω αφημμενη
[ ]νεεϲ δὲ πόλιν ἔπηπον
[ ]. . . . . . αϲ Δαΐφοβόν τ' ἄμα
[ ]. ν οἰμώγα δ' [ἀπ]ὺ τ̣ε̣ί̣χοϲ
[κα]ὶ παίδων ἀΰτα
[ ]ο̣ν πέδιον κάτη⌋χε.
[λ]ύ̣⌋ϲϲαν ἦλθ' ὀλόαν ἔχων
[ ] ἅ̣γ. ναϲ⌋ Πάλλαδοϲ ἀ θέων
[ ]ϲ̣ι θε⌋οϲύλαιϲι πάντων
[ ]τα ⌊μακάρων⌋ πέφυκε.
[ ]ϲ̣ι δ' ἀμφο⌋ῖν παρθενίκαν ἔλων
[ ] παρεϲτάκο⌋ιϲαν ἀγάλματι
[ ] ὀ Λόκροϲ οὐδ' ἔδε⌋ιϲε
[ ]. οϲ πολέμω δότ̣ε̣ρ[ρ]⌋αν
[ ]ν· ἀ δὲ δεῖνον ὐπ' [ὄ]φρ⌋υϲι ‖
ϲμ[⌊πε]λ̣[ι]δνώθειϲα κα̣τ οἴνοπα]μέλοντο δὲ ἕνεκε̣ν̣[. .]
α̣. . ⌊πόν]τ̣ο̣[ν] ἐκ δ' ἀφάντοι[ϲ]α̣ι τὴν πόλιν καὶ [
ἐξαπ[ίναϲ] ⌊ἐκύκα θυέλλαιϲ·]τ̣οῦ Αἴ(αντοϲ) ὅρο(ϲ) ἐ̣π̣ι̣λ̣ε̣γ̣ό̣μ̣ε̣ν̣[ον
αἴδη. [⌊]φ̣ [ ]χάριν ἀνδρὸϲ ο̣ϲ̣[
ἰραιϲ. [
Αἴαϲ ἀχ. [
ἄνδρο̣ϲ̣[
. . μ̣ο̣[
. . ε̣ρ̣ . [
ἔβαϲκε[

3 αὔ]χενι Merkelbach ‖ 4 καὶ μάν κ'] Merkelbach: ἦ μάν κ'] Kassel: καί κ' ὤϲ] Page 5 αἰ τὸν θεοβλ]α̣β̣έ̣ντα Page 6 ἴϲωϲ κε] Page: οὔτω κε]? 7 πραΰτέρα]ϲ Page: ληοτέρα]ϲ Merkelbach 8 ἴρηα δ'] vel γέραιρα δ'] Merkelbach 9 ἄγαλμ'] Merkelbach: ἴκτωρ] Page 10 κάτηχ'] Merkelbach: κάτιϲδ'] Page; ἐπαππένα Lobel: ἀπαππένα P. Colon.; γενήω Merkelbach: γενείω P. Colon. 11 δυϲμέ]νεεϲ Merkelbach, πόλη'? 12]υ̣π̣ Merkelbach 13 ἔπεφν]ο̣ν (e.g.) Merkelbach 14 fort. ἔλαμψε κα]ὶ 15 Δαρδάνι]ο̣ν Page, Treu; incipit P. Oxy. 2303 fr. 1 (*a*) et (*b*) 16 init. Merkelbach;]υ̣[.]αν̣η̣αθοα̣. [.] ἔχων P. Oxy. 17 ἐϲ ναῦον (e.g.) Page, Kassel 18 φώτεϲ]ϲι vel ἄνδρεϲ]ϲι Merkelbach: θνάτοι]ϲι vel βρότοι]ϲι Page 19 δεινοτά]τα Merkelbach: πικροτά]τα vel αἰνοτά]τα Page, *S&A* 283–4; μακάρων om. P. Colon. 20 χέρρεϲ]ϲ̣ι Page *S&A* 283: ἐν χέρ]ϲι West 21 ϲέμνωι] Page: θέαϲ] Kassel: ἄλλωϲ] Merkelbach 22 ὔβριϲϲ'] Page, *S&A* 283: ἔρυϲϲ'] Merkelbach, Kassel: ἔξηλκ'] West: fort. ἤιϲχυνν' 23 παῖδα Δ]ί̣ο̣ϲ̣ Page, *S&A* 283: τὰν βάρ]ε̣ο̣ϲ̣ West 24 γόργωπι]ν Page, *S&A* 283 ‖ 25–31 obelis notantur 25 ϲμ[ύχει Treu: ϲμ[έρδνον Merkelbach 26 ἄϊξ[ε πόν]το[ν Page, *S&A* 283 28 desinit P. Oxy. 30 'α vel η' Merkelbach 32 'Statt μ kann vielleicht auch αρ gelesen werden. Vor der Lücke ein runder Buchstabe, ε oder ϲ oder ο' Merkelbach 33 ad init. fort. π Merkelbach; in tertio loco ϲ vel ε

παννυχια[
πρωτοις̣[
δεινα. . [
ἄϊξε πόν[τον
ὦρσε βίᾳ[ν
. . . ς̣ι̣ς̣ε̣[
πάντᾱ πε̣[
. . .]τ̣ο̣ν̣ο̣ς[
δ̣[. .]. ροcενος[
ουδ̣ω̣δεκαμ[
ζώει μὲν ω. [
ἀτανβροτ. [
ὠυρραδον. [
ἔπει κεληττο̣[
. . . . . ς̣ωπ[

37 fort. κ̣[Merkelbach 44 ‘ου vel δυ’ Page 46 ‘α vel η’ Merkelbach
In scholio sub v. 24 posito, v. 3] τοῦ Αἴ(αντος) ὅρο(ς) ἐ̣π̣ι̣λ̣ε̣γ ό̣μ̣ε̣ν̣[ον Lloyd-Jones et Rea:]χουα. ο̣ρ̣° ἐ̣π̣ι̣λεγ ο̣μ̣ε̣ν̣[Merkelbach

Translation

‘. . . (putting to shame?) those who (had acted?) unjustly. We must put a (noose?) about his neck and destroy him by stoning . . . it would have been far better for the Achaeans if they had killed (the man who offended against the gods); then as they sailed past Aegae ‖ they would have found the sea (more gentle) . . . the daughter of Priam in the temple (clasped) the image of Athene, giver of much booty, holding its chin, while the enemy were besetting the city. (They killed) . . . and Deiphobus, and a wail of grief (went up) from the wall, and the cry of children filled the whole (Dardanian) plain. (Ajax) came in deadly madness to the temple of holy Pallas, her who of all the immortal gods is (most terrible) to sacrilegious (mortals). And with both hands the Locrian seized the girl as she . . . stood by the image and (ravished?) her, not fearing (the daughter of Zeus) who gives strength in war, (her of the fearful eyes?). But she went and (her eyes burned?) terribly beneath her brows, and she (darted) over the wine-dark sea and suddenly stirred up the hidden blasts of the storm . . .’

I

When he discussed the Oxyrhynchus fragment at *S&A* 283, Page classed the piece among the ‘Non-Political Poems’. We now see that although it narrates a story from ancient myth, it does so in order to furnish an example that is meant to illustrate a modern happening.

The citation of examples from past history to throw light on actual events[3] is common already in the *Iliad*; among many instances, it is enough to recall how Achilles in urging Priam to eat with him reminds him that even Niobe in the hour of her deepest sorrow was persuaded to eat (24. 602). Choral lyric took over the practice from the epic; in Alcman's Louvre *Partheneion* (fr. 1) the story of the defeat of the sons of Hippocoon is used to point a moral (36 f.),[4] and in later poets of this genre there are innumerable instances. In monodic lyric there are naturally fewer cases, but even here the practice is attested. Sappho in a famous poem (fr. 16) illustrates her declaration that the most beautiful thing on earth is what one loves by pointing to the case of Helen, who sacrificed everything for the sake of Paris.[5] Alcaeus (fr. 38) supports his advice to Melanippus to ‖ drink and enjoy life while it lasts by recalling the example of Sisyphus, whose cleverness earned him a respite from death, but not for long.[6] If we had more long fragments of Alcaeus, I suspect that other mythological narrations in his poetry would turn out to be there in order to furnish examples; for instance, the story of how Helen left her husband and caused the Trojan War in fr. 283 and the comparison between Helen and Thetis in fr. 42 both look as if they were meant to point a moral. Had we more long fragments, the gap between 'political' and 'non-political' poems might be bridged in other instances also, as it has been in this. As in Sappho fr. 16, this poem seems to have returned towards its end to the modern instances which the example was meant to illustrate, for at l. 47 there seems to be an undoubted reference to Pittacus. That seems not to have been Alcaeus' invariable practice in such cases; the poem partly preserved in fr. 42 ends with the end of the myth, as the coronis shows.

Commentary

1. The poem begins in the middle of a sentence, and the first line cannot be restored with safety while we remain ignorant of what went before it. I see no need to assume the existence of a superlative form *αἰϲχυν[νέϲ]τατα* (Merkelbach, p. 85); what stood in the text was most probably a participle, either active or, less likely, passive.

[3] Many instances are listed and discussed in the invaluable Basle dissertation of Robert Oehler, a pupil of Von der Mühll, *Mythologische Exempla in der älteren griechischen Dichtung* (Aarau, 1925). See also Ed. Fraenkel, *Aeschylus, Agamemnon* (Oxford, 1950), ii. 461, 470.

[4] H. Patzer, *Die Anfänge der griechischen Tragödie* (Wiesbaden, 1962), 104 f., makes some good observations about Alcman in this connection, and about the history of this *topos* in choral lyric and in tragedy. On Pindar see C. M. Bowra, *Pindar* (Oxford, 1964), 290 f., 304 f.

[5] See Page, *S&A* 129, for other possible instances in Sappho.

[6] See Bowra, *Greek Lyric Poetry* (2nd edn., Oxford, 1961), 162 and Ed. Fraenkel, *Horace* (Oxford, 1957), 185.

2. Merkelbach thinks that the first word of the second line meant something like 'a muzzle' or 'a halter'.[7] Unfortunately neither *φιμόν*, *πάccαλον*, nor *κλοιόν* nor any of the commoner words meaning 'yoke' fits, and he can only offer *φορβῆν*, supposed to be a form of *φορβειάν*, which means a horse's feeding muzzle which is attached to a manger. Merkelbach takes *αναγκα* to stand for the dative, and supplies the main verb of the sentence from what precedes. This may be right, but *ἀνάγκα* as nominative would make sense; at fr. 249 it is taken in this way by Page, *S&A* 196–7, and this is how Page prints it here. Merkelbach seems to think that the puzzling scholion indicates that *αναγκα* here means 'noose' or 'fetters'; but even if he is right in taking the word to be a dative, this sense is not necessarily relevant here, even if the author of the scholion thought so. It is quite possible that the scholion referred not to this word at all but to whatever word is missing at the beginning of l. 2. Note that in the scholion *δεcμον* (Page) and not *δεcμα* (Merkelbach) seems to ‖ have been written. At the beginning of the scholion Page prints *την αιcχυνην*: he tells me he would now accept *την αγχ̣ο̣ν̣η̣ν̣*.

2–3. If Merkelbach is right in taking the sense to be 'We should place a halter about his neck and stone him', the verb at the end of l. 3 must have meant 'to crush', 'to punish', 'to kill', or else have been a verb that together with *λαβολίωι* would yield the sense 'to stone'. The only verb that occurs to me that might do is *'παλοίαν*. *ἀπαλοῖαν* occurs at *Il.* 4. 522 of the severing of tendons and at Callim. fr. 260. 1 of breaking off a horn (see *HSCP* 72 (1968), 33 f. [= *Greek Comedy, Hellenistic Literature, Greek Religion, and Miscellanea*, ch. 14, p. 141]), but is used by Nonn. *D.* 9. 320 in the sense of 'destroy'. But I do not feel confident that Alcaeus would have used the epic form of this verb, and offer it only with many reservations. *λαβολίωι* (glossed by *λευcμῶι*) is new, but in view of *λάβολοc* in fr. 68. 3 (where it is used with gerundival sense, like *λιθόλευcτοc* at Callim. *epigr.* 41. 5 Pf. and Alexander Aetolus fr. 3. 12 Powell) it is not surprising.

4. Kassel's *ἦ μάν κ'*] suits the space, and also, I think, the style, better than Merkelbach's *καὶ μάν κ'*]: the latter combination of particles is not so far attested in the Lesbian poets. Clearly the sense is that the Greeks would have been better off if they had stoned Ajax the son of Oileus for having violated Cassandra in the temple of Athene, for they would not then have encountered the famous storm which scattered their returning fleet, leading to the death of Ajax and delaying the return of many others. See below, pp. 49 f.

5. Page's *αἰ τὸν θεοβλ]α̣β̣έ̣ντα* suits the sense; for this verb, compare

[7] Hirzel, op. cit. (n. 12), 243–4, points out that victims of stoning were sometimes tied up so that they could not escape; thus in Philostr. *Heroicus*, 10. 7, Palamedes has his hands bound.

Aesch. *Pers.* 831 λῆξαι θεοβλαβοῦνθ' ὑπερκόμπωι θράcει, where Broadhead is right against Groeneboom in taking the sense as active, 'damaging the gods', rather than passive.

6. Page suggests ἴcωc κε but I slightly prefer οὕτω κε: the simplicity of the latter seems to me more like Alcaeus than the irony of the former. On Aegae, see below, p. 51.

7. Whether we adopt Page's πραϋτέρα]c or Merkelbach's (somewhat less attractive) ληοτέρα]c, a usage is in question which is not uncommon in prayers, as when the Chorus of the *Oedipus Coloneus* (1482) say to the unknown *daimon* whose wrath they fear, ἐναιcίου δὲ coῦ τύχοιμι. One implores a divinity that one may find him ἐναίcιοc, ἵλεωc, πρᾶιοc; which inclines me to prefer Page's guess.

8. Cassandra was not a priestess, so I do not like the suggestion ἴρηα. Sappho fr. 44. 16 has the form Περάμοιο and Alcaeus fr. 42. 2 has Πεϱράμωι, but it would not be safe to conclude that Alcaeus really wrote Περάμω in this place. Alcaeus might well use the epic as well as the vernacular form of such a name: at fr. 42. 11 he has Νηρεΐδων, though Sappho (fr. 5. 1) has Νηρήϊδεc.[8] ||

9. πολυλάϊδοc looks back to Athene's Homeric epithet of ἀγελείη; cf. ληῖτιc in the Doloneia (*Il.* 10. 460).

10. ἀπαππένα glossed by ἀφημμένη is surprising. First, Lobel (ap. Merkelbach, p. 87) must surely be right in suggesting that ἐπαππένα instead of ἐφημμένη was intended; in similar contexts one often finds such verbs as ἀπαρτωμένη 'hanging from', but a compound of ἅπτομαι 'I touch', would have ἐπι- and not ἀπο-. Next, Merkelbach explains the form by comparing ὄππατα, the Lesbian equivalent of ὄμματα. But that is a noun, and here we are dealing with a verbal ending, which we should not expect to suffer this mutation; none of the forms set out by E.-M. Hamm, *Grammatik zu Sappho und Alkaios* (ADAW 1951) 21, §τ30, which Merkelbach refers to, is parallel to this, nor do fr. 130. 10 and Sappho fr. 24(*b*). 5 serve to confirm it. Lobel and Page both think the word must be corrupt for the form with double μ, and it is hard not to agree with them. Or was this form an analogical absurdity forced into the text by grammarians?

The Lesbian form is γένηον, as Merkelbach remarks (cf. Alcaeus fr. 120. 9); γενείω is probably an error. 'The word must, on this evidence, mean "chin" in Lesbian, not "beard",' says Page (ap. Merkelbach, p. 87). In other dialects also it can mean either 'chin' or 'beard', as LSJ rightly say; but I know no other place where it is used of the chin of a *female* person.

11. The lengthening of the iota of πόλιν is surprising. Lobel suggests

[8] See I. Kazik-Zawadzka, *De Sapphicae Alcaicaeque elocutionis colore epico* (Wrocław, 1958), 43–4.

πόλιν⟨*δ'*⟩, thinking, presumably, of the two places in the *Iliad* where the expression *ἔφεπον κρατερώνυχας ἵππους* has the indirect object *Πατρόκλωι*. But these do not prove that *ἐφέπω* can be treated as if it were a verb of motion; in this place the sense must be 'the enemy were besetting the city', and *πόλινδ'* will not do. Page suggests inserting after *πόλιν* a *τ'* which would couple it with *Δαΐφοβον* in l. 12; but even if we leave out of account the gap which precedes that name, the sense favours a strong stop after *ἔπηπον*. Merkelbach quotes K. Strunk for two attempts to explain the lengthening by linguistic arguments; but supposing either could be accepted (and I find neither compelling), why do we find this isolated instance? I am baffled by this phenomenon, but observe that a poet who uses the genitive form *πόληος* (fr. 41. 18) might be capable of using the accusative form *πόληα* which occurs at [Hes.] *Scut.* 105. Callim. 43. 60 must have taken it from early epic.

In view of the occurrence of *ἔπηπον* here, the supplement *'πέπ[οισαι* at fr. 45. 5 becomes more plausible.

12. What stood at the beginning of this line? Merkelbach interprets the traces at the beginning as *υ̣π̣* but says that the apparent *υ* may be a grave accent and the apparent *π* perhaps *ϲι*; in the gap, there may be room for only one rather than two letters. Before *αϲ* there is a small circle in the middle of the line, 'rather small for *β* or *ρ*'. I should expect to find here the names of Trojans killed together with Deiphobus; perhaps *αϲ* was the end of ‖ the accusative plural of a patronymic, or else *παῖδαϲ* followed a genitive in *-οιο*.

13–14. Merkelbach suggests *δέδαε*, Page *ὄρωρε*; neither will do, since a perfect of this kind is equivalent to a present, and the historic present is not used in early lyric: see J. Wackernagel, *Vorlesungen über Syntax*, i^2 (Basle, 1926), 163, and other literature recently listed by R. Führer, *Formproblem-Untersuchungen zu den Reden in der frühgriechischen Lyrik* (Zetemata 44 (Munich, 1967), 93–4). I find it hard to think of an adequate supplement; one might consider *ἔλαμψε*, if only the word were used in this way in lyric as it is in tragedy.

Lobel and Page quote *Il.* 22. 447 *κωκυτοῦ δ' ἤκουϲε καὶ οἰμωγῆϲ ἀπὸ πύργου*. One is also reminded of Xenophon's memorable account of how the news of Aegospotami reached Athens: *ἐν δὲ ταῖϲ Ἀθήναιϲ τῆϲ Παράλου ἀφικομένηϲ νυκτὸϲ ἐλέγετο ἡ ϲυμφορά, καὶ οἰμωγὴ ἐκ τοῦ Πειραιῶϲ διὰ τῶν μακρῶν τειχῶν εἰϲ ἄϲτυ διῆκεν, ὁ ἕτεροϲ τῶι ἑτέρωι παραγγέλλων* (*Hell.* 2. 2. 3). Did the thing really happen, or did Xenophon remember what he knew from the poets ought to happen?

14–15. Kassel well compares *Il.* 16. 78 f. *οἱ δ' ἀλαλητῶι / πᾶν πεδίον κατέχουϲι* and Aesch. *Pers.* 426 f., *οἰμωγὴ δ' ὁμοῦ / κωκύμαϲιν κατεῖχε πελαγίαν ἅλα*: note also Eur. *Tro.* 556 f. *βοὰ κατεῖχε περγάμων ἕδραϲ* and Hdt. 1. 111 *κλαυθμῶι κατείχετο οἶκοϲ*. *ἀϋτή* in the *Iliad* regularly refers to the cry of fighting warriors. Behind the present passage seems to lie *Od.* 14. 264–8 = 17. 433–7:

> αἶψα μάλ' Αἰγυπτίων ἀνδρῶν περικαλλέας ἀγροὺς
> πόρθεον, ἐκ δὲ γυναῖκας ἄγον καὶ νήπια τέκνα
> αὐτούς τ' ἔκτεινον. τάχα δ' ἐς πόλιν ἵκετ' ἀϋτή.
> οἱ δὲ βοῆς ἀΐοντες ἅμ' ἠοῖ φαινομένηφι
> ἦλθον· πλῆτο δὲ πᾶν πεδίον πεζῶν τε καὶ ἵππων
> χαλκοῦ τε στεροπῆς.

Δαρδάνι]ον πέδιον will do well to signify what in the *Iliad* is called πεδίον Cκαμάνδριον (2. 465), or πεδίον Ἰλήϊον (21. 558), or Τρωϊκὸν πεδίον (23. 464; cf. 10. 11). In the *Iliad* the Dardanians are a special people, the subjects of Aeneas and Anchises, but later the word is a mere synonym for 'Trojan'.

16. Compare *Il.* 9. 304–5:

> νῦν γάρ χ' Ἕκτορ' ἕλοις, ἐπεὶ ἂν μάλα τοι σχεδὸν ἔλθοι
> λύσσαν ἔχων ὀλοήν.

In the case of Ajax, as in that of Hector, the 'madness' in question was to prove even more 'deadly' to its possessor than to others.

17. It is odd that elsewhere the epithet ἁγνά does not seem to be applied to Athene before the paean of Lamprocles (*PMG* 735). It is not a prerogative of virgin goddesses, for in the *Odyssey* it is applied to Persephone (11. 386) as well as Artemis, and in the Demeter hymn it is applied both to Persephone (337) ‖ and to Demeter (203, 429). Both Lesbian poets use it of the Charites (Alcaeus fr. 386, Sappho frr. 53, 108. 8). It is often applied to temples, as in the Elean hymn to Dionysus (*PMG* 871. 1–3) and at Alcman fr. 14(*b*); here it is obviously appropriate.[8a]

18. Lobel, P. Oxy. XXI, p. 87, points out that θεοσύλης is shown to be an old Ionic word by Hipponax fr. 118. 1 Masson (= P. Oxy. 2176. 1), where l. 14 of the commentary supports his supplement θεό[συλιν against θεο[ισεχθρὴν (Ed. Fraenkel, *Kleine Beiträge zur klassischen Philologie* (Rome, 1964, i. 242–3); cf. Alcaeus fr. 59(*b*), where ἱερόσυλος probably glossed the same word. The synizesis, rare as it is in Alcaeus, is not abnormal; cf. Lobel, Cαπφοῦς Μέλη (Oxford, 1925), pp. lxiii f.[9] συλάω is the *vox propria* for the spoliation of the gods; see K. Latte, *Kl. Schr.* (Munich, 1968), 416 f. Apart from the regular use for the robbing or seizing of property, it is also used of the seizure of persons (as in the Delphian manumission inscriptions quoted by W. Schulze, *Kleine Schriften* (2nd edn., Göttingen, 1966), 163), so that the seizure of

[8a] [See R. Parker, *Miasma* (1983), 147 f.]
[9] Cf. Kazik-Zawadzka, op. cit. (n. 8), 21.

Cassandra, just as much as a robbery from the temple, could be accounted an act of *θεοσυλία*. Still, this word, like *ἱεροσυλία*, doubtless served as a general term for sacrilege.

Among the various possible supplements *φῶτες*]*ϲι* is specially attractive because of its contemptuous note, but it is far from certain.

19. Alcaeus fr. 237. 1 calls Eros *δεινότατος θέων*. In Homer the verb *φύω*, even in the perfect, always conveys the suggestion of growth; this seems to be the earliest place in which *πέφυκε* means little more than 'is'. [Cf. Callim. fr. 260. 41 Pf. (= *SH* 288) *βαρὺς χόλος αἰὲν Ἀθήνης*.]

20. Lobel, P. Oxy. XXI, p. 87, shows that the dual is not abnormal. To the Lesbian instances of *παρθενίκα* given by Page (Alcaeus fr. 45. 5, Sappho fr. 44. 15) add now Sappho, P. Oxy. 2357 fr. 4. 4.

21. *παρεστάκοισαν* is the normal Lesbian equivalent of *παρεστηκυῖαν*: see Page, *S&A* 284.

22. Verbs meaning 'drag' are regular in this kind of context; note *Il.* 6. 465 *ἑλκηθμός* and 21. 62 *ἑλκηθείσας . . . θύγατρας*. Eur. *Tro.* 70 uses *ἕλκειν* of this very incident, and Lycophron's Cassandra says of herself *ἑλκυσθήσομαι* (358). Still, I favour *ἤισχυνν'* or *ὕβρισσ'*. With *οὐδ' ἔδεισεν* compare *Il.* 13. 623–4 and *Od.* 22. 39. In Quintus, Ajax 'shames' Cassandra (*ἤισχυνεν* 13. 422)[10] and Athene says of Ajax *οὐδ' ἔδδεισεν ἐμὸν μένος* (14. 438).

23. Page quotes Hes. *Op.* 356 *ἅρπαξ . . . θανάτοιο δότειρα*: there is also Pind. fr. 109. 3, where *στάσις* is called *πενίας δότειρα* and Nic. *Alex.* 612. *πολέμω δότερραν* probably means 'giver of strength in war', as in Pind. *Paean* 4. 40 *πόλεμον Διὸς* means 'Zeus' might in war'.

24. Perhaps it is fortuitous, but *γοργωπός* and *γοργῶπις* are so far attested in no author earlier than Aeschylus. With *ὑπ' ὄφρυσι* compare *Il.* 15. 607 f. *τὼ δέ* ‖ *οἱ ὄσσε* / *λαμπέσθην βλοσυρῇσιν ὑπ' ὀφρύσιν*. The context seems to demand that the word following *ὄφρυσι* should be a verb, and the only verb beginning with *σμ-* that seems at all likely to be suitable is *σμ*[*ῦχεν* which Treu has suggested; note *κενεαὶ γὰρ ὑποσμύχονται ὀπωπαί* at Ap. Rhod. 2. 445. But see p. 47 below.

25. The *Suda* s.v. *ὑποδράξ* quotes

> *ἡ δὲ πελιδνωθεῖσα καὶ ὄμμασι λοξὸν ὑποδρὰξ*
> *ὀσσομένη*

τουτέστιν ὠχριάσασα καὶ ὑποβλεψαμένη διὰ τὴν ὀργήν. Pfeiffer rightly prints the quotation as Callim. fr. 374; Hecker's conjecture that all hexameter fragments quoted in the *Suda* without an author's name belong to the *Hecale* has not been controverted in a single instance, despite the large amount of new material discovered since his time (see

[10] Cf. Triphiodorus 647–8:

> *Κασσάνδρην δ' ἤισχυνεν 'Οϊλῆος ταχὺς Αἴας*
> *Παλλάδος ἀχράντοιο θεῆς ὑπὸ γοῦνα πεσοῦσαν.*

R. Pfeiffer, edn. of Callimachus, ii (Oxford 1953), Prolegg. ad Fragm., pp. xxxiii–xxxiv and on fr. 238. 21). Reitzenstein and Wilamowitz thought the person referred to was the angry Medea (see Pfeiffer ad loc.). But the great rarity of the word *πελιδνωθεῖcα*, together with the fact that Alcaeus applied it to the angry Athene, supports Pfeiffer's suggestion that the passage comes from the crow's description of Athene's anger with her; see *HSCP* 72 (1968), 141–2 [= *Greek Comedy, Hellenistic Literature, Greek Religion and Miscellanea*, 149].[10a]

26. *ἄϊξε* is conjectured here by Page. *ἀΐccω* is, as he observes, often applied to the movements of the gods in Homer. He cites *Il.* 4. 78, where it is said of Athene that *ἤϊξεν ἐπὶ χθόνα*: four times (2. 167, 4. 74, 7. 19, 22. 187) her action is described in the formulaic line *βῆ δὲ κατ' Οὐλύμποιο καρήνων ἀΐξαcα. οἴνοπα πόντον*—in Homer usually preceded by *ἐπὶ*, once by *εἰc*—is yet another Homeric expression.

26–7. Compare *Od.* 5. 292. 3 *πάcαc δ' ὀρόθυνεν ἀέλλαc / παντοίων ἀνέμων*; 11. 400 *ὄρcαc ἀργαλέων ἀνέμων ἀμέγαρτον ἀϋτμήν*; 14. 254 *ὄρcαc ἀργαλέων ἀνέμων ἐπὶ πόντον ἀήταc*; 24. 110 *ὄρcαc ἀργαλέουc ἀνέμουc καὶ κύματα μακρά*. For *ἐξαπίναc* compare *Od.* 12. 288 *ἤν πωc ἐξαπίνηc ἔλθηι ἀνέμοιο θύελλα*; *Il.* 9. 6 *Βορέηc καὶ Ζέφυροc . . . ἐλθόντ' ἐξαπίνηc*; 17. 57 *ἐλθὼν ἐξαπίνηc ἄνεμοc cὺν λαίλαπι πολλῆι*.

ἀφάντοι[c is taken by Page to mean 'dark'; he compares Parmenides 9. 3 *φάεοc καὶ νυκτὸc ἀφάντου*. This may be right, but I think it likelier to mean 'invisible', 'lurking', like the Latin *latentes*; compare Soph. *Phil.* 296–7, where Philoctetes, describing how he rubbed two stones together to make fire, says *ἐν πέτροιcι πέτρον ἐκτρίβων μόλιc / ἔφην' ἄφαντον φῶc*.

κυκάω is not used of winds in Homer, but it is applied to the action of Achilles upon the Scamander at *Il.* 21. 235, 240, 324 and to the sea inside Charybdis at *Od.* 12. 238, 241; Sappho has it in a different sense at fr. 137. 4.

25–31. Lines 25–31 have the obelus against them, and Merkelbach points out that it probably stood against l. 24 also. Page infers that the lines obelized were wrongly placed and guesses that ll. 32–7 represent the stanza that should have ‖ followed ll. 20–3. 'We may have to suppose', he writes, 'that 32 was one of the obelised lines, or that 32 represents a second writing of 24, after the error involved in 25–31 had been detected' (ap. Merkelbach, p. 91). 'This is supported', he argues, 'by the fact that there is no Greek word beginning *cμ* which could have made sense between *ἃ δὲ δεῖνον ὐπ' ὄφρυcι* and *πελιδνώθειcα*: *cμέρδνον* could not stand together with *δεῖνον* and *cμέρδναιc* (for *-αιcι*) would be contrary to the dialect.'

30. The word following *Αἴαc* must, as Merkelbach remarks, have

[10a] [See now A. S. Hollis, *Callimachus*, Hecale (1990), pp. 41–4 and 240; in the latter place, for *D. P. Fowler*, read *R. L. Fowler*.]

been some case of the word Ἀχαιοί. Perhaps the sense was that Athene brought ruin upon the Achaeans for the sake of one man's crime. Note the scholion that stands below l. 24, and compare Lycophron 365–6:

ἑνὸϲ δὲ λώβηϲ ἀντί, μυρίων τέκνων
Ἑλλὰϲ ϲτενάξει πᾶϲα τοὺϲ κενοὺϲ τάφουϲ.

and Triphiodorus 650

ἀνθ' ἑνὸϲ Ἀργείοιϲιν ἐχώϲατο πᾶϲιν Ἀθήνη.

34–5. Page observes that ll. 34–5 might well follow l. 25, thus:

] ἀ δὲ δεῖνον ὑπ' ὄφρυϲι
] πελιδνώθειϲα, κὰτ οἴνοπα
ἔβαϲκε πόντον, ἐκ δ' ἀφάντοιϲ
παννυχίαιϲ ἐκύκα θυέλλαιϲ.

Although the simple verb βάϲκειν occurs elsewhere only in the imperative, Sappho fr. 21. 7 has ἀμφιβάϲκει, so that the articulation appears probable. But the conjecture is not recommended by the fact that, if it is accepted, θυέλλαιϲ has to carry two epithets. According to most authorities, Athene was helped by Poseidon, who raised the storm which scattered the Achaean fleet. Did ll. 34–5 refer to his action?

37–9. Merkelbach (p. 93) suggests 'e.g. [κὰτ οἴνοπα] ἄϊξε πόν[τον, ἐκ δ' ἀφάντων] ὦρϲε βία[ν ἀνέμων ...'. Perhaps; but the repetitions in the text are not so startling as to compel us to believe that something like this has happened. Still, ὦρϲε is just the verb in which we should expect the action of raising the winds to be described. See above, on ll. 26–7, and compare *Il.* 12. 253 ὦρϲεν ... ἀνέμοιο θυέλλαϲ and *Od.* 5. 109; for the winds, cf. *Od.* 5. 478 and 19. 440 ἀνέμων ... μένοϲ ὑγρὸν ἀέντων; note also Hes. *Op.* 625, *Theog.* 869.

41. πάνται occurs at Sappho fr. 44. 28 and fr. 65. 9.

47. ὠὔρραδον is puzzling. Pittacus is called τὸν Ὕρραον ... παῖδα at fr. 129. 13. Callim. *epigr.* 1. 2 Pf. calls him παῖδα τὸν Ὑρράδιον and this form occurs so often in the grammarians (see Pfeiffer ad loc.) that it must have been found in Alcaeus (so A. S. F. Gow, in *The Greek Anthology: Hellenistic Epigrams*, ed. Gow–Page (Cambridge, 1965), 205, on his *epigr.* 54. 2). Perhaps ωυρραδον is a mistake for ωυρραδιον. But in that case what is the initial omega doing? ‖ ωὐρράδ⟨ι⟩ον might be part of some such phrase as ωὐρράδιον γένεθλον, 'O offspring of Hyrras'.

48. Merkelbach recalls that the boat (ἀκάτιον) supplied by Mnemon for the return of Myrsilus occurs in a commentary on Alcaeus (fr. 305 i 17–18). But that can have nothing to do with this passage, and we cannot know whether κέλητο[here means 'boat' or 'horse'. For all we know, Ajax as he swam to the rock on which he eventually perished may have bestridden a plank as Odysseus did: ἀμφ' ἑνὶ δούρατι βαῖνε, κέληθ' ὡς ἵππον ἐλαύνων (*Od.* 5. 371).

II

The commentary shows the pervasive influence of the epic on the poem's style and diction; 'the Epic theme, here as elsewhere, brings features of Epic style with it', as Page wrote with only the Oxyrhynchus fragment before him (*S&A* 285). The rape of Cassandra had been described in the *Iliou Persis* and the storm in the *Nostoi*, as well as in the *Odyssey* (4. 495 f.).[11] The crime of Ajax led to the greatest catastrophe that befell the Achaeans during the whole Trojan episode; by transforming their chief helper among the gods into a deadly enemy, it plunged them in the very moment of their triumph into disaster. Alcaeus could have chosen no more signal example of the general truth that it is often better for a community to destroy a member of it who has incurred the anger of the gods, in order that its other members may not share his fate.

The punishment of stoning, which according to Alcaeus the Achaeans would have been wise to apply to Ajax, was particularly appropriate to the perpetrators of crimes likely if unpunished to bring disaster upon the whole community. Rudolf Hirzel in a masterly paper[12] showed that its origins go back to an early period of society, and that its special aptness for punishing such crimes lay in the fact that every member of the community could take a part in it, so that the collective guilt could easily be purged by its infliction. Originally, indeed, it was not so much a means of inflicting death as a means of extruding a member from his share in the community, as in the Ionian ceremony the φαρμακοί were extruded. In the *Iliad* Hector tells Paris that he deserves stoning (3. 56–7), and epic and tragic poetry abound with references to that punishment.[12a] ‖

There can be little doubt that in the *Iliou Persis* the Achaeans came near to stoning Ajax for his offence against Cassandra. The famous painting of the scene soon after the sack of Troy with which Polygnotus decorated the Cnidian Lesche at Delphi showed Ajax standing by an altar with a shield, ὀμνύμενος ὑπὲρ τοῦ ἐς Κασσάνδραν τολμήματος. Cassandra was sitting on the ground and holding the image of Athene, and Agamemnon and Menelaus, wearing helmets, were administering

[11] For the rape of Cassandra in literature, see Juliette Davreux, *La Légende de la prophétesse Cassandre* (Bibl. Liège 94, 1942), 42 f.; for a short bibliography, see Page, *S&A* 285 n. 1.

[12] 'Die Strafe der Steinigung', *Abh. Leipzig*, 27 (1909), 225 f.

[12a] [See D. Fehling, *Ethologische Beiträge auf dem Gebiet der Altertumskunde* (1974), 59 f.]

the oath to Ajax (Paus. 10. 26. 3). A few chapters later, Pausanias mentions that Ajax and Odysseus were enemies because Odysseus advised the Greeks to stone Ajax for his crime against Cassandra (10. 31. 2).

What oath was Ajax swearing? Carl Robert was certainly wrong in thinking[13] that he was swearing to redeem his offence by instituting the famous tribute of the maidens whom his native country of Opuntian Locri later used to send to Troy. In a trial according to Homeric justice the defendant would deny his guilt on oath; thus during the quarrel that arose out of the chariot-race during the funeral games of Patroclus, Menelaus challenges Antilochus to deny on oath that he has cheated (*Il.* 23. 581 f.). Ajax presumably took an oath of this nature; such an oath might be seconded by friends of the swearer, who would take a similar oath on his behalf.

It seems that some standard account described a trial of Ajax for his crime against Cassandra; perhaps Odysseus was the prosecutor. Carl Robert long ago made it probable[14] that the main authority followed by Polygnotus in his Iliou Persis was the cyclic epic of the same name. In Polygnotus' painting (see Plut. *Cim.* 4) in the Stoa Poikile, Pausanias (1. 15. 2) saw 'the kings assembled on account of the crime of Ajax against Cassandra'. That too indicates that an early standard authority described a trial scene, and this was in all probability the *Iliou Persis*. The most detailed account of the episode which we possess occurs in a work more than a thousand years later than the *Iliou Persis*, in the second of the *Refutationes* ascribed to Libanius and in the reply to it contained in the third of the *Confirmationes* attributed to that author.[15] Calchas ‖ tells the Achaeans that Athene is angry with them, and Ajax is put on trial; Odysseus prosecutes (*Ref.* 2. 9), but Ajax escapes by taking refuge at an altar. An assembly is held, and Menelaus insists on sailing for home at once, while Agamemnon stays behind to attempt to placate Athene. That last detail is given also in Proclus' summary of the *Nostoi* attributed to Agias,[16] and it seems probable that the account given in the *Iliou Persis* and the *Nostoi* is being followed.

Further there is an indication that the story of Ajax's trial was told

[13] *Die griechische Heldensage*, iii (Berlin, 1923), 1268. The right view is taken by A. C. Pearson, *The Fragments of Sophocles*, i (Cambridge, 1917), 11. For the oath in Homeric legal procedure, see K. Latte, 'Der Rechtsgedanke im archaischen Griechentum', *Kl. Schr.* 237. [See *The Justice of Zeus* (2nd edn. 1983), 202 n. 110.]

[14] *Hallisches Winckelmannsprogramm*, 17 (1893); see pp. 74 ff., esp. p. 79.

[15] Libanius, ed. Förster, viii (Leipzig, 1915), 128 f. and 150 f. The author was probably the 5th-c. sophist Nicolaus of Myra; see *RE* xii (1925), 2520 s.v. Libanios and (1936), 448 s.v. Nikolaos 21.

[16] *Homeri Opera*, ed. T. W. Allen, v (OCT, Oxford, 1946), p. 108; E. Bethe, *Der troische Epenkreis* (Darmstadt, 1966), 36 (= *Homer, Dichtung und Sage* (Leipzig, 1922), ii. 182); A. Severyns, *Recherches sur la Chréstomathie de Proclos*, iii: *La* Vita Homeri *et les sommaires du Cycle*, i (Bibl. Liège 132, 1953) Procli Fragmenta, p. xi.

by a commentator on the particular poem of Alcaeus which we are now discussing. In *CR*, NS 15 (1969) [R49], 72 I combined frr. 108 and 84 of the work on early lyric published as P. Oxy. 2506 (in Part XXIX of the series) to give the following result:

ἀ]λλὰ καὶ τὸ[ν] ὑπὸ τῶν [
Αἴ]αντα κατη⟨ι⟩τιαμένο[ν
]α [ἐπὶ] τῆι Καϲϲάνδρα[ι
κ]αὶ γὰ[ρ] ạὐ̣τ[ὴν τῶι ἀ]γάλμ̣[ατι

It seems likely that Ajax was said to have been 'accused' by the Achaeans for having committed a crime (e.g. *παρανομήϲαντα*) against Cassandra, and that the writer went on to describe how he seized her as she clasped the image of Athene. The word 'accused' indicates that this author knew the story of the trial of Ajax, which as we have seen probably occurred in the *Iliou Persis*.

The storm struck the fleet as it was passing Aegae (l. 6); [Alcaeus may have understood Aegae in Lesbos, but the original story must have concerned] the place in Euboea variously called Aige, Aigai, and Aigaia and identified by Stephanus of Byzantium with Carystus.[17] In the Towneleian Scholia to the *Iliad* (13. 21), this is said to be the Aegae to which Poseidon repairs from Samothrace to harness his chariot before travelling to Troy; the same view is taken by Strabo 386 (cf. 405). The unidentified author of a work called *Nesias*, 'the book of islands', cited in *Σ* A.R. 1. 1165 (p. 105 Wendel) takes this to be the place from which the name of the Aegean Sea || derives.

In the *Odyssey* Ajax meets his end at the 'Gyraean rocks', which are regularly located in the neighbourhood of Mykonos, Delos, and Tenos.[18] But Proclus in his summary of the *Nostoi* puts the storm and the death of Ajax near Caphereus, the southern promontory of Euboea; and most later authorities either give this version only or contaminate this version with the other in which the neighbourhood of Delos is involved. Bethe in his book on the Trojan epics[19] assumes that Euboea came into the story only with the legend of Nauplius, who avenges the death of his son Palamedes at the hands of the Achaeans by lighting the beacon which lured them on to the jagged cliffs of the coast near Caphereus; he supposes that Proclus wrongly stated that the *Nostoi* mentioned Euboea through an accidental memory of the other

[17] See W. Spoerri in *Lexikon des frühgriechischen Epos* (Göttingen, 1955–), 241 f. s.v. *Αἰγαί*. [See R. L. Fowler, *Phoenix*, 42 (1988), 100–1 n. 17.]

[18] See F. H. Sandbach, *CR* 56 (1942), 63 f.

[19] Op. cit. (n. 16), 131 (= *Homer, Dichtung und Sage*, ii. 279). The reprinting of this work, invaluable for its edition of the fragments of the epics in question, makes it necessary to remind the reader that Bethe's thesis that 'The Little Iliad' was simply a name for the *Aethiopis*, *Iliou Persis*, and *Nostoi* together was refuted by Severyns, *Rev. phil.* 49 (1925), 162 f.

version. But now that we know that Alcaeus mentioned Aegae, it becomes virtually certain that Proclus is correct. In the *Odyssey* the storm overtakes the Greek near Mykonos; but what is true of the *Odyssey* is not necessarily true of the *Nostoi*. If in later writers the location in Euboea is found together with the Nauplius story, that does not prove that the two went together from the start; the *Nostoi* might have made the storm strike near the Euboean coast without bringing in Nauplius. In fact Bethe himself (p. 132) observes that Nauplius was mentioned in the *Nostoi*; Pausanias (1. 22. 6) mentions that his sons helped Aegisthus against Orestes, which Bethe (p. 132) thinks he may have found in that epic. Although Athene and Poseidon caused the storm, so that Nauplius' wrecking activity is superfluous, the two stories are by no means inconsistent. For all we know, the story that the Achaeans, buffeted by the storm, saw what seemed to be a welcoming light and were lured straight on to the most dangerous rocks of the Mediterranean may be as old as the *Iliou Persis*.

5(85)

Λυκαιχμίαις

THE text of the new commentary on Alcaeus (P. Oxy. 3711. 32) indicates that the correct reading of Alcaeus 130. 25 is *λυκαιχμίαις*, and not *λυκαιμίαις* ('wolf-thickets'), as Page thought.[1] Haslam suggests that *λυκαιχμίαις* is nominative, with the meaning either of 'wolf-man' (on the basis of the gloss *λυκόβροτος* in Hesych. *λ* 1369) or some (otherwise unparalleled) cognate of *αἰχμή*, 'wolf-spearman'. Presumably then, these lines would emphasize Alcaeus' isolation; the wolf or wolf-man who lives alone, apart from men.[2]

But it is also possible that *λυκαιχμίαις* is dative plural, as Haslam would prefer. As Page observed, Alcaeus also uses the short dative at the end of a line later in this poem (130. 30) and also in 308(*b*). 2;[3] cf. also Anacreon 356(*b*). 4–5 *καλοῖς / ὑποπίνοντες ἐν ὕμνοις* and 378. 1 *πτερύγεσσι κούφηις* (*κούφαις* codd.).[4] The word would then mean 'wolf-battles', rather than 'wolf-warriors', like *ἱππαιχμίαις*, 'horse-battles' *Σ* Pind. *Nem.* 1. 23 (iii. 14 Dr.); the lines would describe not Alcaeus' loneliness, or his kinship to the werewolves of myth, but his war tactics. In [Eur.] *Rhes.* 208 ff. Dolon puts on a wolf-skin for practical (rather than symbolic) reasons, to make his tracks four-footed and so hard for the enemy to follow; cf. the stratagem used by Josephus against Vespasian in Jotapata, in which his men, in order to get by the guards, 'covered their backs in skins (*νάκεσιν*), so that if someone saw them he should have the illusion that he was dreaming about dogs' (*Bell. Jud.* 3.

[1] M. W. Haslam, P. Oxy. 53 (1986), 123; cf. D. L. Page, *Sappho and Alcaeus* (1955), 205.

[2] R. A. Buxton, 'Wolves and Werewolves in Greek Thought', in J. Bremmer (ed.), *Interpretations of Greek Mythology* (1986), 63; cf. C. Mainoldi, *L'Image du loup et du chien dans la Grèce ancienne* (1984), 128; L. Gernet, 'Dolon le loup', in *Anthropologie de la Grèce antique* (1968), 164.

[3] Page (n. 1) 208. The passages cited in this connection by E.-M. Hamm, *Grammatik zu Sappho und Alkaios* (1957), 147 and A. M. Bowie, *The Poetic Dialect of Sappho and Alcaeus* (1981), 112 ff. add nothing of significance. There is real reason to doubt three, and perhaps four, of the six cases where Bowie suggests that the short dative is a strong possibility. At Sappho 2. 14 both Lobel–Page and E.-M. Voigt print Bergk's *ἄβρως*; at 55. 3 all print Fick's *δόμωι* and at 81. 1 his *ἐράτοις*; at 160 Lobel–Page express doubt about the text and Voigt and Campbell place a crux against *τέρπνα*; at Alcaeus 333 all print Lobel's *ἀνθρώπω*.

[4] For alleged instances of the short dative in the iambographers, see H. Lloyd-Jones, *Females of the Species* (1975), 81–2, on Semon. 7. 70 W.

Zeitschrift für Papyrologie und Epigraphik, 68 (1987), 9–10 (with M. R. Lefkowitz).

192).[5] The wolf's best trick is his resemblance to a dog, 'the most savage to the gentlest of animals' (Plato, *Sophist.* 231 A); his methods always involve cunning, so that wolf-warfare is secret and deceptive, the antithesis of open and regulated hoplite warfare. But, as Pindar says in *Pyth.* 2. 82 ff., the wolf's methods are at least courageous; unlike those of the 'crafty ‖ citizen' (δόλιος ἀςτός) who flatters everyone and weaves his deception, the wolf is a friend to his friends, and 'towards an enemy like an enemy I will run under like a wolf, from time to time treading on crooked paths'.[6] Thus the Onomacles—whoever he was—to whom Alcaeus compares himself was not 'a type of exile or hermit or lone-wolf'[7] but an active fighter, not in conventional combat (φεύγων τ[ὸν] π[όλ]εμον, P. Oxy. 3711. 33–4, cf. 130. 26]ον [π]όλεμον), but in guerrilla warfare; as he says in the final lines of the stanza, 'it is better not to renounce rebellion' (ςτάςιν γὰρ . . . οὐχ †ἄμεινον ὀννέλην, 26–7).

[5] Cited by Porter in his edition of the *Rhesus* (1916). Cf. also the behaviour of the German *ferialis exercitus* with black shields and painted bodies described by Tac. *Germ.* 43; Gernet (n. 2), 161 n. 4.

[6] Buxton (n. 2), 64; W. Burkert, *Homo Necans* (1972), 102–3.

[7] Page (n. 1), 204.

6(60)

More about Antileon, Tyrant of Chalcis. Solon fr. 33 and Aristophanes, *Equites* 1042–4

> Aristotle, *Pol.* 5. 10. 3, 1316ᵃ29 ἀλλὰ μεταβάλλει καὶ εἰc τυραννίδα τυραννίc, ὥcπερ . . . , καὶ εἰc ὀλιγαρχίαν, ὥcπερ ἡ ἐν Χαλκίδι ἡ Ἀντιλέοντοc, καὶ εἰc δημοκρατίαν, ὥcπερ. . . .
>
> Alcaeus 296 (P2) a 8 (end of poem)
>
> μᾶλλόν]κ̣' ἄξιοc Ἀντιλέοντ[οc[1] ὅδ'] ἦc ἀπυδέρθην.

Oxford — P. MAAS

[1] ἀντὶ λέοντ[οc edd.: correxi. On tyrants in Euboea cf. E. Ziebarth, *Inscr. Graec.* xii. 9, p. 147.

I HAVE reproduced a characteristically concise note by Paul Maas ('How Antileon's Tyranny Ended', *CR*, NS 6 (1956), 200 = *Kl. Schr.* (Munich, 1973), 4). Here are two additions:

1. Solon fr. 33 West

> οὐκ ἔφυ Σόλων βαθύφρων οὐδὲ βουλήειc ἀνήρ.
> ἐcθλὰ γὰρ θεοῦ διδόντοc αὐτὸc οὐκ ἐδέξατο·
> περιβαλὼν δ' ἄγραν ἀγαcθεὶc οὐκ ἐπέcπαcεν μέγα
> δίκτυον, θυμοῦ θ' ἁμαρτῆι καὶ φρενῶν ἀποcφαλείc.
> ἤθελον γάρ κεν κρατήcαc, πλοῦτον ἄφθονον λαβὼν
> καὶ τυραννεύcαc Ἀθηνέων μοῦνον ἡμέρην μίαν,
> ἀcκὸc ὕcτερον δεδάρθαι κἀπιτετρίφθαι γένοc.

We do not know Antileon's date. But since Alcaeus took it for granted that his hearers would know the story, it would not be especially surprising if Solon did also. If Solon had a known story in mind, the speech which he puts into the mouth of his imaginary critic gains something in effectiveness.

2. Ar. *Eq.* 1036–44 (the Paphlagonian is expounding an oracle to his master Demos):

Classical Philology, 60 (1975), 197.

ΠΑ. ὦ τᾶν, ἄκουσον, εἶτα διάκρινον, τόδε.
Ἔστι γυνή, τέξει δὲ λέονθ᾽ ἱεραῖς ἐν Ἀθήναις,
ὃς περὶ τοῦ δήμου πολλοῖς κώνωψι μαχεῖται
ὥς τε περὶ σκύμνοισι βεβηκώς· τὸν σὺ φυλάξαι,
τεῖχος ποιήσας ξύλινον πύργους τε σιδηροῦς.
ταῦτ᾽ οἶσθ᾽ ὅ τι λέγει;
ΔΗ. μὰ τὸν Ἀπόλλω ᾽γὼ μὲν οὔ.
ΠΑ. ἔφραζεν ὁ θεός σοὶ σαφῶς σώιζειν ἐμέ.
ἐγὼ γὰρ ἀντὶ τοῦ λέοντός εἰμί σοι.
ΔΗ. καὶ πῶς μ᾽ ἐλελήθεις Ἀντιλέων γεγενημένος;

Σ Ἀντιλέων: οὗτος πονηρὸς κωμωιδεῖται καὶ πολυπράγμων. προειρήκει δὲ ὁ Κλέων ἀντὶ τοῦ λέοντος αὐτῶι εἶναι φύλαξ. VΕΓΘΜLh (*Scholia in Aristophanem*, Pars I, Fasc. II (Groningen, 1969): *Scholia vetera in Equites*, ed. D. Mervyn Jones et *Scholia Tricliniana in Equites*, ed. N. G. Wilson, p. 231).

One needs no very wide acquaintance with the ancient commentaries on Aristophanes to know that the explanation in the scholia has no authority. Explanations of the identity of *κωμωιδούμενοι* are very often guesses, based solely on the text of the play itself. If the Antileon whom Demos has in mind is none other than the tyrant of Chalcis, the joke is not without point.

I regret my failure to equal Maas's matchless brevity.

7(77)

Pindar[1]

By no means every good poet can be credited with a master mind. To deserve such a designation, a poet does not need to have a philosophy, or to have versified the philosophy of others; he does need to have produced a considerable body of verse which communicates a distinctive vision of the world, conveyed with great imaginative power. If this canon is applied with strictness, only great poets will be found to satisfy it.

The poet I have chosen to speak about has a name familiar to most educated people. Since the Renaissance several famous poets have imitated him, or have believed that they were doing so; for example Ronsard and Chiabrera, Cowley and Gray, the young Goethe and Hölderlin. There are many translations, including good modern ones[2] by Richmond Lattimore, Maurice Bowra, and recently Frank Nisetich, and much scholarly work has been devoted to Pindaric interpretation. But since the Renaissance comparatively few people with a working knowledge of the Greek language have studied Pindar with serious attention; and among these only a minority would have claimed, stating their honest opinion, that he deserved to be considered as a master mind in the sense that I have just described.

This is partly because Pindar has always had a reputation for being difficult. One cannot deny that he is difficult, but the difficulty has been exaggerated; his style and language, once superficial awkwardnesses have been overcome, are hardly as difficult as those of Sophocles. His text is better preserved than that of any of the great tragedians; and though some of his poems are written in complicated metres, about half are written in ‖ dactylo-epitrite, whose main features can be set out in half a page.[3] The main difficulties in the way of literary appreciation lie in the conventions of the genre he writes in: the

[1] References are to the Teubner text edited by B. Snell and revised by H. Maehler (vol. i, 1971; vol. ii, 1975). [Maehler has now revised vol. ii in his own name (Leipzig, 1989).]

[2] Lattimore, *The Odes of Pindar* (1947; 2nd edn. 1976); Bowra, *The Odes of Pindar* (1969); Nisetich, *Pindar's Victory Odes* (1980) (with useful explanatory material).

[3] See P. Maas, *Greek Metre* (1962), 40–1.

Proceedings of the British Academy, 68 (1982), 139–63 (Lecture on a Master Mind).

sometimes abrupt transitions from one topic to another are often hard to grasp and the unity of the poems has proved so hard to comprehend that many scholars have denied that they have any unity at all. We need to learn to read Pindar not only in the sense of how to construe him, but in that of how to view each poem and each part of each poem in the light of the tradition it belongs to; and it is only very lately that scholars have begun to give readers the assistance that they need.

Readers would not have been scared away by Pindar's difficulty if they had been convinced that the effort needed to understand him would be rewarded; but for many years the people best qualified to help them have not encouraged them to make that effort. The great period of German scholarship which ended, roughly speaking, with the fall of the empire did much for the editing of the text and of the extensive ancient commentaries, and for the philological and historical elucidation of Pindaric problems. But its principal figures complained that Pindar's world was strange to modern minds, and that he lacked sympathy with the new science and philosophy that in his time was active in the eastern part of the Greek world.[4] They did not regard him as the equal of the Attic tragedians, who in their view stood for progress while Pindar was the poet of an aristocratic world doomed to rapid extinction. Even Gilbert Murray, whose feeling for poetry strongly responded to Pindar's verse, complained that Pindar was 'a poet and nothing more';[5] he would not have credited him with being a master mind. A living scholar who has done much for Pindar, Bruno Snell,[6] writes that in Pindar's time Attic tragedy was requiring that justice should be done on earth, and in consequence was making many demands not only of men but of the gods, but that Pindar kept himself far from any such thoughts, which seemed to him presumptuous.

If a great poet belongs to a world remote from our own and ‖ gives expression to its outlook, that is no reason why we should not respond to the greatness of his poetry; it is enough to think of Homer, or of Dante. The notion that, because the world of Pindar's aristocratic patrons was visibly perishing before his death, therefore his poetry is somehow less valuable than that of poets belonging to communities whose prosperity was then beginning scarcely deserves respect. Neither does the complaint that a lyric poet shows no interest in science and philosophy: if Keats reproached Newton with having explained the

[4] Representative opinions are quoted in my article 'Modern Interpretation of Pindar: The Second Pythian and Seventh Nemean', *JHS* 93 (1973), 109 f. [= ch. 9]. I hope the reader of that piece will pardon a few repetitions in this, made necessary by my subject.

[5] In *The Literature of Ancient Greece* (1897; repr. 1966), 112 (cited on p. 110 of the article referred to in n. 4).

[6] *Die Entdeckung des Geistes* (4th edn. 1975), 94; cf. the English version by T. G. Rosenmeyer, *The Discovery of the Mind* (1953), 89.

mystery of the rainbow, shall we refuse to read Keats's poetry on that account? In fact the tragedians see life in terms of the traditional religion scarcely less than Pindar does, for all the play Euripides may make with the new rhetoric and with casual allusions to the new speculations.

Many people are prejudiced against Pindar for political and social reasons. He was an aristocrat,[7] belonging to one of the great clans whose members were spread over different civic communities. His own city, Thebes, belonged to the division of the Greek race called Aeolian, and like most Dorian and Aeolian communities was ruled by an aristocracy; at the time of the Persian invasion of 480, geography determined that Thebes had about as much chance of resisting the invader as Denmark in 1940 or Czechoslovakia in 1945. Many of those who commissioned Pindar to celebrate their victories in the great games were noblemen; some were even monarchs, like the rulers of the Greek colonies in Sicily or in Cyrene. In his poetry, Pindar often stresses the importance of breeding; learning is no use, he says, unless nature has made one apt to learn. A distinguished ancient historian, perhaps not especially skilful in linguistic or aesthetic matters, seems to think it morally wrong for anyone to admire Pindar's poetry.[8]

We should not now make the mistake of exaggerating the differences between the Dorian, Aeolian, and Ionian divisions of the Greeks. The poetry of the archaic period, like its art, cut across these barriers; thus Pindar's older contemporary Simonides and his nephew Bacchylides came from the Ionian island of Ceos, but this did not prevent them from investing their choral lyric poetry with the thin patina of West Greek dialect which was customary in a genre belonging traditionally to western Greece. Pindar did not write poems only for individuals: he was sometimes commissioned ‖ by communities, including, on more than one occasion, democratic Athens. He often proclaims his sympathy with his patrons of the moment, and several times praises the Dorian institutions observed in Syracuse and Aegina; yet he nowhere, if his text is understood correctly, makes a declaration of political allegiance, as Yeats did when he wrote a poem for General O'Duffy and his Blueshirts. Indeed, W. S. Barrett has made it seem probable that in the Twelfth Olympian he saluted the new régime in Himera which had replaced one controlled by his former patrons, the rulers of Akragas.[9] The truth is that he was not a romantic individualist, like Yeats, but a poet operating within the terms of fixed religious and poetical conventions, and like the tragedians he is concerned less with

[7] See M. R. Lefkowitz, *HSCP* 67 (1962), 230 f.

[8] M. I. Finley, *Aspects of Antiquity* (1968), 43 f.

[9] See his article in *JHS* 93 (1973), 23 f.

what is temporary and accidental in human life than with what is permanent and fundamental.

The preponderance in Pindar's extant work of odes written to celebrate athletic victories is due to accident. His poems were collected by the scholars of Alexandria in seventeen books. Only six of these consisted of poems written for men; these were the four books of victory odes for athletes, with one book of enkomia and one of dirges.[10] The remaining eleven books consisted of poems written for the service of the gods, most of them what we should call hymns of one sort or another. Of this large output only the victory odes have survived virtually entire. But we have enough fragments of Pindar's other poems, some from quotations by other authors and others from papyri discovered in the sands of Egypt, to have a very fair notion of their character. It is clear that the religious background, the use of myth, and the general outlook expressed in the victory odes were, like their language, style, and metre, similar to those found in the other poetry of Pindar.

The convention of the victory ode was already established well before the date of the Tenth Pythian, the earliest poem of Pindar whose date we know. The great poet Simonides, born in 556, more than thirty years before Pindar, must long before then have been an established writer in this genre. Almost all surviving victory odes were written to honour victories won in the four great games held in connection with Panhellenic religious festivals. The Olympic Games at Olympia in the western Peloponnese and the Pythian Games at Delphi were held every four years, and the ‖ Nemean Games at Nemea in the Argolid and the Isthmian Games on the Isthmus of Corinth were held every two years. It is generally assumed that each victory ode was performed, with an accompaniment of dance and music created by the poet, by a chorus of men, trained if circumstances permitted by the poet himself; the performance took place either at the scene of the games soon after the victory, or later at the victor's home. Most remarkably, the poet speaks through the chorus in the first person singular, in his own person.[11] This is not the case in all of Pindar's poetry: in his poems for choirs of maidens, for example, the chorus speaks in its own character. The lost poems of which we know most are the paeans, a kind of lyric poem in honour of Apollo or Artemis or both, but occasionally extended to the service of the other gods. In some paeans the chorus speaks in its own person, but in others the poet

[10] See *Vita Ambrosiana* in *Scholia vetera in Pindari carmina*, ed. A. B. Drachmann, i (1903), 3. 6 f.; cf. P.Oxy. 2438. 35 f.; for a discussion of the discrepancy between these authorities, see I. Gallo, *Una nuova biografia di Pindaro* (n.d., 1976?), 72 f.

[11] See M. R. Lefkowitz, art. cit. (n. 7), 177 f.

speaks. Was every victory ode performed by a chorus? Till lately it was believed that the lyric narrative poems on epic themes composed by the sixth-century Sicilian poet Stesichorus were performed chorally; but now that we know that several of them had many more than a thousand lines it seems likelier that the poet or another chanted them to the accompaniment of the lyre, as rhapsodes did the poetry of Homer.[12] One of Pindar's victory odes, the Fourth Pythian, is three hundred lines long, and largely consists of epic narrative; I wonder whether this, and perhaps also other victory odes, was not performed, as Pindar's encomia must have been, by a single person.

A victory ode normally contains certain recurring elements. It praises a particular victor on a particular occasion, and usually his family and his city also; it usually contains a modicum of general reflection upon human life in relation to the gods, and often the narration of a myth judged to be appropriate to the occasion. But these elements occur in different orders, and are linked together by different methods of transition. Since August Boeckh[13] early in the nineteenth century laid the foundations of modern Pindaric scholarship, the unity of the victory ode has been the subject of much controversy.

For the greater part of that century, most scholars believed in ‖ its essential unity. Boeckh himself distinguished between an objective unity determined by the occasion for which the ode was written and a subjective unity dictated by the poet's personal preoccupations. His collaborator Ludolf Dissen introduced the notion of a 'Grundgedanke', or central thought; he and many others did harm by trying to boil down the content of whole poems into painfully trite and jejune summaries. At the end of the century this produced a violent reaction, whose main proponent was Wilamowitz.[14] Sharply rejecting the idea that a victory ode had any unity, he and his followers concentrated on the philological exegesis of the text and on its historical aspect, taking comparatively little interest in the construction of the poems and the poet's art. In 1928 Wolfgang Schadewaldt[15] so far modified this trend as to offer a refined version of Boeckh's objective and subjective unity. Objective unity was determined by what Schadewaldt called the 'programme' of the ode, the praise of the victor and his home and family and other historical elements; to these were added the general

[12] See M. L. West, *CQ*, NS 21 (1971), 302 f.; M. W. Haslam, *QUCC* 17 (1974), 33, with n. 53.

[13] *Pindari opera quae supersunt* (2 vols., each in 2 parts, 1811–21); the commentary on the Nemeans and Isthmians is by Ludolf Dissen, who later published his own commentary on the whole of Pindar, *Pindari carmina quae supersunt cum deperditorum fragmentis selectis* (1830; 2nd edn., with additions by F. G. Schneidewin, 1843).

[14] Especially in *Pindaros* (1922). On the Pindaric studies of Wilamowitz, see L. E. Rossi, *ASNP*, 3rd ser., 3 (1973), 119 f.

[15] *Der Aufbau des pindarischen Epinikions* (1928, repr. 1966).

reflection and the narration of the myth; and upon this material the poet imposed other matter derived from his own thoughts and feelings. A different kind of neo-unitarianism has been advocated by Hermann Fränkel,[16] who found the main content of each ode to lie in its contribution to the general picture of the divine government of the world and the values which it generates that is built up by the poet's work as a whole. In his detailed exposition of Pindaric poems Fränkel has proved one of the best modern critics; but to say that the individual ode, composed for performance on a particular occasion, must be explained in terms of its relation to the entire corpus, like different pieces of a jigsaw puzzle that are given away in different packets of some product, is an odd way of dealing with the specific problem of the poem's unity. Other kinds of neo-unitarianism, based upon a more or less vaguely conceived symbolism, seem to me to have been a good deal less effective.

A new trend started in America during the early sixties, one not unconnected with Milman Parry's work on the Homeric formula, and indeed with the general movement towards 'structuralist' ‖ analysis of poetry and other things. In 1962 Elroy R. Bundy[17] brought out detailed studies of two Pindaric victory odes, the Eleventh Olympian and the First Isthmian. He strongly insisted on the importance of understanding the conventions of the epinician genre, showing how often formal devices that conform to standard patterns are used at corresponding places in different poems in order to effect transitions. In several places he and his followers have been able to resolve in terms of such devices problems which earlier scholars had tried to explain by assuming the presence of personal or historical allusions for which there was no independent evidence. For this purpose Bundy invented a formidable technical terminology, oddly recalling the edition published at Wittenberg in 1616 by Erasmus Schmid, who analysed each ode in terms of the categories prescribed in the rhetorical handbooks of the early Roman Empire. As an Ariadne's clue to the unity of the victory ode, Bundy offered the principle that its main purpose is to praise the victor, and that all that it contains ultimately subserves that purpose.

Two years after Bundy's studies there appeared a survey of Pindaric criticism from the time of Boeckh, in which another American scholar,

[16] *Dichtung und Philosophie des frühen Griechentums* (1950; 2nd edn. 1962); English translation by M. Hadas and J. Willis as *Early Greek Poetry and Philosophy* (1973); see also his 'Pindar's Religion', *Die Antike*, 3 (1927), 39, reprinted in W. M. Calder III and J. Stern, *Pindar und Bakchylides* (Wege der Forschung 134, 1970), 232 f.

[17] *Studia Pindarica* I, II (University of California Publications in Classical Philology, 18/1–2 (1962), 1–34 and 35–92 [repr. 1986]).

David Young,[18] forcefully argued for a not dissimilar approach; he later followed it with studies of four odes, contending that each is 'a unified, meaningful work of literary art'. Young objects that Bundy recognizes no unity in the epinician beyond mere linear continuity, and concentrates too narrowly upon the theme of praise. He himself uses his grasp of the traditional themes and the conventional rhetoric of the genre to explain the complex but coherent working-out of each poem's argument. Since then a number of scholars in Europe as well as in America have adopted a similar approach.

The new movement has done a great service to the understanding of Pindar by emphasizing the importance of convention in his art, and so helping to correct the vast damage done by the fatal conjunction of nineteenth-century historicism with nineteenth-century Romanticism. The Romantics held all poetry, and particularly lyric poetry, to be the spontaneous outpouring of the poet's sincerest feelings; a genuine poet felt an overmastering urge to ‖ express in verse his own most deeply held convictions. In many ways they appreciated the Greek tragic and lyric poets better than their eighteenth-century precursors, whose understanding of the poetry of archaic and classical Greece was limited by their preoccupation with the rules of their own classicism; and it was natural that they should assume that the Greek poets were like themselves, gravely underrating the importance of form and convention in their work. Wilamowitz in his famous book on Pindar, published in 1922, when he was 74 years of age, adopted a biographical approach; so did Maurice Bowra as late as 1964,[19] never doubting that Pindar was a lyric poet like Yeats, expressing his personal attitude to the world and to those with whom he might have dealings. Both relied upon the data furnished by the ancient lives of Pindar and the ancient commentaries upon his work, adding further matter based on modern inference from the actual poems. But as Mary Lefkowitz in particular has shown,[20] little reliance can be placed upon the biographical traditions about the early Greek poets; much of them has no securer basis than jokes from comedy later taken seriously by owlish grammarians or frivolous speculations based upon the poets' actual works. The most solid part of the ancient commentators' evidence for Pindar's life is chronological: lists of the victors at Olympia and Delphi were available, so that most Olympian and Pythian odes are dated. But the

[18] First published in *The Minnesota Review* 4 (1964), 584 f.; reprinted with some changes in Calder and Stern, op. cit. (n. 16), 1 f.; cf. *Three Odes of Pindar: A Literary Study of Pythian 11, Pythian 3 and Olympian 7* (*Mnemosyne*, Suppl. 9, 1968); *Pindar, Isthmian 7, Myth and Exempla* (*Mnemosyne*, Suppl. 15, 1971).

[19] *Pindar* (1964).

[20] See *The Lives of the Greek Poets* (1981), 57 f.; also 'Pindar's Lives', *Classica et Iberica* (*Festschrift Marique*) (1975), 71 f., and other articles listed in *The Lives of the Greek Poets*, 179.

dates of Nemean and Isthmian odes are matter for conjecture; and the attempts made by stylometric methods to guess at the date of poems whose date we do not know have not, so far at least, had very much success.

Romanticism as well as historicism has derived encouragement from the ancient commentators, who seem to operate with a theory of poetic inspiration deriving, in the last resort, from Plato. The poet writes in frenzy, they assume, as though improvising; when he employs the traditional device of apologizing for a digression by saying that in his excitement he has been carried far out of his path, they innocently take his word for it. Since the Renaissance a vast amount of misunderstanding has been fostered by the second ode of Horace's fourth book, in which he warns a would-be imitator of Pindar of the hazards of the attempt, depicting Pindar's verse as a rushing torrent, 'bound by no laws'. In fact Pindar scrupulously obeys many laws, not only those of || a strict form and metre, the same pattern of strophe, antistrophe, and epode being repeated many times in the same poem, but the conventions of the genre which it is the great merit of the new movement to have made us properly aware of. In doing so it has placed new weapons in the hands of those who would defend the essential unity of the victory ode, and since it started anti-unitarianism has been very much on the defensive.

But the new trend has its dangers, and we must guard against them. Bundy himself, as I know from talks with him in 1969, six years before his untimely death at only 51, regretted the forbidding dryness of his revolutionary studies, and was planning work which would do better justice to Pindar as a poet. Inevitably his approach has been adopted by a number of those humdrum scholars who are forever on the lookout for a mechanical recipe for getting results, and these have created a dreary scholasticism, treating an ode as a collection of commonplaces strung together by a few stock devices. In fact the commonplaces were originally chosen because of their relation to reality, and both they and the traditional devices are used as the context demands, and are skilfully adapted by the poet according to his needs, so that to talk of 'formulas' in this connection is out of place. The clue to understanding lies, indeed, in the knowledge that above all else the poet aims to praise the victor; but Pindar was a human being writing to please other human beings, and those who wish to understand him will neglect the historical dimension only at their peril. He does honour to the victor by placing the moment of felicity which his triumph gives him in the context of his historical situation against the background of the permanent situation of the world as the gods govern it.

This is not the only danger that confronts the new unitarianism. A second is that of exaggerating the significance, sometimes real enough, of the repetitions of words and ideas within a poem;[21] this is sometimes combined with an exaggerated notion of the importance of symbolism in Pindar's work.[22] A third danger is that of using excessive ingenuity[23] to try to prove the myth narrated in ‖ a given poem to be relevant to that poem's theme; in many cases we must be content to observe that the myth illustrates Pindar's view of heroism and its relation to the gods only in a general way. To have unity, a Pindaric ode does not need to have a streamlined unity.

In the light of these considerations, let us consider the recurrent elements of the epinician ode. First, there is the religious background. Pindar is above all else a religious poet, and much of the failure to understand him has been bound up with the failure to understand the Greek religion that pervades his poetry. That failure has encouraged the Romantic error of taking Pindar's religion for a personal construction of his own. For a believer in the archaic Greek religion, to praise a mortal man was a hazardous enterprise. The gods are immortal, and alone enjoy true happiness; men are the creatures of a day, and are subject to all the misfortunes which the day may bring. The gods govern the universe, understandably enough, in their own interest, and with little regard for that of men; the various creation-myths make men an accidental or a casual element in the scheme of things. The gods in Pindar's poetry are most vividly realized as individual personalities, and something of the same concreteness invests even the personified abstractions which so often figure in his work. But these gods work through real forces in the world, so that to deny their existence would be pointless, since the reality of the forces which they stand for cannot be denied.

No man can be pronounced fortunate till he is dead; at any moment an unforeseen catastrophe may overwhelm the happiest among mankind. But men may enjoy moments of felicity; these are dependent on the favour of the gods, and come most of all to men descended from the unions with mortal women which the gods allowed themselves

[21] It is a failing of the useful introduction to Pindar and Bacchylides by M. R. Lefkowitz, *The Victory Ode: An Introduction* (1976). [With regard to Aeschylus, see now West, *Gnomon*, 59 (1987), 195 f.]

[22] This combination has spoiled the interpretation of the Seventh Olympian put forward by David Young, who has done much for Pindaric studies; see the review of his *Three Odes of Pindar* (cited at p. 145, n. 2 above) by H. Maehler, *Gnomon*, 42 (1970), 441 f.

[23] This fault is exemplified in what is in many ways a good and useful book, A. Köhnken's *Die Funktion des Mythos bei Pindar* (1971); see the somewhat too severe notice by S. Radt, *Gnomon*, 46 (1974), 113. For an instance of this kind ‖ of thing, one may note the treatment of *Pyth*. 12. 10 by Köhnken, *BICS* 25 (1978), 92 f.; in order to force the text into close conformity with the gnomic utterance at the end, he is even driven to argue that *δυϲπενθέϊ ϲὺν καμάτωι* must refer to the distress experienced by Perseus, and not to the Gorgons lamenting for their dead sister.

during the heroic age. Such men's heroic nature makes them capable of becoming rulers, warriors, athletes, or poets. Nature is vitally important, and without natural aptitude education is in vain. But natural gifts by themselves are not enough; hard labour and endurance are required, and the need for them is quite as strongly stressed as that for natural gifts. Not that even the conjunction of natural ability and strenuous effort can command success; in the last resort, all mortal triumph || is dependent on the favour of the gods. The gods may grudge success to any mortal; the divine envy which may make them do so is conceived by Pindar not as mere spite, but as a facet of divine justice, whose workings mortals are not always able to perceive.

Any man who praises other men must remain always conscious of the danger of provoking envy, not only divine but human. Envy is a force which any man who strives after any kind of *areta*, excellence, must fight; of this the man who praises others must remain aware. Hesiod, the greatest Boeotian poet before Pindar, had stressed the community between kings and poets, both of whom derive authority from Zeus, kings directly, but poets through Apollo and the Muses.[24] Conscious of his own *areta* as an inspired poet, Pindar has a deep sense of community with the men of action, rulers, warriors, and athletic victors, whom he praises; like them he has to struggle against envy, not simply that of rivals, but that which any man who praises others has to strive to overcome. If the poet can contrive to vanquish envy and to place the victory in its setting in the world of gods as well as men, then the victor's brief moment of felicity can in a limited but real sense be made eternal.

To the modern mind it may seem strange, even ridiculous, that athletes should be coupled with rulers and warriors as bearers of *areta*. That is not how the archaic Greeks saw the matter. A victory in the games, with their strong religious associations, carried vast prestige; athletes endured arduous training of a kind that fitted them for war as well as triumph in the games. The Aeginetans, for whom Pindar wrote more odes of victory than for any other community, won the prize for valour awarded by the Greek army after Salamis; the Sicilian tyrants, whatever modern moralizers may say about their ideology, saved the Greeks of Sicily from the Carthaginians and Etruscans. Again, the conventional references to the customary fee should not lead us to suppose, as ancient comic poets did in jest and later grammarians in all seriousness, that Pindar and Simonides were especially mercenary.[25] One proof of the noble man's nobility, according to the poetic

[24] *Theogony* 75–103.

[25] See Lefkowitz, *The Lives of the Greek Poets*, 50 f.

convention, was his readiness to spend his money generously in assuring that his fame would live.

Yeats lived at a time when there was no generally accepted picture of the world to serve as background for his poetry, so that he was compelled to construct one for himself out of very disparate materials. Pindar, like Dante, inherited the picture of the world || fashioned by the religion of his people. The greatest error of Romantic criticism has been to treat this as though it were a personal philosophy or religion of the poet's own making; we can see that it is common to him and to other writers in the genre, both to the older Simonides, of whom we know little, and to the younger Bacchylides, of whom we possess large parts of fourteen victory odes. If we make allowances for the differences of genre and avoid the pervasive error of reading ideas into ancient tragedy that are not there, we can see that it is not very different from the religious outlook that we find in Aeschylus or Sophocles.[26]

This Romantic misconception has been fostered by the force and clarity of Pindar's many statements in the first person, which give the reader nurtured on Romantic verse the irresistible impression of being in contact with a powerful and independent poetic personality. That is in a sense true; but we shall not appreciate Pindar's poetry correctly unless we are aware of how much he has in common with other writers in the same genre, with Greek archaic poetry in general, and with all believers in archaic Greek religion. Presented by a poet whose imagination has grasped its tragic truth with utter honesty and clarity, the austere world of early Greek religion comes across to us in all its beauty and with all its hardness.

Against this backcloth the poet must place the victory won by his client of the moment. The odes commonly start with an elaborate prelude, ultimately deriving, it would seem, from the invocation of deities with which cult-hymns began. Sometimes the poet invokes a god, but more often a personified abstraction, or the hero or heroine who personifies a place; he may address his own Muse, who is about to perform in honour of the victor; he may start with general reflection or with praise of the victory; or he may describe his ode by means of an elaborate simile, likening it to a cup or bowl ceremonially offered by one man to another or to one of the splendid treasuries which wealthy communities built at the great religious centres to contain their dedications.

Nothing is more indicative of the distorting effect exercised by Romanticism upon Pindaric criticism than the proneness of critics to

[26] See in general my book *The Justice of Zeus* (2nd edn. 1983).

assume that the mention of the victory and the enumeration of other victories won by the victor and by other members of his family must have been irksome duties which the poet resented and was eager to get out of the way as soon as possible in order to give expression to his personal preoccupations. In fact the victory is the starting-point and epicentre of the entire poem, and the other ‖ victories of the family must be mentioned too, since they help to demonstrate that these men possess that nobility of nature which makes effort fruitful and wins the divine favour which alone can bring success. In the hands of an indifferent poet this part of the poem might indeed turn out monotonous, but Pindar displays the greatest virtuosity in the avoidance of this danger. With typical Greek realism he frequently remarks that not all the members even of the greatest athletic families possess equal gifts; talent often skips a generation or so. What counted for the Greeks was victory, and Pindar has no compunction in imagining the discomfiture of the defeated rivals of the winner. Not that the odes dwell at all frequently on the incidents of the games, despite the poet's fondness for athletic metaphors.

The character and history of the victors, their families, and the civic communities they belonged to stimulate the poet to recall history as well as legend; to the Greeks of his time the boundary between the two was vague and almost non-existent. Naturally the part of the poems dealing with the victors takes on a special interest when they are great personages like Arcesilas, king of Cyrene, Hieron, ruler of Syracuse, Theron, ruler of Acragas, and other members of his family. A poetic convention that is as old as Hesiod allows Pindar to address such people as an equal, and to advise them to persist in their noble ways, advice which the audience knows there is no possibility of their neglecting, any more than Britannia might reject the advice of those who in song urge her to rule the waves.

Most victory odes include the narration of a myth, sometimes short, but sometimes very long. Pindar's narrative technique is masterly; he knows what to leave out, and has a flair for the significant detail. Sometimes it is not easy to perceive the myth's special relevance to the situation; but at other times that relevance is so obvious that one cannot help asking why it is not obvious in every case.

In the Third Pythian, for instance, the myth has a peculiar aptness. This is not a true victory ode, but a poem of consolation addressed to Hieron, who was suffering from a painful illness. Pindar starts by expressing the wish that he could call back from the dead Chiron, the noble centaur, who taught the great healer Asclepius. He tells how Coronis, when pregnant with Asclepius by Apollo, broke the law of the gods by having intercourse with a mortal; Apollo destroyed her, but as the funeral pyre was about to consume her body, snatched the child from her womb and brought him to Chiron for his education. Some poets might have ‖ found it wiser not to mention that in the end Asclepius was persuaded by a great fee to break, like his mother, the law of the gods by raising a man from the dead, so that Zeus destroyed him with the thunderbolt; but Pindar goes on to tell it, as a reminder of man's mortality and fallibility. Later, he recalls to Hieron words of Homer which he interprets to mean that the gods give men two parts of evil for one part of good;[27] even Peleus and Cadmus, the two heroes who married goddesses, met with grievous misfortunes. Human happiness never lasts for long, and a man must adapt himself to whatever fortune the gods send him. Pindar concludes the poem with a reminder that poetry can confer on human triumphs a kind of permanence.

Still more perfect is the applicability of the myth of the First Pythian.[28] The poem honours a Pythian victory of Hieron's chariot, but also celebrates his foundation of the new city of Aitna, founded on the site of Catane, the modern Catania, as a kingdom for Hieron's son Deinomenes. Very near that city lies the mighty volcano from which it takes its name; beneath it, in the legend, but stretching all the way from Etna to Vesuvius, lies Typhos or Typhoeus, the mightiest of the giants who have challenged Zeus. The poem starts with a solemn invocation of the Apolline instrument, the lyre, and a description of the celestial music played to the gods upon Olympus by Apollo and the Muses. Even the savage war-god yields to that music's spell; but the enemies of Zeus are infuriated by its strains. One such is Typhos, and the poet memorably describes his fiery breath, which only a few years before the poem was written had burst forth in a devastating eruption. Pindar prays that he may never displease Zeus, interprets the Pythian victory of Hieron as a good omen for the new city, and prays to Apollo, lord of Delphi, for its prosperity. All human excellence depends upon the gods, and Pindar prays to them for Hieron's welfare, making allusion to his great achievements. Since Hieron is grievously sick, he is compared with Philoctetes, the hero whose poisoned foot did not prevent him from making a great contribution to the Greek victory over Troy. Now Pindar returns to Aitna and his hopes for it; he prays Zeus to keep off the ever present danger from the barbarous enemies of the Greeks, with an allusion to the glorious victory which Hieron had won at Cumae, near to Vesuvius, four years earlier, and coupling ‖ it with the earlier victory of Hieron's brother and predecessor, Gelon, over the Carthaginians and with those of the mainland Greeks over the

[27] *Il.* 24. 527 f., misinterpreted, it would appear, by Pindar; see C. W. Macleod, *Homer, Iliad, Book XXIV* (1982), 133.

[28] See the excellent treatment of this poem by Ed. Fraenkel, *Horace* (1957), 276 f.

Persians. So the poet comes to his final praise of Hieron, contrasting the fame won by his nobility and generosity with the ill repute of an earlier Sicilian tyrant, the cruel Phalaris. Happiness and fame are the two greatest things for mortals. In this great poem, the continuing presence of the giant, a natural enemy of the beauty and order which the Olympian gods stand for, brought low by Zeus but living and always to be reckoned with, mirrors the abiding threat to Sicily from the barbarians; the Greeks of Sicily lived a life as precarious as that of modern Israel. It mirrors also the threat to Hieron's dynasty from its enemies; a few years after the performance, Hieron was dead and the rule of his dynasty was at an end. The myth's significance is even wider, for it relates also to man's position in the world; by great exertions he may merit and enjoy the favour of the gods, but its continuance cannot be guaranteed, and nothing that is human lasts for ever.

But in other cases the relevance of a myth is far from obvious. In the First Nemean, for example, written for the Syracusan general Chromios, Pindar tells how the infant Heracles strangled the snakes sent by Hera to kill him and his brother, and in the Eleventh Pythian, written for the Theban boy Thrasydaios, he tells the story of the murder of Agamemnon and the revenge of his son Orestes. A myth may be relevant in more than one way; so that any legend about Heracles is appropriate to a Nemean victory, because he founded the Nemean games, and any myth connected with Delphi, like the myth of Orestes, is relevant to a Pythian victory. But another reason why these myths are relevant is that both illustrate the strength that goes with true nobility, and the power and justice of the gods; Heracles and Orestes through their valour triumph against overwhelming odds.[29] Just so when Pindar honours an Aeginetan victor, he always recounts one of the many legends about the great heroes descended from the founder of Aegina, Aeacus; all such stories illustrate the fundamental principles that govern the universe of archaic Greek religion.

Already in the seventh-century Spartan choral poet Alcman we find mythic narrative closely linked with gnomic reflection; either a reflection prompted by the situation may be illustrated by a || myth, or else a myth may prompt such a reflection. Taken out of context, Pindar's gnomic reflections might be made to appear as so many commonplaces, despite the striking manner in which they are expressed; but in their proper contexts they do not give that impression, unlike many of the similar reflections in the epinicians of Bacchylides.

Transition between the various elements of the poems is effected by means of various technical devices, most of which Bundy furnished

[29] For good discussions of the First Nemean, see G. A. Privitera, *GIF* 24 (1972), 28 f. and P. W. Rose, *HSCP* 78 (1974), 145 f.; on the Eleventh Pythian, see Young, *Three Odes* (n. 18), 1 f.

with somewhat alarming technical names. It is important to recognize these and to understand the function they perform; but it is equally important to realize that they are never trotted out mechanically, but are cunningly disposed to suit the context and the requirements of the poet's art. There is absolutely no fixed order for the elements of the victory ode, each poem having not only its own metrical pattern but its own structure with its own particular shape; the claims of certain scholars to have 'created a generalized formal model' cannot be substantiated.

It is precisely when we have made full allowance for the element of convention both in Pindar's picture of the universe and in his literary technique that the power of his poetic genius makes the most profound impression. To grasp what is most distinctive in his art we may start by comparing him with his younger contemporary Bacchylides. Bacchylides has often been unfairly blamed for not having been a Pindar. He is an admirable poet, whose smoother and lighter style has its own special quality, one seen to less advantage in his victory odes than in the brief narrative lyric poems which are termed dithyrambs. One may say of Bacchylides what Eliot says of the early work of Yeats, that 'these are beautiful poems, but only craftsman's work, because one does not feel present in them that particularity which must provide the material for the general truth.'[30] What enables Pindar to seize on that particularity is his unusual power to select the detail that will most vividly bring home the scene he is describing or the truth on which he is insisting. His determination to describe such a detail with absolute precision sometimes results in the choice of a manner of expression so unusual as to seem bizarre. In the First Nemean, the infant Heracles grasps by the throat the twin serpents sent by Hera; 'and as they choked time breathed the lives out of their dread bodies'—*ἀγχομένοις δὲ χρόνος / ψυχὰς ἀπέπνευσεν μελέων ἀφάτων* (46). In a poem in which he describes how Heracles dealt with the man-eating horses of the Thracian ‖ king Diomedes,[31] Pindar tells how the hero came to the stables and distracted the attention of the horses by tossing one of their grooms into the manger to be devoured; 'and swiftly a crunching sound rang out through the white bones as they splintered'—*ταχέως δ' ἀράβησε διὰ λευκῶν / ὀστέων δοῦπος ἐρεικομένων.*

The so-called Longinus says of Ion of Chios and Bacchylides that they are 'impeccable, uniformly beautiful writers in the polished manner', but of Pindar and Sophocles that they 'sometimes set the world on fire with their vehemence, for all that their flame sometimes

[30] T. S. Eliot, *Selected Prose*, ed. F. Kermode (1975), 251, from a lecture on Yeats originally delivered in 1940.

[31] Fr. 169. 24; on this poem, see my article in *HSCP* 76 (1972), 45 f. [= ch. 10].

goes out without reason and they collapse dismally'.[32] These two passages, I suspect, are examples of the kind of thing in Pindar that distressed the gifted rhetorician of the early imperial period; his style, like those of Aeschylus and Sophocles, conserves an element of archaic roughness, displeasing to the taste of periods like the eighteenth century, when 'the polished manner' is in fashion, but seeming to a different kind of criticism to give these authors strength and to preserve them from the facile smoothness which is the excellence of writers of the other kind, but is also the quality that prevents their rising to the greatest heights.

Very often in Pindar the significant detail is beautifully expressed, besides being hit off with an aptitude that contributes to its beauty. The holy island of Apollo in the Aegean, Pindar wrote in his hymn to Zeus,[33] 'mortals call Delos, but the blessed ones on Olympus call the star that shines far off in the dark earth'—*ἅν τε βροτοὶ / Δᾶλον κικλήϲκοιϲιν, μάκαρεϲ δ' ἐν Ὀλύμπωι / τηλέφαντον κυανέαϲ χθονὸϲ ἄϲτρον*. Bacchylides could never have imagined how Delos must appear to the gods looking down on earth from heaven. The birth of Athena is a favourite subject of art as well as literature; it was memorably depicted on the eastern pediment of the Parthenon. Pindar in the Seventh Olympian describes how Hephaestus split his father's skull with his brazen axe and how Athena leapt up, uttering her warcry, 'and Heaven shuddered at her, and Mother Earth'—*ἁνίχ' Ἁφαίϲτου τέχναιϲιν / χαλκελάτωι πελέκει πατέροϲ Ἀθαναία κορυφὰν κατ' ἄκραν / ἀνορούϲαιϲ' ἀλάλαξεν ὑπερμάκει βοᾶι, / Οὐρανὸϲ δ' ἔφριξέ νιν καὶ Γαῖα μάτηρ* (35 f.).

In the Thirteenth Olympian Pindar tells the Corinthian story of Bellerophon, who made many vain attempts to catch the winged horse Pegasus, 'till Pallas brought him a golden bridle, and his ‖ dream was changed to waking, and she said, "Are you asleep, prince, son of Aiolos? Come, take this to make horses love you! Sacrifice a white bull to the Tamer, and show him this!" So much as he dozed he heard the maiden say, with her blue aegis in the dark.' After duly sacrificing to Poseidon Bellerophon succeeds; once mounted, he cannot resist showing off in typically Greek fashion, and does a war-dance in the sky—*ἀναβαὶϲ δ' εὐθὺϲ ἐνόπλια χαλκωθεὶϲ ἔπαιζεν*.[34] Effects of this kind are not easily appreciated in translation.

Arnold wrote of Homer 'that he is eminently rapid; that he is eminently plain and direct, both in the evolution of his thought and in the expression of it, that is both in his syntax and in his words; that he is

[32] 33. 5; see D. A. Russell, *'Longinus' on the Sublime* (1964), 159.

[33] Fr. 33c.

[34] *Ol.* 13, 65 f.; the last line quoted is l. 86.

eminently plain and direct in the substance of his thought, that is in the matter and ideas; and finally, that he is eminently noble.'[35] When one first reads that famous judgement, it seems somewhat paradoxical; how can the notably grand style of Homer, carrying so great a weight of ornament and prone to such elaborate verbal effects, be called rapid, plain, direct? But when one has considered Homer's style with care, distinguishing, so to speak, the flesh from the bones of his poetry, one must acknowledge the justness of the description. Pindar even more obviously carries a great weight of ornament, and has in addition the quality imposed by his resolute search for precision that makes him at times seem almost baroque; yet I contend that Arnold's words are just as true of Pindar as of Homer, as the longer passages I shall now examine are designed to show.

At the end of the Fourteenth Olympian, an exquisite short poem for a boy victor from Orchomenos, where the Graces had their cult, Pindar remembers the boy's dead father. 'Go, Echo', he writes, 'to the black-walled house of Persephone to bring the glorious news to his father; see Kleodamos and tell him that in Pisa's famous valley his son has crowned his locks with the feathers of splendid prizes.'

μελαντειχέα νῦν δόμον
Φερcεφόνας ἔλθ', Ἀχοῖ, πατρὶ κλυτὰν φέροιc' ἀγγελίαν
Κλεόδαμον ὄφρ' ἰδοῖc' υἱὸν εἴπηιc ὅτι οἱ νέαν
κόλποιc παρ' εὐδόξοιc Πίcαc
ἐcτεφάνωcε κυδίμων ἀέθλων πτεροῖcι χαίταν. (20 f.)

In the Tenth Nemean, for the Argive wrestler Theaios, the myth || is placed, unusually, at the end. From a catalogue of Argive legends—Pindar uses the device of *praeteritio*, making a series of brief mentions of stories he does not intend to tell—he comes to the victor and his long list of victories. It is not surprising that this family should produce great athletes, for the victor's ancestor Pamphaes once had the privilege of entertaining the Dioscuri, Castor and Polydeuces. In Sparta they share with Hermes and with Heracles the direction of the games, and they take good care of righteous men; the sons of gods may be relied on. Now follows the story of the last battle of the Dioscuri against their cousins and enemies, the sons of Aphareus, Idas and Lynceus (55 f.):

Turn and turn about, they spend one day with their father Zeus, and one beneath the caverns of the earth, in the hollows of Therapna, each fulfilling a fate like the other's; for it was that which Polydeuces chose, rather than to be altogether a god and live in heaven, after Castor had been slain in battle. Idas, angry over some affair of cattle, had gashed him with the point of his brazen

[35] Matthew Arnold, *On the Classical Tradition*, ed. R. H. Super (1980), 102. The quotation is from the lectures *On Translating Homer*, given at Oxford in 1861.

spear. Looking down from Taygetos, Lynceus saw them seated inside the trunk of an oak; for his was the most piercing eye of all men upon earth. With swift feet they came up at once, and in a moment did the dreadful deed. And the Apharetids suffered sorely at the hands of Zeus; for at once Leda's son came in pursuit; and they took their stand near their father's tomb. From it they seized a funeral monument, a wrought stone, and hurled it at the chest of Polydeuces. But they failed to crush him or to drive him back; and darting forward with his swift javelin he drove the bronze into the ribs of Lynceus. And Zeus hurled upon Idas the sooty thunderbolt, and alone they were burned together; it is hard for men to contend with those whose power is greater.

And quickly back to his mighty brother came the son of Tyndareus, and he found him not yet dead, but breathing in great gasps. Shedding hot tears he groaned, and cried aloud, 'Father, son of Cronos, what release shall there be from sorrow? Grant me too death together with him, lord! A man's honour is gone when he has lost those who belong to him; few among mortals can be trusted in hard times to take a share of labour!' So he spoke, and Zeus came to meet him, and uttered these words, 'You are my son; but this man is mortal seed, sown when the hero came after to your mother. Come, I give you the choice; if you wish to escape death and hated old age, and live on Olympus with me and with Athena and with Ares of the black spear, you have a right to. But if you fight for your brother, and have a mind to share alike in all things, you may live half the time below the earth, and half in the golden halls of heaven.' After that speech he did not hesitate in his decision, but opened the eye of Castor of the brazen corselet, and made him speak again.

I hope that even that bald rendering, which can give no notion || of the greatness of the poetry, can make the reader who does not know the language aware that Pindar, often censured for grandiloquence, is capable of great simplicity and directness. Arnold's words about Homer apply in every detail to this passage.

Pindar's last dated poem is the Eighth Pythian, performed in 446 BC, when he will have been well over seventy. It was written for a wrestler called Aristomenes from the small island of Aegina, visible from Athens, inhabited by Dorians under aristocratic leadership. Its people had been notably active in trade, in sea-voyaging, and in war; Pindar seems to have had a special link with them, for he wrote more victory odes for them than for any other people, no less than fourteen out of the fifty whose recipients are known to us. Since Aeacus, son of Zeus by the nymph Aegina, was father both of Peleus, father of Achilles, and of Telamon, father of Ajax, the island could claim some of the greatest fighting heroes of Greek legend; before the battle of Salamis, not far from Aegina, and in legend ruled over by Telamon, their statues were carried round the fleet, and gave great encouragement to the men about to fight.[36] Telamon helped Heracles to conquer Troy in the time of Priam's father, king Laomedon; Achilles, his son Neoptolemus, and Ajax all took part in the siege of Troy described in the *Iliad*; and both wars are depicted in the sculptures from the temple of Aphaia, on Aegina, now in Munich. These sculptures illustrate Pindar's odes for Aeginetan victors, just as those from the temple of Zeus at Olympia illustrate the myth of Pelops narrated in the First Olympian, for in poem after poem Pindar dwells on the rewarding themes supplied by the glorious deeds of Aeacus' descendants.

At the start of the Eighth Pythian, Pindar invokes Hesychia, 'Tranquillity'. It is a commonplace of the epinician genre that the victor wins peace, repose, tranquillity as the reward of his mighty labours, so that this choice is natural enough. But that may not be the only reason for it; tranquillity is the opposite of *polypragmosyne*, that readiness to interfere in other people's business that was thought characteristic of the imperialist democracy of Athens.[37] In 446, the year of this poem, Aegina had been for twelve years under Athenian domination; later, after the Peloponnesian War had broken out in 431, the Athenians would expel its entire population.||

Gentle Tranquillity, daughter of Justice who makes cities great, you who hold the keys supreme of counsels and of wars, receive for Aristomenes the honour due to his Pythian victory. For you know how to give and how to take a gentle touch with flawless instinct. But when someone has welded ungentle rancour into his heart, you strike back hard against the power of your enemies, and put their insolence in the bilge. Porphyrion was one who did not realize the folly of provoking you. What one gains is most precious when he from whose house one carries it is willing.

In time violence has been the ruin even of the mighty boaster; Typhos of Cilicia did not escape its consequence, neither did the king of the giants; but they were brought low by the thunder and by the arrows of Apollo, he who received the son of Xenarkes with kindness when he came from Kirrha garlanded with the leaves of Parnassus and with Dorian revelry.

Porphyrion and the king of the giants are the same person;[38] the allusion is to the battle in which the giants challenged the might of the Olympians and were overwhelmed by them, a favourite theme of archaic art and of much poetry, now mostly lost. In the First Pythian we have found the presence under Etna of Typhos, the most formidable enemy of Zeus, not to be unrelated to the ever-present danger to

[36] Hdt. 8. 64, 2; cf. 83–4.

[37] See H. T. Wade-Gery, *JHS* 52 (1932), 224–5 = *Essays in Greek History* (1958), 265–6; cf. V. Ehrenberg, *JHS* 67 (1947), 46 f.

[38] So O. Schroeder, *Pindars Pythien* (1922), 70; cf. F. Vian, *La Guerre des géants: Le mythe avant l'époque hellénistique* (1952), 197.

the Greeks of Sicily from their barbarian enemies. The strong stress laid on the power of self-defence possessed by Tranquillity 'who holds the keys supreme of counsels and of wars' must cause us to wonder whether here too the poet may be glancing at the present situation of the community to which his client belongs; it may be significant that here also Typhos finds a mention.

Not far from the Graces has fallen the lot of the island of the righteous city, which has a part in the far-famed deeds of glory of the sons of Aeacus; solid has been her fame from the beginning; she is sung of as the nurse of heroes supreme in many victorious contests and in many fierce battles; and the same quality is manifest in her men.

Now Pindar uses a standard device to move from the praise of the victor's city to that of the victor.

But I have no leisure to dedicate the whole long story with the lyre and with sweet singing, lest satiety should come and irk us. Let my art give wings to the matter now in hand, the debt I owe you, boy, the newest of noble things. In your wrestling you follow in the tracks of your mother's brothers; at Olympia you do not shame Theognetos, nor at the Isthmus the victory of Kleitomachos, brave and strong. By bringing glory to the ‖ Meidylidai you earn the praise which once the son of Oikles spoke in riddles, when he saw his son and his friends' sons standing fast in battle at seven-gated Thebes, when the Epigonoi came from Argos on that second march.

The son of Oikles is the warrior and prophet Amphiaraus, one of the seven champions who accompanied Adrastus, king of Argos, in his famous and disastrous attempt to restore Polyneices to the throne of Thebes; and the allusion leads up to the brief myth of the poem. In the next generation Adrastus and the sons of the Seven, known as the Epigonoi, launched a new attack; among them were such heroes as Diomedes and Sthenelus, known from the *Iliad*, and this time the attack met with success.

Thus he spoke as they fought: 'By nature a valiant spirit from the fathers shines upon the sons. Clearly I see Alcman plying his shield with its sinuous snakes, foremost in the gates of Cadmus. He who suffered in the disaster long ago now enjoys a better omen, the hero Adrastus. But in his own family his fortune will be the opposite; he alone of the Danaan army shall gather the bones of his son, and by a fortune sent from the gods shall return to the broad streets of Abas with his host unscathed.' So spoke Amphiaraus; and I myself rejoice to crown Alcman with flowers and to sprinkle him with song; for he is my neighbour and the guardian of my possessions, and on my way to earth's navel, famed in song, he met me and prophesied by the art belonging to his house.

Amphiaraus belonged to one of the great prophetic families of heroic Greece; he had not been killed, but had been swallowed up by the earth, and in his shrine at Oropos, on the borders of Boeotia and Attica, continued to prophesy. It would seem that Alcman or Alcmaeon, like his father, was the object of a hero-cult, that Pindar lived near his shrine and had deposited property there for safe-keeping, in accordance with a common practice, and that he believed himself to have received a prophecy from the hero. Perhaps he had a waking vision, but it seems likelier that he dreamed a dream, for at the shrine of Amphiaraus inquirers slept in the sanctuary in the hope of a dream in which they would receive their answers. The simple significance of the myth is made explicit by the poet; it lies in the commonplace, often expressed in victory odes, that nature implants nobility in the sons of noble fathers. In the earlier battle the Seven had all perished, but Adrastus survived, carried to safety by the wonder-horse Arion; in the attack of the Epigonoi all the sons of the Seven survived, but Aigialeus, son of Adrastus, was killed by the defending commander Laodamas, son of Eteocles, who was himself killed by Alcman. As so often, the myth carries || a reminder that even the greatest human triumphs are seldom undiluted by an element of sadness.

Now the poet directly addresses Apollo, lord of Delphi:

You, Far-Darter, lord of the glorious temple that welcomes all in the hollows of Pytho, granted him there the greatest of delights, but first at home in your festival you brought him a gift eagerly to be snatched up, that of the pentathlon. King, I pray you to deign to look upon all my goings according to a harmony! The band of revellers with sweet song has justice standing near; and I pray that the gods grant unstinted favour to your fortunes.

As often, gnomic reflection follows the prayer for divine favour; that prayer is necessary, because without such favour, no success is possible for men.

For if a man possesses good things that have come without long effort, many think that he is clever among fools and builds up his success by schemes based on right judgements. But such things do not lie within the power of men; it is the god who gives them, now putting one on top and bringing down the other, now the reverse. Enter the contest with right measure in your heart![39] You have the honours you won at Megara, and in the lowland of Marathon, and with three victories you have prevailed in Hera's games in your own country.

Thus the victor's wins in local games are rapidly dispatched. The Greeks of Pindar's time did not share English notions of sportsmanship.

Upon four bodies you came down with fierce intent; for them no equally

[39] In l. 78 I have followed A. Turyn, *Pindari carmina cum fragmentis* (1944, 1948; 2nd edn. 1952), in accepting Bergk's conjecture; cf. *Paean* 2. 34, and also *Ol.* 3. 48 and *Pyth.* 2. 34.

delightful homecoming was decreed in the Pythiad, nor did sweet laughter arouse joy when they returned to their mothers; in dark passages they cower, avoiding their enemies, stung by their disaster.

From the gloom of the defeated Pindar passes to the happiness of the victorious; but he is constantly aware of the dangers that attend all praise of human effort, and comes back immediately to the fragility of human happiness.

He who has some new glory for his lot, revelling in great luxury, flies on wings given to his mighty deeds by hope. In a moment delight flowers for men, and in a moment it falls to the ground, shaken by a stern decree. Humans are creatures of a day;[40] what is one of them? what is he not? The dream of a shadow. But when the Zeus-given glory comes, a bright ‖ radiance lies on men, and life is sweet. Aegina, dear mother, convey this city in freedom with the aid of Zeus and mighty Aeacus and Peleus and valiant Telamon and with Achilles.

Aegina, the eponymous heroine, mother of Aeacus by Zeus, is begged 'to convey this city in freedom'; with her are joined Zeus himself and three of the five mighty heroes of the family. In the light of this we must surely acknowledge that the invocation of Tranquillity and the reminder of the fate met by the barbarous giants who challenged the Olympians, though fitting perfectly the ordinary requirements of an epinician ode to honour the victory of Aristomenes, cannot have been felt by the audience to be irrelevant to the dangerous situation in which Aegina at that time found herself. But the poem is infinitely remote from the political pamphlet against democracy and imperialism which a certain kind of criticism would make of it; it ends with reflections that have the widest possible relevance to the conditions of human life.

Ὁ δὲ καλόν τι νέον λαχὼν
ἁβρότατοϲ ἐπὶ μεγάλαϲ
ἐξ ἐλπίδοϲ πέταται
ὑποπτέροιϲ ἀνορέαιϲ, ἔχων
κρέϲϲονα πλούτου μέριμναν. ἐν δ' ὀλίγωι βροτῶν
τὸ τερπνὸν αὔξεται· οὕτω δὲ καὶ πίτνει χαμαί,
ἀποτρόπωι γνώμαι ϲεϲειϲμένον.
ἐπάμεροι— τί δέ τιϲ; τί δ' οὔ τιϲ; ϲκιᾶϲ ὄναρ—
ἄνθρωποι. ἀλλ' ὅταν αἴγλα διόϲδοτοϲ ἔλθηι,
λαμπρὸν φέγγοϲ ἔπεϲτιν ἀνδρῶν καὶ μείλιχοϲ αἰών.
Αἴγινα φίλα μᾶτερ, ἐλευθέρωι ϲτόλωι
πόλιν τάνδε κόμιζε Δὶ καὶ κρέοντι ϲὺν Αἰακῶι
Πηλεῖ τε κἀγαθῶι Τελαμῶνι ϲύν τ' Ἀχιλλεῖ.

[40] The manuscripts are surely right in reading *ἄνθρωποι* in l. 96; everything that lies between *ἐπάμεροι* and *ἄνθρωποι* is a parenthesis; otherwise *ἐπάμεροι* can hardly be construed.

Despite the loss of about three-quarters of Pindar's work, the victory odes together with the fragments are enough to justify the claim that he has produced a body of poems that communicate a distinctive vision of the world, conveyed with great imaginative power. Since the victory odes constantly show gods and men in close relation to each other, it may be fortunate that we possess them, rather than, say, four books of hymns of various kinds; for their subject-matter is well calculated to display Pindar's tragic view of human life.[41] Seeing the victor and his circumstances ‖ always against the background of archaic Greek religion, and being always aware of the difficulty which a believer in it must find in praising any mortal, Pindar, by using the dark colour of the greater part of men's existence to bring out the brightness of that small part of it which is lit up by what he calls 'the Zeus-given glory', achieves a magnificent chiaroscuro. He is grateful to the gods for all the splendour of the world, and for the occasional favours that they show to men; but he never forgets the precariousness of human life, and knows that men have no right to count on the continuance of such favours. He shows himself aware of one client's sickness, another's exile, the losses in battle suffered by the family of another, the threat to Sicily from the barbarians and to Aegina from the Athenians; he reminds the victors that not even the greatest heroes of the past were exempt from misfortune. In the Second Olympian and in several of his dirges, Pindar speaks of a happy life in the afterworld granted to certain souls who have lived out three lifetimes without committing an injustice; Peleus, Cadmus, and Achilles are among them. Here the Homeric notion of Elysium as the haven of a fortunate few seems to be combined with beliefs of the sort which Pindar's contemporaries associated with the name of Orpheus; and yet he faces up to a realistic view of the limitations of human life, with complete freedom from illusion and complete absence of false comfort. In the last resort, what makes his poetry so powerful is his power to present, to use Eliot's words once more, 'that particularity which must provide the material for the general truth'; in that respect, and not in that alone, Pindar resembles Dante.

[41] In the *Times Literary Supplement* for 26 Feb. 1982, p. 207, Sir Stuart Hampshire defends state subsidies for the arts. 'Glory', he writes, 'unlike mere success or fame, lives on in the individual's mind and in national consciousness; it is the natural reward of athletes and heroic soldiers in battle, and, most of all, of artists and of poets and of composers. A nation that is unwilling to give solid ‖ material support and encouragement to the probable sources of its glory will have a dispirited, nameless history, a dim existence on the margin of things. It still seems to me odd that almost any Frenchman is willing to acknowledge this truism, and very few Englishmen; perhaps it is because puritan ethics, which originally included hostility to the Court and its ways, survives as hostility to all rewards that are not earned through merchandise sold in the market.' Pindar would have agreed.

8(84)

Pindar and the Afterlife*

No early Greek poet is more aware than Pindar of the mortality of man and of the impermanence of all human things. For brief moments of their brief lives, certain men may be irradiated by the splendour that comes from Zeus; but even in his hour of triumph, a man must remember that his limbs are mortal, and that at the end of all things his covering will be earth.‖

That makes it remarkable that in certain poems Pindar voices beliefs about what happens after death which are at variance with those

* The following abbreviations have been used:

Burkert, *GRAKE* [= *GR*]	W. Burkert, *Griechische Religion der archaischen und klassischen Epoche* (Stuttgart etc., 1977) [= *Greek Religion: Archaic and Classical* (1985)].
Burkert, *LSAP*	W. Burkert, *Lore and Science in Ancient Pythagoreanism* (Cambridge, Mass. 1972) = 2nd edn. *Weisheit und Wissenschaft: Studien zu Pythagoras, Philolaos und Platon* (Nuremberg, 1962).
Bremmer	Jan Bremmer, *The Early Greek Concept of the Soul* (Princeton, 1983).
Claus	David B. Claus, *Toward the Soul: An Inquiry into the Meaning of ψυχή before Plato* (New Haven, 1981).
Cole	Susan G. Cole, *GRBS* 21 (1980), 223 ff.
D.–K.	H. Diels und W. Kranz (eds.), *Die Fragmente der Vorsokratiker*, 3 vols. (Berlin; 6th edn., Berlin, 1951–2).‖
Guthrie	W. K. C. Guthrie, *Orpheus and Greek Religion* (London, 1935; 2nd edn. 1952).
Hampe	R. Hampe, 'Zur Eschatologie in Pindars zweiter olympischer Ode', in *Ἑρμηνεία: Festschrift Otto Regenbogen* (Heidelberg, 1952), 46–65.
Kern, *OF*	O. Kern (ed.), *Orphicorum Fragmenta* (Berlin, 1922).
KRS	G. S. Kirk, J. E. Raven, and M. Schofield, *The Presocratic Philosophers* (Cambridge, 2nd edn. 1983; 1st edn. by Kirk and Raven, 1957).
Linforth, *AO*	I. M. Linforth, *The Arts of Orpheus* (Berkeley, 1941).
Nilsson, *GGR*	M. P. Nilsson, *Geschichte der griechischen Religion*, i (Munich, 1940; 2nd edn. 1955; 3rd edn. 1967).
Rohde, *Psyche*	E. Rohde, *Psyche* (Tübingen, 1894; 2nd edn. 1898, of which subsequent edns. are reprints) = English version by W. B. Hillis (1925).
van Leeuwen	*Pindarus' Tweede Olympische Ode*, ed. J. van Leeuwen (2 vols.; Assen, 1964).
West, *OP*	M. L. West, *The Orphic Poems* (Oxford, 1983).
Wilamowitz, *Pindaros*	U. von Wilamowitz-Moellendorff, *Pindaros* (Berlin, 1922).
Zuntz, *Persephone*	G. Zuntz, *Persephone* (Oxford, 1971).

Pindare, Fondation Hardt, Entretiens sur l'antiquité classique 17 (Vandœuvres-Geneva, 1985), 245–83.

which were generally held among his contemporaries and which the great body of his work appears to presuppose. The evil are punished and the good rewarded; men's souls are reincarnated in other bodies; and certain chosen persons are transported to the Islands of the Blest, and that not simply because they are the children or the favourites of the gods, but because they have lived out three lifetimes without committing an injustice. The principal account of this belief is in the Second Olympian Ode, performed in 476 BC in honour of the Olympic chariot victory of Theron, ruler of Acragas. Acragas was the home of Empedocles, at that time about twenty years of age, who was later to put forward a remarkable theory of reincarna‖tion, certainly influenced by Pythagoras; and it was not very far distant from Thourioi, in whose neighbourhood some of the gold tablets buried with dead persons to help them in the underworld and often held to reflect Orphic or Pythagorean beliefs have been discovered. Much of Pindar's account of the next world is distinctively heroic and Pindaric, and Roland Hampe in a fine article has done his best to stress this fact; but there is a solid residuum of matter that shows an unmistakable affinity with what is considered Orphic or Pythagorean. Did Theron belong to an Orphic or Pythagorean sect? Wilamowitz was among those who have contended that Pindar was voicing the beliefs of his patron, not his own; but Rohde and others have insisted that Pindar's manifest sincerity and the calm assurance with which he puts forward the beliefs in question make it impossible that he himself should not have entertained them.[1]

That problem has been endlessly discussed; so also has that of the exact nature of the beliefs about the next world that are in question. As lately as 1971, the whole matter was treated with great learning and with impressive amplitude by Günther Zuntz in his fine book *Persephone*.[2] This work contains first a detailed history of the cult of Demeter and Persephone, the great goddess of Sicily, to whom according to Pindar Zeus gave that island as a present;[3] then a new edition of Empedocles' poem *Katharmoi*, with useful contributions to the constitution and the understanding of its text; and finally the best and most useful edition so far of the texts preserved upon the gold leaves from Thourioi and other places, with full commentary and discussion.‖

But since Zuntz wrote new research has thrown light on some of the problems he discussed, and even more significantly new evidence bearing on those problems has been discovered. David Claus and Jan

[1] *Pindaros*, 251 f.; Rohde, *Psyche*, ii. 204 f. = English version, 414 f.

[2] See the review by W. Burkert, *Gnomon*, 46 (1974), 326 f.

[3] *N.* I 14; cf. L. R. Farnell, *The Cults of the Greek States*, iii (Oxford, 1907), 65; 337; 375; Zuntz, *Persephone*, *passim*.

Bremmer have published important investigations of the early history of the concept of the soul.[4] Fritz Graf[5] has made a valuable study of the Orphic poetry relating to Eleusis and its mysteries; Martin West[6] has given us a masterly treatment of Orphic poetry in general, with a bold attempt to assign each of the poetical fragments in Kern's collection to that one of the six Orphic theogonies to which it belongs; and various aspects of these problems have been discussed in various places by the eminent scholar who has done more than any man living for the study of Greek religion, Walter Burkert.[7] In 1974 was discovered the earliest and most interesting of all the gold leaves, that found at Hipponion, the Roman Vibo Valentia.[8] In 1978 the Russian scholar A. S. Rusyaeva published the inscriptions from three small ‖ tablets of bone found on the site of Olbia,[9] the Greek colony on the Dnieper, where as we know from the story of the Scythian prince Scyles told by Herodotus (4. 76 ff.), the worship of Dionysus was popular in early times; and these are of much significance both for Dionysiac worship and for early Orphism. Zuntz in the wake of Wilamowitz had treated Orphism with a robust scepticism, regarding the gold tablets as Pythagorean rather than as Orphic and pointing to the absence from them of any reference to Dionysus as an indication of this fact. But as Pentheus found, Dionysus has a way of turning up when and where you least expect him, and in the gold tablet from Hipponion and the bone tablets from Olbia he has done so in a way somewhat disconcerting for the learned author of *Persephone*. That makes it necessary for certain questions that have often been asked with regard to the Second Olympian and certain Pindaric fragments to be revived again. For the understanding of these poems the questions

[4] Claus and Bremmer: see the list of abbreviations, above.

[5] F. Graf, *Eleusis und die orphische Dichtung Athens in vorhellenistischer Zeit* (Berlin, 1974); for Pindar's treatment of the legend of Heracles' initiation, see H. Lloyd-Jones, *Maia*, NS 19 (1967), 211 f. [= ch. 11].

[6] West, *OP*.

[7] Burkert, *LSAP*; 'Le laminette auree. Da Orfeo a Lampone', in *Orfismo in Magna Grecia. Atti del Quattordicesimo Convegno di studi sulla Magna Grecia* (Naples, 1975), 81 f.; 'Orphism and Bacchic Mysteries. New Evidence and Old Problems', in W. Wuellner (ed.), *The Center for Hermeneutical Studies*, Colloquy 28 (1977), 1–8 and 31 f.; *GRAKE* 436 f. [= *GR* 296 f.].

[8] G. Foti and G. Pugliese Carratelli, *PP* 29 (1974), 91 f.; R. Merkelbach, *ZPE* 17 (1975), 8–9; M. L. West, *ZPE* 18 (1975), 229–36; M. Marcovich, *ZPE* 23 (1976), 221–4; M. Gigante, *PP* 30 (1975), 223–5; H. Lloyd-Jones, ibid., 225–6; G. Pugliese Carratelli, ibid., 226–31, and *PP* 31 (1976), 458–66; G. Zuntz, *WS*, NF 10 (1976), 129 ff.; *SEG* xxvi. 1139; xxvii. 226 bis; W. Burkert, 'Neue Funde zur Orphik', in *Informationen zum altsprachlichen Unterricht*, ii/2 (Graz, 1980), 27 ff.; A. Henrichs, in B. F. Mayer and E. P. Sanders (eds.), *Jewish and Christian Self-Definition*, iii (London, 1982), 154; Cole, 223 ff. (with bibliography); Robert Parker, *Miasma* (Oxford, 1983), 286 f.; R. Janko, *CQ*, NS 34 (1984), 89 ff.

[9] See F. Tinnefeld, *ZPE* 38 (1980), 67 ff.; W. Burkert, art. cit. (n. 8), 36; M. L. West, *ZPE* 45 (1982), 17 ff.; *OP* 17 ff. and see Index s.v. Olbia.

are of great importance, so that I must ask you to forgive me for detaining you for some time in the strange world of early Orphic, Pythagorean, and Dionysiac speculation about the human soul and its destiny in the world it enters after death.

Let us first glance briefly at the Second Olympian[10] itself, so as to consider how the mention of beliefs about the afterlife fits into the framework of the poem. Addressing the songs that rule the phorminx, the great lyre, the poet asks what god, what hero, what man is to be || honoured; the god is Zeus, the lord of Olympia, the hero is Heracles, the founder of its games, the man is Theron, victor in the Olympic chariot-race. Theron is just in the regard he shows for guests; he is the support of Acragas; he is the fine flower of a family whose members bear auspicious names, keeping the city straight. With much labour they secured the sacred habitation of the river of Acragas, and were the eye of Sicily; and their fated lifetime came bringing wealth and gratitude, thanks to their genuine excellences. Next comes a solemn prayer to Zeus, lord of Olympus and of Olympia, to take pleasure in the song and to secure the ownership of the land that has been their fathers' to Theron's heirs. What has been done, in justice and injustice, not even Time, the father of all things, can render undone; but may there be forgetfulness, conjoined with happy fortune; for good things in which we delight can subdue malignant pain so that it perishes, when fate given by a god sends up prosperity on high.

We know virtually nothing of how the Emmenidai achieved their power. It is usual for Pindar to lay stress upon the pain and labour that have gone to the winning of the triumphs which he celebrates; but it would not be wise to assume that his mention of their struggles was nothing but the reiteration of a commonplace. It is also usual for Pindar to say that the final triumph causes its winner to forget his pain, so that it would be equally unwise to assume that the reference to the past and the impossibility of altering it must hint at something sinister which Theron might wish forgotten. Pindar could, I think, speak of Time the father of all things without having in mind the special || importance of Time in Orphic theogonies;[11] and he could speak of the divine fate as sending up prosperity from the earth without being

[10] Douglas Gerber's material in his *A Bibliography of Pindar* (Americ. Philol. Assoc. 1969), 19 f. and María Rico's from *Ensayo de Bibliografía Pindárica* (Madrid, 1969), 87 f. may be supplemented from Luigi Lehnus (ed.), Pindaro. *Olimpiche: Traduzione, commento, note e lettura critica* (Milan, 1981), 27; since then the ode has been commented on by G. Kirkwood (ed.), *Selections from Pindar* (Chico,|| Calif., 1982), 61 f. and by J. K. and F. S. Newman, *Pindar's Art: Its Tradition and Aims* (Hildesheim, 1984), 162 f. (a work with whose method I am out of sympathy); cf. L. Gernet, in the posthumously published essay at *Les Grecs sans miracle* (Paris, 1983), 70 ff. Detailed treatment and a vast doxography is provided by van Leeuwen (see the list of abbreviations, above).

[11] But see West, *OP* 103 ff. and the last sentence of n. 82 on p. 110.

directly conscious of the notion that wealth is sent up from the nether regions by the chthonic deities.[12]

Now the poet illustrates his view of human fortune from the history of the House of Laios, from which Theron claimed descent through Polyneices. Great though their sufferings were, the daughters of Cadmus achieved fame; their sorrows were outweighed by greater goods. Semele, though destroyed by the lightning, lives among the gods, loved by Pallas, by Zeus the father, and by her son; Ino lives an immortal life beneath the waves; of Agave Pindar says nothing. Mortals cannot know when they will die, or when they will finish the day with happiness still undiminished; the tides bring now delight, now labour to men; and so Fate who sustains the fortune of the house may bring not only god-given prosperity but also pain once more. Oedipus fulfilled Apollo's prophecy and slew his father, and the Erinys saw it and caused his sons to slay each other; but Polyneices when he fell left behind him Thersandros, who triumphed at Thebes with the Epigonoi.

So Pindar comes to the formal praise of the Olympic triumph and of the other victories of Theron and his brother; it is proper, he declares, that a descendant of this family should be praised in song. Success brings relief from sorrows; wealth adorned by deeds of valour brings the right moment to act in one thing after another, giving sensations that bring delight (*ἁβροτέραν* Stadtmüller, Wilamowitz); wealth is a star seen from afar, true radiance for a ‖ man. To appreciate the significance of Pindar's praise of wealth, one must remember that the Greeks thought wealth honestly come by to be god-given, so that wealth deserved praise hardly less than excellence.

But Theron has not only wealth, but knowledge of what happens in the next world that will serve him well. 'If a man has wealth and knows the future', the poet continues in a sentence whose anacoluthon should not perplex the reader; and then follows the section of the poem which must be examined in some detail, since I cannot translate or summarize these words without explaining why I interpret them in the way I do (ll. 56–60):

> *εἰ δέ νιν ἔχων τις οἶδεν τὸ μέλλον,*
> *ὅτι θανόντων μὲν ἐνθάδ' αὐτίκ' ἀπάλαμνοι φρένες*
> *ποινὰς ἔτεισαν, τὰ δ' ἐν τᾶιδε Διὸς ἀρχᾶι*
> *ἀλιτρὰ κατὰ γᾶς δικάζει τις ἐχθρᾶι*
> *λόγον φράσαις ἀνάγκαι.*

Let us deal first with the problem of the meaning of *ἀπάλαμνοι φρένες*. Many scholars, taking it for granted that the sentence must be

[12] See Ed. Norden (ed.), *P. Vergilius Maro, Aeneis, Buch VI* (Berlin, 1927³), 38 f.; against, Hampe, 46 f.

taken closely with the sentence which follows it, so that the dead pay the penalty for the crimes which someone judges, have taken the word *ἀπάλαμνοι* to mean 'wicked'.[13] They agree that in general the meaning of the word is not 'wicked', but 'helpless', standing to *παλάμη* as the commoner synonym *ἀμήχανοc* does to *μηχανή* (cf. *Ol.* 1. 59, where Tantalos' *ἀπάλαμον βίον* is not a wicked life): but they have cited four passages in which they argue that its meaning must be 'wicked'. Let us consider these passages, one by one:

1. Eur. *Cyc.* 597–8

> *χώρει δ' ἐc οἴκουc, πρίν τι τὸν πατέρα παθεῖν*
> *ἀπάλαμνον·* (Canter: *ἀπαλλαγμὸν* L) ||

There is indeed a danger that the Cyclops may do 'something bad' to Silenus; but the literal meaning of *τι . . . ἀπάλαμνον* is 'something which he cannot help', just as at *Il.* 8. 130 *ἀμήχανα ἔργα* means 'things which cannot be helped', or 'things about which nothing could be done'; cf. Hes. *Theog.* 589 *δόλον αἰπύν, ἀμήχανον ἀνθρώποιcιν.*

2. Solon fr. 27. 11–12 West

> *τῆι δ' ἕκτηι περὶ πάντα καταρτύεται νόοc ἀνδρόc,*
> *οὐδ' ἔρδειν ἔθ' ὁμῶc ἔργ' ἀπάλαμνα θέλει.*

Here the *ἔργ' ἀπάλαμνα* are the equivalent of the Homeric *ἀμήχανα ἔργα*: no doubt these things are bad, but the literal meaning is not 'bad' but rather 'impossible', in the sense in which one says that a particular person is 'impossible', meaning that he will not see reason.

3. Theognis 279–82

> *εἰκὸc τὸν κακὸν ἄνδρα κακῶc τὰ δίκαια νομίζειν,*
> *μηδεμίαν κατόπιcθ' ἁζόμενον νέμεcιν·*
> *δειλῶι γάρ τ' ἀπάλαμνα βροτῶι πάρα πόλλ' ἀνελέcθαι*
> *πὰρ ποδόc, ἡγεῖcθαί θ' ὡc καλὰ πάντα τιθεῖ.*

Once again, the *ἀπάλαμνα . . . πολλά* are like the *ἀμήχανα ἔργα* in Homer.

4. Theognis 481 (of a drunken man): *μυθεῖται δ' ἀπάλαμνα.* Here as in all these passages the reference is to things which are *ἀμήχανα*, and which are so called because nothing can be done about them; to reduce the meaning of the adjective to the simple 'bad' is not permissible.

In the Pindaric passage, the word is not applied to 'things'; it stands in agreement with the noun *φρένεc*. With that noun it is most unlikely

[13] Tycho Mommsen, *Annotationis criticae supplementum ad Pindari Olympias* (Berlin, 1864); cf. Wilamowitz, *Pindaros*, 248; W. Schadewaldt, *Hellas und Hesperien*, i (2nd edn., Zürich, 1970), 159 ('der hienieden Hingeschiedenen frevelhafte Herzen'); D. L. Page, *Sappho and Alcaeus* (Oxford, 1955), 315.

to mean 'minds you can do nothing about'; the natural sense would be, as Rohde ‖ saw,[14] 'minds that have no resource, feeble minds'. The spirits of the dead are sometimes said to have no φρένες whatever, as at *Il.* 23. 104, or at *Od.* 10. 493, where Tiresias alone is said to have his φρένες ἔμπεδοι; they are ἀμενηνὰ κάρηνα. At Aesch. *Cho.* 157 the dead Agamemnon has a φρήν which is ἀμαυρά, like the χθονία φρήν of the dead kings of Cyrene at *Pyth.* 5. 101. The sense must be that when men die here, their feeble minds at once pay the penalty; the penalty consists in their minds becoming feeble, that is to say, in death.

If the penalty which those who die pay consists in death, it is not the same as the penalty paid by the souls whose crimes are judged according to the following sentence. Indeed the sense appears to be that all who die here immediately pay a penalty; then someone who pronounces sentence with cruel necessity judges crimes committed in this realm of Zeus. The expression chosen reminds one that Hades also is a realm of Zeus; so that if one wishes to guess at the identity of the judge, it is natural to think first of the subterranean Zeus known to us from *Il.* 9. 457 and from Aesch. *Supp.* 156–7 and 230–1. But what do those who die pay the penalty for? I shall return to that later.[15]

'But the good receive an easier life than ours, with nights equal to ours and having the sun for equal days,‖ never vexing earth nor the sea's water with their hands; but those who have rejoiced in keeping their oaths live a life free from tears at the court of gods enjoying special honour; but those others endure toil not to be contemplated' (*Ol.* 2. 61–7). Some scholars, of course, take the view that nights and days in the underworld are said to be of equal length, and it has been argued that this is indicative of a specially just dispensation. But in fr. 129, from a dirge, Pindar says that the sun shines for those in the world below during our night, and the interpretation which I have preferred harmonizes with that.[16] Boeckh took τιμίοις θεῶν (ll. 65–6) to mean 'with those honoured by the gods'; but surely it means 'gods who have special honour', and refers doubtless to Persephone and those about her.

[14] Rohde, *Psyche*, ii. 208 n. 3 = English version, 442 n. 35.

[15] Albert Henrichs, to whom I am indebted for helpful comments on this paper, has already illustrated the concept of ποινή as a retribution for spilled blood by citing Hdt. 2. 134. 4; 3. 14. 5; 7. 134. 2; 136. 3, Antiphon, *Or.* 2 (*Tetralogia A*) 4 (δ), 11, and Kern, *OF* 32 d and 32 e. 4, and also *Ol.* 2. 57 and fr. 133. He thinks that the penalty consists in some sort of accountability after death rather than in death itself; but to me it seems likelier that all men make atonement to the goddess by their deaths, but that she accepts atonement only from the good.

[16] Thus Rohde, *Psyche*, ii. 210 n. 1 = English version, 444 n. 38; this is the natural way to take the words, and has the advantage of making them correspond with fr. 129. But Wilamowitz, *Pindaros*, 249 says, 'an ewig gleichen Sonnentagen', and L. Woodbury, *TAPA* 97 (1966), 597 ff. has argued that equal days and nights imply a just dispensation. Most have followed Wilamowitz (see van Leeuwen, 181 f.). On the differences between the afterlife of the Second Olympian and that of the θρῆνοι to which frr. 129 and 131 belong, see Fr. Solmsen, *Hermes*, 96 (1968), 503 f.

'And they who have had the strength to remain three times on each side and to keep their soul free altogether from unjust acts go along the road of Zeus to the tower of Kronos; here the breezes of Ocean blow about the Islands of the Blest. Flowers of gold blaze, some on land, growing from the splendid trees, and others fed by the water; with clusters of these they join hands and weave crowns, in the upright counsels of Rhadamanthys, whom the great father has as a companion ready by his side, the husband of Rhea who has the highest throne of all. Peleus and Cadmus are numbered among them; and Achilles was brought there by his mother, after she had won over the heart of Zeus with her prayers—he who brought down Hector, the matchless || firm pillar of Troy, and did Cycnus to death, and the Ethiopian child of Dawn' (ll. 68–83).

The dirge of which fr. 129 is part also described the agreeably idle existence led by the good in Hades; it too speaks of golden fruit, together with crimson meadows and the enjoyment of riding, gymnastics, and board-games, and the delicious scent of incense from continual sacrifices. Most notably, it mentioned rites of initiation (fr. 131a *ὄλβιοι δ' ἅπαντεϲ αἴϲαι λυϲιπόνων τελετᾶν* (Wilamowitz: *λυϲίπονον τελετάν* codd.); and that reminds us that in the Third Olympian, which is the companion poem to the Second, Pindar speaks of the Emmenidai as 'guarding with pious purpose the *τελεταί* of the blessed ones' (41).

But for a chosen few there is a higher destiny in store, transportation to the Island of the Blest, located in Oceanus.[17] The place is mentioned in the *Odyssey*, where it is said that Menelaus as son-in-law of Zeus will be conveyed there, and the whole race of heroes finishes up there in Hesiod's *Works and Days*. But here Pindar mentions only a select number of heroes, not including Menelaus. Peleus and Cadmus both married goddesses, and both are often mentioned by Pindar as great heroes who despite some reverses enjoyed singular felicity; the same is true of Achilles, whose afterlife is usually located in his own private island of Leuce in the Pontos. Rhadamanthys[18] is often placed in Elysium, a place sometimes identified with the Islands of the Blest, as well as being frequently a judge in Hades, so that there is no need to appeal to his Boeotian connections to explain his presence; as a known pattern of virtue he is an appropriate inmate for a resort to which virtue, or at least abstinence from injustice, can gain admittance. Kronos presides over the Islands in Hes. *Op.* 173*a–e*,|| lines which figure in two papyri but not in a third, nor (except for 173*a*, in another place) in the medieval tradition. He was notoriously more indulgent to men

[17] See Nilsson, *GGR* 324 ff.; Burkert, *GRAKE* 300 ff. [= *GR* 194 ff.].

[18] See L. Malten, 'Elysion und Rhadamanthys', *JDAI* 28 (1913), 35 ff.

than his successor Zeus, and his associations with the Titans, who are often connected with the origins of men, may not be irrelevant.

Pindar now breaks off his disquisition on the life after death with a transitional formula specially appropriate to the present context: 'Beneath my arm are many swift arrows in my quiver, which speak to those who can understand; but in general they need interpreters' (ll. 84–6). This is not the only place where Pindar compares himself to an archer and his words to arrows. But though 'a word to the wise' is a common enough expression—Bacchylides 3. 85 says *φρονέοντι cυνετὰ γαρύω*, with no mysterious overtones—it is hard not to be reminded of such formulas as *φθέγξομαι οἷc θέμιc ἐcτί, θύραc δ' ἐπίθεcθε βέβηλοι* (Kern, *OF* 245. 1–2) or *ἀείcω ξυνετοῖcι· θύραc δ' ἐπίθεcθε βέβηλοι* (ibid., 334). Can it be that the hearer is being warned that the preceding passage contains certain allusions which will be clear only to persons familiar with certain secret doctrines known to those acquainted with the mysteries?

The thought that his poem is hard for some people to interpret brings Pindar to another of his poetical commonplaces, that of the contrast between the naturally gifted person and the industrious learner: 'Wise is he who knows much (i.e., understands much) by nature; but learners shriek loudly like a pair of crows against the godlike bird of Zeus' (ll. 86–8).[18a] Ancient commentators saw here an allusion to Pindar's supposed rivals, Simonides and Bacchylides, which recent scholarship has rightly rejected. But how is one to explain the dual *γαρύετον*? Bergk's alteration of this word to *γαρυέτων* has been adopted by several scholars during the last few years; but Kirkwood seems right in pointing out that one would expect the form || *γαρυόντων*.[19] A different explanation has occurred to me. The dual was originally designed to signify not so much the number two as the concept of duality or the notion of a pair belonging together; thus in Homer it is constantly used of things that form a natural pair, like hands or thighs, or of brothers, spouses, or friends who regularly act together. It is a known fact that crows are often seen in pairs, and A. de Jongh in 1865[20] was the first to adduce two passages in Aratus which seem likely to be relevant. At *Phaen.* 966–9 we read

καί που κόρακεc δίουc cταλαγμοὺc
φωνῇ ἐμιμήcαντο cὺν ὕδατοc ἐρχομένοιο·
ἤ ποτε καὶ κρώξαντε βαρείῃ διccάκι φωνῇ
μακρὸν ἐπιρροιζεῦcι τιναξάμενοι πτερὰ πυκνά.

[18a] [See now G. W. Most, *CQ*, NS 36 (1986), 304 f.]

[19] *CQ*, NS 31 (1981), 240 ff. and later *Selections from Pindar*, 35 f.

[20] *Pindari carmina Olympia* (Utrecht, 1865), ad loc.; on this way of using fable, see M. R. Lefkowitz, *HSCP* 73 (1969), 55 n. 13.

The same surprising combination of plural and dual is found at 1021–3 of the same poem:

> καὶ χῆνες κλαγγηδὸν ἐπειγόμεναι βρωμοῖο
> χειμῶνος μέγα σῆμα, καὶ ἐννεάγηρα κορώνη
> νύκτερον ἀείδουσα, καὶ ὀψὲ βοῶντε κολοιοί.

Jackdaws, of course, are the other kind of bird to which Pindar compares his detractors; at *Nem.* 3. 82 he writes κραγέται δὲ κολοιοὶ ταπεινὰ νέμονται.

Now Pindar orders his θυμός to aim its bow at the mark; the arrows that bring fame are to be aimed at Acragas. Swearing a solemn oath,[21] Pindar declares that in a hundred years the city has given birth to no man who has done more kindnesses to his friends than Theron. The poet ‖ cuts short his encomium with a mention of satiety provoked by the envy of small-minded persons; none can count the grains of sand, and who can enumerate the joys Theron has brought to others?

The stress laid upon Theron's kindness and generosity to others is most notable; he is designated by the word εὐεργέτας, the very word which later became the standard term for the Hellenistic king as 'benefactor'.[22] Pindar avoids the crudity of claiming directly that Theron is likely after death to share the bliss of Peleus, Cadmus, and Achilles; yet the whole burden of the poem suggests that after the many hardships of a noble and generous life he is worthy to be granted no ordinary reward.

Let us now return to the section of the poem, beginning at l. 61, that is concerned with the life after death. I have argued that the first sentence means that when men die, they at once pay a penalty when their wits become feeble, as the wits of the dead are commonly thought to be. I have also argued that the reference cannot be to a punishment ordained by the judge who in the following sentence is said to judge crimes committed upon earth. In that case, for what offences can the penalty of death be paid?

Socrates in Plato's *Meno* quotes a passage of Pindar (fr. 133), perhaps from a dirge, that has been much discussed. It may be rendered as follows: 'The souls of those for whom Persephone accepts atonement for her ancient grief, she sends back to the sun above in the ninth year; from them rise up great kings and men of mighty strength or great in wisdom, and for the rest of time they are called by men holy heroes.'

[21] Van Leeuwen, ii. 522 finds fault with Wilamowitz for saying that Bacchylides was readier with oaths than Pindar; but see Bacchyl. 5. 42, with H. Maehler's n. (*Die Lieder des Bakchylides*, i: *Die Siegeslieder*, ii (Leiden, 1982), 99).

[22] On the notion of εὐεργέτης in Pindar, see Hampe, 48 f.

Zuntz (*Persephone*, 86) writes that the reference remains obscure; 'It must be', he writes, 'to a myth which || accounted, in a manner unknown to us, for the fact that the soul, which is "from the gods", ever lost its divine status.' According to West (*OP* 110 n. 82), 'If Pindar thought of these souls as having begun their career as fallen gods'—a supposition based, like that of Zuntz, upon the assumption that Empedocles can guide us here—'the "ancient grief" for which they atone should be their original offence (perjury or bloodshed, as Emp. B 115?).'

Another theory which has been put forward, and which has attained considerable popularity, is based not upon the analogy offered by Empedocles, but on the conjecture that a myth narrated in an Orphic theogony can supply the answer. First put forward by Paul Tannery in 1899,[23] it was revived by Salomon Reinach in 1922, and was argued for in detail by H. J. Rose in 1936 and again in 1943; it has won the approval of the sceptical Linforth, of Burkert, and of Robert Parker in his important book *Miasma*. According to this theory, the 'ancient grief' of Persephone is the grief caused her by the Titans, who slaughtered, cooked, and devoured her son Dionysus.

The Orphic Theogony in question is that known as the Eudemian Theogony, much of whose contents were later incorporated in the Rhapsodic Theogony. Much of our evidence for its contents, which is most fully set out by Linforth and West,[24] comes from authors of the imperial || period, but the facts are clear enough; West dates the Eudemian Theogony [in the late fifth century. But the myth of the slaughter of Dionysus is older].[24a] The original Dionysus was the son of Zeus by his own daughter Persephone. While still a child he was decoyed with the offer of toys by the evil Titans, who slew him, cut him into pieces, boiled the pieces, and stuck them upon spits and then roasted them. The one part that survived was the heart, which Athena rescued and brought to Zeus, who swallowed it,[25] and soon after had intercourse with Semele, so that Dionysus was born again. The Titans were destroyed by Zeus with his thunder; from the soot that rose from their charred ashes mankind came into being. Man thus partakes of the

[23] P. Tannery, *RPh* 23 (1899), 129; S. Reinach, *RA*, 5th ser., 9 (1919), 162 f.; *Cultes, mythes et religions* (Paris, 1922), 61 f.; H. J. Rose, 'The Ancient Grief', in *Greek Poetry and Life: Essays presented to Gilbert Murray* (Oxford, 1936), 79–96, and *HTR* 36 (1943), 247–50; cf. Linforth, *AO* 348 f.; Burkert, *GRAKE* 443 [= *GR* 298]; R. Parker, *Miasma*, 299 f.

[24] Linforth, *AO* 307 f.; West, *OP* 140 f.; M. P. Nilsson, *HTR* 28 (1935), 221 f. = *Opuscula selecta*, ii (Lund, 1952), 673 f.; on the question of what purpose the myth was designed to serve, see West, *OP*, ch. 5 *passim*, and M. Détienne, *Dionysos mis à mort* (Paris, 1977), English version, *Dionysos Slain* (Baltimore, 1979); L. J. Alderink, *Creation and Salvation in Ancient Orphism* (Chico, 1981).

[24a] [See F. Graf, *Gnomon*, 57 (1985), 588–9.]

[25] According to Hyginus, *fab.* 167, Zeus swallowed the heart; cf. A. Henrichs (ed.), *Die Phoinikika des Lollianos* (Bonn, 1972), 69 f. and West, *OP* 162 f.

nature of the Titans, who though divine were evil, but also of the nature of Dionysus, son of Zeus and of Persephone.[26]

The Eudemian Theogony told the story of Zeus' birth in Crete, his nursing by Ida and Adrasteia, and his protection by the Curetes which is familiar from Hellenistic poetry.[27] Dionysus too was born in Crete and was guarded by the Curetes, a story known from so early a source as the *Bacchae* of Euripides. Early Cretan religion, with its god who died and was reborn each year, was in its original form distinct from Greek religion, and was adapted to it only with some difficulty. The great god might be called Zeus, in which case his father must be Kronos and his mother must be Rhea, who could be equated with the Great Mother, as in the *Bacchae*, where despite the part played by the Semele story, the importance of Dionysus' divine mother cannot be concealed (see E. R. Dodds, *Euripides, Bacchae* (2nd edn., Oxford, 1960), on 78–9). But the great god of ‖ Crete might also be equated with Dionysus, in which case his mother was Persephone. For the purpose of the cult myth in which Dionysus arrives at Thebes, his mother must be the daughter of the local king, to wit Semele. But even Semele was originally an earth goddess, as the etymology of the name reveals; the fundamental fact is that Dionysus must be the son of an earth goddess. Zeus may be equated with Hades, as he is in the ninth book of the *Iliad* (457) and in the *Supplices* of Aeschylus (156 ff.; 230–1); so might Dionysus, as he is by Heracleitus (22 B 15 D.–K.). The chorus of Cretan initiates of Idaean Zeus in Euripides' *Cretes* (fr. 472 Nauck^2 = 79 Austin) mentions 'the thunder', in connection with 'the raw feasts of night-wandering Zagreus'. Aesch. fr. 228 Radt said that Zagreus was the son of Hades; in the post-Homeric epic *Alcmaeonis* (fr. 3) he was addressed as *θεῶν πανυπέρτατε πάντων*. Callimachus speaks of Dionysus Zagreus (fr. 43, 117; cf. frr. 517 and 643); he and Euphorion (fr. 13 Powell) both told the story of Dionysus' murder by the Titans. In Crete there was a tomb of Zeus; at Delphi there was a tomb of Dionysus,[28] from which he was annually 'roused up' during the winter month Daidaphoros. When the Eudemian Theogony made Apollo take the remains of Dionysus from Crete to Delphi and bury them there, it was alluding to this belief; this reminds one of the tradition, featured in the *Homeric Hymn to Apollo*, that Delphi was founded by some Cretans. Our information about Epimenides, set out with great clarity by West, *OP* 45, is infuriatingly vague; but was poetry put out under his name the source of these Cretan elements in the theogonies?[29] ‖

[26] But see Burkert, *GRAKE* 253 [= *GR* 163].

[27] See Guthrie, 108 f. and West, *OP* 131 f.

[28] See West, *OP* 146 f.

[29] Pointing out that there was a Cretan element in the population of Gela, Nancy Demand, *GRBS* 16 (1975), 347–57, has conjectured that the beliefs of Theron regarding the afterlife had a Cretan origin.

Zuntz (p. 86 n. 3) complains that he 'cannot visualize Pindar, or Theron, indulging in theological abstrusities of this kind'. But it must be agreed that Pindar often took pains to familiarize himself with myths and beliefs cherished by his patrons that cannot have been widely known. It is certain enough that in the Second Olympian and other poems Pindar alluded to beliefs of the kind usually called Orphic or Pythagorean, and since he did this it would hardly be surprising if he made use of myths pertaining to those beliefs. We do not have to suppose that the allusion to the 'ancient grief' was as cryptic to the ancient hearer as it is to us, since for all we know the poem which contained it may have made the matter quite explicit, and in any case the poem may have been performed somewhere where the myth was known to many persons present. The allusion in the Second Olympian to the paying of a penalty would indeed have been obscure; only we must remember that Pindar, using language that recalls that of the mysteries, says that the arrows of his poetry had a voice for those who understood.

Zuntz in the same place complains that 'there is in ancient Akragas, and most of Sicily, a marked scarcity of evidence for any cult of Dionysus . . . and without Dionysus this whole "Orphism" becomes non-existent'. As Susan Cole[30] has shown, there is rather more evidence for Dionysiac religion in Magna Graecia than Zuntz allows. But for Dionysus to be important in a cult connected with Persephone that was familiar in a certain region it would not be necessary for Dionysus to have a cult of his own in that neighbourhood; at Eleusis there is no denying the importance of Iakchos, who is commonly identified with Dionysus. Scholars have long been puzzled by the invoca‖tion of Dionysus by the chorus of Sophocles' *Antigone* (1115 ff.):

πολυώνυμε, Καδμείας ἄγαλμα νύμφας
καὶ Διὸς βαρυβρεμέτα
γένος, κλυτὰν ὃς ἀμφέπεις
Ἰταλίαν, μέδεις δὲ
παγκοίνοις Ἐλευσινίας
Δηοῦς ἐν κόλποις, ὦ Βακχεῦ . . .

Ἰταλίαν was changed to *Ἰκαρίαν* by Unger, to *Κιδαρίαν* by Bergk, to *Φυταλίαν* by Seyffert; Gerhard Müller in 1967 approved Unger's conjecture, and Dawe in 1979 conjectured *Οἰχαλίαν* and put it in the text. But it would seem that after saluting Dionysus as the son of Semele and Zeus, the chorus names him as the ruler of two localities where he was specially connected with *τελεταί*, first Italy and then Eleusis.

[30] *GRBS* 21 (1980), 234 f.

But what reason have we for thinking that Dionysus was connected with *τελεταί* in Magna Graecia? Three years after the appearance of Zuntz's splendid book, the most interesting of all the gold tablets was found at Hipponion, in the territory of ancient Locri.[31] Its last two lines (15–16) read as follows:

> *καὶ δὴ καὶ cὺ πιὼν ὁδὸν ἔρχεαι ἅν τε καὶ ἄλλοι*
> *μύcται καὶ βάκχοι ἱερὰν cτείχουcι κλ(ε)εινοί.*

Here we find the allusion to Dionysus whose absence led Zuntz to pronounce that the gold tablets might be termed 'Pythagorean', but should not be termed 'Orphic'. In the wake of Wilamowitz, Zuntz is somewhat disdainful about anything 'Orphic'; he seems unable to forget Plato's mock||ery of the absurd pretensions of the Orpheotelestai, and its echo in Theophrastos' character of the Superstitious Man. But the term 'Orphic' is of very wide extension, and to separate what is 'Orphic' from what is 'Pythagorean' is by no means easy. Not only did Epigenes, apparently a fourth-century writer,[32] name certain Pythagoreans as the authors of Orphic poems, but Ion of Chios,[33] writing during the fifth century, said the same thing of Pythagoras himself.

Next, the bone tablets from Olbia, a place where we know Dionysiac cult to have existed in the sixth century, bear the inscription ΔΙΟΝ, pretty clearly standing for *Διόνυcoc*, and also a word which certainly begins with the letters ΟΡΦΙΚ and which may be ΟΡΦΙΚΟΙ.[34] If it is, we have the earliest *safe* instance of that term as the name of a group of persons; even if it is not, we have some kind of allusion to Orphism. The same tablets have *βίoc θάνατoc ἀλήθεια ψεῦδoc*: tables of opposites are regarded as Pythagorean, even though these particular examples remind us less of the Pythagorean mathematical tables of opposites preserved by Aristotle than of various pairs of opposites coupled by Heracleitus.[35]

We must also consider the famous passage of Herodotus dealing with this topic; of the two versions in the manuscripts, the longer is surely to be preferred, as Burkert seems to me to have established. Herodotus writes (2. 81) that in the refusal to allow woollen garments to enter temples or to be buried with the dead, the Egyptians *ὁμολογέουcι . . . τοῖcι Ὀρφικοῖcι καλεομένοιcι καὶ Βακχικοῖcι, ἐοῦcι δὲ Αἰγυπτίοιcι καὶ*

[31] See n. 8 above; Susan Cole has shown that the word *βάκχοι* implies a Dionysiac element. Cf. the archaic inscription from the cemetery of Cumae *οὐ θέμιc ἐν|τοῦθα κεῖcθ|αι μὲ τὸν βε|βαχχευμέ|νον* (Ed. Schwyzer (ed.), *Dialectorum Graecarum exempla epigraphica potiora* (Leipzig, 1923), n. 792). [See now W. Burkert, *Ancient Mystery Cults* (1987), 21 f., 87; and the Addendum below, pp. 105 f.]

[32] See Kern, *OF* 222 and cf. West, *OP* 9 f.

[33] See Kern, *OF* test. 248 (Ion 36 B 2 D.–K.); cf. Linforth, *AO* 111 and West, *OP* 7, 9.

[34] See the full discussion by West, *OP* 17 f. [; cf. Burkert, op. cit. (n. 31), pp. 46 f.].

[35] See Burkert, *LSAP* 51 f.; cf. Heraclitus Ephesius 22 B 21, 48, 62, 67, 76–7, 126 D.–K.

Πυθαγορείοιϲι. Parker, following ‖ Burkert, has well written that 'the traditional tug-of-war between pan-Orphism and pan-Pythagoreanism has given way of late to a recognition that coincidences between the two doctrines are probably more important than divergences.[36]

If I am right in taking the penalty paid by all the dead in the Second Olympian to be identical with the atonement accepted by Persephone from certain dead persons in the other poem, which Wilamowitz (*Pindaros*, 252) rightly warned should not be assumed necessarily to have been a dirge, then the dead from whom the goddess accepted the atonement were presumably the good. The souls of these, according to the poem quoted in the *Meno*, are sent back to the world by the goddess after eight years; the period recalls the nine-year exile endured, according to Hesiod, *Theog.* 793 ff., by gods who have committed perjury. From these good souls, Pindar says, come great kings, men of mighty strength—one may think of athletes, but as Hampe (63) says surely not only of athletes—and men great in wisdom, who will later be called heroes. The doctrine found in the Second Olympian would seem to be a further refinement of this, made by combining the Homeric and Hesiodic notion of the Islands of the Blest for selected heroes with a doctrine of paradise as the reward of three successive lives free from injustice that must derive from an Orphic or a Pythagorean source.[37] The mention of the kings, strong men, and wise men has often called to mind ‖ the fragment in which Empedocles (31 B 146 D.–K.) writes that the souls of the wise end by becoming prophets, poets, doctors, and leaders. Empedocles, if the usual dating is correct, will have been about twenty when the Second Olympian was first performed. But his work should be used with great caution in guessing at the nature of the doctrine known to Pindar, for like Pindar's his was an original and independent mind.

Empedocles does not speak of the *psyche*, but of a *daimon* who has been imprisoned in the body as a punishment for certain crimes, very possibly for having eaten flesh, and may eventually win release from the cycle of successive reincarnations. Pindar speaks, indeed, of the human soul as 'from the gods' in one of the fragments from a dirge (fr.

[36] *Miasma*, 290; cf. Burkert, *GRAKE* 445 [= *GR* 298], and see Burkert's discussion of the Herodotean passage, *LSAP* 127 f.

[37] The exact implications of *ἐϲτρὶϲ ἑκατέρωθι μείναντεϲ* (*Ol.* 2. 68–9) have been much argued over. Tycho Mommsen, op. cit. (n. 14), 30, argued for one life on earth, then a stay in Hades, then a second life in the world above followed by translation to the Islands of the Blest (cf. H. S. Long, *A Study of the Doctrine of Metempsychosis in Greece from Pythagoras to Plato* (Diss. Princeton, 1948), 29 f.; Hampe, 63). But Rohde, *Psyche*, ii. 212 n. 2 (= English version, 445 n. 32),‖ thought it meant three lives on each of the two sides; K. von Fritz, *Phronesis*, 2 (1957), 86–7, and D. MacGibbon, *Phronesis*, 9 (1964), 5–11, with various refinements, take views similar to Rohde's. [The poet's words do not seem to me to furnish sufficient evidence to decide the question; but the greater simplicity of Mommsen's view surely makes in its favour.]

131b). 'The body of all', this fragment says, 'follows mighty death; but the *αἰῶνος εἴδωλον* is still alive; for that alone is from the gods, and it sleeps while the limbs are active, but while they sleep it reveals in many dreams the coming decision of pleasant and of unwelcome things.' The word *αἰών* is coupled with *ψυχή* at *Il.* 16. 453 *λίπηι ψυχή τε καὶ αἰών*: elsewhere in Homer it means 'life', but after Homer 'lifetime'. Homer calls the spirits of the dead *εἴδωλα καμόντων*, and they are usually imagined as images of the living man; thus *αἰῶνος εἴδωλον* means 'image of life'.[38] Jan Bremmer[39] seems to me to have shown that ‖ the notion of a soul which not only survives the living man, but is able while he lives to leave the body at intervals, in the way described by Pindar and illustrated also by such stories as Herodotus tells of Aristeas and Hermotimus, and in a striking passage of the Hippocratic *De victu* in which it is described in terms not unlike those used by Pindar, is not a fifth-century innovation; using comparative material, much of it from Sanskrit sources, he argues that the notion of such a detachable soul is common to several Indo-European cultures from an early date. Bremmer distinguishes between the 'free soul', in Greek *ψυχή*, and the 'body soul', denoted by such words as *θυμός, μένος, νόος*: whether 'soul' is an adequate word to describe the latter may be disputed.

Oddly enough Bremmer does not raise the question of when the soul first came to be called 'immortal', or 'divine'. I know of no assertion that the soul comes from the gods earlier than Pindar fr. 131b; but as Burkert has written 'the mystery cults bring man the hope of escaping death and joining the gods; and it is an easy step from this to the doctrine that man is of divine descent and returns at death to his place of origin'.[40] The Greeks had various accounts of the creation of man, none of which makes very ambitious claims for him.[41] Hesiod (*Op.* 119 ff.) asserts that the immortals made him; but the particular immortals usually ‖ associated with the creation are the Titans, and particularly Prometheus. The Titans had been vanquished by Zeus, and were from several points of view less respectable than other

[38] See Claus, p. 117.

[39] See n. 4 above; Bremmer, 51 cites the Hippocratic passage. In this connection we may note that A. Dihle, 'Totenglaube und Seelenvorstellung im 7. Jahrhundert vor Christus', in *Jenseitsvorstellungen in Antike und Christentum: Gedenkschrift für Alfred Stuiber* (*JbAC*, Erg.-Bd. 9, 1982), 9 f. has argued that the Nekyia of the ‖ *Odyssey* reveals more than one conception of the *psyche*. Like Dihle's consultant Dr William Furley (see p. 19), I am not impressed by the argument that Anticleia's soul reveals different preoccupations from those of male heroes, since as she was a woman and Odysseus' mother, this is what I should expect. But when Dihle remarks 'Daß ein Dasein wie das der Psyche des Orion oder des Tantalos Grundlage einer Belohnung oder Bestrafung der auf der Welt vollbrachten Taten sein kann, liegt auf der Hand', he is making an observation of great interest.

[40] *LSAP* 359.

[41] See Zuntz, *Persephone*, 365; S. G. F. Brandon, *Creation Legends of the Ancient Near East* (London, 1963), 189.

Olympians; but they were certainly divine, and a divine pedigree that made man descended from them was better than none at all. Empedocles made man a fallen *daimon*, and the gold tablets, the earliest of which has been dated about ten years after the production of the Second Olympian, make a similar claim.

Let us take a look at the tablets and consider their relation to what Pindar says about the life after death. They were placed in graves so that the departed might take advantage of his membership of the group to which he belonged to establish his credentials as one of those who in Pindar's terminology are called 'the good'. Men die, according to the physician Alcmaeon of Croton,[42] who is often thought to have had Pythagorean connections, because they cannot fit the beginning on to the end; and Pindar wrote that the man who had been initiated at Eleusis was fortunate, because he knew the end of life and knew the Zeus-given beginning (fr. 137). The tablet would enable the soul of the dead initiate to begin again soon after his end.

The Hipponion tablet,[43] which has been dated about 465 BC, contains a fuller and better version of the poem ‖ found with variations on B1, from Petelia, south-east of Thourioi, B2, from Pharsalos in Thessaly, and B3–8, all from Eleutherna in central Crete; B1 and B2

[42] 24 B 2 D.–K. = *KRS*, fr. 455 (p. 347).

[43] See R. Janko, art. cit. (n. 8), for a careful attempt to reconstruct the archetype. My purpose requires only that I give a text of H, a text of the B tablets (n. 44, based on Zuntz, *Persephone*, 368), and a text of the A tablets, also based on Zuntz (n. 46).

Μναμοςύνας τόδε δῶρον· ἐπεὶ ἂν μέλληιςι θανεῖcθαι
εἰς' Ἀΐδαο δόμους εὐήρεας, ἔςτ' ἐπὶ δεξιὰ κρήνην
πὰρ δ' αὐτὰν ἑςτακῦ⟨ι⟩' ἁ λευκὰ κυπάριςςος
ἔνθα κατερχόμεναι ψυχαὶ νεκύων ψύχονται.
ταύτας τᾶς κράνας μηδὲ ςχεδὸν ἐγγύθεν ἔλθηις ‖
πρόςθεν δ' εὑρήςεις τᾶς Μναμοςύνας ἀπὸ λίμνας
ψυχρὸν ὕδωρ πρόρεον· φύλακες δ' ἐπύπερθεν ἔαςι.
οἱ δέ ς' ⟨ἐπ⟩ειρήςονται ἐν⟨ὶ⟩ φρεςὶ πευκαλίμ[ηιςιν
ọṭ⟨τ⟩ι διεξερέεις Ἄϊδος ςκότος ἠερόεντος.
εἶπον· "ὑὸς Γαίας καὶ Οὐρανοῦ ἀςτερόεντος,
δίψαι δ' ἠμ' αὖος καὶ ἀπόλλυμαι· ἀλλὰ δότ' ὦ[κα
ψυχρὸν ὕδωρ προρέον τῆς Μνημοςύνης ἀπὸ λί[μνης."
καὶ δή τοι ἐρέουςιν ὑποχθονίωι βαςιλῆϊ.
καὶ δή ςοι δώςουςι πιεῖν κείνας [ἀπὸ λίμνας.
καὶ δὴ καὶ ςὺ πιὼν ὁδὸν ἔρχεαι ἅν τε καὶ ἄλλοι
μύςται καὶ βάκχοι ἱερὰν ςτείχουςι κλε⟨ε⟩ινοί.

1 *δῶρον* Ll.-J. (1975), cf. A5, 3: *ἠρίον H*: see Janko, 92, for other recordings 2 Two lines seem to have been conflated 3 There are no white cypresses in this world; but why should they not be imagined in the next? 8 Suppl. Ll.-J. (1975) 9 [M. L. Lazzarini, *ASNP*, 3rd ser., 17 (1987), 332 reads ΟΡΟΕΕΝΤΟϹ, and A. C. Cassio, ibid. 333 f. conjectures *ἠερόεντος*. See Lazzarini, ibid. 329 f.] 10 *Γαίας*: *Βαρέας* ed. pr., approved by Janko 13 Lazzarini: *καὶ δή τοί ς' ἐλέουςιν ὑποχθόνιοι βαςιλῆες H* 14 *κείνας* Ll.-J.: *τᾶς Μναμοςύνας H*: *ταύτας* Marcovich 15 *καὶ ςὺ πιών* Luppe, Gil, Gallavotti (see Janko, p. 89 n. 1) 16 *κλε⟨ε⟩ινοί*: I do not favour the conjecture *κλυτάν τε* (B. Feyerabend, *RhM* 127 (1984), 4 f.).

are assigned to the fourth century and B3–8 to the third century BC. Zuntz, on p. 368, gives a composite version, giving all the variants, of the poem on the first two B tablets, which must be compared with the tablet from Hipponion (H);[44] tablets 3–8 (see Zuntz, p. 362) are not relevant to my purpose.‖

The soul arrives in the underworld desperately thirsty, 'dry with thirst'. Near the entrance is a fountain, where according to H4, souls may be seen getting refreshment (*ἔνθα κατερχόμεναι ψυχαὶ νεκύων ψύχονται*). *Ψύχεcθαι* is cognate with *ψυχή*, so that refreshment is a basic need of the soul; but these souls are getting it from the wrong place, for this is the fountain of Lethe, the source of forgetfulness. If it drinks of this fountain, the soul will forget the knowledge gathered in its previous existence; it was the mark of the superior soul, such as that of Pythagoras or later Empedocles, that it was able to remember its former lives. The soul of the initiate must drink not of this fountain, but of the fountain of Mnemosyne. H starts with the difficult and unmetrical words *Μνημοcύνης τόδε ἠρίον* ('this is the tomb, the monument, of Mnemosyne'). Metre and sense will be restored if we suppose that *Μνημοcύνης τόδε δῶρον*, a phrase which is found in the much later tablet A5. 3, was what stood in the original poem.

The fountain of Mnemosyne is guarded by watchers, who when the initiate approaches will question him as to his identity and origin. In reply, he is to say that he is a ‖ child of Earth and Heaven. The gods, of course, were children of Earth and Heaven; but so were the Titans, and

[44] 1. B1 *εὑρήcεις δ' Ἀΐδαο δόμων ἐπ' ἀριcτερὰ κρήνην*
B2 *εὑρήcεις Ἀΐδοιο δόμοις ἐνδέξια κρήνην*
2. B1 *πὰρ δ' αὐτῆι λευκὴν ἑcτηκυῖαν κυπάριccον·*
B2 *πὰρ δ' αὐτῆι λευκὴν ἑcτηκυῖαν κυπάριccον·*
3. B1 *ταύτης τῆc κρήνης μηδὲ cχεδὸν ἐμπελάcειαc*
B2 *ταύτης τῆc κρήνης μηδὲ cχεδόθεν πελάcηιcθα.*
4. B1 *εὑρήcεις δ' ἑτέραν, τῆc Μνημοcύνης ἀπὸ λίμνης*
B2 *πρόccω δ' εὑρήcεις τὸ Μνημοcύνης ἀπὸ λίμνης* ‖
5. B1 *ψυχρὸν ὕδωρ πρόρεον· φύλακεc δ' ἐπίπροcθεν ἔαcιν.*
B2 *ψυχρὸν ὕδωρ πρόρεον· φύλακεc δ' ἐπύπερθεν ἔαcιν.*
5a. B2 *οἱ δέ c' ⟨ἐπ⟩ειρήcονται ὅ τι χρέοc εἰcαφικάνειc·*
5b. B2 *τοῖc δὲ cὺ εὖ μάλα πᾶcαν ἀληθείην καταλέξαι·*
6. B1 *εἰπεῖν· Γῆc παῖc εἰμὶ καὶ Οὐρανοῦ ἀcτερόεντοc·*
B2 *εἰπεῖν· Γῆc παῖc εἰμὶ καὶ Οὐρανοῦ ἀcτερόεντοc·*
7. B1 *αὐτὰρ ἐμοὶ γένοc οὐράνιον· τόδε δ' ἴcτε καὶ αὐτοί.*
B2 *Ἀcτέριοc ὄνομα·*
8. B1 *δίψηι δ' εἰμὶ αὔη καὶ ἀπόλλυμαι· ἀλλὰ δότ' αἶψα*
B2 *δίψηι δ' εἰμ' αὖοc· ἀλλὰ δότε μοι*
9. B1 *ψυχρὸν ὕδωρ πρόρεον τῆc Μνημοcύνης ἀπὸ λίμνης.*
B2 *πιεῖν ἀπὸ τῆc κρήνης.*
10. B1 *καὐτοί ⟨coι⟩ δώcουcι πιεῖν θείηc ἀπ[ὸ κρή]νηc.*
11. B1 *καὶ τότ' ἔπειτ' ἄ[λλοιcι μεθ'] ἡρώεccιν ἀνάξει[c] . . .*

B1 had apparently three more verses.

so is even man; *ἓν ἀνδρῶν, ἓν θεῶν γένος· ἐκ μιᾶς δὲ πνέομεν / ματρὸς ἀμφότεροι*, Pindar says at the beginning of the Sixth Nemean (1–2). Then the soul must explain that it is perishing of thirst, and must ask for water from the Lake of Memory. Now the rulers beneath the earth will take pity on the soul, and will allow it to drink; after that it will be free to tread the sacred road previously trod by other *μύσται καὶ βάκχοι*. The sacred road reminds us of the 'sacred way' at Eleusis and of Pindar's road of Zeus; but the road here spoken of is trodden by all the good, and not only by the select company that finds the 'mystic path' to Rhadamanthys.[45]

A1–4 were found in the necropolis of Thourioi; the first three come from a tomb assigned to the middle of the fourth century BC, the other from one perhaps as late as about 300 BC; A5, found at Rome, is about six hundred years later, and bears the name of its owner, one Caecilia Secundina.[46] First, the soul claims to be pure and to come from the

[45] H. Lloyd-Jones and P. Parsons (eds.), *Supplementum Hellenisticum* (Berlin, 1983), no. 705; see H. Lloyd-Jones, *JHS* 83 (1963), 93–4 [= *Greek Comedy, Hellenistic Literature, Greek Religion, and Miscellanea*, ch. 16, p. 186 f.]. Burkert made this observation; see Cole, p. 231 n. 28; and cf. Hegesippus, *AP* 7. 545. 1–2 = Gow–Page, *HE* 1913–14.

[46] A1:

1 *ἔρχομαι ἐκ κοθαρῶν κοθαρά, χθονίων βασίλεια,*
2 *Εὐκλῆς Εὐβουλεύς τε καὶ ἀθάνατοι θεοὶ ἄλλοι·*
3 *καὶ γὰρ ἐγὼν ὑμῶν γένος ὄλβιον εὔχομαι εἶμεν.*
4 *ἀλλά με μοῖρ' ἐδάμασσε [καὶ ἀθάνατοι θεοὶ ἄλλοι]*
καὶ ἀστεροβλῆτα κεραυνῶι.
5 *κύκλου δ' ἐξέπταν βαρυπενθέος ἀργαλέοιο,*
6 *ἱμερτοῦ δ' ἐπέβαν στεφάνου ποσὶ καρπαλίμοισι,*
7 *δεσποίνας δ' ὑπὸ κόλπον ἔδυν χθονίας βασιλείας.*
[ἱμερτοῦ δ' ἐπέβαν στεφάνου ποσὶ καρπαλίμοισι]
8 *"ὄλβιε καὶ μακαριστέ, θεὸς δ' ἔσηι ἀντὶ βροτοῖο."*
9 *ἔριφος ἐς γάλ' ἔπετον.* ||

4 et 7 dittographias del. Kaibel   4 *κεραυνῶι* Zuntz: *κεραυνον* tab.

A2:

1 *ἔρχομαι ἐκ καθαρῶν καθαρά, χθονίων βασίλεια,*
2 *Εὖκλε καὶ Εὐβουλεῦ καὶ θεοὶ δαίμονες ἄλλοι·*
3 *καὶ γὰρ ἐγὼν ὑμῶν γένος εὔχομαι ὄλβιον εἶναι·*
4 *ποινὰν δ' ἀνταπέτεισ' ἔργων ἕνεκ' οὔτι δικαίων·*
5 *εἴτε με μοῖρ' ἐδάμασσ' εἴτ' ἀστεροπῆτι κεραυνῶ(ν).*
6 *νῦν δ' ἱκέτης ἥκω παρ' ἁγνὴν Περσεφόνειαν*
7 *ὥς με πρόφρων πέμψηι ἕδρας ἐς εὐαγέων.*

2 *θε⟨ῖ⟩οι*? (cf. A3. 2)   6 *ἀγαυὴν* Diels   7 *ἕδρας εἰς εὐαγεόντων* Diels

A3:

1 *ἔρχομαι ἐκ καθαρῶν καθα⟨ρά, χθ⟩ο⟨νίων⟩ βασίλεια,*
2 *Εὖκλε [υα] καὶ Εὐβουλεῦ καὶ θεοὶ ὅσοι δαίμονες ἄλλοι.*
3 *καὶ γὰρ ἐγὼν ὑμῶν γένος ⟨ὄλβιον⟩ εὔχομαι εἶναι· [ὄλβιον]*
4 *ποινὰν δ' ἀνταπέτεισ' ἔργων ⟨ἕνεκ'⟩ οὔ τι δικαίων·*
5 *εἴτε με μοῖρ' ⟨ἐδάμασσ'⟩ εἴτ' † εροπητι [κη] κεραυνο †.*
6 *νῦν δ' ⟨ἱ⟩κ⟨έτης⟩ ἥκω παρ' ἁ⟨γνὴν⟩ Περσεφ⟨όνειαν⟩*
7 *ὥς με πρόφρων πέμψηι ἕδρας ἐς εὐαγέων.*

2 *θεοὶ ὅσοι*] *θεῖοι*?

pure; this will presumably imply that the bearer of ‖ the tablet has been initiated and has observed certain purificatory rules. It addresses the Queen of the Chthonioi, who is obviously Persephone, but also Eukles, Eubouleus, and other immortal gods.‖

Eukles, as Zuntz (p. 310) remarks, has been compared with *Κλύμενος*,[47] a familiar name of Hades, and the Euklos mentioned in the Oscan tablet from Agnone may indeed be Hades. That suggested to Olivieri the trinity Persephone, Hades, Dionysus. But can Eubouleus be Dionysus? He is Dionysus in the *Orphic Hymns*, where he is equated with Plouton and is called a son of Demeter and Dysaules; but more often Eubouleus, or Euboulos, is a title of Hades, though it is also a title of Zeus. Zuntz is emphatic in warning us against the identification suggested by Olivieri. But as he says the name Eubouleus is an euphemism; so is the name Eukles; and it is highly characteristic of the language of the mysteries, at Eleusis in particular, to denote awe-inspiring deities by obscure and euphemizing aliases. What is significant is that the tablets name one female deity, the Queen of the Chthonioi, together with two male deities, thus giving us the trinity of one female and two male persons which we should expect. That matters more than the attempt to establish an exact equation between the names of the trinity's male members and other known gods.

As in the B poems, the soul goes on to claim divine ancestry, but in a less specific way, for it says, 'for I too claim to be of your blessed race'. A2 and A3 go on to claim to have paid the penalty for unjust actions, 'whether Fate subdued me or the lord of lightning with his bolt'. A1, however, omits the claim to have paid the penalty, going straight from the claim to divine birth to say, 'But Fate subdued me, and the lord of lightning with his bolt.' This must surely remind one of Pind. fr. 133,

A4:

1 *ἀλλ' ὁπόταμ ψυχὴ προλίπηι φάος Ἀελίοιο,*
2 *δεξιὸν † εcοιαcδεεϲτ † ⟨ἰέ⟩ναι πεφυλαγμένον εὖ μάλα πάντα.*
3 *χαῖρε παθὼν τὸ πάθημα τὸ δ' οὔπω πρόcθ' ἐπεπόνθειc·*
4 *θεὸc ἐγένου ἐξ ἀνθρώπου· ἔριφοc ἐc γάλα ἔπετεc.*
5 *χαῖρ⟨ε⟩ χαῖρε· δεξιὰν ὁδοιπόρ⟨ει⟩*
6 *λειμῶνάc τε ἱεροὺc καὶ ἄλcεα Περcεφονείαc.*

1 and 5 suppl. Zuntz

A5:

1 *ἔρχεται ἐκ καθαρῶν καθαρά, χθονίων βαcίλεια,*
2 *Εὔκλεεc Εὐβουλεῦ τε, Διὸc τέκοc ἀγλαά· ἔχω δὲ*
3 *Μνημοcύνηc τόδε δῶρον ἀοίδιμον ἀνθρώποιcιν.*
4 *"Καικιλία Cεκουνδεῖνα, νόμωι ἴθι δῖα γεγῶcα."*

Tablet C is not relevant to my purpose (see Zuntz, 344 ff.); neither is the new tablet published by J. Breslin, *A Greek Prayer* (Pasadena, Calif., Ambassador College, 1977) and transcribed by R. Merkelbach, *ZPE* 25 (1977), 276, which adds nothing to our knowledge.

[47] See Lasos fr. 1 Page = *PMG* 702.

where Persephone is said to accept atonement for certain souls, and also of the Second Olympian, where those who die are said at once to pay the penalty.‖

The mention of the lord of lightning leads Zuntz (p. 316) to suggest that all tablets bearing this particular verse were placed in the graves of persons killed by lightning. It has long been known that early religion attributed a peculiar sanctity to persons and places struck by lightning; yet deaths by lightning are not so frequent that one would expect three of the few tablets discovered to contain a verse written to refer to them. Also, one remembers that according to the myth recounted in the Eudemian Theogony, the Titans were punished for the murder of Dionysus by being destroyed by Zeus' lightning; and it seems to me highly probable that it was this episode to which the soul of the initiate here refers.

Against this interpretation Zuntz (p. 312) has objected that in l. 3 (καὶ γὰρ ἐγὼν ὑμῶν γένος ὄλβιον εὔχομαι εἶναι) one would expect not ὑμῶν but ϲοῦ. But as the parallel in the other group of tablets indicates, in saying 'I am a child of earth and heaven' the soul is not claiming descent from Persephone, but is rather claiming to share the descent from Earth and Heaven which the Titans had in common with the gods. The soul, he complains, could hardly expect to curry favour with Persephone by claiming descent from the murderers of her child; but as the soul points out it has now paid the penalty, and Dionysus as well as the Titans was its ancestor.

A2 and A3 conclude with a couplet in which the soul declares that it comes as a suppliant to mighty Persephone, so that she may graciously send it to the seats of the blessed ones. But A1 offers a different conclusion, five lines long. The first three say, 'I have flown out of the grievous cycle of deep mourning, and have attained with swift feet to the longed-for crown, and have crept beneath the lap of the Mistress, Queen of Earth.'

Zuntz (pp. 320 f.) remarks that, since life with its uncertainties is sometimes compared to a cycle one cannot ‖ assume without a definite indication that we have here a reference to a cycle of successive incarnations, like that described by Empedocles; but since Diogenes Laertius [8. 14], probably following Aristoxenos, says that Pythagoras spoke of the 'wheel of Ananke', and since there is a line of Orphic poetry (Kern, *OF* 229–30) κύκλου τε λῆξαι καὶ ἀναπνεῦϲαι κακότητοϲ, he admits that κύκλοϲ was 'a Pythagorean term for life on earth conceived as a series of doleful incarnations'. It seems certain that the word has this connotation here.[47a]

[47a] [See R. Renehan, *CP* 80 (1985), 147.]

In the next line, 'I have attained to the longed-for crown', the metaphor must be from the garland that was the prize in an athletic contest. Then the soul claims to have found a refuge with Persephone; one is reminded of the tragic metaphor by which children are said to take refuge under their mother's wings as though she were a hen.

A1 continues with words presumably put into the mouth of the goddess, 'Happy and blessed one, you shall be a god instead of a mortal.' The divinity here claimed is asserted on behalf of the *daimon* that is the equivalent of the human soul with far greater conviction and far greater eloquence by Empedocles; Pindar, we remember, says in fr. 131b that the human soul comes from the gods.

A1 concludes with a line of prose, containing the mysterious expression, 'I am a kid who has fallen into milk'. The riddling use of a proverbial expression recalls other sayings associated with the mysteries; among countless attempts at explanation, one may single out R. S. Conway's suggestion[48] that it was a proverb indicating that an apparent disaster has resulted in unparalleled good fortune. But maybe there was no suggestion of an apparent disaster, and the phrase simply meant that the initiate was extremely fortunate.||

My conclusion is that the tablets give a picture of the beliefs behind initiation rites that has more in common with what we learn from Pindar than has so far been allowed. Now that we have a rough general notion of the Orphic or Pythagorean beliefs that were entertained not only in South Italy, but in other places also, we can go back to Pindar and consider how he may have adapted these notions to his own literary purposes. The tablets, which are designed simply to help the soul of the initiate to present his credentials, distinguish the common souls who drink of the fountain of forgetfulness from the souls of the initiates who drink at the fountain of memory; but they give no notion of the existence of the special third category whom Pindar places in the Islands of the Blest. Neither does the comparatively simple scheme which we see in operation here give any indication, despite the reference to the 'grievous cycle of deep mourning', of the series of incarnations, starting from plants and animals and ending with the fallen *daimon*'s return to its divine status, described by Empedocles.

Pindar describes with sympathy the easy life in Hades of those who have been good throughout a single life, and in fr. 129 we seem to have a more detailed picture of that life and of the pleasures which it offered. But his special interest is reserved for the inhabitants of the Islands of the Blest, dwelling in their island in Oceanus with its golden flowers, with Kronos and with Rhadamanthys. The Orphic or Pytha-

[48] *Bull. John Rylands Library*, 1933, 76, cited by Guthrie, 179. [See pp. 106 f.]

gorean scheme with its moral element has so far influenced Pindar that he admits those who have lived out three lives without injustice; but he prefers to think in terms of the Islands of the Blest known from Homer and Hesiod, where one could find the great heroes whose deeds he loves to celebrate, like Cadmus, Peleus, and Achilles.

Theron's great benefits to others, it is delicately || implied, though not directly stated, make it possible that after death he may attain this privileged position. But Pindar writes of his patron and his family and their fortunes in a manner far removed from any complacent optimism. Tracing their history from the time of Cadmus and his daughters, and dwelling on the sad episode of Oedipus and his sons, he lays the strongest stress on the sorrow which alternates with the happiness of even those human beings who attain to most felicity. Theron himself has lived a life of many vicissitudes, it is made clear, though without details, in the first triad of the poem. Between the initiate and his future life there lies, in any case, the bourne of death; all humans must pay the penalty to Persephone. If the beliefs about the afterlife which figure in this poem can properly be called 'Orphic', then Orphism as Pindar has presented it is as far from the vulgar Orphism of the practitioners described by Plato as the Christianity of Paul Claudel or T. S. Eliot is removed from that of the vulgar practitioners of that religion. His depiction of the highest bliss is couched in heroic terms; where Empedocles thought of it as a reward for purity, Pindar prefers to think of as a reward for heroism.

Let us return to the question mentioned at the start; did Pindar 'believe' in this doctrine of an afterlife, or did he merely work into his poem allusions to a belief cherished by his patron Theron? It is a question calculated to perplex Christians and persons whose outlook is conditioned by Romanticism. Wilamowitz was not a Christian, but it has justly been remarked that the very title of his great book on Greek religion indicates that its author has been brought up as a Lutheran. However, in dealing with Greek religion, the question as to what people *believed* is not always the right question to ask. The Greeks did not believe their myths in the same literal way in which Christians, until comparatively lately, believed theirs, although in our time || the growth of scepticism has forced Christians more and more to fall back on the defence of their myths as myths that was first elaborated late in the eighteenth century. Again, since the decline of Romanticism it has become easier to understand that a great poet can write great poetry about images or ideas which have captured his imagination even without being in a Christian or a Romantic sense totally 'sincere'.

The kind of 'Orphism' we find in the Second Olympian was not confined to Magna Graecia, but existed in several parts of Greece. It may not have been so different from what one might call Eleusinian Orphism, with which Pindar was certainly familiar,[49] as is usually assumed; at Eleusis also we find Dionysus, and at a critical stage of the celebration of the mysteries the hierophant proclaimed that the Mistress had given birth to a holy child, Brimo to Brimos, presumably with reference either to Iakchos or to Ploutos.[50] It would appear that Theron was an initiate of a mystery cult; the Third Olympian (l. 41) refers to the Emmenidai as 'guarding the *τελεταί* of the blessed ones with pious purpose'. Even if Pindar had no part in such a cult and did not share its special interests—and in Greek religion it was hard to honour all gods in equal measure—one can understand that while he was composing a poem in honour of a family that subscribed to it, a family, moreover with which he seems to have had a specially close personal connection, this like the Eleusinian cult may have kindled his imagination. As he presents it, the eternal bliss of the elect has a heroic colour; and nothing in his poem minimizes the finality and solemnity of death, the penalty that all mortals must render to Persephone.||

DISCUSSION

M. Hurst: Chacun sera reconnaissant à M. Lloyd-Jones d'avoir hardiment repris en mains une vieille question à la lumière de documents nouveaux. Il y a ceux qui, dans un livre sur Pindare cherchent du nouveau sur la 2e *Pythique*, mais il y a aussi ceux qui cherchent du nouveau sur la 2e *Olympique*: ces derniers ne seront pas déçus, sans doute. Du sens de certains mots jusqu'à la vision de l'énoncé 'mythique' complexe de cette ode, vous nous offrez de quoi réfléchir à nouveau et peut-être de quoi changer d'avis.

M. Gerber: I am intrigued by the interpretation you offer for *ἀπάλαμνος* and *Ol.* 2. 51–8, but I should like to see a more detailed explanation of the syntax such an interpretation involves.

M. Lloyd-Jones: I would say that *ἀπάλαμνοι φρένες ποινὰς ἔτεισαν* was equivalent to *αἱ φρένες αὐτῶν τὰς ποινὰς ἔτεισαν ἀπάλαμνοι γιγνόμεναι.*

M. Pòrtulas: Le professeur Lloyd-Jones a soulevé la question des origines de l'homme. J'avoue ne pas le comprendre quand il dit que 'Hesiod asserts that the immortals made him.' Sans doute les dieux ont-ils créé les races d'or, d'argent et de bronze. Mais la race de fer? Hésiode ne dit pas quelle est son origine. D'où ma conviction qu'une des singularités de l'orphisme est précisément d'avoir donné à ce sujet un enseignement. La religion grecque ne disposait pas d'une anthropogonie communément acceptée, ce qui a frappé les historiens des religions. Ils ont donc insisté sur la singularité, à cet égard, de l'orphisme.

[49] See F. Graf, op. cit. (n. 5); on Pind. fr. 346, probably from a Pindaric poem which mentioned the initiation of Heracles by Eumolpos, see H. Lloyd-Jones, art. cit. (n. 5) [= ch. 11].

[50] See Burkert, *GRAKE* 430 [= *GR* 288].

Deuxième remarque: la question des rapports entre l'orphisme et le pythagorisme mérite, sans doute, une discussion plus poussée. Je me limiterai toutefois à signaler qu'on a de bonnes raisons de croire qu'elle était surtout de nature politico-sociale: les pythagoriciens ont constitué un mouvement à caractère fermé, avec le dessein de concilier le refus ‖ orphique de la vie de la *πόλις* avec leur propre intervention dans les affaires politiques. Il serait fallacieux de se borner, pour traiter de ce problème, à des points de vue étroitement philologiques. Je suis reconnaissant à M. Lloyd-Jones d'avoir évité cet écueil.

M. Lloyd-Jones: Je me sens conforté, Monsieur, par votre approbation.

M. Köhnken: I have learned much about Orphism as a background for interpreting *Ol.* 2 and agree with your explanation of *ἀπάλαμνος* (l. 57) and *γαρύετον* (l. 87). Concerning the relationship between *Ol.* 2 and frr. 129 ff. I should like to ask whether the fact that the accounts of the afterlife given in both places are basically the same does not argue against the assumption that in *Ol.* 2 Pindar is voicing the beliefs of his addressee.

M. Lloyd-Jones: The poem may not have contained quite the same motif, but it might in any case have been written for the same people and so have referred to the same beliefs.

M. Köhnken: Ich weiß nicht, ob die übliche Auffassung der Partie Ol. 2, 30 ff. (*ἤτοι βροτῶν γε κέκριται πεῖρας οὔ τι θανάτου, οὐδ' ... ὁπότε ... τελευτάσομεν· ...*), auf die Sie kurz Bezug nehmen (S. 251): 'Mortals cannot know when they will die, or when they will finish the day with happiness still undiminished', richtig ist. Spricht nicht der Kontext und das auf die vorhergehenden *paradeigmata* der Kadmostöchter Semele und Ino zurückweisende *ἤτοι* (30; vgl. zur Funktion dieser Partikel Pyth. 12, 13) eher dafür, den Satz folgendermaßen zu verstehen: 'für die Sterblichen ist also wirklich eine Todesgrenze[51] durchaus nicht festgelegt und (es ist) auch nicht (bestimmt), wann wir den friedlichen Tag in unzerstörbarem Glück beenden werden', d.h. das Dasein ist für die Sterblichen mit dem Tode nicht zu Ende, wie das Beispiel von Semele ‖ und Ino zeigt, die nach ihrem (schrecklichen) Tod für alle Zeit glücklich 'weiterleben' (25 von Semele *ζώει ... ἀποθανοῖcα* sie 'lebt', obwohl sie 'gestorben' ist; vgl. 29 f. von Ino *βίοτον ἄφθιτον / ... τὸν ὅλον ἀμφὶ χρόνον*; s. auch 24 *κρεccόνων πρὸc ἀγαθῶν* im Mythos und 33 *ἀτείρει cὺν ἀγαθῶι* in der allgemeinen Aussage: die Sentenz ist im Hinblick auf die vorhergehenden Exempla formuliert). Die vorgeschlagene Erklärung scheint mir gestützt zu werden durch Parth. 1. 14–15: *ἀθάναται δὲ βροτοῖc ἁμέραι, cῶμα δ' ἐcτὶ θνατόν* und fr. 131b 1–2: *cῶμα μὲν πάντων ἕπεται θανάτωι ... ζωὸν δ' ἔτι λείπεται αἰῶνοc εἴδωλον* und sie würde zu Ihrer Interpretation der Verse 56 ff. passen.

M. Lloyd-Jones: I agree.

M. Reverdin: Nous voici à nouveau sur un terrain solide, et je m'en réjouis. Il est dangereux de divaguer à propos de l'orphisme, du pythagorisme. C'est pour parler avec Rabelais, 'sasser et beluter son temps en ce bas monde'!

M. Lloyd-Jones: The history of people's attitudes to 'Orphism' and its

[51] Der Ausdruck *πεῖραc ... θανάτου* (31) im Anschluß an Od. 22, 41: *ὑμῖν ... πᾶcιν ὀλέθρου πείρατ' ἐφῆπται* (vgl. Il. 6, 143).

problems in our times take the form of a Hegelian triad; excessive confidence lends to exaggeration (such people as V. Macchioro inventing an Orphic church with clergy, dogmas, and sacraments); then comes the sceptical reaction (Wilamowitz, Linforth); finally a cautious middle view.

Mme Lefkowitz: Professor Lloyd-Jones's analysis illustrates once again the importance of the patron–poet relationship. Unlike the Romantic poet, who preferred to draw his inspiration from nature, the ancient poet wrote for and was influenced by his patron, often with remarkable results, like *Ol.* 2.

M. Lloyd-Jones: It may be significant that the person for whom this poem was written belonged to the family of patrons with which there is the best reason to believe that the poet was on intimate terms.‖

In such a case, whether someone 'believed' is perhaps not the most important question; Catholics may find this easier to understand than Protestants.

M. Hurst: Vous avez évoqué prudemment le rapport avec Empédocle. Ne pensez-vous pas qu'on puisse aller un peu plus avant et lire dans la description de l'île des bienheureux le schéma des quatre éléments (72: *αὖραι* / *φλέγει* — 73: *χερcόθεν* / *ὕδωρ*)?

M. Lloyd-Jones: It is an ingenious suggestion, and may conceivably be right; but I am wary of reading back Empedoclean notions into Pindar.

ADDENDUM (1989)[52]

K. Tsantsanoglou and G. M. Parássoglou, 'Two Gold Lamellae from Thessaly', *Ἑλληνικά (Δημοcιεύματα τῆc Ἑταιρείαc τῶν Μακεδονικῶν Cπουδῶν)*, 38 (1987), 3–17 have now published two copies of the same poem from gold tablets, shaped like ivy-leaves, found in December 1985 by Athanasios Tzafalias near the Thessalian village of Petroporos, near the site of the ancient Pelinnaion, known to us from Pindar's Tenth Pythian Ode. 'The publication of the new tablets', Albert Henrichs has written (in 'Between Country and City: Dimensions of Dionysos in Athens and Attica', forthcoming, which I am much obliged to him for having shown me), '. . . marks a turning-point in the study of these controversial texts'. The first editors have made a good job of the publication; but I should like to add a footnote to my article of 1981. First, I reproduce their text.

a

νῦν ἔθανες | καὶ νῦν ἐγ|ένου, τριcόλβ|ιε, ἄματι τῶιδε. |
εἰπεῖν Φερ ʽcʼ εφόν|αι ̣ς̣ʼ ὅτι Β⟨άκ⟩χιος αὐτὸc | ἔλυcε.
τα{ι}ῦροc | εἰc γάλ(α) ἔθορεc.
αἶ|ψα εἰc γ⟨ά⟩λα ἔθορεc. |

[52] I am obliged to the first editors for useful discussion, and also to Professor James Barr, Dr Robert Parker, and Dr Christine Sourvinou-Inwood.

(κ)ριὸϲ εἰϲ γάλα ἔπεϲ⟨εϲ⟩. |
οἶνον ἔχειϲ εὐ|δ⟨α⟩ι̣μονατιμν|
κὰπ(ι)μέν|ε̣ι ϲ' ὑπ̣ὸ̣ | γῆ̣ν τέ|λεα ἅϲ⟨ϲ⟩α|περ ὄλ|βιοι ἄλ|λοι.

3 *γάλ(α)*: *γαλδ* 5 *(κ)ριὸϲ*: *χριοϲ* 7 *κὰπ(ι)μένει· καπυμενει*

b

νῦν ἔθανε⟨ϲ⟩ | καὶ νῦν ἐ|γένου, τριϲόλ|βιε, ἄματι | ⟨τῶι⟩δ̣ε̣.
⟨ε⟩ἰπ̣ε̣ῖν Φερ|ϲεφό⟨ναι ϲ'⟩ ὅτι Βά⟨κ⟩χιο|ϲ αὐτὸϲ ἔλυϲε. |
ταῦροϲ εἰ⟨ϲ⟩ γάλα ἔ|θορ⟨ε⟩ϲ.
κριὸϲ εἰϲ γάλ⟨α⟩ | ἔπεϲε⟨ϲ⟩.
οἶνον ἔ|χειϲ εὔδαι|μον|τιμ|μν

1–2. First, the deduction from ll. 15–16 of the Hipponion tablet (see *EH* 31 (1985), 264 = 93 above) that the mysteries to which the poem is related are those of Dionysos is now placed beyond all doubt. On the connection of the Dionysiac mysteries with the afterlife, see W. Burkert, *Ancient Mystery Cults* (1987), 21 f. and *passim*; R. Merkelbach, *ZPE* 76 (1989), 15 f. has pointed out that by being initiated into the mysteries of Dionysos the dead person must have made atonement to Persephone for her 'ancient grief'.

The first editors suggested that *τριϲόλβιε* had been substituted for *μάκαρ*, vocative, or for *θεὸϲ*, used as a predicate; also, it might have been substituted for *μάκαρ*, not vocative but predicative, and used as a synonym for *θεόϲ*. In favour of either of the last two alternatives one may cite, with the first editors, A1. 8 (Zuntz, p. 301) *ὄλβιε καὶ μακαριϲτέ, θεὸϲ δ' ἔϲηι ἀντὶ βροτοῖο* and A4. 4 *θεὺϲ ἐγένου ἐξ ἀνθρώπον* (Zuntz, p. 329). But Dr Robert Parker points out that if *νῦν ἐγένου* means 'you have been born on this day', we obtain a powerful contrast with *ἔθανεϲ*, and this favours the first of the three alternatives. The substitution of the unmetrical *τριϲόλβιε* for *μάκαρ* indicates that we may expect to find that the original poem has not been copied accurately, but that we must expect to find changes, either deliberate or accidental.

The dead person is to tell Persephone, who is undoubtedly the *χθονίων βαϲίλεια* of the first line of A1, A2, and A3, that Dionysos himself has released him; having been initiated (cf. H16 *μύϲται καὶ βάκχοι*), he has paid the penalty for the goddess's ancient grief and has been released from the cycle (see *EH* 31 (1985), 259 f. and 276 f. = pp. 89 and 100 above).

3–5. These lines, as the first editors observe, obviously recall the mysterious words at A1. 9 and A4. 4 *ἔριφοϲ ἐϲ γάλα ἔπετον*. As to the significance of this formula, Zuntz, pp. 324–7 and I at *EH* 31 (1985), 276 = p. 101 above) were equally in the dark.

But in 1914 A. B. Cook, *Zeus*, i. 675–6 had offered an explanation of the formula which must now be reconsidered. 'We must, I think,' he wrote, 'start from the fact, noted by Dr. Frazer (*Anthropological Essays Presented to Edward Burnett Tylor*, Oxford, 1907, 151 f.) that semi-civilised folk relish meat boiled in milk.' This practice is familiar to us from the biblical injunction 'Thou shalt not seethe a kid in his mother's milk' (Exod. 23: 19 and Deut. 14: 21); its prevalence in the Mediterranean area, both in ancient and in modern times, is amply demonstrated by M. Haran, 'Seething a Kid in its Mother's Milk', *Journal of Judaic Studies*, 30 (1979), 23–35 and 'Das Böcklein in der Milch seiner Mutter und das säugende Muttertier', *Theologische Zeitschrift*, 41 (1985), 135–59 and Othmar Keel, *Das Böcklein in der Milch seiner Mutter und Verwandtes* (Göttingen, 1980). (All these works were brought to my attention by Professor James Barr.)

In the myth that lies at the back of these mysteries, the Titans boiled the flesh of the dismembered Dionysos in a cauldron. 'I should conjecture', wrote Cook, i. 675, 'that there was a version of the Dionysus-myth in which the god boiled in a cauldron and subsequently devoured was done to death not as a bull but as a kid. I am further inclined to think that his worshippers, by way of identifying themselves with him, took the name of "kids" and actually pretended to be seethed like him in a cauldron.'

The mention of a bull in l. 3 is unsurprising in that no animal was more frequently identified with Dionysos (see e.g. E. R. Dodds, *Euripides, Bacchae* (2nd edn. 1960), p. xviii), but since mature animals are not boiled in milk, it may seem to make against Cook's explanation. So may the mention of the ram in l. 5, if indeed the ram is mentioned, as it certainly appears to be.

At this point, let us consider what is meant by l. 4. ΑΙΨΑ is written, but *αἴξ*, considered as a possible emendation by the first editors, is undoubtedly correct. In magical and religious texts corruption is especially frequent; those who copied them were often ignorant of, or indeed indifferent to, their proper meaning, since for them such a text was what the early Romans called a *carmen*, whose magical power was not affected by inaccuracy. The goat had a very special relation to Dionysos (see e.g. Walter F. Otto, *Dionysos: Mythos und Kultus* (1933), 155–7). 'The most usual victim was a bull', Dodds writes (loc. cit.), speaking of the Bacchic *ὠμοφαγία*. '... We hear also of *ὠμοφαγίαι* of wild goats or fawns'; and goats often figure in cult-inscriptions as sacrificed to Dionysos. Nor is the male animal always specified: on the Erchia calendar (F. Sokolowski, *Lois sacrées des cités grecques* (1969), no. 18 *Α* 44–51 and *Δ* 33–40), Semele and Dionysos are each to receive the sacrifice of a goat (*αἴξ*) on 16 Elaphebolion, the day of the City Dionysia, and according to Pausanias 9. 8. 1 at Potniai a goat was sacrificed to Dionysos Aigobolos as a surrogate for a human being.

In l. 5 A has ΧΡΙΟϹ, which the first editors very reasonably emend to κριὸϲ, which is B's reading. The ram, like the bull, is a symbol of strength and virility, and often occurs as a sacrificial beast. True, it seems not to be offered to Dionysos, though W. Luppe, *ZPE* 76 (1989), 13 has reminded us that the god turned himself into a ram in the *Dionysalexandros* of Cratinus (see R. Kassel and C. Austin, *PCG* iv (1983), 140–1). However, the ram is often offered to Persephone, as Dr Christiane Sourvinou-Inwood reminds me; she points out that in the calendar of Nikomachos (Sokolowski, op. cit., Suppl. (1962), no. 10. 64) and in the calendar of the Marathonian Tetrapolis (Sokolowski, op. cit., no. 20 *B* 44) Kore receives the sacrifice of a ram, and recalls the ram sacrifice at Eleusis (see W. Burkert, *Homo Necans* (1972), 311 = p. 282 of the English version, nn. 39–40 (for other instances, see F. Graf, *Nordionische Kulte* (1985), 194 and 282 n. 97) and the rams' heads at the corners of the Peisistratean Telesterion (see G. Mylonas, *Eleusis and the Eleusinian Mysteries* (1961), fig. 21). Dr Sourvinou-Inwood tentatively suggests that the bull and the goat, both of which are mentioned with the verb ἔθορεϲ, are offered to Dionysos, but that the ram, mentioned with the verb ἔπεϲεϲ, is sacrificed to Persephone. Her suggestion is attractive; but it is worth mentioning the possibility that ΧΡΙΟϹ and ΚΡΙΟϹ are mistakes for χοῖροϲ. The pig was of course one of the commonest sacrificial beasts, and was sometimes offered to Dionysos; in the fourth-century Coan calendar (Sokolowski, op. cit., no. 151 *A* 44–5) a pig (χοῖροϲ) and a kid are sacrificed to Dionysos Skyllitas. If this were right, the mention of the noble beast, the bull, would be followed by that of the two commonest and cheapest sacrificial beasts. The bull, perhaps, would be offered only by the wealthy; humbler initiates would be content with one of the two humbler beasts, like Horace with his lamb (*carm.* 2. 17. 32; see Nisbet and Hubbard ad loc. and the literature they cite).

Of course the bull and ram would not have been boiled in milk; presumably the original formula referring to a kid, which is found at A1. 9 and A4. 4, was later adapted to include the name of other sacrificial beasts. In any case, Cook's suggestion that the kid of A1. 9 and A4. 4 represents a surrogate sacrifice offered by the initiate is likely to be right, and if so it is applicable also to the beasts mentioned in the new tablets.

6. This line is puzzling. Dionysos is indeed a god of wine; but what is wine doing here? Should we divide εὐδαίμον' or εὔδαιμον? and what does **(A)TIM(N)M** represent? Hardly, I think, ἄτιμον, which does not seem to suit the context.

The two letters at the end might suggest μνημοϲύνηϲ or μνήμηϲ. At

B1. 4 (Zuntz, p. 359) the dead person will find in Hades the spring of Memory, from which cold water flows, which he must ask the guardians to let him drink of; cf. B2. 4 (Zuntz, p. 361), and see Zuntz, p. 353 for a composite text of B1 and B2. Was the dead person said in this poem to have metaphorical wine (after all, a Dionysiac substance) from the spring of Memory? If so, one might tentatively conjecture something like *οἶνον ἔχεις, εὔδαιμον, ἀπὸ μν[ήμης τινὰ λίμνης*. Henrichs (pers. comm.) suggests that *οἶνον* could refer to the banquet of the blessed in the underworld; but I prefer my own proposal.

7. The first editors read *κἀπιμένει ς'*, 'and there await you beneath the earth *τέλεα* such as the other blessed ones have'; they think *ἔχουςι* may be supplied from *ἔχεις* in l. 6. W. Luppe, *ZPE* 76 (1989), 13 f. rightly finds that this gives very peculiar syntax; he tentatively suggests a radical alteration, *καὶ ςὺ μὲν εἶς ὑπὸ γῆν τελέςας ἅπερ ὄλβιοι ἄλλοι*. But the first editors offer also the alternative suggestion *κἀπιμένεις* (or *κἀπιμενεῖς*). This may be right, and I do not see that it would oblige us to read *τελέςας* or *τελέςας'*, 'having performed the rites which the other *ὄλβιοι* had performed'. As they themselves say (p. 15), *τέλεα* may refer not only to rites but to goals or prizes, or to dignities or honours. The first editors point out to me that on this view one would still need to supply something like *ἔχουςι*, which would not be entirely easy; but since *κἀπιμένεις* has a personal subject, it is a little easier to supply a verb whose subject would be *ὄλβιοι ἄλλοι* than it would be to supply one after *κἀπιμένει ς'*, whose subject would be *τέλεα*.

9(53)

Modern Interpretation of Pindar The Second Pythian and Seventh Nemean Odes

SINCE I was taken, thirty-five years ago, to hear E. R. Dodds lecture on the *Bacchae*, his work has been one of my chief sources of inspiration. My Sather Lectures form a kind of commentary on his, and if I sometimes disagree with him, or see things from a different point of view, that will not prevent the understanding reader from seeing how greatly I admire him and how much my work owes to his. His inaugural lecture at Oxford was called 'Humanism and Technique in Greek Studies';[1] and no great scholar of our time, except perhaps Rudolf Pfeiffer, has kept so perfect a balance between the two. In that lecture Dodds, as the circumstances of the time required, pleaded for more attention to the content, as against the form, of ancient writings; and throughout his career he has applied his masterly technique to just those problems of the ancient world which are of most interest and importance to the modern. But he has always borne in mind that a scholar who hopes to throw light upon such problems must do all he can to master the technique of his profession.[2] Both in his humanism and in his technique, he offers an example from which all classical scholars of our time can profit.

To me Pindar seems one of the greatest Greek and also one of the greatest European poets. But some would dispute this proposition; and I believe that many even of those who would assent to it in reality admire him less than other great poets who seem to me to be his equals. There are two main reasons why Pindar has received less than justice. One is that he is believed to have a narrow and restricted outlook, which is often unfavourably compared with that of the great tra-

[1] Oxford, 1936.

[2] In 1964 he wrote, 'Textual criticism must always remain an essential part of the training of any literary scholar; and even the ordinary man who will never be a scholar ought to know something of the history of the texts he reads and the methods by which they were established. . . . It would be wrong in principle, even if it were possible in practice, to exclude textual criticism completely from the undergraduate course . . .' (*Proc. Class. Assn.* 61 (1964), 19). Then as in 1936, Professor Dodds stressed most what most needed stressing at the time.

Journal of Hellenic Studies, 93 (1973), 109–37 (volume of honour of E. R. Dodds).

gedians; the other is that he is difficult. The question of whether Pindar's outlook is narrow I shall treat comparatively briefly here; I shall not try here to describe that outlook at any length, though I may do so later. In comparing it with that of the tragedians, I shall be able to save space because of having already, in my book *The Justice of Zeus*, written about the *Weltanschauung* of the early Greek poets. Next, I shall pass to the difficulties which Pindar presents to modern readers. The difficulty with which I shall be most concerned will be that of eluding the dangers inherent in the Romantic and historicist approach to Pindar which until eleven years ago was adopted in virtually all modern treatments and which still comes most naturally to most readers, including several distinguished scholars. In 1962 Elroy L. Bundy[3] raised issues of central importance not only for the historical understanding but also for the literary appreciation of the poet, and so initiated a critical discussion which is still proceeding. My best way of contributing to this discussion within the space at my disposal will be to give a specimen treatment of two poems. I have chosen two poems of great difficulty, the Second Pythian and the Seventh Nemean, and I have chosen these particular poems because they will best serve to illustrate my attitude to the new tendency in Pindaric studies. The Second Pythian has been discussed very little in recent years, the Seventh Nemean very often; my treatment of the latter will therefore be comparatively brief. The problems of the Second Pythian seem to me to a large extent soluble when attacked from the starting-point recommended by Professor Bundy; the Seventh Nemean, on the other hand, will serve to show that the consequences of treating a poem in blind obedience to a theory erected into a dogma may || be unfortunate. To deal adequately with the questions raised by Bundy would need a large book; all I can do here is to draw attention to a few of them.

Most of my readers will at least pay lip-service to the statement which I made above, that Pindar is one of the greatest Greek, and one of the greatest European poets. But for English-speaking readers he has never meant as much as any of the three great tragedians; and many who would accept my claim would hesitate to agree that he was as great a poet as Aeschylus or Sophocles. Despite all changes of fashion since it was published in 1897, Gilbert Murray's judgement expressed in the history of Greek literature which he published at the age of thirty-one[4] would be echoed by many readers even now. 'Often in thinking over the best pieces of Pindar', Murray wrote, '—the majestic

[3] See below, p. 121.

[4] *The Literature of Ancient Greece* (1897; repr. 1956), 112.

organ-playing,[4a] the grave strong magic of language, the lightning-flashes of half-revealed mystery—, one wonders why this man is not counted the greatest poet that ever lived, why he has not done more, mattered more. The answer perhaps is that he was a poet and nothing else.' 'The thing that he loved', Murray says later, 'was real heroism. But he could not see it out of its traditional setting; and when the setting was there, his own imagination sufficed to create the heroism.' Pindar, he repeats later, 'is nothing but a poet. There is little rhetoric, no philosophy, little human interest; only that fine bloom—which he calls ἄωτον—which comes when the most sensitive language meets the most exquisite thought . . .'. Although some of the most influential scholars in England and America disagreed—Jebb[4b] and Gildersleeve in the nineteenth, Bowra and J. H. Finley in the twentieth century—I think most modern readers of Pindar would underwrite this judgement.

In Germany the case is different. By far the greatest contribution to the scholarly explanation of Pindar has been made by Germans; and for German poets of the great period, Pindar has been important as he has never been for French, Italian, or English poets. Although they claimed to write Pindaric odes, Ronsard and Chiabrera, Cowley and Gray,[5] are further from Pindar than Goethe,[6] and further still than Hölderlin.[7] But even in Germany the acceptance of Pindar as a poet has, for the most part, been less than whole-hearted. 'Pindar', wrote Eduard Schwartz in 1902,[8] 'ist weder ein reicher noch ein gefälliger Dichtergeist. Der Kreis seiner Gedanken ist streng geschlossen, der Pomp seiner Sprache steif, die Formen seiner Poesie konventionell; die Sache, der er diente, war längst nicht mehr lebendig und berechtigt zu dauern.' Wilamowitz was in many ways a sympathetic interpreter of Pindar. Yet he could write in the closing chapter of his *Pindaros* of 1922:[9] 'Seine Welt ist uns ganz fremd; ihre Sitten, ihr Dichten und

[4a] Why do people always mention organ-playing in connection with Pindar? Is it simply a reminiscence of Tennyson's famous line about Milton? Or is it because Jebb rendered Browning's *Abt Vogler* (*Translations into Greek and Latin Verse* (2nd edn. 1907), 2 f.) into the metre of the Fourth Pythian Ode?

[4b] *JHS* 3 (1882), 144 f. = *Essays and Addresses* (1907), 41 f.

[5] See G. Highet, *The Classical Tradition* (1949), 230 f. (with notes); D. M. Robinson, *Pindar, a Poet of Eternal Ideas* (1936), is useful for its information about modern translations and imitations, if for nothing else. On Ronsard, see in particular I. Silver, *The Pindaric Odes of Ronsard* (1937). Most Pindarizing poetry since the Renaissance has less to do with the real Pindar than with the Pindar of Horace, *Odes* 4. 2; see E. Fraenkel, *Horace* (1957), 435.

[6] See E. Grumach, *Goethe und die Antike*, i (1949), 226 f.; cf. A. Lesky, *Gesammelte Schriften* (1966), 633 f.

[7] See M. B. Benn, *Hölderlin and Pindar* (The Hague, 1962), useful for its bibliography.

[8] *Charakterköpfe aus der antiken Literatur* (4th edn. 1956), 32.

[9] P. 463; the opening words seem to echo A. Croiset, *La Poésie de Pindare* (1880), iii: 'Ce monde dorien dont il est la dernière grande voix est bien plus loin de nous qu'Athènes et que l'Ionie; le

Trachten ‖ für uns reizlos, wenn nicht anstößig. Er selbst ist kein reicher Geist. Von Macht und Größe des Vaterlandes weiß er nicht, nichts von Fortschritt in irgendwelcher Richtung. Weder die Erkundung der weiten Erde noch die Lösung der tausend Rätsel reizt ihn, die uns die Natur um uns und in uns ergibt. Wissenschaft ahnt er nicht; träte sie ihm nahe, würde er sie als gottlose Torheit abweisen. All das Große, an dem unsere Seele hängt, ist ionisch-attisch; aber von diesem Wesen mag er nichts wissen; nicht nur Odysseus, Homer selbst ist ihm antipathisch. Fremdartig ist uns selbst seine Kunst, nicht minder in dem, was ihm eigen ist, wie in dem gegebenen Stile.' These words were written in 1922; they may be supplemented by a quotation from a living scholar who has done great services to Pindar. 'Während Pindar in Theben wirkt', writes Bruno Snell,[10] 'bildet sich in Attika ein ganz neues Verhältnis zur Welt. Die Tragödie erhebt den Anspruch, daß es gerecht auf der Welt zugehen müsse, and stellt damit mancherlei Forderungen an die Menschen, aber auch an die Götter, ohne sie stets erfüllt zu finden, — und dabei verstummt das Rühmen. Pindar hält sich bewußt von solchen Gedanken fern, die ihm Anmaßung sind. Wohl ändert er gelegentlich einen Einzelzug der überlieferten Sage, der ihm den Glanz des Göttlichen zu trüben scheint, aber er verfällt keinem Zweifel an der Ordnung und Schönheit des Lebens, so schwach und hinfällig auch das Irdische sein mag, und er fühlt sich nicht bemüßigt, am Bestehenden ändern zu wollen. In vornehmer Gelassenheit nehmt er die Welt, wie sie ist, — die trotz allem Dunkel durchwebt ist von den goldenen Fäden der Himmlischen.' All these writers consider Pindar a great poet; but would any of them hold him to be the equal of Aeschylus or Sophocles? I doubt it.

Let us consider what these criticisms amount to. Pindar was a Theban; he belonged to a community of Dorian origin. Thebes was ruled by an oligarchy, and Pindar himself is commonly held to have been an aristocrat; most of his patrons were of West Greek origin, some were kings or tyrants, and all the rest appear to have been noblemen. Scholars like to remind us that Pindar's world was perishing even before his death, as though this were not true of other great poets and as though it somehow diminished the value of his poetry. Pindar's

fond et la forme, les idées et l'art nous en sont étrangers.' That Homer is 'antipathetic' to Pindar is evidently deduced from *Nem.* 7. 20 f. The deduction is not justified; it is normal for a poet to say that other poets sometimes tell lies, whereas he himself tells the truth (cf. Hes., *Theog.* 24–9, and see below, pp. 141 f.). At *Isthm.* 4.37 f. Homer appears under a different aspect. The admiration and gratitude which we owe to Wilamowitz cannot prevent me from remarking that this passage shows all the arrogant and ponderous philistinism of a Berlin public monument of the Wilhelmine era.

[10] *Die Entdeckung des Geistes* (4th edn. 1975), 94. Does tragedy demand that justice should be done in the world? In a sense, yes; but in that sense Pindar also does; see p. 136 below.

outlook, it is felt, was narrowly restricted; and it is unfavourably contrasted with the progressive and optimistic attitude to the world held to characterize the Attic tragedians. By itself the fact of having been a Theban and an aristocrat or a sympathizer would hardly be sufficient to prove the point; but evidence of a restricted outlook has been discovered by some scholars in Pindar's actual works. Can any defence be offered?

First, the cultural differences which divided the west Greeks from the east Greeks in Pindar's time have been, we must agree, considerably exaggerated. It could not be contended that such differences existed in the field of art; and it is doubtful whether west and east Greek poets felt that they belonged to different worlds. It appears from comedy that the poems of Alcman, Stesichorus, and Ibycus were well known at Athens.[11] Poets like Simonides, Bacchylides, and Pindar himself were active all over the Greek world; and in the Syracuse of Hieron Pindar and Aeschylus were both welcome. Pindar's dithyramb for Athens was acceptable to the Athenians.[12] Doubtless here as elsewhere the best connoisseurs of poetry were the best educated; but are we to suppose that at Athens only Megacles and a few noblemen[13] could enjoy Pindar's poetry? The leading Greek poets were not felt to be the exclusive property of a particular city, but belonged to a panhellenic tradition going back to Homer and Hesiod.

Science, philosophy, and rhetoric were, it is true, at this time mainly an Athenian and || Ionian preserve, although the contribution of some Sicilian cities of west Greek origin and the careers of some individuals such as Hippias of Elis should remind us that the boundaries could be crossed. But Pindar was a poet, and a poet who wrote for the most part choral lyric, and it is not reasonable to criticize him for neglecting contemporary knowledge. His mention of the eclipse, in the tradition of Archilochus, gives a poet's reaction to that event;[14] whether Pindar as a private person felt like this about it we have no means of knowing. In a poem, he said that philosophers and scientists 'reaped an unprofitable harvest of their wisdom'. From the standpoint of the religious outlook common to the early Greek poets, this is true, and anyone who maintains that Aeschylus or Sophocles would have been shocked by these words does so at his peril. Here, for once, Pindar may profitably be compared with a Romantic poet; when Keats protested against Newton's treatment of the rainbow, he was not trying to refute Newton, but expressing his attitude as a poet.

[11] See Ar. *Lys.* 1297 f.; *Pax* 798 f.; *Thesm.* 161.

[12] See frs. 76 f.; cf. Wilamowitz, *Pindaros*, 272 f.; Bowra, *Pindar*, 142–3.

[13] *Pyth.* 7 is for Megacles, *Nem.* 2 for Timodemos of Acharnai.

[14] *Paean* 9.1 f.

At this point one may remark that the tragedians themselves show a good deal less interest in contemporary speculation than is commonly assumed. Many scholars have asserted that Aeschylus and Sophocles show an acquaintance with pre-Socratic thought; not one has proved it. Euripides makes fairly numerous allusions to contemporary science and philosophy; but how far does this affect the outlook on the world that forms the background of his dramas?

Do his social and political beliefs in fact affect Pindar's poetry so as to render his outlook narrow and parochial? We do not in fact know that Pindar was an aristocrat; Hermann Fränkel has reminded us[15] that since at *Pyth.* 5. 76 the 'I' may perfectly well refer to the leader of the chorus, Pindar's descent from the clan of the Aegidae cannot be taken as certain. What is certain is that in his surviving poetry he makes no political pronouncements. Most of the surviving poems are written in honour of west Greek patrons, and not surprisingly he often speaks with respect of the ancient institutions of the Dorians. He repeatedly gives expression to his loyalty to his own city and his acceptance of her way of life. But where does he decry other forms of government? When he says, 'I find fault with a tyrant's lot', he is not attacking tyranny, but disclaiming the wish to be a tyrant, as Archilochus had done before him;[16] when he refers to the popular assembly of a democratic state as 'the noisy host',[17] it is unsafe to assume that the description is meant to be pejorative. Of course his poems contain expressions of sympathy with his friends and patrons. Despite the delicate position of his own city, he applauds, in praising Aeginetans or Athenians, or Spartans, the Greek victory over the Persians;[18] he sympathizes in praising Syracusans with the Sicilian Greeks in their struggle with the Carthaginians and Etruscans;[19] and in all probability the Eighth Pythian shows his sympathy with Aegina when she was threatened with destruction. But none of these sympathies involve him in a statement of political principle.

Such a statement would have been alien to the whole tradition of the Greek poets. In the Greece of Pindar's time, poets concerned them-

[15] Imre Mueller, *Quomodo Pindarus chori persona usus sit* (Diss. Freiburg, Darmstadt, 1914), 29 f.; H. Fränkel, *Dichtung und Philosophie des frühen Griechentums* (2nd edn. 1962), 485 n. 2 (neglected by Bowra, *Pindar*, 100 (cf. 152); K. Latte, *GGA* 207 (1953), 40 = *Kl. Schr.* 723 agreed with Fränkel. [But see M. R. Lefkowitz, *HSCP* 67 (1962), 230 f., who has shown that 'I' means Pindar here, as everywhere in the Epinicians.]

[16] See David C. Young, *Three Odes of Pindar* (*Mnemosyne*, Suppl. 5, 1968), 9 f. (on *Pyth.* 2. 53).

[17] At *Pyth.* 2. 87 ὁ λάβρος cτρατὸc does not mean 'the brute multitude', as Bowra, *Pindar*, 137 says. The word λάβροc is applied by Homer to winds (*Il.* 2. 148; *Od.* 15. 625) and water (*Il.* 21. 271; *Od.* 15. 293), then to speech (*Il.* 23. 474, 478–9); here it refers to the noise made by the assembly.

[18] *Isthm.* 5. 48; *Nem.* 2. 13; *Pyth.* 1. 75 f.; see J. H. Finley, *HSCP* 63 (1958), 121 f. (though I have doubts about his interpretation of *Isthm.* 8).

[19] *Pyth.* 1. 72; *Nem.* 9. 27.

selves with the temporary problems of the day only so as to set them against the backcloth of the permanent conditions of existence. Adherence to the rule of 'the few' or 'the many', belief in the desirability of a more or less restricted suffrage, did not affect a man's outlook on the gods and their ‖ government of the universe or on their attitude to human life. Pindar's attitude to these cannot be said to be restricted by his partiality for a narrow social group, nor does it differ markedly from the attitude of other persons who were not 'oligarchs'.

Pindar does, it is true, frequently insist that true excellence, in war, poetry or sport, is that which comes 'by nature'.[20] He denies true poetic genius to those crow-like persons who have 'learned', presumably by mimicry, and lift up their voices against the bird of Zeus, the true poet.[21] The true poet has been taught, like Phemius in the Odyssey, by a god.[22] Just so the gift of warlike or athletic excellence is divine, and recurs in certain families favoured by the gods. As a human action is suggested by a god, but is still the action of its human author, who cannot disavow responsibility, so any great aptitude possessed by a human being is the gift of a god, but the recipient of such a gift must second it by making the greatest effort in his power. This is true not only of the gift of poetry, but of all divine gifts of excellence. In his references to athletic success, Pindar puts the greatest emphasis on the hard work which it demands, however great the victor's natural endowment.[23]

Likewise he makes it clear that athletic skill, although it often occurs in members of the same family, does not occur with regularity. Like everything else in human life, according to Pindar, it is subject to the law of change; it may skip a generation and descend from grandfather to grandson, or it may reappear not in a victor's son, but in his nephew.[24] Pindar repeatedly stresses that no amount of skill combined with practice can in itself command success; success is not possible without the favour of the gods.[25] What is true of victory in the games is true of every kind of human triumph; Pindar's view of athletics is only one aspect of his view of life. The gods sometimes exalt men, but sometimes push them down; not even the life of great heroes is free of misfortune. Pindar's view of human life is not egalitarian. But it is in no way mean, narrow, or trivial.

[20] See Bowra, *Pindar*, 5 f.

[21] *Ol.* 2. 86 f.; cf. *Pyth.* 9. 100 f.; *Nem.* 3. 80 f.

[22] *Od.* 22. 347–8. For Hesiod (*Theog.* 81 f.), the poet is he whom the Muses have honoured from his birth.

[23] See the index of A. Köhnken's book, *Die Funktion des Mythos bei Pindar* (see p. 122 below), s.v. *Mühe*; cf. Bowra, *Pindar*, 172 f.

[24] See *Nem.* 6. 9 f.; 11.39 f.

[25] See e.g. *Pyth.* 1.41 f.; cf. H. Gundert, *Pindar und sein Dichterberuf* (1935), 19; Bowra, *Pindar*, 67.

I shall not in this paper attempt a full account of the view of human life which forms the background to Pindar's poetry, though I shall say a little more about it later. Instead, I shall say a word about the tragedians, with whom he is so often unfavourably compared.[26] Were they indeed progressive, optimistic, rational, unlike the reactionary, pessimistic, irrational Pindar?

In the case of Sophocles, the proposition is not easy to sustain. The Sophoclean hero's *aretē* leads him to transgress the bounds set to human aspirations by gods who, in the Greek sense, must be considered just, but whose justice consists not only in punishing mortals for their crimes against each other but in maintaining a universal order in which men have only a humble place and limited prerogatives. Sophocles has sometimes, not without reason, been found to have an affinity with Pindar.

But what of Aeschylus? Had he not a progressive, optimistic outlook? The Aeschylean Zeus has granted to men that 'grace that comes by violence' which consists in the assurance that his justice will ultimately punish their crimes against each other. In comparison with what the gods of other religions offer, that is a somewhat limited concession. The notion of Aeschylus as a progressive optimist is based largely on the final stages of the Oresteian trilogy. But what happens at the end of the *Oresteia*? The problem that confronts Athene and the Areopagus resembles the paradoxes of Zeno. Orestes would have been punished had he not avenged his father; yet now he must be punished, according to the Erinyes, guardians of justice, for having killed his mother. Only an arbitrary solution is possible, and such a solution is offered by Athene, basing her decision on the accident of her own ‖ unique parentage. That solution she is able to impose by persuasion with a judicious modicum of threats; but the Erinyes are not defeated or dishonoured. Whether Aeschylus is 'for' or 'against' the reform of the Areopagus is not easily decided, nor is it, for our present purposes, particularly important; what matters is the firm insistence, strongly underlined by the close correspondence between the words of the Chorus in the central stasimon and Athene's charge to the Areopagus before they deliberate upon their verdict, that in the government of state and universe alike the element of 'what is formidable' is valuable and must always be preserved. The Athenians are assured, as they had been by Solon, that their city and her institutions will continue to flourish, through the favour of Athene; but they are by no means encouraged to imagine that this state of affairs will lead them to any sort of political or religious utopia. At the end of the Prometheus

[26] For the views taken in this section, see my book *The Justice of Zeus* (Sather Lectures 41, 2nd edn. 1983).

trilogy Zeus, in all probability, granted to mortals a share in justice in the form of the 'grace that comes by violence'. But Aeschylus did not forget that it is Zeus, and not Prometheus, who rules the universe. Like all Greek poets of the archaic and classical periods, he was well aware of the conditions of human life laid down by the gods; and he did not imagine that the people of his own city were exempt from them.

But Euripides, it will be argued, is surely quite a different case; he, with his comparatively numerous signs of acquaintance with contemporary science, philosophy, and rhetoric, must be a rationalist and a progressive. Thirty years ago this view was very widely held; the first steps in undermining it were taken by Professor Dodds, in the famous article called 'Euripides the Irrationalist'.[27] If Euripides knows the views of modern thinkers, it does not follow that he wrote his plays to recommend them. Like Aeschylus and Sophocles, Euripides was a tragedian, concerned to present such conflicts as arise in the course of human life rather than to recommend his own solutions of contemporary problems. Divinity in his works, far from appearing as benevolent and understandable, is inscrutable and incalculable; the justice of the gods is done, but it is not the same as human justice. In essentials Euripides' general outlook on the world seems to me very close to that of Aeschylus and Sophocles. Nor do I find that, when allowance has been made for the difference of genre and poetical tradition, the world outlook of any of the tragedians is significantly different from that of Pindar. That is why the view that the tragedians must be superior, because they are more intelligent and more progressive, seems to me totally mistaken.

People fail to appreciate Pindar not only because they find him not to be progressive, but because they find him difficult. He is indeed difficult, but the nature of the difficulty is not always appreciated. His style and language present grave problems; but they are hardly as difficult as those of Sophocles, who is confidently read by many people who will not dare tackle Pindar. That would be the case even if Pindar's text were as corrupt as that of Sophocles, whereas in fact it is a good deal better preserved. The source of many of the greatest difficulties is not the language itself, nor the corruption of the text, but the conventions of the genre and the sometimes abrupt transitions from one topic to another which these entail. Many readers who can see the beauty of Pindar's style and language find it hard to read a Pindaric poem as a whole. We need to be taught how to read Pindar not only in the sense of how to construe him, but in that of how to view each poem and each section of each poem in the light of the tradition to which it

[27] *CR* 43 (1929), 97 f. = *The Ancient Concept of Progress* (1973), 78 f.

belongs; and in this few books written in English are of much help to us.

How hard it is to understand the relation of each section of a Pindaric poem to the whole is shown by the controversies over 'the unity of the epinikion' which have continued since the beginning of the nineteenth century.[28] Throughout that period, it was generally ‖ agreed that the epinikion was a unity; but many of the attempts to show in what the unity consisted were somewhat crude and arbitrary. Late in the century began the anti-unitarian reactions, whose main theorist was A. B. Drachmann[29] and whose main practical exponent was Wilamowitz. Drachmann argued that the traditional elements of the epinikion were too diverse to allow of its having any real unity; Wilamowitz in a series of publications extending from the treatment of the Sixth Olympian Ode in his *Isyllos von Epidauros* of 1886 to his *Pindaros* of 1922 analysed the poems on anti-unitarian lines.

Wilamowitz had set himself the vast task of surveying Greek civilization in its totality. He aspired to combine the tradition of exact interpretation of literary texts derived from Gottfried Hermann with the newer kind of scholarship, making use of monuments, inscriptions, and the data of comparative linguistics as well as literary evidence, whose pioneers had been Boeckh, Welcker, and Karl Otfried Müller. That Wilamowitz made a tremendous contribution to the understanding of Greek literature as well as other Greek studies is beyond dispute. Yet his approach, like that of all scholars, can be seen with the passage of time to have its limitations.[29a] Not all can feel sympathy with his robust brand of Romanticism; and the historicism which so strongly conditioned his approach to the study of antiquity can now be seen to have its dangers.

'The century which lies between the Napoleonic wars and the Great War', said Professor Dodds in his inaugural lecture,[30] 'was, on the whole, dominated, despite important exceptions, by the ethic of Romanticism'; the Oxford contemporary of Eliot and friend of Auden and MacNeice was one of the first English classical scholars to point out to his colleagues that this had ceased to be the case. The Romantics

28 A useful, though sometimes tendentious survey of the controversy is given by David C. Young, 'Pindaric Criticism', first published in the *Minnesota Review*, 4 (1964), 584 f., and later reprinted with some ‖ changes in W. M. Calder III and J. Stern, *Pindar und Bakchylides* (Wege der Forschung 134, 1970), 1 f. See also ch. i of W. Schadewaldt, *Der Aufbau des pindarischen Epinikions* (Schr. der Königsberger Gelehrten Gesellschaft, Geisteswissenschaftliche Klasse, 5/3, 1928), 259 f. (repr. Darmstadt, 1966, and B. A. van Groningen, *La Composition littéraire grecque* (2nd edn. 1960), 331 f.

29 *Moderne Pindarfortolkning* (1891).

29a See U. Hölscher, *Die Chance des Unbehagens* (1965) [and my introduction to Alan Harris's translation of Wilamowitz's *History of Classical Scholarship* (1982).].

30 Op. cit. (n. 1), p. 16.

held poetry, in particular lyric poetry, to be the spontaneous outpouring of the poet's true feelings; a genuine poet feels an overmastering urge to express in verse his own most genuinely held convictions. No wonder that the nature of poetical conventions, as Housman remarked with reference to textual interpretation, was better apprehended during the eighteenth than during the nineteenth century. Such conventions, the Romantics thought, might be important for the understanding of the sophisticated verse of the seventeenth and eighteenth centuries which they rejected; they found it hard to believe that they could be significant for the work of those Greek poets with whom they felt a special affinity, and for whose understanding they did so much. Even in the case of Horace Romantic critics have had difficulty in appreciating the full importance of conventions: all too often they have taken as a personal statement what is simply the expression of a *locus communis*, and they have greatly exaggerated the value of the works as a source for the biography and personal attitudes and opinions of their author. Horace, one might suppose, is a writer in whose work the element of convention and sophistication cannot easily be underrated; yet this was done, and traces of the delusion can be seen even in the learned, intelligent, and humane study of him by the late Eduard Fraenkel.[31] How much more excusable, and how much more misleading, is their misunderstanding of Catullus, an author they found it fatally easy to take for one of themselves! With early Greek lyric, likewise, the Romantics felt a special affinity, by no means altogether without reason; how much better, in many ways, they understood both Greek lyric and Greek tragedy than had their immediate predecessors, or any generation since the revival of learning! Yet this very sympathy with the poets carried the germ of a serious misunderstanding. Pindar, who on the surface seems to have so little in common with ‖ sophisticated poets like Horace or Pope and so much with the lyric poets of the Romantic age, was bound to be distorted by a Romantic interpretation.

Hardly less powerful than the distorting influence of Romanticism was that of historicism. Wilamowitz and his contemporaries insisted that a poem was as much a historical document as a historical work, an inscription, or a monument; so it is, provided that even in assessing its value as historical evidence one takes proper account of its artistic character and purpose. In the climate generated by the historicism of the nineteenth century, it was dangerously easy to try to extract from a poem more in the way of historical information than its actual content justified.

[31] *Horace* (1957); contrast the introduction of R. G. M. Nisbet and M. Hubbard to their commentary on Book 1 of the *Odes* (1970), and see G. W. Williams, *PBA* 56 (1972), 432–3.

Even now, fifty years after its first appearance, Wilamowitz's *Pindaros* is an exciting book to read. Like all his works, particularly those of his astonishing last decade, it is full of life and colour; the author's comprehensive learning often brings illumination; and many of the new interpretations he advanced in it still hold their ground. Its influence has been enormous. Despite its element of Crocean aestheticism, Gennaro Perrotta's *Saffo e Pindaro* (1935) is dominated by Wilamowitz; and Sir Maurice Bowra's *Pindar* (1964) moves almost entirely within the lines which Wilamowitz marked out. An important advance was made by Wolfgang Schadewaldt, who in 1928 published a study of the various recurrent elements in the epinikion and the way in which they are combined;[32] Illig[33] did much to illuminate Pindar's narrative technique; and Hermann Fränkel in the course of a series of brilliant interpretations[34] has made a specially valuable contribution to the understanding of Pindar's religious outlook; all three, like Bruno Snell,[35] adopt in general an approach like that of Wilamowitz. An attempt to demonstrate the unity of the epinikion in terms of a more or less hazily conceived 'symbolism' has been in my judgement notably unsuccessful; from the works of G. Norwood[36] and his followers most readers have returned despairingly to the anti-unitarians.

A trend which is immeasurably more interesting than that begun by Norwood started in 1962 with the publication of detailed studies of the Eleventh Olympian and First Isthmian Odes[37] by Elroy L. Bundy, a professor in the University of California at Berkeley. Bundy strongly insisted on the importance of understanding the conventions of epinician poetry, and supported his contention by showing in the chosen poems how often a *locus communis* may be shown to occur in different poems, but in corresponding places and with a similar function. The somewhat formidable technical terminology which he has invented to indicate the special purposes of each formal device is curiously reminiscent of the edition published at Wittenberg in 1616 by Erasmus Schmid, who analysed the odes in terms of the categories prescribed in the rhetorical handbooks of the imperial period. Bundy sharply reminded his readers that the main purpose of an epinician ode is to praise the victor in whose honour it is written; and he emphatically

[32] See n. 28.

[33] L. Illig, *Zur Form der Pindarischen Erzählung* (1932).

[34] Op. cit. (n. 15), 483 f.; cf. 'Pindars Religion'; *Die Antike*, 3 (1927), 39 f. = Calder and Stern, op. cit. (n. 28), 232 f., and his review of Schadewaldt, *Gnomon*, 6 (1930), 1 f. = *Wege und Formen des frühgriechischen Denkens* (2nd edn. 1960), 350 f.

[35] Op. cit. (n. 10), 118 f.

[36] *Pindar* (Sather Classical Lectures 19, 1945).

[37] *Studia Pindarica*: I, *The Eleventh Olympian Ode*; II, *The First Isthmian Ode* (Univ. of California Publications in Classical Philology, 18 (1–2 (1962) [repr. 1986]).

warned them against taking it for granted that all or most difficulties of interpretation are to be explained by supposing the existence of a personal or historical allusion.

Since 1962 Bundy has most unfortunately published nothing; but his methods have been applied, with varying degrees of fidelity, by a number of followers. David C. Young has written a valuable sketch of the history of Pindaric criticism,[38] and has applied his own ‖ principles in four detailed studies of individual odes.[39] Young combines a modified version of Bundy's approach with more questionable elements derived from the school of symbolists that springs from Norwood. Erich Thummer in his commentary on the Isthmian Odes[40] has applied, I fear somewhat mechanically, principles like those of Bundy; he has prefaced to the work a useful collection of *loci communes* arranged under general headings. W. J. Slater, the author of a valuable new Pindaric lexicon, has offered some detailed interpretations on Bundy's lines, mostly useful but at times exaggerated.[41] Adolf Köhnken has shown that the advice on kingship given in the last part of the First Pythian Ode is offered not to the young Deinomenes but to the mature Hieron;[42] since Hesiod's time, at least, it had been a convention that a poet, inspired by the Muses, might address a ruler in such a tone without impertinence. The same writer has set out to show how the myths of six epinician odes can be shown, on Bundyite principles, to subserve the unity of the whole, and has carried through the work with considerable success.[43]

Bundy has beyond doubt done an immense service to the understanding of Pindar by reemphasizing the importance of convention in Pindaric art, by warning us against the naive romantic assumption that lyric poetry must involve unreserved self-revelation, and by exposing as gratuitous assumptions a number of supposed allusions invented by scholars in order to explain the difficulties of the text. At the same time, it is possible to pursue this approach too far, as some of Bundy's

[38] See n. 28. In passing, let me inform the learned author of something that will give him pleasure. He complains (see Calder and Stern, op. cit. (n. 28), p. 52) that only the Italian reviews of F. Dornseiff's *Pindars Stil* of 1921 were unfavourable. He should see the review of P. Maas in *Zeitschrift für Ästhetik und allgemeine Kunstwissenschaft*, 16 (1922), 407 f., for a judgement closely in accordance with his own.

[39] *Three Odes of Pindar: A Literary Study of Pythian 11, Pythian 3* and *Olympian 7* (*Mnemosyne* Suppl. 9, 1968); admirably reviewed by H. Maehler, *Gnomon*, 42 (1970), 441 f.); *Pindar, Isthmian 7, Myth and Exempla* (*Mnemosyne* Suppl. 15, 1971).

[40] *Pindar: Die Isthmischen Gedichte* (2 vols., 1968); M. M. Willcock, *CR*, NS 21 (1971), 336, supplies an interesting review, though he might have pointed out more of the mistakes the book contains.

[41] *CQ*, NS 19 (1969), 86 f.

[42] *Hermes*, 98 (1970), 1 f.

[43] Op. cit. (n. 23).

followers have already shown. The Ariadne's clue to the understanding of an epinician ode is, as Bundy has insisted, the realization that its main purpose is to praise the victor. But the victors whom Pindar celebrated lived at a particular time in a particular society. Each had his own place in that society and in the history of that time; some held an important place there. It was therefore inevitable that some echo of events in that world should be found in Pindar's poetry. Many of his historical allusions are acknowledged facts, as no serious scholar would dispute; and it is natural that there should be others whose existence, owing to the poverty of our information, is not so easy to demonstrate. In some cases Bundy's followers have gone too far. Like most new approaches to the study of great poetry, the new approach to Pindar does not offer by itself a key to the solution of all difficulties; rather, it must be judiciously combined with other principles and other methods. The recognition that certain *loci communes* recur at frequent intervals in choral lyric, as in other Greek poetry, must not encourage people to suppose that Pindar, any more than other Greek poets, automatically strung together a selection of 'formulae', a word which should be carefully avoided in this connection. We should indeed guard against assuming that the poetic personality that appears in Pindar's writings corresponds at all points with the poet's private character; and we should be more hesitant than our predecessors when it comes to explaining difficulties in the poems by means of factual assumptions resting on no other evidence. But we should not forget that Pindar was a human being writing for other human beings in a particular and individual historical and social environment. We can best elucidate and appreciate the poet's art by using our understanding of his principles of composition to see how he applies these principles in the special circumstances offered by a given context.

So let us turn to the most puzzling of the Sicilian odes, the Second Pythian. Its occasion and its date are wrapped in mystery.[44] Aristophanes[45] of Byzantium placed it among the ‖ Pythian Odes, and Apollonius the Eidographer followed him; but there is no certainty that it belongs there. Callimachus, perhaps following Zenodotus, called it a Nemean; perhaps he meant that it belonged with the group of κεχωριϲμέναι, or unclassified odes, that stood at the end of the Nemeans in the ancient editions. Callistratus, the pupil of Aristo-

[44] The best statement of the problems is that of P. Von der Mühll, *MH* 15 (1958), 215 f. [= *Ausgewählte kleine Schriften* (1976), 165 f.] 'Dici vix potest quantum in eo expediendo laboraverint interpretes', wrote Hermann of this poem in 1834 (*Opuscula*, vii. 115).

[45] For the opinions of ancient scholars, see A. B. Drachmann, *Scholia Vetera in Pindari Carmina*, ii ‖ (*Scholia in Pythionicas*), 31; for the contributions of the scholars named to Pindaric scholarship, see J. Irigoin, *Histoire du texte de Pindare* (1952), chs. iv–vii and R. Pfeiffer, *A History of Classical Scholarship from the Beginnings to the Hellenistic Age* (1968), s.nn.

phanes, called it an Olympian; so did Ammonius, the pupil of Apollonius. Timaeus called[46] it a *θυςιαςτική*, or a *θυςιαςτήριος*, meaning, I suppose, an ode designed to accompany a sacrifice.

The poem celebrates a victory of Hieron in a chariot race. Von der Mühll, it is true, has argued[47] that because Artemis and Hermes are said (9 f.) to give Hieron glory whenever he yokes his horses, therefore this ode celebrates no definite occasion, but merely praises Hieron because he is always victorious. But the claim (4) to bring news of 'the earth-shaking four-in-hand', the statement (6) that Hieron has crowned Ortygia and the specific pronoun *κείνας* in the phrase 'those mares with flashing reins' (7) all help to show that this was no exception to the general rule that an epinician ode celebrates a victory. But what victory?

The only victories in chariot races at the great Panhellenic Games which we know Hieron to have won were won at the Pythian Games of 470 and at the Olympic Games of 468. Some scholars have thought that this ode celebrated a victory which was not won at the Panhellenic Games at all. Boeckh[48] thought it was won at the Iolaeia at Thebes, and found many to agree with him, Farnell[49] thought of local games at Syracuse. Such suggestions cannot be refuted, any more than they can be established; though the manner in which the victory is spoken of hardly suggests a minor occasion. Wilamowitz[50] and Schadewaldt[51] held that this poem, like the First Pythian Ode, celebrated the Pythian victory of Hieron's chariot in 470. D. S. Robertson[52] suggested that it was written to honour Hieron's Olympic chariot victory of 468, praised in the third ode of Bacchylides; Bowra[53] by a different set of arguments reached the same conclusion.

There has been similar uncertainty about the date. Boeckh put the poem as early as 477; Schroeder[54] hesitated between 475 and 471; Von der Mühll argued for 475; Wilamowitz and his followers, like Robertson and Bowra, placed it later, as their view of the occasion demanded.

The content of the poem also has been the subject of controversy. The exact significance of the myth of Ixion has been debated; but there has been even more dispute over the last part of the poem, from l. 72 to the end. Most modern scholars regard it as a remonstrance addressed to

[46] *FGrH* 566 F 141.

[47] Loc. cit. (n. 44), 220 [= 172].

[48] See his commentary, ad loc.

[49] ii. 119. P. M. Fraser, *Ptolemaic Alexandria* (1972), i. 452, asserts as though it were a known fact that the ode was written to honour a success in local games.

[50] *Pindaros*, 285 f.

[51] Op. cit. (n. 1), 325 f. = 67 f.

[52] *PCPS* 127–9 (1924), 35.

[53] *HSCP* 48 (1937), 1 f. = *Problems in Greek Poetry* (1953), 66 f.; cf. *Pindar*, 410.

[54] *Pindars Pythien* (1922), 1–3.

Hieron, who is exhorted not to believe certain persons who have slandered the poet at his court; some have agreed with the ancient commentators that these persons must have been rival poets. Instead of recording the various opinions which have been expressed, I wish to examine the text, section by section, and while guarding against excessive confidence in this or any other single line of approach, to use as a clue the principle reaffirmed by Bundy that an encomium is designed, in the last resort, to praise the victor.

'Syracuse of the mighty city, precinct of Ares mighty in war, nurse divine of men and horses that delight in iron, to you I come from bright Thebes, bringing this song, bearing news of the earth-shaking four-in-hand in which Hieron, mighty charioteer, crowned Ortygia with a garland that shines from far—Ortygia, seat of Artemis of the river, with whose aid he tamed with gentle hands those mares with flashing reins. For the archer ‖ maiden and Hermes of the contest with both hands give him radiant glory, when he yokes the might of his horses to the polished seat and to the chariot that guides the bridle,[55] calling upon the mighty god who wields the trident' (1–13).

It is worth remarking that this praise of Syracusan valour leaves little to be said for the argument that the absence of any direct reference to the victories of Himera and Cumae indicates an early date. Next we may notice that Pindar's home was Thebes, so that his sending his poem from Thebes does not prove that the victory it celebrates was won there. Hieron's home was Syracuse, and his crowning of Ortygia does not show, as Farnell imagined, that his victory was won at Syracuse. The poem was written, it would appear, in honour of a specific victory; but where or when that victory was won, neither the prologue nor any other part of it gives us any means of knowing. In the absence of anything like adequate evidence, I would prefer to suspend judgement, a thing Pindaric scholars do too seldom in matters of chronology.[55a] Schadewaldt, who follows Wilamowitz in taking the poem to be not an epinikion in the strict sense but a poetical epistle, thinks Pindar felt obliged to give it the standard form of an epinikion.[56] I do not see how Schadewaldt knows that it is a poetical epistle, and not an epinikion proper.

Hieron himself, the poet says, trained the horses. Does that point to a time in his life before he became afflicted with the stone and other troubles? It would be unwise to deny that royal persons in ancient Syracuse, as in modern England, may have occupied themselves with

[55] On the sense of πεισιχάλινα, see D. S. Robertson, *Studi in onore di L. Castiglioni*, ii (1960), 801 f.

[55a] I suspect that Robertson and Bowra are right about the occasion, but do not think their arguments amount to proof.

[56] Op. cit. (n. 28), 326–8 = 68–70.

such matters. But it would be equally unwise to rule out the possibility that Pindar may have spoken of Hieron as having trained the horses, even though he had taken little or no part in their training;[57] so that I see no evidence for dating here.

The poem continues: 'Different men have wrought for different kings a sweet song to reward their prowess' (14–15). That is a transitional formula of a kind often used by Pindar;[58] but it also touches directly on the theme of gratitude which lies at the heart of the poem. First comes the example of Cinyras as a good man rewarded by the praise which he deserves; then follows the similar example of Hieron. 'Often the voices of the Cyprians resound in praise of Cinyras, whom golden-haired Apollo loved, the favoured priest[59] of Aphrodite; it is gratitude, I think, in regard[60] for some kindly action, that impels them. And you, son of Deinomenes, the maiden of Locris in the west sings of before her house,[61] she who after desperate struggles in war through your might has cast her first untroubled glance' (15–20).‖

Cinyras, the Cyprian king mentioned in the *Iliad* as a guest-friend of Agamemnon, was proverbial for his wealth.[62] That is why he is mentioned at *Nem.* 8. 18. That and the favour shown him by Apollo, who in one place is said to have been his father,[63] make him a specially apt instance here. But as the text shows he is mentioned as a benefactor who has received his due reward in praise.

Then follows an unmistakable historical allusion. But what gratitude

[57] Wilamowitz changed his mind over this question, as Ed. Fraenkel, *Horace* (1957), 172 n. 3 pointed out.

[58] See Bundy, *SP* 1. 7 f.

[59] In an ingenious article (*RCCM* 2 (1960), 30 f.), Anna Morpurgo Davies argued that at 1. 17 the word *κτίλος* meant not 'pet' or 'favourite', as is usually supposed, but 'ram', as in its earliest instances. A ram has a place in the cult of Carneian Apollo, but none in that of Paphian Aphrodite; a sacred animal in that cult would be, I regret to say, the pig. Even if Hes. fr. 323 M.–W. is regarded as a doubtful instance, the sense 'tame' is as early as Empedocles (B 130. 1. D.–K.).

[60] *'Oπιζομένα* is middle; see H. Gundert, op. cit. (n. 25), 141 n. 364.

[61] Does *πρὸ δόμων* refer to the singer's house, or to Hieron's? and what kind of occasion is envisaged? Satyrus, *On the Demes of Alexandria*, P. Oxy. 2465, fr. 2, col. ii, 12 f., giving directions for procedure on the occasion of a religious procession in honour of the goddess Arsinoe Philadelphos, writes: *οἱ δὲ] βουλόμενοι θύειν Ἀρσιν[όηι Φιλα/δέ]λφωι θυέτωσαν πρὸ τῶν ἰδ[ίων οἴ/κ]ων ἢ ἐπὶ τῶν [δω]μάτων* . . . The restoration *ἰδ[ίων οἴ/κ]ων* is due to W. Morel (ap. Lloyd-Jones, *Gnomon*, 35 (1963 [R41], 453) who compares Chariton 3. 2. 15 (in a similar context) *ἔθυεν ἕκαστος πρὸ τῆς ἰδίας οἰκίας*; one might also compare Eur. *Bacch.* 68–70, where Dodds is surely right to punctuate *τίς ὁδῶι τίς ὁδῶι; τίς μελάθροις*; (cf. A. J. Festugière, *MH* 9 (1952), 244, cited on p. 75 of the 2nd edn. of Dodds's commentary). These passages seem to suggest that the house may be the singer's own, and that the occasion envisaged may be that of some kind of religious procession, perhaps resembling the performance of one of Pindar's own poems. The scholia have *προελθοῦσα . . . τῶν οἴκων*, rightly.

[62] *Il.* 11. 19 f.; cf. Tyrtaeus 9.6.

[63] ΣTheocr. 1. 109. His connection with Apollo may have arisen from his name. Eustathius on *Il.* loc. cit. derives it from the Hebrew word for 'harp', but a Greek would naturally have thought of *κινυρός*, etc.

of the Locrians to Hieron does the poet have in mind? According to a scholion on the passage,[64] the Locrians had been threatened by Anaxilas of Rhegium, and Hieron had sent his brother-in-law Chromios to warn him off. This event would have to lie between Hieron's accession to power in 478 and Anaxilas' death in 476; it is usually placed in 477.[65] Those who have believed the scholion and have assumed that the allusion must be to a recent event have found here evidence for a date about 476. But Robertson[66] has pointed out that there was another occasion on which Hieron came to the help of the Locrians. A scholion on l. 38 says that the Locrians were threatened by the tyrants Anaxilas and Kleophron, but were saved by Hieron. Now according to Justin (21. 3) the Locrians were once attacked by Leophron, tyrant of Rhegium; so that the Kleophron of the scholion is presumably the son and successor of Anaxilas. Diodorus (11. 66) places the restoration of the sons of Anaxilas not long before Hieron's death in 467/6; but the chronological inexactitude of Diodorus is notorious. Robertson thought Leophron's attack on Locri took place in 471 or 470. There is a third possible explanation. Hieron's saving of the Locrians from Anaxilas is known to us only from the scholion I have mentioned and from another on *Pyth.* 1. 98;[67] the latter cites no historian as his authority, but the comic poet Epicharmus in his *Nasoi* (fr. 98 Kaibel). But even if this is the truth, Pindar may have had no such thing in mind. Like all the inhabitants of South Italy and Sicily, the Locrians had good reason to be grateful to the conqueror of the Etruscans at Cumae in 474. Why should the Locrian maiden be specified? Locri was a home of poets, as the poet of the Tenth Olympian was aware.[68] Can we be sure that it was not this great achievement that the Locrian maiden praised? It might have been the theme of poets for many years after its occurrence. To sum up, the allusion to Locrian songs in praise of Hieron is uncertain, and its value in dating the poem very limited.

Now follows the myth of Ixion (21 f.): 'And by the command of the gods, they say, Ixion speaks these words to mortals as he whirls every way on the flying wheel: "Requite your benefactor with a kind return and do him honour." He learned this in truth; for after he had attained a pleasant life with the children of Kronos, he could not long sustain his happiness, when with mind deranged he desired Hera, who is set

[64] Drachmann, op. cit. (n. 45), ii. 38.

[65] e.g. by H. Berve, *Die Tyrannis bei den Griechen*, i (1967), 150.

[66] Op. cit. (n. 52).

[67] Op. cit. (n. 45), p. 18.

[68] See *Ol.* 10. 14; fr. 140b, 1–6; for the alleged Locrian origin of Stesichorus, and for the Locrian poet Xenocritus, see C. M. Bowra, *Greek Lyric Poetry* (2nd edn. 1961), 82–4.

apart for the mighty joys of Zeus' bed; but pride drove him to arrogant folly; and soon he suffered what he deserved and got trouble beyond that of others. Two crimes brought him grief; he was the first hero to introduce among mortals the shedding of kindred blood, not without treachery; and once in the capacious chambers of Zeus he attempted Hera!'[69]

I cannot agree with the scholar who a few years ago argued that Ixion is not imagined ‖ as speaking these words, but merely as bringing them to mind by his example.[70] Virgil is explicit in the similar case of Phlegyas (*Aen.* 6. 18 f.):

> Phlegyasque miserrimus omnis
> admonet, et magna testatur voce per umbras:
> 'discite iustitiam moniti et non temnere divos.'

Heyne thought that Virgil had this passage of Pindar in mind;[71] certainly it is an instance of the same *locus communis.* The words Ixion utters illustrate a common theme in Pindar, that of the duty to show gratitude and the criminality of ingratitude; Bundy[72] has exhaustively catalogued the instances.

A subsidiary *gnome*—what Bundy would call a 'gnomic foil'—follows next (34 f.). 'A man should look on the measure of all things at his own level. His lawless coupling cast him into all kinds of misfortune; he too was the victim of such actions;[73] for in ignorance the man lay with a cloud, embracing a sweet falsehood. In shape it was like the daughter of Kronos, matchless among the heavenly ones; the hands of Zeus made it to ensnare him, a beautiful cause of disaster. He brought on himself his binding to the four-spoked wheel, his own destruction; in fetters not to be escaped from he received the message that comes to all. Remote from the Graces, she bore him an arrogant son, solitary, like

[69] We may notice in the passage the regular vocabulary of the concepts of Hybris and Ate: *ἔμαθε* (cf. Homer's *παθὼν δέ τε νήπιος ἔγνω* and Aeschylus *πάθει μάθος*; see, if documentation is needed, H. Dörrie, *Abh. Mainz*, 1956/5, 306 f.): *μαινομέναις φραςίν*: *ὕβρις* drives Ixion to *ἄτη*: *παθὼν ἐοικότα*. As in Aeschylus' trilogy about him, Ixion figures as an example of ingratitude punished by the justice of Zeus.

[70] D. Kuijper, *Mnemosyne*, 4th ser. / (1963), 162 f.

[71] On Virgil's knowledge of Pindar, see Lloyd-Jones, *Maia*, NS 19 (1967), 229 [= ch. 11, p. 187], Eduard Fraenkel pointed out to me that my warning against underestimating the width of Virgil's reading echoed his own at *Philol.* 87 (1932), 247–8 = *Kleine Beiträge zur kl. Philologie*, ii. 178–9. See A. Seidel, *De Vergili studiis Pindaricis* (Diss. Breslau, 1925); cf. L. P. Wilkinson in *Forschungen zur römischen Literatur* (Festschrift Büchner, (1970), 286 f.

[72] *SP* II. 86–91.

[73] The passage has perplexed commentators; see e.g. G. Tarditi, *PP* 2 (956), 197 (whose own suggestion does not convince). But Snell and Turyn rightly read *ποτὶ καὶ τὸν ἵκοντ'*. The iota is shortened by the omission of the augment, and the sense of *ἵκοντ'* is that listed under II 2 in LSJ s.v. *ἱκνέομαι*, in which anger, suffering, and other undesirable things are said to 'come upon' or 'visit' people. The *καὶ* is like that in such phrases as *οἶνος καὶ Κένταυρον* . . . (*Od.* 21. 295); cf. *Pyth.* 3. 55.

his mother, having no honour among men or in the realm of the gods; she reared him and called him Centaur.[74] He mated with Magnesian horses on the spurs of Pelion, and there was born a horde prodigious, like both parents, like the mother below, but above like the father.'

Schadewaldt, after Hermann and others, has admirably shown how the myth of Ixion illustrates the motive of ingratitude, the omnipotence of Zeus, and the inevitable punishment of the ungrateful man by Zeus' justice; I need not repeat what he has said.[75] What follows will show that in this poem the myth is perfectly appropriate to the central theme.

The power of the gods that is attested by any prodigious thing or happening supplies a transitional formula very often used by Pindar.[76] 'The god accomplishes his every purpose in accordance with his hopes, the god who catches the winged eagle and passes the dolphin of the sea, and makes a proud mortal bow, while to others he gives success[77] that never grows old' (49 f.).

Then follows what seems an abrupt transition to a more perplexing passage (53 f.). 'But I must avoid the sharp tooth of slander. For I have seen from far off the scolding Archilochus in perplexity grow fat upon abusive enmities; and to be wealthy while one attains one's proper fate is the best thing wisdom offers.'‖

In considering the relevance of this to what precedes it, we must start from the word *χρεών* in l. 53. Bundy speaks of 'the countless appeals to propriety in introductions, transitions, and conclusions', and adds that 'all such words and phrases attach to the laudator's *χρέος* and to the appropriate manner of discharging it' (*SP* II 73). Why does the poet say that he must avoid evil-speaking? Evil-speaking is the opposite of praise; doers of good like Cinyras and Hieron deserve gratitude, of which praise is a manifestation.[78] Pindar must not be like Archilochus, a great poet of the past who savagely abused his enemies. A famous passage of Archilochus runs, 'My heart, my heart, confounded by perplexing cares, . . . take your stand opposite the enemy and defend yourself against your foe'.[79] Archilochus, then, is mentioned because of

[74] P. Von der Mühll, *MH* 25 (1968), 221 [= *AKS* 242 f.] showed that Pindar derived *Κένταυρος* from *κεντεῖν* and *αὔρα*. On the significance of the myth of the centaurs, see G. S. Kirk, *Myth: Its Meaning and Function in Ancient and Other Cultures* (Sather Classical Lectures 40, 1970), 132 f.

[75] Op. cit. (n. 28), 328–30 = 70–2; following Wilamowitz, *Pindaros*, 289, he draws attention to the echo of the prelude of Hesiod's *Works and Days* here; cf. Young, *Pindar, Isthmian 7* (cited in n. 39), 37 n. 125. J. Martin, *REG* 82 (1969), pp. xxii f. thinks that Pindar is urging Hieron not to behave like Ixion.

[76] See Bundy, *SP* I 2–3, 9–11.

[77] *κῦδος* means rather more than 'fame'; see H. Fränkel, *Wege u. Formen* (n. 34), 71 n. 2.

[78] Bundy, *SP* II 56 f. illustrates this common notion; with *Isthm.* 1. 41–51 he compares *Isthm.* 5. 24–32; see also Bacchyl. 5. 187–97 (cited by him on p. 60), *Pyth.* 9. 96–9 and *Nem.* 7. 61 f.

[79] Fr. 128 West; P. Friedländer ap. Gundert, op. cit. (n. 25), 141 n. 367, thought Pindar had this in mind when he wrote these words.

his attacks upon his enemies; but how can Pindar then go on to say, just as he does at the beginning of the Fifth Pythian Ode, that wealth when it is granted by divine will and not gained unjustly is the best thing wisdom has to offer! That translation of the difficult l. 56 has not, it is true, been accepted by all scholars; but it is certainly recommended by the order of the words, as Eduard Fraenkel with his special expertise in word-order pointed out.[80] The clue to the transition lies in the word *πιαινόμενον*, which must in this context imply not simply that Archilochus had many enemies, but that he made money, or somehow profited, out of his enmities.[81] Whether Archilochus made money from his poetry is uncertain; Pindar certainly did, and may have assumed that Archilochus did the same; at least Archilochus utilized his enmities for the purposes of his poetry. The nature of the context enjoins this interpretation, which well suits the natural meaning of the verb. Archilochus, then, is introduced as an instance of an evil-speaker, but the transition from him to Hieron is managed by contrasting the source from which they derive their wealth or profits.

We come to Hieron for the section of the poem that contains the main laudation of the victor[82] (56 f.). 'You can in truth point to this with your generous mind, lord who commands many well-walled cities and their people.[83] And if anyone says that any other in Greece among those before him was greater in his possessions or in his honour,[84] he struggles unavailingly with empty mind. I will embark in a garlanded vessel[85] as I proclaim his excellence. The support of youth is valour in grievous wars; and from this also I say you won boundless glory, battling among the charging cavalry and among foot-soldiers; and the counsels of riper age allow me without hazard to praise you on every count.'

Many modern scholars seem to believe that to enumerate the

[80] Ap. Schadewaldt, op. cit. (n. 28), 72–3 n. 2. D. Gerber in a learned article has argued for this view (*TAPA* 91 (1960), 100 f.). R. W. B. Burton, *Pindar's Pythian Odes* (1962), 120, thinks that *σοφία* here must mean poetry; that would remove the point of contrast between Archilochus and Hieron. The same prejudice that 'Pindar seldom, if ever, uses the word to mean wisdom in a general sense' has been responsible for the repeated misinterpretation of *Nem.* 7. 17, where *cοφοὶ* means not 'poets' but 'wise men who are willing to pay poets to preserve their fame'; see n. 118 below.

[81] A wrong interpretation of *πιαινόμενον* has often been defended by citing Bacchyl. 3. 68, where a pi was written by the second corrector against the verb in the clause *ὅcτιc μὴ φθόνωι πιαίνεται*. But there too it is presumably implied that the envious profit from their envy, unless H. Richards and J. Schöne were right to read *Ϝιαίνεται*.

[82] In Bundy's terminology the 'crescendo'; see *SP* I 17.

[83] To infer anything about the date of the poem from this manner of addressing Hieron strikes me as most unreasonable; but Burton, op. cit. (n. 80), i. 15 follows Wilamowitz, *Pindaros* (1922), 285.

[84] For this kind of solemn asseveration, see Bundy, *SP* II 59 n. 60.

[85] Ships were, on festal occasions, decorated with flowers; see L. Deubner, *ARW* 30 (1933), 73 [= *Kl. Schr.* 123], and F. Bömer, *Ovids Fasten*, ii. 42.

triumphs of the victor and to praise him was for Pindar a disagreeable necessity, to be got over as quickly as possible. It is the great merit of Bundy to have reminded us that in an epinician ode the praise of the victor must be of central importance. Here this solemn praise is uttered with reference to ‖ a principle which Pindar in his works enunciates in various ways and amplifies with various details; that if a man has wealth and also *arete*, if he has wealth which is not ill-gotten, but is his by the will of the gods, then he deserves glory, which it is the duty of other men not to grudge him and the task of the poet to perpetuate. Hieron has wealth, honourably won; in wealth and honour, he is unsurpassed. In praise of his *arete*, the poet will set out on a voyage on a ship whose prow is garlanded; Hieron has the proverbial combination of valour in war and wisdom in counsel; therefore, his glory is deserved and the poet's task in celebrating it is easy.

At this point Pindar wishes the victor well; he says *χαῖρε*. This word can mean 'Farewell' as well as 'Fare well', and since less than ten lines from the end of the Third Nemean Pindar similarly salutes Aristocleides, it has been argued that the real end of the ode comes here, and that ll. 72–96 are a kind of appendix. The passage of the Third Nemean[86] that is in question is like this in more ways than one. There too the poet tells the victor he is sending him his poem; there too he describes it, mentioning that it is in Aeolic measure; in the Nemean the *aulos*, in the Pythian the *phorminx* is to accompany the performance. In the Nemean an apology for lateness in delivery leads the poet to contrast himself, the eagle, with the chattering daws, before uttering his final praise of the victor's glory. In the Second Pythian also the poet after saluting the victor contrasts himself—in this case in the character of the honest man, the straight speaker—with others, this time the flatterers and deceivers. The eagle and the daws occupy three or four lines, the straight speaker and the deceivers twenty-four; but I do not think that the difference in length between the two passages need be specially significant.

'This song I am sending over the grey sea, like Phoenician merchandise;[87] graciously look upon the Castor-song in Aeolian strains, granting it favour for the sake of the seven-stringed lyre' (68 f.). What

[86]
χαῖρε, φίλος. ἐγὼ τόδε τοι
πέμπω μεμιγμένον μέλι λευκῶι
σὺν γάλακτι, κιρναμένα δ' ἔερς' ἀμφέπει,
πόμ' ἀοίδιμον Αἰολίςςιν ἐν πνοαῖςιν αὐλῶν,
ὀψέ περ. ἔςτι δ' αἰετὸς ὠκὺς ἐν ποτανοῖς,
ὅς ἔλαβεν αἶψα, τηλόθε μεταμαιόμενος,
δαφοινὸν ἄγραν ποςίν·
κραγεταὶ δὲ κολοιοὶ ταπεινὰ νέμονται.

[87] Cf. *Nem*. 5. 3.

does Pindar mean by *Καςτόρειον*? The same term seems to be employed in the First Isthmian (17), where Pindar wishes to celebrate Herodotus in a Kastoreion or a hymn for Iolaus. Castor and Polydeuces are commonly connected with horses and with chariots, so that it seems reasonable to suppose that a Castor-song was one in honour of a victory in a horse-race or a chariot-race, and to compare *Ol.* 1.102, 'I must crown him with a horse-melody in Aeolic song'.[88] The term 'Aeolic' (for which compare also *Nem.* 3. 79) will refer not to the metre, but the music.[89] It seems, then, that this description could apply to any poem in honour of a horse-race or a chariot-race. But which poem does Pindar mean?

The ancient commentary says that he meant a hyporcheme, from which we have more than one quotation.[90] Is this a mere guess, or does the commentator speak from knowledge? We cannot be certain. Some moderns have thought he meant the First Pythian, others the First Olympian; such suggestions cannot be confirmed, but neither can they be refuted. Several scholars have maintained that it is the Second Pythian Ode itself. Gildersleeve, no mean grammarian, warns us that 'Pindar's use of *μέν* and *δέ* is so tricky that the Kastoreion is not necessarily different' from the Second Pythian. We cannot be sure that he is wrong; and the comparison with *Nem.* 3. 76 f. made above might encourage a believer in his view.[91]‖

This brings us to the concluding passage which has been the cause of so much controversy. Pindar has asked Hieron to give a gracious reception to his poem or poems. He goes on to say, 'Be such as you are according to my words'.[92] Instead of adding to the lengthy discussion of this sentence, I shall simply say that I see no difficulty in taking *οἷος ἐςςὶ* to go with *μαθών* as well as with *γένοιο*. Pindar has praised Hieron; let Hieron now reveal in action the character the poet has described. The word *μαθών* does not imply that without Pindar's description Hieron would not have known his own character; it means much the same as *ἀκούςας* would mean in this context. We should guard against

[88] Plutarch's statement (*Lyc.* 22) that a Kastoreion was a Spartan marching-song played on the aulos is not necessarily relevant to a Kastoreion as Pindar uses the term.

[89] See M. I. Henderson, *New Oxford History of Music* i. (1957), 382–4. We cannot know whether the linguistic Aeolisms detected in the poem (see B. Forssman, *Untersuchungen zur Sprache Pindars* (1966), 13–20) have anything to do with its allegedly Aeolic character.

[90] Fr. 105 S.–M. = 121 Turyn.

[91] Heyne, Boeckh, Christ, and Schroeder in their commentaries, Von der Mühll, *RhM* 72 (1918), 307 [= *AKS* 189] and art. cit. (n. 44), 218 n. 13 [= *AKS* 170 n. 13; cf. *MH* 15 (1958), 141 f. = *AKS* 174 f.] and B. Forssman, op. cit. (n. 89), 19 f. are among those who have agreed with him.

[92] Wilamowitz took this view at *SBPAW* 1901, 1313 f., but retracted it at *Pindaros* 291. See the excellent statement of it by Schadewaldt, op. cit. (n. 1), 331 = 73. So also Farnell, iii. 128 f., and E. Des Places, *Le Pronom chez Pindare* (1947), 72.

assuming that Pindar thinks that without his exhortation Hieron would not behave nobly. Hieron always does behave nobly (cf. *Pyth.* 1. 86 f.).[92a]

From the mention of Hieron's noble nature, Pindar goes on to distinguish between immature minds, whose sense of values is perverted, and the sound judgement of the ripe intelligence. 'An ape is ever[92b] handsome among children, handsome; but Rhadamanthys enjoys happiness, because he has reaped the blameless harvest of his good sense, and takes no joy in deceptions in his heart within him, such joy as comes ever to a mortal through the schemes of whisperers. A plague irresistible to both sides are the speakers of slanders, like altogether to the character of foxes' (73 f.).

Is the monkey mentioned because of its imitations, suggesting the behaviour of flatterers, or because of its tricks? We do not need to ask this question. Pindar says that children think a monkey handsome; clearly the point of the contrast between them and Rhadamanthys is that a mature intelligence sees things as they are and judges rightly. But we are told that Rhadamanthys takes no joy in deceptions; such joy comes through the schemes of whisperers; and they are slanderers, who resemble foxes. Many scholars have assumed that Hieron is being warned not to believe slanderers; but this is only a very indirect inference from the text before us. The just and wise man, of whom Rhadamanthys in the type, has been contrasted with the slanderer; and this is the theme on which the poet now dilates.

Line 78 as it is transmitted hardly makes sense; in the sentence κέρδει δέ τί μάλα[93] τοῦτο κερδαλέον τελέθει, the dative κέρδει, apparently there simply to lend emphasis to κερδαλέον, would be most unusual. The most attractive attempt at emendation is Huschke's[94] substitution for κέρδει of κερδοῖ, a Sicilian word for 'vixen'. But κέρδει, because of the presence of κερδαλέον, may have displaced some altogether different word; the best editorial procedure would be to mention Huschke's conjecture, but to place a crux against κέρδει in the text. Anyway, the general sense is clear; the poet is asking what profit comes to slanderers through their deceptions.

'For though another equipage is suffering grievously at sea, I escape sinking, like a cork above the surface of the deep' (79 f.).[95] We must be

92a See A. Köhnken, *Hermes*, 98 (1970), 1 f.

92b With Tycho Mommsen I would put the comma after αἰεί.

93 The word-order suggests that the μάλα is part of the interrogation; cf. πώμαλα.

94 I. G. Huschke, in August Matthiae's *Miscellanea Philologica*, 1 (1803), 30. Huschke was a pupil of Heyne at Göttingen, and later professor at Rostock; see Bursian, *Geschichte der classischen Philologie in Deutschland* (1883), 641 f.; cf. Schroeder, *Pindars Pythien*, 119 f. Boeckh, ii/2. 250, calls him 'vir elegantissimus et acutissimus'.

95 Cf. *Dithyramb* 1. 16 (fr. 70a S.–M.) φυγόντα νιν καὶ μέλαν ἕρκος ἅλμας: not that this makes it quite certain that ἅλμας goes with ἕρκος and not with ἀβάπτιστος.

careful not to assume that the use of the first person singular here means that the poet must be speaking of himself. It can hardly be denied that in Pindar 'I' is very often [almost] equivalent to *τις*,[96] and it would be most unsafe to take it for granted that he means himself rather than the honest victim of slanderers in general.‖

'It is impossible', the poet continues (81 f.), 'for a crafty citizen to utter a word of power among noble men; none the less, throughout his life[97] he greatly flatters all. I have no part in his impudence; may it be mine to show friendship for a friend, and against an enemy I, as an enemy,[98] will run in like a wolf, treading differently at different times in winding paths.'[99] Again we must be careful not to take it for granted that the first person is autobiographical; more probably 'I' stands simply for the *εὐθύγλωσσος*, the honest, plain-spoken man who is contrasted with the cunning slanderers. That is fully confirmed by what follows (86 f.). 'Under every rule the straight speaker excels, when there is one master, when the noisy crowd and when the wise[99a] guard the city.'

Now follows a difficult transition. 'We must not quarrel with the god, who now upholds the power of one party, now gives great might to others. But not even this contents the mind of the envious, but they stake their claims according to an excessive measure and driving that stake[100] into their own hearts inflict a grievous wound, before they have attained what they are striving for. It is best to accept and bear lightly the yoke upon one's neck; but kicking against the goad makes one's path slippery. May I give pleasure to noble men in my dealings with them!'

Men should not resist the god, Pindar is saying, as Greek moralists so often say; they should not struggle against the laws of human life which the gods impose. They should know that human prosperity, at

[96] See David C. Young, *Three Odes* (n. 39), 58 f. and Köhnken, *Der Mythus* (n. 23), 209 f.

[97] *ἄγαν* offends against sense and responsion, and no emendation is convincing. Boeckh's *ἀγὰν*, 'bend', is in the words of Burton, op. cit. (n. 80) 'not very convincing of the movements of a fawning spaniel'; Heyne's *ἄταν* may derive some colour from the use of *ἀντιδιαπλέκει* at Aeschines 3. 28 but gives the verb a most unusual object. J. G. Griffith ap. Burton (loc. cit. n. 3) suggests *ἆμαρ*: but then *πάγχυ* would stand oddly between this object and its verb.

[98] In early Greece it was part of standard morality to do good to one's friends and evil to one's enemies; see my book *The Justice of Zeus*, 40.

[99] Schadewaldt, op. cit. (n. 28), 68 = 326 wished to amend *πατέων* to *πατέονθ'*: against, see H. Fränkel in his review (quoted in n. 34), 359.

[99a] See above, p. 115.

[100] This translation is designed to bring out Pindar's play on the similarity of *ἑλκόμενοι* and *ἕλκος*. I agree with Gildersleeve and Farnell (both ad. loc.) and with A. C. Pearson, *CQ* 18 (1924), 156–7, in accepting the view stated in the scholia that *στάθμας* means not a balance but a measuring-line, and that *ἑλκόμενοι* means 'drawing out for oneself'. The envious persons have just been mentioned, and it is easy to supply them as subject, so that it is wrong to change *τινος* to *τινες*. Pindar may well speak of 'an excessive kind of measuring-line'. [For another view, see G. W. Most, *AJP* 108 (1987), 571 f.]

the best, does not last for long, so that the envious hatred of one human being against another is inevitably futile. But this knowledge does not content envious humans; measuring their claims by a standard that demands too much, they cause grief to themselves, not being content with the knowledge that the prosperity of others which angers them will not be permanent. It is best to accept the conditions laid down by the gods for human life; it is best not to kick against the pricks.

At the end the poet prays that he may give pleasure to noble[101] men. That may be taken as a general statement, implying that it is a duty to give praise where praise is due. But in this context it has the further function of expressing the poet's eagerness to give praise to Hieron, and to combat those envious persons who might deny him the glory that is his due.

The whole concluding passage of this poem, it seems to me, is fully understandable if we suppose that Pindar is dilating on a common theme of encomiastic poetry, that of the duty of men, and particularly poets, to give great men proper credit for their benefits to others and to abstain from envy. This accords with the significance of the myth of Ixion, which as the poet clearly tells us is introduced as an example of the dangers of ingratitude. In saying this, I am not attempting to reduce the content of the poem to a series of dry and barren formulas. On the contrary, I believe that the best reward of an increased understanding of the conventions of encomiastic lyric and Pindar's way of handling its standard themes will lie in the insight we attain into the poet's manner of adapting a given *locus communis* to the requirements of a particular situation.||

The frequency in Pindar's poetry of the theme of envy and ingratitude[102] is to be accounted for by the requirements of the religion in which he and his audiences believed. For the believers in that religion, to praise a mortal man is a difficult and even a dangerous action. Real felicity belongs only to the gods; mortal men, even those favoured by the gods, are granted only certain moments of true happiness; these quickly pass, and will be followed by misfortune, and in the end death is inevitable. That makes it highly desirable for the favoured mortal to ensure that his brief hours of happiness and glory attain a kind of immortality; this can only be achieved through the action of a poet, who through the gifts of Zeus' son Apollo and his daughters the Muses can preserve human achievements from oblivion.

[101] The prayer at the end of *Ol.* 1 is obviously similar.

[102] See Bowra, *Pindar*, 186–7; Thummer, op. cit. (n. 40), i. 80–1; Köhnken, op. cit. (n. 23), Index s.v. 'Neid'. I have not seen W. Steinlein, *Φθόνος und verwandte Begriffe in der älteren griechischen Literatur* (Diss. Erlangen, 1944), cited by Thummer, op. cit., p. 80 n. 56. S. Eitrem, 'The Pindaric Phthonos', *Studies Presented to D. M. Robinson*, ii (1953), 531 f., has some useful references.

In praising the victor, the poet must abstain from any utterance that might seem to infringe the privileges reserved for the gods; that might bring down upon his patron that 'divine envy' which guards Dike, the order of the universe, against the encroachments of mortals who try to rise above their proper sphere. It is not only the envy of the gods that mortals must beware of; they are inevitably menaced by the envy of their fellow mortals. Envy is the force against which the poet who would praise his patron has to struggle, and in order to assure his patron's fame the poet must do battle against his detractors; this battle is a common theme of Pindar's poetry, and the mention of it serves to guard against the greater danger of divine envy of his patrons' greatness. Unstintingly, without envy, the poet must give glorious deeds their due; in doing so, he must struggle against the envy of true greatness shown by the common run of men.[102a]

Pindar's lyric poetry, like the tragedy of the great tragedians and like all great poetry, is concerned far less with the accidental particulars than with the permanent conditions of human life. Yet the theme of envy seems to be touched on with special frequency and special gravity when the patron who is being celebrated is rendered by his wealth and power specially prone to the attacks of envy. We know that Hieron, like Theron and Arcesilas of Cyrene, had in fact many powerful enemies, who before long were to bring down the dynasty; and this fact can hardly be quite irrelevant to the emphasis upon this theme in the poems which Pindar addressed to him. But to understand Pindar's treatment of that theme, we must understand the religion which supplied his poetry with its background.

Since the time of the ancient commentators it has been generally believed that the concluding passage of the Second Pythian alludes to happenings at the court of Hieron.[103] Some modern scholars, it is true, have questioned the statement, found in the scholia and believed by many moderns, that Pindar is defending himself against his poetical rivals, Simonides and Bacchylides.[103a] They have pointed out that nothing in the text compels us to suppose that these poets are in Pindar's mind; and that the ancient scholars may have imagined it

[102a] That envy is not to be avoided by those who aim at greatness: *κρέϲϲον γὰρ οἰκτιρμοῦ φθόνοϲ* (*Pyth.* 1. 85; cf. Aesch., *Agam.* 939 *ὁ δ' ἀφθόνητοϲ οὐκ ἐπίζηλοϲ πέλει*). The scene of the tapestries in the *Agamemnon* well brings out the Greek religious attitude to *φθόνοϲ*. The great man is bound to provoke the *φθόνοϲ* of men; that does not matter, provided he does nothing to provoke that of the gods. Steinlein, op. cit. (n. 72), according to Thummer, loc. cit., 'vom Neid als etwas Erwünschtem spricht', while E. Milobenski, *Der Neid in der griechischen Philosophie* (1964), argues that that view of it is not consistent with Pindar's. But Pindar, like Aeschylus, thought that *φθόνοϲ* was both to be desired and to be avoided.

[103] Bowra, *Pindar*, 135, believes this.

[103a] See, for example, Mary R. Lefkowitz, *HSCP* 73 (1968), 55 n. 13 [and *Lives of the Greek Poets* (1981), 57 f.].

simply because they knew that Bacchylides celebrated Hieron's victory in the Olympic chariot-race of 468, and did not know Pindar to have done the same.[104] The nearest approach to any positive evidence in favour of the theory comes from the Second Olympian Ode, where Pindar seems to say that *two* noisy crows shriek against Zeus' eagle; || but the lightest possible alteration converts the dual to a plural imperative,[105] so that the number two vanishes. But though not all modern scholars believe in the allusion to the Cean poets, almost all suppose that Pindar is warning Hieron against believing the slanders which certain enemies have uttered against him. Their writings show how the text can be twisted to yield this implication. But I believe that I have shown that the most natural interpretation, strongly reinforced by the parallels supplied by a consideration of Pindar's treatment of similar themes throughout his works, makes it absolutely unnecessary to have recourse to any speculations of this kind. That Pindar in a poem written to celebrate a chariot victory, or at any rate to praise a patron, should have chosen to tack on at the end a complaint against his own private adversaries seems to me a notion wholly inconsistent with all that we know about encomiastic poetry and its canons, about the nature of patronage, and about the poetic practice of the ancient world.[106] The belief originated in the idle speculations of the ancient scholars; but during the nineteenth century it was fostered by the conjunction of historicism and Romanticism. Historicism tried to explain all difficulties in terms of historical allusions, and when history was not known encouraged scholars to invent it; and Romanticism took

[104] See e.g. Burton, op. cit. (n. 80), 126 f.

[105] Bergk's alteration of γαρύετον to γαρυέτων is accepted by Schroeder and Snell [= Maehler], though not by Farnell, Bowra, or Turyn; for a doxography of the problem, see J. van Leeuwen, *Pindarus' Tweede Olympische Ode* (1964), 245 f. [But see ch. 8, p. 88.]

[106] F. Mezger, *Pindars Siegeslieder* (1880) deserves honourable mention here. On p. 50 he writes, 'Die Erklärung der Ode ist mit ganz besonderen Schwierigkeiten verknüpft, so dass es nicht zu verwundern ist, wenn hier die Ansichten der Ausleger noch mehr als sonst auseinander gehen. Nur darin stimmen sie alle überein, dass sie in dem Gedichte eine Reihe von Warnungen und Ermahnungen erblicken, welche der Dichter mit seinem gewohnten Freimuth an den zum Missbrauch seiner Macht hinneigenden Fürsten richte. Und gerade dies anzunehmen ist sehr bedenklich. Denn es hat sehr wenig Wahrscheinlichkeit für sich, dass sich ein Fürst, noch dazu von dem Stolze Hierons, dazu hergegeben haben sollte, dass ihm ein Dichter an seinem Ehrentage unter grösstem Pompe vor tausenden seiner Untertanen den Text lesen durfte, und dass er ihn dafür noch reichlich bezahlte. *Hieron wollte von Pindar Lobgedichte und Siegesgesänge, und Siegesgesänge und Lobgedichte hat Pindar ihm auch gedichtet.*' As Young, ap. Calder and Stern op. cit. in n. 28, 26, remarks, 'Mezger often made remarkably perceptive and unique observations about what was happening in the poems.' But when he gets to the detailed interpretation of the final section, Mezger is led by the 'ungewöhnliche Gemutserregung' of ll. 79–85 to conclude that Pindar is defending himself against his enemies.

Bundy, *SP* I 4 n. 15 has written, 'I believe that this ode, on which I am preparing a monograph, contains nothing personal to Pindar.' I agree, and hope that this article will provide a statement that will do duty at least until Bundy's monograph appears. Every serious student of Pindar looks forward with keen expectancy to the resumption of Bundy's extremely valuable Pindaric studies.

it for granted that the true poet poured out what was in his heart, even when he was executing a commission for a great ruler for which he was to receive a fee.[107]

The dangers of an uncritical application of the new approach can best be illustrated from the recent handling of the difficult problems presented by the Seventh Nemean Ode. This was written at an unknown[108] date to honour Sogenes, son of Thearion, an Aeginetan victor in the boys' pentathlon at the Nemean Games.‖

The ancient commentary on this poem tells us that Aristarchus and his pupil Aristodemus believed that in it Pindar tried to excuse himself for an uncomplimentary reference to Neoptolemus made in a Paean he had written earlier.[109]

That statement was not taken seriously by all nineteenth-century scholars.[110] But when in 1908 Grenfell and Hunt published large fragments of the Paean in question, its truth seemed to have been confirmed. Wilamowitz in an article written immediately after the publication of the Paean and again in his *Pindaros* of 1922[111] held that Pindar in the Seventh Nemean was defending the account of the end of Neoptolemus which he had given in the Paean; Schadewaldt[112] in 1928 agreed with him.

[107] In correcting this tendency, we must resist the temptation to react too far in the opposite direction. Thus when Pindar writes in a poem for the Alcmaeonid Megacles, who was ostracized in 487/6 and won his Pythian victory in 486, that he feels sorry that the victor's great deeds have been requited with envy (*Pyth.* 7. 18 f.), it would be stupid not to acknowledge that the use made of the theme of envy had some relation to the actual fact. Thummer, op. cit. (n. 40), i. 72 n. 48, after drawing attention to the facts, writes, 'Wir halten es deshalb für methodisch richtig, auch hier, wo eine konkrete historische Situation, auf die sich die Stelle beziehen könnte, vorhanden ist, die Aussage des Dichters aus der Topik, in die sie hineingehört, zu deuten und lediglich mit der Möglichkeit eines konkreten Bezuges zu rechnen.' That seems to me over-cautious.

[108] The dates given in the scholia (ιδ' = 547 and κδ' = 527) are impossible, and no certain emendation has been offered. Wilamowitz and other scholars argue for a date about 485; others, less numerous, follow Hermann in putting the poem in about 467. See E. Tugendhat, *Hermes*, 88 (1969), 385 n. 1; S. Fogelmark (below n. 115), nn. 30 and 31. Fogelmark argues for a later date. I share his scepticism about the arguments for a later date ‖ advanced by W. Theiler, 'Die zwei Zeitstufen in Pindars Stil und Vers', *Schriften der Königsberger Gelehrten Gesellschaft*, Geisteswissenschaftliche Klasse, 17/4 (1941), 255 f. = *Untersuchungen zur antiken Literatur* (1971), 148 f. But I find his own arguments (derived from the incidence of colour-words and from the nature of references to Zeus and Apollo) still less convincing than I find Theiler's. A style like Pindar's offers uncommonly few reliable indications of date, and few of the many attempts to date poems whose dates have not been preserved by means of the Olympian and Pythian victor lists can command great confidence. Any serious attack upon the problem would have to take into account far more factors than even Theiler, let alone Fogelmark, has considered.

[109] See Σ on 70 (*Scholia vetera in Pindari carmina*, ed. A. B. Drachmann, iii. 126 (on Aristarchus); Σ on 94a (ibid. 129. 4); Σ on 123a (ibid. 134. 6 f.); Σ on 150a (ibid. 137. 3). See H. Fränkel, *Hermes*, 89 (1961), 387 f.

[110] See Tugendhat, op. cit. (n. 108), 386 n. 1; Hermann, for instance, thought this explanation was an invention of the ancient commentators.

[111] *SBPAW* 1908, 328 f. = Calder and Stern, op. cit. (n. 28), 127 f.; *Pindaros*, 159 f.

[112] Op. cit. (n. 28).

In 1960 the relation between the Paean and the Seventh Nemean was re-examined in a careful and judicious article by Ernst Tugendhat.[113] Tugendhat showed that the differences between the narration about Neoptolemus in the paean and that in the epinikion are not such as to prove that Pindar is defending his earlier treatment of the story. Rather, he is asserting that he in his poetry speaks the truth; this is something he does in many of his poems, but here he does it with special emphasis, and accompanies his assertion with a fresh treatment of the story of Neoptolemus, designed to win approval in Aegina. Tugendhat holds that only one short passage (ll. 102–4) contains an unambiguous reference to the earlier treatment of the Neoptolemus story and the trouble it had caused. He shows with great skill how the handling of the matter of Neoptolemus is combined with the praise of the victor, both in relation to the pervading theme of human destiny, which ordains for all men a mixture of good and bad fortune, and that of poetry, which can perpetuate the good.

Here, if anywhere, we seemed to have a certain example of a personal and historical allusion by the poet. That view was challenged by Bundy in the first of his two papers of 1962. '*N.* 7, a straightforward enkomion,' he wrote, 'has been canonized ... as the poet's personal apology for offensive references to Neoptolemus in the ode we now possess fragmentarily as *Pa.* 6';[114] the authors of the scholia, he adds, had only the odes to aid them. Bundy's opinion has been justified in more detail by Erich Thummer in four pages of the introduction of his commentary on the Isthmian Odes of 1968; by W. J. Slater in an article of 1969; and now at greater length by Adolf Köhnken, in the useful book, published in 1971, in which he has examined the function of the myth in six of Pindar's epinicians. The received opinion has been defended by G. F. Gianotti, by C. P. Segal, and by S. Fogelmark.[115]

The best way of surveying this controversy will be by examining the Seventh Nemean, section by section, not giving a detailed commentary, but concentrating upon matters relevant to the central problem. Like several Pindaric odes, this one opens with the invocation of a divinity; usually it is not one of the great gods, but a minor deity or a ‖ personified abstraction which is specially relevant to the present moment. Here it is Eileithyia, goddess of childbirth.

'Eileithyia, who sits beside the deep-thinking Fates, child of mighty

[113] Op. cit. (n. 108).

[114] *SP* I. 4; cf. 29, n. 70; and see J. Fontenrose, 'The Cult and Myth of Pyrrhos at Delphi', *University of California Publications in Classical Archaeology*, 4/3 (1960), 223 n. 14.

[115] Thummer, op. cit. (n. 40), i. 94 f.; Slater, op. cit. (n. 41), 91 f.; Köhnken, op. cit. (n. 23), 37 f.; G. F. Gianotti, *RFIC* 94 (1966), 385 f.; Ellen Wüst, *Pindar als geschichtsschreibender Dichter* (Diss. Tübingen, 1967); C. P. Segal, *TAPA* 98 (1967), 431 f. (with bibliography); S. Fogelmark, *Studies in Pindar, with particular reference to Paean VI and Nemean VII* (1972).

Hera, hear me, bringer to birth of children. Without you we cannot look upon the light nor upon black night to get our share in your sister, Hebe of the lovely limbs. We do not all draw breath for lives that are alike; yoked as we are by destiny, we are constrained by different fates. With your aid Thearion's son also was singled out for excellence, and the glory of Sogenes is sung of among the men of the pentathlon' (1–8).

The ancient commentators inquired why Eileithyia was invoked here. The five different reasons they offer are discussed by Hermann Fränkel in an article of 1961.[116] He singles out as correct the fourth of these: that Pindar always praises those whose excellence is inborn or natural rather than those whose excellence is due to learning, and Eileithyia has equipped Sogenes to become a great athlete from his birth. This may be part of the truth, but it is not the most important part. First, the first of the explanations in the scholia is worth noting. This is that Pindar is playing on the victor's name.[117] Sogenes is compounded from the root *cω-*, meaning 'save' or 'preserve', and the root *γεν-* meaning 'family' or 'birth'; Eileithyia is called *γενέτειρα τέκνων*, which in this context must mean 'bringer to birth of children'. The ancient commentator thinks this a frigid pun; but Pindar, like other early poets, attributed a mystic significance to names. When in a passage quoted in the following scholion he plays on the derivation of the name of his patron Hieron from the word for 'holy', he is not being merely frivolous. The play on names serves to introduce the goddess, but it by no means exhausts the significance of her choice. She was from early times connected with Hera, goddess of marriage and mother of Hebe, goddess of youth and youthful vigour; she was from early times linked with the Fates, who spin out the destiny of men given them at their birth. The mention of these connections, and of the special functions of the goddess, normal in the invocation of a divinity, lead Pindar straight to what is a frequent topic in his poetry, the varied destinies of men; this is to be a central theme of the present ode, serving

[116] Op. cit. (n. 109), 391 f.

[117] See Σ on I a quoted on p. 116 of Drachmann's third volume, l. 19: *ἔνιοι δέ φαcι πρὸc τοὔνομα τοῦ Σωγένουc παρειλκύcθαι τὴν Εἰλείθυιαν· εἶναι γὰρ αὐτὴν cωγενῆ τινα διὰ τὸ τὰ γεννώμενα ἀναcώιζειν· τὸν οὖν Πίνδαρον ψυχρευcάμενον πρὸc τοὔνομα τῆc Εἰλειθυίαc μεμνῆcθαι*. Paul Maas drew Schadewaldt's attention to the possibility that Eileithyia might have been recalled to Pindar by the victor's name (Schadewaldt, op. cit. 297 = 39 n. 2). He does not quote the scholion. G. Coppola, *Introduzione a Pindaro*, 51, who also takes this view, does quote it. The next scholion rejects the idea on the ground that Pindar plays on names only when an actual homonymy exists, as at fr. 105. 2 (*ζαθέων ἱερῶν ὁμώνυμε*) or fr. 120. 1, where Pindar alludes to the homonymy of Alexander of Macedon and Alexander or Paris, son of Priam. The argument is not a good one; note the opening of Bacchyl. 6: *Λάχων Διὸc μεγίcτου λάχε φέρτατον πόδεccι / κῦδοc* . . . Some doubt attaches to the etymologies of Amenas from *αἰεὶ μένειν* at *Pyth.* 1. 67 (cf. 1. 64) and of the Kronion from *χρόνοc* at *Ol.* 10. 49 f. with which Pindar is credited by J. H. Quincey, *RhM* 106 (1963), 144–5 and 146 (in support of the second, Quincey might have quoted Pherecydes of Syros; see M. L. West, *Early Greek Philosophy and the Orient*, 10). For other etymologies in early poetry, see Pfeiffer, *SB der Bayer. Akad.* 1938, p. 9 n. 2 and Ed. Fraenkel on Aesch. *Agam.* 682.

to link its parts together. Then follows the first mention of the victor and his triumph, and from here, at the beginning of the antistrophe (9 f.), the poet moves to the praise of the victor's native island.

'For he lives in the song-loving city of the men of Aegina with their clashing spears; they are most eager to nurse a spirit that strives in contest.' 'Song-loving' as an epithet of Aegina takes Pindar straight to a central theme of all his poetry, that of the unique power of poetry to preserve the memory of great deeds. 'And if a man triumphs in action, he gives honey-sweet occasion to the streams of the Muses. For mighty deeds of valour lie in darkness if they lack songs; and in one way only we know of a mirror for noble actions, if by means of Memory with her bright veil they find a reward for labour in verses whose singing brings fame.'‖

Now follow words that have been misunderstood even in recent times.[118] 'Wise men know when the third wind is coming, and avarice does not distort their judgment. Rich and poor alike come to death's mark.' The third wind is the wind that will stir up the third and most formidable of three successive waves; the same notion is in Plato's mind when in a famous passage of his *Republic*[119] he compares the problems presented to Socrates by the formidable task of justifying the communistic elements in his ideal state to three successive waves, the last being the greatest, which he must stand up to. 'Wise men' in Pindar often means 'poets', but the wise men here are clearly not poets; they are men who have done noble deeds, and are not prevented by avarice from spending money to ensure that the memory of these deeds will live in poetry.

Next Pindar proceeds to the most famous instance of fame conferred by poetry. 'I believe that the fame of Odysseus is more than what he suffered because of the sweet words of Homer; for upon his falsehoods and his winged art there lies a majesty; and skill beguiles us, deceiving us with tales.' In this passage there is one word missing; and it would be possible to restore it in such a way that the falsehoods upon which there

[118] e.g. by Thummer, op. cit. (n. 40), i. 97 n. 77. The right explanation was given by G. Fraccaroli, *Le ode di Pindaro* (1894), 586; O. Schroeder, *Sokrates*, NF 1 (1913), 532 (cf. his Appendix of 1923, 522); by A. Puech in his Budé text of Pindar; by H. Fränkel in his review of Schadewaldt, cited in n. 34; by G. Méautis, *Pindare le Dorien* (1962), 51; by Carlo Pavese, *QUCC* 2 (1966), 109 f.; by G. F. Gianotti, op. cit. (n. 115), 388 f.; by Tugendhat, op. cit. 401 n. 5; by A. Setti, *Studia Florentina A. Ronconi sexagenario oblata* (1969), 410–11; by Köhnken, op. cit. 44 with n. 41 [and by D. C. Young, *TAPA* 101 (1970), 638 f.]. There is no excuse whatever for anyone who gets it wrong; p. 468 of Slater's *Lexicon to Pindar* (1969) contain several instances of *σοφός* and *σοφία* in a general sense.

[119] 472 A. *τριταῖον ἄνεμον* here must mean 'the third wind', not 'the third day's wind'. For this sense of the adjective, see Barrett on Eur., *Hipp.* 275; the context shows that this is the sense here. [C. Carey, *A Commentary on Five Odes of Pindar* (1981), 142 objects that 'there are no waves in our sentence', but since *κύματα* are created by winds (note *Il.* 2. 396–7 *κύματα . . . παντοίων ἀνέμων*), the sense is clear enough (cf. 31).].

lies a majesty were those told by Odysseus in the *Odyssey*, and not those of Homer. But the context shows clearly that this is wrong.[120] The Muses told Hesiod (*Theog.* 27) that they could tell lies resembling the truth, and also tell the truth when they pleased. Pindar claims very strongly that he in his poetry tells the truth; but he does not claim that all poets do so; and the point here is that poetry can confer fame even on those who do not deserve it.

The thought of the deceptive power of words and the fallibility of human judgement gives the transition to the praise of Ajax. In all the poems Pindar wrote for Aeginetans, he praises Aeacus, son of Zeus and Aegina and his descendants, his sons Peleus and Telamon, his grandsons, Achilles and Ajax, and his great-grandson Neoptolemus; in this ode, Ajax and Neoptolemus are praised.[121] 'The heart of most men is blind; for if it could have seen the truth, mighty Ajax would not, in anger over the arms, have driven his smooth sword through his heart—Ajax, who after Achilles was strongest of those whom the escort of Zephyrus' breath sped in the swift ships straight to Ilus' city to bring back his wife to fair-haired Menelaus.' In similar fashion Pindar in the Eighth Nemean (24 f.) mentions the fate of Ajax as an instance of the power of envy, passing next to what is for him the kindred topic of deceit.

But in the Seventh Nemean the praise of Ajax is a kind of parenthesis; it does not divert the poet from the motive of fame as a bulwark against oblivion. 'But the wave of Hades comes upon us all, and falls upon the obscure and on the famous; and honour belongs to those whose tender fame the god makes grow when they are dead.'

The expression 'the wave of Hades' seems to take up the metaphor of 'the third wind'. Just as rich and poor alike perish, so do the famous and the unknown; the emendation in ‖ Snell's text which would substitute for 'the obscure and the famous' 'him who does not and him who does expect it' has nothing to be said for it.[122] Honour belongs to those whose fame the god increases; the supernatural agent is the god, without whose aid the human agent, the poet, could do nothing.

Now comes what seems a sudden transition to the second hero descended from Aeacus who is praised in the poem, Neoptolemus, son

[120] In l. 22 a syllable is missing; most editions refer *οἱ* to Homer (see E. Des Places, op. cit. (n. 92), 31; and follow Hermann in reading *ψεύδεσί οἱ ποτανᾷ ⟨τε⟩ μαχανᾷ*. Köhnken has argued for Erasmus Schmid's ⟨*γε*⟩ in preference to ⟨*τε*⟩ since he thinks the lies should be those of Odysseus and not those of Homer.

[121] Thummer's belief that Pindar in this poem has to praise Neoptolemus with special emphasis because he is only a minor hero (op. cit., i. 94) shows a total failure to understand what the Aeacidae mean for Pindar that I find amazing.

[122] The point is well dealt with by Köhnken, p. 66 (with nn. 141–2). For *δοκεῖν* meaning 'to be celebrated', cf. Soph. *OT* 1191–2. [Cf. D. C. Young, *CSCA* 4 (1971), 249 f.; Maehler now keeps *δοκέοντα*.].

of Achilles. At the beginning the text is uncertain; I give a literal rendering of what is printed in Snell's edition: 'It was to bring help that he came to the mighty navel of broad-breasted earth—and Neoptolemus lies in the land of Pytho—after he had sacked Priam's city, over which the Greeks had laboured. And when he sailed away he missed Scyros, but came to Ephyra. And he ruled briefly in Molossia; but the race ever honours him for that royalty.[122a] And he left to visit the god, bringing treasures from the first-fruits of Troy; and there as he came against him in a battle over meat a man thrust at him with a knife. And his Delphian hosts were deeply grieved. But he paid a debt to fate; for it was ordained that for the future there should be one of the Aeacidae in the ancient grove, by the god's house with its fine wall, and that he should live there to see justice done at the processions for heroes with their many sacrifices.'

At this point we must consider what is said about Neoptolemus in what we have of the Sixth Paean. The passage about him (98 f.) forms part of a long section in praise of Apollo. Apollo, the poet says, postponed the fall of Troy by killing Achilles; Apollo pitted his strength against Hera and Athena. But even Apollo could not save Troy, since not even Zeus overrules the decrees of Fate; the Achaeans sent for Achilles' young son, Neoptolemus, and he took the city. 'But he never again saw his kind mother, nor the horses of the Myrmidons in his father's fields, as he urged on the bronze-helmed throng. He came to the Molossian land near Tomaros, and did not escape the winds, nor the far-darter with the broad quiver. For the god had sworn that because he had slain the aged Priam after he had leapt upon the altar at the hearth, he should not come to his loving home, nor to old age; and he slew him as he fought with the attendants over the honours due to him in the precinct the god loves beside earth's broad navel.'

Is Tugendhat (p. 391) right in saying that this is not a modification of what is said in the Nemean, but a significantly different story? Let us compare the two. In the Paean, Neoptolemus offends Apollo by the atrocious act of killing the aged Priam while he was a suppliant at the altar of Zeus which stood in the courtyard of the palace. In both stories, Neoptolemus sails away meaning to reach Thessaly and is diverted to Molossia, only in the Paean the god causes the winds to drive him off his course. In both accounts, Neoptolemus perishes in a brawl with the Delphians over the division of sacrificial meats. Only it is significant that in the Paean Apollo himself is the killer,[123] in the

[122a] The word-order indicates that *οἱ* is attributive, not ethic-possessive; see Des Places, op. cit. (n. 92), 31–2.

[123] At *Paean* 6. 119 we must read *κτάνεν*: see S. L. Radt, *Pindars zweiter und sechster Paian* (1958), 169. J. Fontenrose, op. cit. (n. 114), 223 n. 14, summarizing Bundy's unpublished views, says that this passage does not show that Apollo killed Neoptolemus 'directly', because the Seventh Nemean

Epinikion a man; Pindar strongly hints at the story, recorded as early as Pherecydes in the fifth century, that this was the Delphian hero Machaireus, son of Daitas. The Paean says nothing about Neoptolemus' reason for coming to Delphi, but the scholia mention several reasons given by others. One was that he planned to plunder the temple, another that he came to demand that the god pay him an indemnity for having killed his father. The epinikion says that he came to bring the god an offering from the Trojan booty. It claims that the Delphians were deeply ‖ distressed about the incident, which is certainly not the impression one derives from the account given by Euripides in his *Andromache*. It explains that it had to happen, so that one of the Aeacidae could remain permanently near the temple to supervise religious processions. Unlike the Paean, the epinikion alludes to the claim of the fifth-century kings of Molossia to be descended from Neoptolemus.

To sum up, there is only one difference of hard fact, but that is in the all-important matter of who killed Neoptolemus. The story in the Paean is what we would expect in a poem written in honour of Apollo for a Delphic ceremony; the story in the epinikion is exactly what we might expect the same poet to make of the same tradition in a poem written in honour of an Aeginetan and for performance in Aegina. The second story is wholly consonant with the theory that it was introduced in order to remove offence caused by the first. But so far we have found nothing in the poem that obliges us to believe this; and we shall have to go further with our examination of the ode before we can decide the question.

At the beginning of the section I have just translated, there is a difficult problem of textual criticism and interpretation. I have accepted the emendation which makes the poet say that someone came to Delphi bringing help.[124] Who was that person? Presumably Neoptolemus; but if so, there are two difficulties. First, the introduction of Neoptolemus is strangely abrupt: 'it was to bring help that he came—and Neoptolemus lies in Delphi—after he had taken Troy'. Tugendhat argues that we are not made to wait long for elucidation; but I find the sudden transition to Neoptolemus somewhat disconcerting. Secondly, in what sense did Neoptolemus come to Delphi 'bringing help'? To

'shows that for Apollo a temple attendant was the killer'. But even if we grant the highly questionable proposition that Pindar always believed in a single version of a mythical event and stuck to it throughout his works, it is perfectly clear that in the Paean Pindar made Apollo responsible for the hero's death.

[124] Farnell placed a full stop after the words *τιμὰ δὲ γίνεται / ὧν θεὸc ἁβρὸν αὔξει λόγον τεθνακότων*, and changed *βοαθόων* to *βοαθοῶν*. Snell reads *βοαθοῶν τοι παρὰ μέγαν ὀμφαλὸν εὐρυκόλπου μόλεν χθονόc—ἐν Πυθίοιcι δὲ δαπέδοιc κεῖται—Πριάμου πόλιν Νεοπτόλεμοc ἐπεὶ πράθεν*, ... [See G. W. Most, *Hermes*, 114 (1986), 262–71 for an ingenious new suggestion.]

bring offerings to the god is very oddly described as 'bringing help'; the phrase would be appropriate only if Delphi had been on the point of going bankrupt. Thirdly, Snell's text involves taking the words ἐν (Πυθικοῖcι . . . κεῖται as a parenthesis, which though not impossible is somewhat awkward.

The scholia contain the reading ἔμολον, which suggested to Hermann the reading μόλον. Suppose that βοαθοῶν is right; then the words will mean 'It was to bring help that I came to Delphi', and they will refer to Pindar.[124a] To bring help to whom? The preceding sentence is 'Honour belongs to those whose tender fame the god increases when they are dead.' The god does so by causing poets to sing their praises; and it happens that Pindar more than once uses such terms as 'come to the help of' to describe his relation to the patrons whose great deeds he protects against oblivion. In the Thirteenth Olympian (96 f.) he says, 'I have gladly come as an ally to the Muses on their glorious thrones and to the Oligaithidai (his patron's family); the word ἐπίκουροc meaning 'ally', is also the epithet of the road of words which in the First Olympian (110) he claims to have discovered. In the Ninth Olympian (83) he has come in friendship and goodness to aid the Isthmian garlands of Lampromachos; and in the very Sixth Paean (10–11) which has been thought relevant to our poem he has come to Delphi to protect the citizens and his own honour against the reproach of inability to act.

This interpretation was put forward in 1967 by Ellen Wüst and C. P. Segal. It makes the punctuation distinctly smoother; but if we follow them in taking ἐπεὶ as temporal, as others have always taken it, κεῖται will go ill in the apodosis to a temporal clause with aorist verb. Suppose we take ἐπεὶ as causal. Then the following words will mean, 'And Neoptolemus lies in the land of Pytho, since he took Troy . . . and on his way back missed Scyros, but came to Ephyra'. After the causal ἐπεὶ, we should expect the explanation of how he came to Delphi to follow soon; but the poet launches into an account ‖ of Neoptolemus' travels, and when he finally comes to explain how he got to Delphi, the ἐπεὶ-clause has been abandoned.[125]

The causal ἐπεὶ introduces not simply the statement that Neopto-

[124a] Fraccaroli, op. cit. (n. 118), 588 n. 2 read μόλον and took it as referring to Pindar; see his note. [Thus Turyn and Snell–Maehler.]

[125] 'Die Griechen gebrauchen ἐπεί sehr häufig auch da, wo der grundangebende Satz nicht einen untergeordneten Teil des Hauptsatzes bildet, sondern vielmehr die Geltung eines mit γάρ beigeordneten Hauptsatzes hat, in welchem Falle im Deutschen durch *denn* übersetzt wird': Kühner–Gerth ii. 461. So no one need find it odd that a sentence whose main verb is κεῖται is followed by an ἐπεί clause whose main verb is aorist. Cf. *Ol.* 6. 27; at *Ol.* 1. 26 the misunderstanding of a causal ἐπεί as temporal led to ludicrous results (see J. Th. Kakridis, *Hermes*, 63 (1928), 475 = Calder and Stern, op. cit. (n. 28), 188 f. = Μελέτες καὶ ἄρθρα (1971), 66 f., for the right interpretation).

lemus took Troy, but the whole explanation of how he came to Delphi.

If this interpretation is correct, the problem of whether the ancient commentators were right in thinking Pindar wished to justify himself to the Aeginetans because of the paean is solved, for the occasion on which Pindar came to Delphi to bring help to Neoptolemus in the sense of helping to preserve his fame can refer only to the occasion of the paean's performance.[126] But even without this argument I believe that the necessity of supposing there to be a reference to the paean can be established.

Now follows a difficult passage, taken by Wilamowitz and Schadewaldt to be part of Pindar's self-defence. 'For justice, whose name is beautiful, three words will be enough. The witness is not false, Aegina, who stands over your achievements and those of the descendants of Zeus. I make bold to say it, moving along a road of words that is my own, with power in respect of shining deeds of glory.' My rendering of the last sentence needs a good deal of justification.

When Pindar says, 'For justice . . . , three words will be enough', there is no use in trying to work out what the three words are; in Greek 'three', like our expression 'two or three', can simply mean 'a few'. People who have believed that the Seventh Nemean cannot be explained without reference to the Sixth Paean have assumed that when Pindar says that three words will be enough for justice, he must mean that three words will be enough to justify him against the charges resulting from the Paean. In the lines that follow, Pindar undoubtedly has his own art in mind. That becomes clear once we raise the question of who is the witness to the achievements of Aegina mentioned in the next sentence. Is the witness Apollo, or is it Neoptolemus, or is it Pindar? The presence of Neoptolemus at Delphi does not make it easier for Apollo, or for Neoptolemus himself, to bear witness to the achievements of the Aeginetans. But Pindar does bear witness to their achievements; and in several passages he speaks of his relation to those he praises in these terms or others very like them. At the beginning of the Fourth Olympian, the Seasons have sent him to be witness of supreme achievements, by which he means those of the victors in the Olympic Games; in the Second Partheneion (38 f.), the leader of the chorus claims to be a faithful witness of the good deeds of Agasikles. In the Fourth Isthmian (28), 'testimonies of the measureless glory of men dead or alive' must refer to poetry; and in the Sixth Olympian (20), Pindar declares that he will bear witness upon oath to the admirable qualities of Agesias.

[126] It is strange that Thummer (i. 96 n. 72) has not seen this, although he too reads *βοαθοῶν . . . μόλον*.

The witness is surely Pindar;[127] and when Pindar claims that three words will be enough for justice, he is vindicating his own poetic honour.

Before going back to l. 48, let us consider the difficult sentence beginning in l. 50. *θρασύ μοι τόδ' εἰπεῖν* must, I think, mean, 'I make bold to say this'. I do not think that the 'road of words' can be the predicate of the indirect statement; the pronoun 'this' would then be obscure. Can the words mean, 'I make bold to say this, that I have at home a road of words ...'? The absence of the infinitive of the verb to be seems to me to make against this view. I have underlined the possibility that the words *ὁδὸν λόγων* are an internal accusative depending on the verb *εἰπεῖν*, so that the sense is, 'I make bold to say ‖ this (moving along), a road of words ...'.[128] But I think it likelier that Wilamowitz and Schadewaldt are right in emending *ὁδὸν κυρίαν* to *ὁδὸς κυρία*.

Whatever the exact syntax, the general sense is clear; in speaking of a 'road of words from home that has power in respect of shining deeds of glory', Pindar is speaking of his own poetic art.

The notion of the road or path of poetry is not rare in Pindar. In the first Olympian (110), he says that he will find a helping road of words to praise Hieron; that is the most relevant of many similar expressions. For the notion of a road of words that is valid in respect of glorious deeds, we may recall the opening words of the parodos of Aeschylus' Agamemnon (104 f.), where the old men who form the chorus declare that they have power to sing of the expedition to Troy because, aged as they are, they still possess the gift of song. *οἴκοθεν* means that Pindar does not depend on others, but relies on his own gift, his own 'fountain of immortal verses' (*Pyth.* 4, end).

In this place Pindar is vindicating his own poetic art. Does it follow that when he says that three words will suffice for justice he means that three words will suffice to do him justice? Immediately before, he has been speaking of Neoptolemus; so that it is only natural for the hearer to suppose that the sense is 'Three words will suffice to do justice to Neoptolemus'. It is quite natural for Pindar then to go on to emphasize his own truthfulness; three words, spoken by one as truthful as himself, will be sufficient. Pindar has praised Neoptolemus, and if he then goes on to say 'Three words will be enough for justice', it is perfectly

[127] See Tugendhat, p. 395, with n. 1.

[128] Frl. Wüst, op. cit. (n. 115), 146, seems to incline to this view. Although I know of no closely similar sentence involving a verb of saying with an expression like *ὁδὸν λόγων* as an internal accusative, I do not find the construction impossible. But see Wilamowitz ap. Calder and Stern, op. cit. (n. 28), 138; (cf. *Pindaros*, 163 n. 1) and Schadewaldt, op. cit. (n. 28), 313 = 55; cf. O. Becker, 'Das Bild des Weges' (*Hermes* Einzelschriften 4, 1937), 74 f. [Against: C. Carey in *Dionysiaca* (1978), 35 f.]

possible that he simply means 'A few words will be enough to do justice to so great a hero'. In that case, there is no need to imagine any reference to the complaint provoked by the Sixth Paean. Taken by themselves, and taken independently of my explanation of the words 'to bring help' in l. 33, the words do not prove the existence of such a reference. But if I am right in thinking that at l. 33 Pindar has insisted that he came to Delphi on an earlier occasion to bring help to Neoptolemus, meaning that he came to praise him, then I find it hard not to take l. 48 as a reference to the complaints, and to suppose that the poet's insistence on his own truthfulness, normal as it is in an epinician ode, has in this place a special appropriateness.

Here Pindar passes from the topic by means of a transitional formula of a common type; one tires of everything, even of honey and of making love. 'But in every action rest is delightful; we grow tired even of honey and of the pleasant flowers of Aphrodite.'

When the poet passes to the familiar statement that different men have different fates, he does so because it will serve him as a foil to the praise of the victor's father that will follow. But the choice as foil of this particular commonplace of gnomic wisdom is specially appropriate in a poem which has begun with the invocation of Eileithyia, who sits beside the Fates, and which has gone on from that beginning to mention the diversity of human fortunes. 'By nature we differ from each other in our lives; one man has one lot, one another; and for one man to win success, gathering up all manner of good fortune, is impossible; I do not know of one to whom Fate has granted in permanence this consummation. But to you, Thearion, she gives a fitting moment of happiness; you have found the courage to do great things, and she has not harmed your judgement.' Though Thearion has found courage to achieve triumph, he has resisted the temptation to commit hybris; he is not one of those whose wits have been taken away by Ate so that he has committed acts of folly.

'I am your friend', the poem continues, 'keeping off dark reproach, as though by bringing streams of water to the man I honour I shall praise true glory; and for the noble this reward is fitting' (61 f.). The ceremonial salutation of the victor as a friend is ‖ paralleled, for instance, in the Third Nemean (76) and in the Second Isthmian (end). Keeping off reproach is as much part of the poet's task as praising the victor; here the end of the Second Pythian supplies the best parallel.

'When he is near the Achaean who lives above the Ionian Sea shall find no fault with me; I trust in his friendship; and among my fellow citizens my glance is clear. I do not overstep the mark, but draw out of my path all violence. May future time as it comes on be kind; and men shall learn whether I speak crooked words out of tune.'

Who is the Achaean living above the Ionian Sea? The immediately following reference to Pindar's fellow citizens, suggests that we have here a polar expression of a familiar type; wanting to say 'no one at all will blame me', the poet first says that no one living very far away will blame him, and then that no one living very near will do so. The earlier mention of the claim of the fifth-century Molossian kings to be descended from Neoptolemus has encouraged the ancient commentator, and also most modern scholars, to think that by 'the Achaean' Pindar means a Molossian, whose approval would help to vindicate the poet from the suspicion of having maligned his king's ancestor. It would be natural to assume that the words *Ἰονίας ὑπὲρ ἁλὸς*, in the mouth of a poet writing in Thebes, or in Aegina, referred to the inhabitants of Sicily or Magna Graecia; but since Dissen commentators have cited passages from historians and geographers to show that people living on the coast bordered by a particular sea may be said to live not indeed *beyond*, but *above* the sea in question. The Molossians—like the inhabitants of Pallene—may well have laid stress on their claim to be descended from the Achaeans who took Troy, and it seems a shade likelier that they are meant than that 'Achaean' is being used in the general sense of 'Greek' and that the Greeks of Magna Graecia are the persons referred to.[129] But the deduction that Pindar was the proxenos, in the technical sense, of the Molossians in Thebes is quite unwarranted.[129a] *προξενία* can very well mean 'friendship', as at *Ol.* 9. 83 and *Parth.* 2. 31 (cf. *πρόξενοι* at *Isthm.* 4. 8 and see Aesch., *Supp.* 420 and Soph. *El.* 1451), and that is likely to be the sense here also.

Pindar has come to Delphi to defend Neoptolemus' fame; he is keeping off reproach and is praising a victor who truly deserves praise; no one in the future—and the future is the best judge of truth[130] will be able to say that he has spoken crooked words. All this could be said without a reference to the Sixth Paean; but if that poem could be

[129] See Σ on 95a (Drachmann, iii. 129); Erasmus Schmid translated 'Ionium *supra* mare habitans'. Heyne differed: 'illum dixit Achivum e Sicilia et Italia, si adesset', and also recognized the polar expression: '*nemo nec in propinquis nec in remotis terris habitantium*: pro his memorat Siculos et Italos'. Dissen (first in Boeckh's edn., ii/2. 430, and later in his own) argued that places on mountains near a sea can be said to be 'above' that sea; he quoted Strabo 324, *ὑπέρκειται δὲ τούτου μὲν τοῦ κόλπου Κίχυρος ἡ πρότερον Ἐφύρα πόλις Θεσπρωτῶν· τοῦ δὲ κατὰ Βουθρωτὸν ἡ Φοινίκη. ἐγγὺς δὲ τῆς Κιχύρου πολίχνιον Βουχέτιον Κασσωπαίων μικρὸν ὑπὲρ τῆς θαλάττης ὄν* . . . : 326 *ἀναμέμεικται δὲ τούτοις τὰ Ἰλλυρικὰ ἔθνη τὰ πρὸς τῶι νοτίωι μέρει τῆς ὀρεινῆς καὶ τὰ ὑπὲρ τοῦ Ἰονίου κόλπου· τῆς γὰρ Ἐπιδάμνου καὶ τῆς Ἀπολλωνίας μέχρι τῶν Κεραυνίων ὑπεροικοῦσι Βυλλίονες κτλ.* Cf. Thuc. i. 46. 4 (of Chimerium) *ἔστι δὲ λιμὴν καὶ πόλις ὑπὲρ αὐτοῦ κεῖται . . . Ἐφύρη*. Against these we may consider the many places where *ὑπέρ* with a word for 'sea' means 'beyond the sea', as at *Od.* 13. 256–7 *πυνθανόμην Ἰθάκης γε καὶ ἐν Κρήτηι εὐρείηι / τηλοῦ ὑπὲρ πόντου*. The latter is the commoner usage.

[129a] Even Thummer (i. 97 n. 82) allows a personal allusion here, wrongly.

[130] Cf. *Ol.* i. 33; 10. 54; fr. 159.

shown to have been in his mind, what he says here would take on an added significance.

With an apology for his delay in doing so, Pindar now comes to the praise of the young Sogenes. That section of the poem (71–131) is not relevant to my present purpose. But we must consider ll. 102 f., where following a prayer for Sogenes' prosperity we find the words 'But my heart will never say that I have savaged Neoptolemus with ruthless words.[131] To repeat the same things three or four times is futility, as when someone idly barks at children "Corinth of Zeus".'‖

I am concerned not with the obscure but in this context obviously relevant proverb at the end, but with the sentence that precedes it. Slater claims that it means 'My heart will declare that it has not injured Neoptolemus with intractable words'; and he explains that this really means 'I claim that I have praised Neoptolemus with fitting words'.[132] If anyone supposes that Pindar can really have meant this, let him consider Slater's attempt at translation. It unjustifiably waters down the sense of *ἑλκύcαι*, which as almost everyone since antiquity till now has realized is in this place a metaphor taken from the 'worrying' or 'dragging about' of a body by dogs; the use of the word to mean this is familiar from the *Iliad*.[133] The basic meaning of *ἄτροποc*[134] is 'not to be turned aside', and hence 'ruthless' or 'remorseless'; it is familiar as the name of one of the Fates. *οὔ ποτε* goes with *φάcει*, not with *ἑλκύcαι*. Every translator has seen all this except Slater, who in his violent attempt to fit the text to the Procrustean bed of his dogma has done to Pindar's words just what Pindar is denying that he has done to Neoptolemus.

If we take the words in their natural sense, we must admit that Pindar is denying having attacked Neoptolemus; and we can hardly help accepting the explanation that the Sixth Paean had given offence in Aegina, so that Pindar had to defend it. Tugendhat after a careful and scrupulous discussion concluded that this was the only sentence in the poem that forces us to accept this explanation.[135]

But as he says when we look back over the poem we see other

[131] *Τὸ δ' ἐμὸν οὔ ποτε φάcει κέαρ / ἀτρόποιcι Νεοπτόλεμον ἑλκύcαι / ἔπεcι.*

[132] Op. cit. (n. 41), 92 f.

[133] *Il.* 22. 33 *cὲ μὲν κύνεc ἠδ' οἰωνοὶ / ἑλκήcουc' ἀϊκῶc*; cf. 17. 557–8; just so Heracles says he will throw the head of Lycus to be *κυνῶν ἕλκημα* (Eur. *HF* 568; cf. Herodotus i. 140, Plato, *Rep*. 539 B, Theocr. 1. 135. Σ on 150A (Drachmann iii. 137. 12) says *τὸ δὲ ἑλκύcαι ἀντὶ τοῦ ἐνυβρίcαι. ἡ δὲ μεταφορὰ ἀπὸ τῶν κυνῶν τῶν ἑλκόντων τὰ cώματα.*

[134] On *ἄτροποc*, see H. Fränkel, art. cit. (n. 109), 385; cf. Tugendhat 404.

[135] Slater's treatment of the passage is criticized by Fogelmark, op. cit. (n. 115), 104 f. In dealing with Slater's complaint that it is odd that Pindar should defend 'a supposed insult in a Paean commissioned for Delphians in Delphi against protests by Aeginetans', Fogelmark might have mentioned that, since the Paean in question contains a long section in praise of Aegina, it would be very odd indeed if it was not known of in that island.

passages in a somewhat different light; otherwise the poet's denial that he has savaged Neoptolemus would come abruptly and almost unintelligibly, in this place. Following Frl. Wüst and Segal, I have argued that at l. 33 Pindar claims to have gone to Delphi to bring help to Neoptolemus. In the light of ll. 102–4, this surely refers to his having gone there on the occasion of the performance of the Paean. If that is right, it will be natural to take the words about justice and the poet's truthfulness at ll. 48 f. in the light of this earlier reference. But even if my view of ll. 33 f. is not accepted, it will not be easy to avoid taking ll. 48 f. in the light of the poet's denial at 102 f. that he has attacked Neoptolemus. Pindar has skilfully contrived to make amends for the alleged slight to Neoptolemus within the framework of an epinician ode which conforms to the normal pattern and in no way lacks unity.

The Sixth Paean, containing as it does, for whatever reason, the praises of Aegina, must have been known on that island; and it would not be surprising if its presentation of Neoptolemus as a savage killer, true as it was to the standard version of the cyclic epics, angered the Aeginetans. It is true that in early times the violent and savage actions of heroes did nothing to diminish their heroic status, or their right to be venerated.[136] Ajax the son of Oileus, whose rape of Cassandra was a main cause of the storm which scattered the Achaean fleet as it sailed home from Troy, received heroic honours in his native state of Locri;[137] the chequered careers of Oedipus and Eurystheus do not diminish their right to worship after death, according to Sophocles and Euripides. Yet as early as the archaic age we discern the tendency to censor stories about heroes in places where they received worship, just as we discern the tendency to censor stories about gods. In Attic tragedy Theseus ‖ appears as a character of blameless virtue; it is hard to recognize in the pattern constitutional monarch of Euripides' *Suppliants* the betrayer of Ariadne or the abductor of Helen and would-be abductor of Persephone. Some of the victims of Theseus were not, in the traditions of their own cities, the brigands and monsters they appear in Attic legend. Cercyon, Sinis, and Sceiron were all half-brothers of Theseus, being sons of Poseidon; the last two were also his relations on the maternal side. In the tradition of his native Megara Sceiron is a highly respected person, a suppressor of brigandage. He married Chariclo, daughter of Cychreus, an honoured figure in Corinthian as well as Megarian legend; their daughter Endeis married no less a hero than Aeacus, and became by him mother of Telamon and Peleus. Megarian historians, Plutarch says, defended Sceiron, 'coming to grips with the tradition, as

[136] On the gradual process through which heroes came to be credited with moral qualities, see Angelo Brelich, *Gli eroi greci: un problema storico-religioso* (1958), especially pp. 225 f.

[137] Σ on *Ol.* 9. 166. See L. R. Farnell, *Greek Hero Cults and Ideas of Immortality* (1921), 293 f.

Simonides put it, and "fighting against time immemorial"'.[138] Had any poet written disrespectfully of Sceiron at Megara, there would have been complaints. Pindar's retraction of insult to Neoptolemus has a precedent in the famous case of the palinode, or rather palinodes, of Stesichorus addressed to Helen. It has been surmised that Stesichorus had to recant because he had displeased the Spartans, for whom Helen was the centre of an important cult.[139] Pindar was writing not long after the Aeginetans, asked by the Thebans for help in battle, had lent them the statues of the Aeacidae. On that occasion the Aeacidae were not successful; but at Salamis they were carried round the fleet, and the Greeks were encouraged by the thought of fighting on the territory of the greatest warrior heroes of their mythology.[140] There is no doubt that in Pindar's time hero-cult was an important factor in religion, and consequently in politics; and it followed that the treatment of cult heroes by poets was matter of legitimate public interest.

Just as Pindar rejects certain discreditable stories about the gods, so he is jealous of the honour of certain heroes. Obliged, in the Fifth Nemean, to touch on the departure from Aegina of Peleus and Telamon, he treads delicately when he comes to their murder of their half-brother Phocus, son of Aeacus by the Nereid Psamatheia. Pindar might, indeed, have avoided the story altogether; still, his caution will have shown his patrons his care for the reputation of their ancestors. In the Third Nemean he dextrously avoids the story that the young Achilles fed upon live animals which he had captured, and also the legend that Peleus led his army into Iolcos through gates each of which had upon it half of Hippolyte, wife of Acastus, who had plotted against his life for having rejected her advances.[141] In the Thirteenth Olympian, he prefers not to describe the end of Bellerophon; in the Seventh Olympian, he skirts the topic of those early Rhodians of doubtful reputation, the Telchines. Here, surely, is a point at which the poetic world of Pindar shared a frontier with the world of contemporary reality.[142]

[138] *οἱ δὲ Μεγαρόθεν ϲυγγραφεῖϲ ὁμόϲε τῆι φήμηι βαδίζοντεϲ καὶ τῶι πολλῶι χρόνωι κατὰ Σιμωνίδην πολεμοῦντεϲ οὔτε ὑβριϲτὴν οὔτε ληιϲτὴν γεγονέναι τὸν Σκίρωνά φαϲιν, ἀλλὰ ληιϲτῶν μὲν κολαϲτήν, ἀγαθῶν δὲ καὶ δικαίων οἰκεῖον ἀνδρῶν καὶ φίλον*, *Thes.* 10. The growth of the Theseus saga has been traced by H. Herter, *RhM* 85 (1936), 177 f. and 193 f. and 88 (1939), 244 f. and 289 f.

[139] See C. M. Bowra, *Greek Lyric Poetry* (2nd edn., 1961), 110 f.

[140] See Hdt. 5. 79 f. and 8. 64 and 83–4. Theseus, like the hero Echetlos or Echetlaios, fought at Marathon (Plut. *Thes.* 35; Paus. 1. 32. 4). See Brelich, op. cit. (n. 136), 91; Rohde, *Psyche* (English version) 136 f.

[141] *Nem.* 5. 13 f.; *Nem.* 3. 34; ibid. 45 f. See D. S. Robertson's article 'The Food of Achilles', *CR* 54 (1940), 177–80, esp. 178 f.; more instances are given by M. C. van der Kolf, *Quaeritur quomodo Pindarus fabulas tractaverit quidque in eis mutarit* (1933), also by Bowra, *Pindar*, ch. 2.

[142] Much of the material included in this article formed part of lectures which I had the privilege of delivering as Alexander White Visiting Professor in the Humanities at the University of Chicago in the Fall Term of 1972. I profited greatly from discussing the problems in question

with Professor Anne Burnett: and I have been greatly encouraged by learning that Professor Mary Lefkowitz, in a chapter of a forthcoming book which she kindly permitted me to read [*The Victory Ode* (1976), 8–41], has concluded, independently of me, that the generally held view that the end of the Second Pythian Ode contains a defence of the poet against slanderers is to be rejected. [Cf. now E. Thummer, *RhM* 115 (1972), 293 f.]

[Since this article appeared, the Seventh Nemean has been discussed in detail by M. R. Lefkowitz, *HSCP* 84 (1980), 24–49 (esp. 39 f.), and both the Second Pythian and the Seventh Nemean by C. Carey, *A Commentary on the Odes of Pindar* (1981) and G. W. Most, *The Measures of Praise: Structure and Function in Pindar's Second Pythian and Seventh Nemean Odes* (Hypomnemata 83, 1985). Ian Rutherford's commentary on Pindar's Paeans will contain a discussion of the Sixth Paean in relation to the Seventh Nemean.]

10(51)

Pindar Fr. 169

'Law the king of all, of mortals and immortals, leads them, making just what is most violent with arm supreme. My witness is the acts of Heracles; for he drove the cattle of Geryones to the Cyclopean portal of Eurystheus, without . . . or wage.'

These words have long been known from Plato's quotation in the *Gorgias*, where Callicles, the defender of *Realpolitik*, uses them to support the right of the stronger to rule over the weaker. The Platonic quotation is supplemented by others in Aristides (*Orat.* 45) and in the apology for Socrates which Libanius composed in the fourth century AD to answer the accusation of the sophist Polycrates written in the fourth century BC. Both later writers cite the poem in the course of debates which echo that in Plato; both follow Plato in their interpretation.

It has long been agreed that Plato—not only in the *Gorgias*, but in other places where he quotes the fragment—grossly misinterprets Pindar. He is simply making use of the poem for his own purposes, as other sophists of the fifth and fourth centuries were accustomed to use the poetry they quoted. Whatever the *nomos* is that justifies the violent act of Heracles, it cannot be the right of the stronger; Pindar was no immoralist after the fashion of Callicles.

But what was the *nomos* in question? The point has long been debated, and the debate has not been ended by Lobel's publication, in 1961, of extensive fragments of the same poem contained in a papyrus of the first or early second century AD.

A full bibliography of the discussion before 1956 is provided by M. Gigante, "*Νόμος βαcιλεύc*", *Ricerche filol.* i (Naples, 1956), 79–92; cf. E. Wolf, *Griechisches Rechtsdenken*, ii (Frankfurt, 1952), 190 f.; E. Thummer, *Die Religiosität Pindars* (Diss., Innsbruck, 1957), 118; and E. R. Dodds, *Plato, Gorgias* (Oxford, 1959), 270 f.[1]

P. Oxy. 2450, *The Oxyrhynchus Papyri*, XXVI, ed. E. Lobel (1961), 141, with Plate XVa and b.

[1] [See now D. E. Gerber, *A Bibliography of Pindar, 1513–1966* (1969), 105 f.]

Harvard Studies in Classical Philology, 76 (1972), 45–56.

D. L. Page, *PCPS*, NS 8 (1962), 49 f.
H.-J. Mette, *Glotta*, 40 (1962), 42 f. ‖
E. G. Turner, *CR*, NS 13 (1963), 269.
M. Treu, *RhM* 106 (1963), 212 f.
B. A. van Groningen, *Gnomon*, 35 (1963), 129.
C. M. Bowra, *Pindar* (Oxford, 1964), 74 f.
C. O. Pavese, *Maia*, 16 (1964), 311 f.
B. Snell, edn. of Pindar, ii (Leipzig, 1964), 122 f. (fr. 169).
M. Ostwald, *HSCP* 69 (1965), 109 f. = W. M. Calder III and J. Stern (eds.), *Pindaros und Bakchylides* (Wege der Forschung 134, 1970), 194 f.
W. Theiler, *MH* 22 (1965), 69 f. = *Untersuchungen zur antiken Literatur* (1970), 192 f.
M. Treu, *Ἐπιcτημονικὴ Ἐπετηρὶc τῆc Φιλολογικῆc Cχολῆc Θεccαλονίκηc*, 9 (1965), 207 f. (esp. 221 f.).
Paola Bernardini, *QUCC* 2 (1966), 185 f.
M. Gigante, *Atti del XI Congresso di papirologia* (Milan, 1966), 286 f.
C. O. Pavese, *HSCP* 72 (1968), 47 f.
W. K. C. Guthrie, *A History of Greek Philosophy*, iii (Cambridge, 1969), 131 f.
H. Schmitz, *Hypsos und Bios* (Berne, 1970), 41 f.

Here is an adaptation of the text printed by Snell:

A′ Νόμος ὁ πάντων βαcιλεύc
θνατῶν τε καὶ ἀθανάτων
ἄγει δικαιῶν τὸ βιαιότατον
ὑπερτάται χειρί. τεκμαίρομαι
ἔργοιcιν Ἡρακλέοc·
col. ii ἐπεὶ Γηρυόνα⌋ βόαc
Κυκ'λώπει⌋ον ἐπὶ π'ρόθυρο⌊ν⌋ Ε⌊ὐρυc⌋θ̣έοc
× (?) – – ∪⟩ κ⌋αὶ ἀπ'ριάταc ἔλαcεν,
– ??] Διομήδεοc ἵππουc
– ? μ]ο̣ναρχον Κ[ι]κόνων
παρὰ] Βιcτο⌋νίδι λίμναι
χαλκοθώρ]ακοc Ἐνυαλίου
∪∪ –] ἔκπαγ'λον υἱόν
∪∪∪] . ι̣αντ̣α μέγαν
– οὐ κό]ρωι ἀλλ' ἀρεταῖ.
∪ – γ]ὰ̣ρ̣ ἁρπαζομένων τεθνάμεν
× –] μάτων ἢ κακὸν ἔμμεναι. ‖
– –]εcελθὼν μέγα
∪ – ν]υ̣κτὶ βίαc ὁδόν
]ρε{ν}, λαβὼν δ' ἕν[α] φ̣ῶ̣[τ]α̣ πεδά⟨ρ⟩cι̣ο̣ν̣
φά[τναιc] ἐν λιθίναιc βάλ[∪ – ∪∪ –

9 ἵππο̣[ι]ἐ̣ν̣α̣ν φρέ[ν ∪– –
καί μ[ιν]. ζον. ταχέως
δ᾽ ἀράβη[cε] διὰ [λ]ευκῶν
12 ὀcτ̣έ[ων] δοῦποc ἐ[ρ]⟨ε⟩ικομένων.
ὁ δ᾽ ἄφ[αρ π]λεκτόν τε χαλκόν
ὑπερη[. .]. ε τραπεζαν
προβά⌞τω⌟ν ἁλυcιωτόν
3 δι᾽ ἑρκ[έ]ων, τεῖρε δὲ cτερ̣εῶ⟨c⟩
ἄλλαν [μ]ὲν cκέλοc, ἄλλαν δὲ πᾶχ[υν,
τὰν δὲ πρυμνὸν κεφαλᾶc
6 ὀδ[ὰ]ξ α[ὐ]χένα φέροιcαν.
.ρ. μ̣ι̣[ ] δ᾽ ὅμωc ε[ ]c᾽ ὕπα. [.|]. θυ. [
πικρο[τά]τ̣α̣ν̣ κ̣λάγεν ἀγγε[λία]ν̣
9 ζαμενε[ ]τυρανν[ ]
ποι]κί̣λ̣ω̣[ν ἐ]κ λεχέω̣[ν ἀπέ]δ̣{ε}ιλ[οc
]ν καθε. [ ]c ῥα̣̈. [
]ιον κακ[ ]
]. ο̣ν̣ ἔ̣[ ]
. . .

col. iii .νατ[ ]ν. [
B′ ἔμολε̣[.]αι παῖδα[∪–
Ἡρακ᾽λ[έ]οc εξα. [.]. [.]ν̣ [ ]
3 τεταγ᾽μένον τουτά. [. . . .]εκατ. [
Ἥραc ἐφετ᾽μαῖc· Cθενέλο[ι]ό μιν
υἱὸc κέ[λ]ευcε⟨ν⟩ μόνον
6 ἄνευ cυ[μμ]αχίαc ἴμεν.
καὶ Ἰόλαο[c ἐ]ν ἑπταπύλοιcι μένω̣[ν τε
Θήβαιc] Ἀμ⌟φιτρύωνί τε cᾶμα χέω[ν
9 –??∪]μιᾶι δ᾽ ἐπὶ θήκαι
]ν καλλικέραc
]άδιc, οὓc οἱ
12]ο̣υ cτρατὸc οὐκ ἀέκ[ων
. . .]. αθ̣[. . . . . .]όντ̣[.]κ[. .]. ᾶι
]φέ[. .]. [ ]ρμα. [ ] ‖
]. ω̣προ̣[ ]λ̣ιμ[ ]ν
3]. νεκα[ ]. πολ̣[
]υρεκα[ ]α̣μον
]. οc[ ]
6]υc᾽ε[ ]. ενον̣
]ελ[ ]. νδέμ̣[
]. [ ]. έκ[
9]. [
. . .

From the papyrus we gain, in whole or in part, almost fifty new verses. The guess that the quotation contained the beginning of the poem seems to be confirmed. We learn that from the brief mention of Heracles' violence against Geryones, Pindar went on to a more detailed account of his attack upon another monster, Diomedes, the Thracian king who fed his mares on human flesh.

As Lobel says (p. 141), the poem seems likeliest to be a dithyramb. Let us go through it, line by line.

1 f. To dignify some particular abstraction by conferring on it Zeus' title of 'ruler of gods and men' seems to have been a poetical commonplace. Pindar calls Time 'father of all things' (*Ol.* 2. 17) and again 'lord, supreme over all the blessed ones' (fr. 33); Heracleitus calls War 'the father of all things and the king of all things' (fr. 53 D.–K. = 29 Marcovich); in a famous speech from Euripides' *Andromeda* Love is addressed as 'tyrant of gods and men' (fr. 136 Nauck^2). Pindar, of course, likes to begin a poem with a striking pronouncement such as this; compare, for instance, the opening of the Sixth Nemean Ode with ἓν ἀνδρῶν, ἓν θεῶν γένοc. (See the end of this article.)

What is the object of the verb ἄγει? Schroeder seems to have thought it was τὸ βιαιότατον: he renders 'lenkt rechtfertigend das Gewalttätigste'. So do Dodds (270: 'conducts the uttermost violence') and Ostwald (117: 'brings on . . . what is most violent'). Pavese (ii. 57) thinks the verb is used absolutely, adducing parallels to show that it may be used absolutely of 'leading' by a divine agency. But in all the passages he quotes it is easy to supply an object; and so it is here. That object is surely to be supplied from θνατῶν τε καὶ ἀθανάτων. Nomos leads all mortals and immortals as he will. Compare *Nem.* 11. 42 καὶ θνατὸν οὕτωc ἔθνοc ἄγει μοῖρα. τὸ βιαιότατον is the object of δικαιῶν, not of ἄγει.

Plato's manuscripts have βιαιῶν τὸ δικαιότατον; so has Libanius, doubtless taking his text from a manuscript of Plato. That cannot be what Pindar wrote:|| βιαιόω occurs nowhere else. Dodds in his commentary on the *Gorgias* (p. 271) has sufficiently refuted the view of Wilamowitz that Plato misquoted Pindar; Theiler's attempt at a defence (69 f. = 192 f.) is not convincing.

δικαιῶν τὸ βιαιότατον just suits the context. δικαιοῦν is a factitive; its form suggests that it means 'makes just'. Turyn (see p. 351 of his edition) objects that in other places it means 'bring to justice' (see e.g. Fraenkel, *Aeschylus, Agamemnon*, ii. 203); but the verb is not common, and there is no reason why Pindar should not use it in its basic sense. It is natural to use βιαιότατον of the action of Heracles; in the *Iliad* he is commonly referred to as βίη Ἡρακληείη. So we arrive at the sense 'making just what is most violent'.

Here we must notice an interpretation put forward by Pavese in an article that provides more useful information to illustrate the text than any other so far published. He translates 'bringing violence to justice'; Law, he thinks, in the person of Heracles, brings to justice the evil deeds of Diomedes.

The obvious objection is that at l. 15 below Diomedes is said to have resisted Heracles 'not in insolence but in *ἀρετά*'. That interpretation is confirmed by the scholion in the papyrus, which is convincingly restored to mean 'not by way of insolence but for the sake of *ἀρετή*. For not to let go of one's own property shows courage, not insolence; and Heracles was acting unjustly in taking it away.'

This objection Pavese tries to circumvent by arguing that the subject of the sentence in l. 15 is not Diomedes, but Heracles. The sentence is indeed fragmentary; and Pavese might be right, were it not that the following words utterly destroy his case. 'For it is better', the sense in ll. 16 f. must be, 'to die in defence of possessions that are being seized than to be a coward'. The supplement *γ]ὰρ* can hardly be avoided, and once it is accepted Pavese's argument collapses.

With *ὑπερτάται χειρί*, 'with arm supreme', Pavese well compares Soph. *El.* 1090 f.:

ζώιης μοι καθύπερθεν
χερὶ πλούτωι τε τῶν ἐχθρῶν ὅcον
νῦν ὑπόχειρ ναίειc.

and Eur. *Bacch.* 899 f.:

τί τὸ coφόν, ἢ τί τὸ κάλλιον
παρὰ θεῶν γέραc ἐν βροτοῖc
ἢ χεῖρ' ὑπὲρ κορυφᾶc
τῶν ἐχθρῶν κρείccω κατέχειν; ||

4. *τεκμαίρομαι ἔργοιcιν Ἡρακλέοc.* For this manner of expression, compare *Nem.* 6. 8 *τεκμαίρει καὶ νῦν Ἀλκιμίδαc.* As often, *ἔργα* implies 'mighty deeds', those of a god or hero. After this sentence, it is easy to see that *ἐπεὶ* in l. 6 is causal.

7. The quotation in the Aristides scholia (iii. 408 Dindorf) has genitives plural; the papyrus has accusatives. Which is right? At l. 20 the responsion gives no help; but l. 47 favours the accusatives, for if we can assume a synizesis, *καἰόλαοc* can respond with *κυκλώπειον*. The accusatives involve us, it is true, in the difficulty of having to scan the last syllable of *πρόθυρον* as long; though unusual, this is not unparalleled. On either view, the long first syllable of *Εὐρυcθέοc* must respond with two shorts; that, again, would be unusual but not unique.

Both Tiryns and Mycenae are 'Cyclopean'. Tiryns is often specially associated with Heracles, but Mycenae is the seat of Eurystheus;

Κυκλώπων ἕδοϲ is Mycenae, according to Hesychius *κ* 4483 Latte. At *Ol.* 13.5 Corinth is called *Ἰϲθμίου πρόθυρον Ποτειδᾶνοϲ*, but here it seems, *πρόθυρον Εὐρυϲθέοϲ* means literally 'Eurystheus' portal'.

8. Boeckh's supplement *ἀναιτήταϲ* long held the field; it is now ruled out by the need to respond with ll. 21 and 48. Plato paraphrases *οὔτε πριάμενοϲ οὔτε δόντοϲ τοῦ Γηρυόνου*, Aristides *οὔτε αἰτήϲαϲ οὔτε πριάμενοϲ*, the Aristides scholia †*ἀναιρεῖται*† *καὶ ἀπριάταϲ*. There is hardly room for Theiler's *αἰτηταϲ*, which Guthrie accepts. Page and Mette have independently thought of *ἀνατεί* from *ἄτη*, 'without harm'. That hardly fits the paraphrases, and is in any case unusual. I prefer Page's other suggestion, *ἀναιτί*. He notes the confusion between *ἄναιτοϲ* and *ἄνατοϲ* at Soph. *OC* 786; *οὔτε αἰτήϲαϲ* might, as he says be a paraphrase of *ἀναιτί*, but certainly cannot be retained.[1]

9. Responsion with l. 22 indicates that this line began with two long syllables; Page supplies them with *κεῖνοϲ καὶ* (*οὗτοϲ καὶ* Snell); he has a main verb (*ἔκλεψε*) in l. 10 and a participle (*δαμάϲαϲ*) in l. 13. Snell starts l. 9 with *καὶ κλυτὰϲ*] assuming the second syllable to be anceps; supplies *τὸν γὰρ* at the beginning of l. 10; and supplies the main verb ('*ἀπάτηϲ*' *sim.*') in l. 13. I prefer Page's arrangement, which features a type of *καὶ* common in similar transitions; cf. e.g. *Od.* 21. 295 *οἶνοϲ καὶ Κένταυρον* ... Lobel's supplements of ll. 11 and 12 are accepted by both, and seem hardly avoidable. *χαλκοθώραξ* is the epithet of Abderus at the beginning of the Second Paean (fr. 52b), and of Enyalios at Soph. *Aj.* 179.

13 responds with l. 26; Page's *δαμάϲαϲ* is best, but if one wants a main verb, Gigante's *ἐδάμαϲϲ'* is better than Snell's *ἀπάτηϲ'*.

14–15 respond with ll. 1–2 and with 41–2. Of the first two letters printed ‖ by Snell, I can make out neither iota nor alpha with confidence. Page reads *Διὸϲ ὑπο*]*ϲ̣τ̣άντα μέγαν* | *παῖδ' οὐ κό*]*ρωι ἀλλ' ἀρετᾶι*. Snell offers *νόον ἀεί*]*ραντα*, Mette and Pavese *χόλον ἀεί*]*ραντα*: their supplements would allow the next line to start *οὐ μὰν* (Snell, Gigante) or *οὐ ϲὺν* (Pavese); though if *νόον ἀεί*]*ραντα* were correct, one might expect the next line to start *ἄντ' οὐ κό*]*ρωι, κτλ.* The general sense is clear; Diomedes resisted Heracles (not Heracles Diomedes; see above), in the words of the scholion as interpreted by Lobel—*οὐκ ἐπὶ ὕβρει ἀλλ' ἀρετῆϲ ἕνεκα. τὸ γὰρ τὰ ἑαυτοῦ μὴ προίεϲθαι ἀνδρείου ἐϲτὶ ... ἀλλ' οὐχ ὑβριϲτοῦ. Ἡρακλῆϲ δὲ ἠδίκει ἀφελόμενοϲ.*

The Bistonian Lake is the usual location of Diomedes, as in Eur. *Alc.* 485; according to Strabo 7 fr. 66 and Aelian, *Nat. Anim.* 15. 25, the ruins of his palace were shown in that locality. Biston was the son of Cicon, according to Philostephanus fr. 7 Müller (*FHG* iii. 30); the

[1] [Since the final syllable must be long, we should spell *ἀναιτεί*; cf. W. Schulze, *Kl. Schr.* (2nd edn. 1966), 656 (H.-S.).]

Cicones lived in Bistonia, according to the Orphic *Argonautica* (77). Diomedes was the son of Ares and Cyllene, according to Apollodorus 2. 5. 8; Enyalios often appears as a synonym for Ares.

16–17 are paraphrased by Aristides *οὐ γὰρ εἰκός φηcιν ἁρπαζομένων τῶν ὄντων καθῆcθαι παρ' ἑcτίαι καὶ κακὸν εἶναι*. Page's *κρέccον* is superior to other supplements; at 18, I prefer Gigante's *πρὸ κτη]μάτων* to Page's *πρὸ χρη]μάτων*, but either may be right.

18–19. Snell has *cτέγος δ'*] in 18 and *κρυφᾶν* in 19, Page *κρύβδαν* in 18, and (*c*)*τέγοc* in 19. Since the reader has been prepared for a narration, asyndeton is in order, and Page's supplements are superior. But I do not care for the compound *νυκτιβίαc* which Page posits in 19, continuing *ὁδὸν* | *ὕβριοc εὗ]ρε*. 'By night found the way of violence' makes good sense: remember *τὸ βιαιότατον* at l. 3, and compare *Ol.* 1. 110 *εὑρὼν ὁδὸν λόγων*. Gigante rightly insists that Heracles would hardly be said to have committed *hybris*. Before *νυκτὶ*, something like Theiler's *ἥρωc* might be right.

20. Lobel's *πεδά[ρ]ϲ̣ι̣ο̣ν̣* suits the sense better than Snell's *πεδάcα̣[ιc*. Its drawback is that it requires us to supply two syllables at the end of l. 47 that are not easy to guess, unless one agrees with Pavese in accepting the semiconsonantal treatment of iota here.

21. Diodorus 4. 15 says that the mangers were of bronze and the halters of iron; Heracles keeps the horses busy till he can unfasten them by the simple expedient of seizing a groom and hurling him into the manger. In l. 22 the accent of the papyrus indicates that *μαινομ]έναν*, or whatever stood here, is an accusative, and I do not see why some make it genitive. Page suggests *ὠμοτάταc* | *ἵππω[ν μαινομ]έναν φρέ[ναc ἆcαι*: he writes *μαινομ]έναν*, but seems to take the word as genitive. I should prefer ‖ something like *ὠμοτατᾶν* (or *λαβροτατᾶν*) | *ἵππω[ν μαινομ]έ̣ν̣α̣ν φρέ[να τέρπειν*: compare *μαινομέναιc φραcίν* at *Pyth.* 2. 26.

23. Page gives *κερά]ιζον*, Snell *ἐλάκ]ιζον* or *λάκ]ιζον*: the metre does not help to decide between the two latter possibilities.

24. Does the sound of bones splintering ring out 'through the white bones', or must we change the accent of the papyrus and accent *διαλεύκων* from *διάλευκοc*? To this adjective Pavese objects that it is found in late prose only, and means not 'white' but 'transparent'. It can mean 'mixed with white' or 'with white spots'; and it could mean 'very white' (see R. Strömberg, *Greek Prefix Studies* (Göteborg, 1946), 161). The latter meaning is expressly attested by Philo *περὶ τοῦ θεοπέμπτουc εἶναι τοὺc ὀνείρουc* 1. 35 (iii. 248 Cohn–Wendland), who says *διάλευκοι μὲν τοίνυν εἰcὶν οἱ τηλαυγέcτατοι καὶ ἀριδηλότατοι, τοῦ δια πολλάκιc ἐπὶ τοῦ μεγάλου τιθεμένου, ἀφ' οὗ διάδηλον καὶ διάcημον τὸ μεγάλωc δῆλον καὶ μεγάλωc ἐπίcημον ἔθοc ὀνομάζειν ἐcτί*. Philo's parallels seem to indicate that there is nothing specifically 'late' about the use of δια- involved in this compound. But I still suspect that Pavese is right in suggesting that 'the sound cracked through the rending bones' is a Pindaric way of saying that 'rending bones cracked'. To me it seems an extremely expressive way of saying it.

26 f. *π]λεκτὸν . . . χαλκόν* seems certain; the 'entwined bronze' is that of the horses' chain. The chain was made of links; elsewhere the form is always *ἁλυcιδωτόc*, and perhaps we should restore it here. Page's *ὑπερή[γν]υε* is far the best supplement, and he must be right in taking *τραπεζαν* to be genitive plural. In l. 29, *δι' ἐρκ[έ]ων* presumably means 'through the stalls'; the horses were tethered by a single long chain of bronze links running along the line of stalls, which Heracles with his mighty strength was able to break, so that he could control all of them.

29. Perhaps *δὲ* should be changed to *τε*. Pavese's notion that *cτερεῶι* may be a mistake for *cτελεῶι*, meaning Heracles' club, is impossible, for had Heracles been using his club, the verb would not be *τεῖρε*. Pavese's own translation—'*stung* them [my italics] with his club'—offers the best refutation of his conjecture. Either Page and Snell are right in reading *cτερεῶ⟨c⟩*, which makes good sense; or, as Page suggests, after Lobel, something has dropped out.

30–2. Each of the horses is busy devouring some part of the unfortunate groom. A cup fragment of the Campana collection by Oltos, adduced by Pavese (first in *Maia*, NS 16 (1964), 311 f.), well illustrates the scene although I do not think the club, visible on the vase, has been mentioned by Pindar. [See now D. C. Kurtz, *JHS* 95 (1975), 171–2.] ‖

Lobel points out that *πρυμνόc* is not found with a defining genitive, and therefore takes *πρυμνὸν* as a noun, with *αὐχένα* in apposition, comparing *πρυμνὸν . . . θέναρος* with *Il.* 5. 339 and *πρυμνοῖc ἀγορᾶc ἔπι* at *Pyth.* 5. 93. This may be right, but one cannot rule out the possibility that *πρυμνόc* may be construed with a defining genitive, as *ἔcχατοc* may.

28. By comparing fr. 2 with ll. 28–35, Lobel obtains the reading:

.ρ. μι|]. ὁμωcε[. .]c' υπα. [.]. θυ. [

Treu boldly conjectures that the person who cried out the bitter news to the fierce tyrant was Eremia; Theiler agrees with him. My courage fails me.

34. Pindar elsewhere has *ἔκλαγξα*, but since Bacchylides has *ἔκλαγον* three times, it is not surprising to find it here.

36. The *ποικίλα λέχη* reminds me that Dio Chrysostom (ii. 150 Dindorf) represents Diomedes as something of a luxury-lover: *ποικίλην εἶχεν ἐcθῆτα καὶ καθῆcτ' ἐπὶ θρόνου πίνων δι' ἡμέρας καὶ τρυφῶν*. Lobel quotes the two passages of Pindar in which Alcmene starts up *ἀπέδιλοc* at the approach of the snakes (*Nem.* 1. 50 and *Paean* 20. 14); he clearly thought also of Aesch. *PV* 135, where the Oceanids tell Prometheus how they have come shoeless in their winged car (or cars).

37–40. I can do nothing here; Pavese has thought of *κάθελε*, which would describe what Heracles did to Diomedes.

41 f. The sentence must have had a subject, and also a genitive going with *παῖδα*: perhaps it may have run over from the preceding epode. Something came upon the son of someone, Heracles being mentioned in the genitive. *παῖδ'* may well have been followed by a short and then a long syllable; that would be consonant with *παῖδ'* C[*ρεωϲ*, and Ares was father of Diomedes. In l. 42, Pavese's ingenious supplement *ἐξάριθ̣*[*μ*]*ο̣*[*ν*—'there came upon the son (of Ares) (a labour) of Heracles appointed outside the number'—depends upon a somewhat doubtful interpretation of the traces, and upon the possibility that the preceding epode contained a neuter word that could indicate a labour, such as *ἄεθλον*. Placing a high point after *τεταγμένον*, Pavese, following a suggestion of Lobel, goes on *τοῦτ' ἄρ̣* [*α δωδ*]*έκατον*. The difficulty here is that the high point after *ἐφετμαῖϲ*, whose existence Pavese denies, does seem to be in the papyrus. There seems to be a reference to the story told by Apollodorus that there were originally to have been ten labours, but that twelve had to be performed because Eurystheus refused to count two in which Heracles had received assistance.‖

47. Iolaus himself was buried in the tomb of Amphitryon (*Pyth.* 9. 81 f.; cf. Σ *Ol.* 9. 98 and Σ *Nem.* 4. 20).

49. *θῆκαι* is presumably the tomb; Pavese suggests that *καλλικέραϲ* in l. 15 is an epithet of beasts sacrificed to the hero.

52. Pavese cites *Nem.* 4. 21: *Καδμεῖοί νιν οὐκ ἀέκοντεϲ ἄνθεϲι μείγνυον*: Gigante conjectures *Κάδμ*]*ου ϲτρατὸϲ*.

Metre. Pavese calls the metre 'aeolic'; Theiler calls it 'free dactylo-epitrites'. Neither name helps much. Here is an analysis in terms of the notation devised for Pindar's poems not in dactylo-epitrite by A. M. Dale, *Collected Papers*, 41-97 (see especially pp. 63 f.).

1	∪ ∪ ∪ – | – ∪ ∪ –	ʳs | d	
2	– – ∪ ∪ – ∪ ∪ –	– d d	
3	∪ – ∪ – | – ∪ ∪ – ∪ ∪ –	∪ s | d d	
4	∪ – ∪ – | – ∪ ∪ – ∪ –	x s | d s	cf. *Ol.* 10, str. 2;
5	– – ∪ – | – ∪ ∪	– s | s	*Pyth.* 8 str. 4.
6	∪ – | – ∪ ∪ – ∪ ∪	∧ s | d s	cf. *Ol.* 1, str. 1.
7	– ∪̲ ∪̲ | – ∪ ∪ – ∪ ∪ ∪̲ – – –	∧ s̄ | d d – | s̄	
8	∪̲ – | – ∪ ∪ – ∪ ∪ – ∪ ∪ –	∧ s | d d | d	

9 – – | (?) ∪ ∪ – ∪ ∪ – – ∧ s̄ | d d –
10 – | (?) ∪ ∪ – – ∪ ∪ – – (d ?) | d
11 ∪ ∪ – | ∪ ∪ ∪ – – ∧ d s –
12 – ∪ – | – ∪ ∪ – ∪ ∪ – s | d d
13 ∪ ∪ – – – ∪ – – – d – s –

Miss L. P. E. Parker (Mrs Edwards), to whom I am much indebted for a discussion of the metre, writes:

Two features in particular strike me. 1. Choriambic phrases preceded by true long (not anceps): e.g. 1, 3, 4, 12 and (probably) 10–11. I can find no other poem in which this happens so frequently, but it does happen, e.g. at *Ol.* 10 str. v. 2 and at *Pyth.* 8 str. v. 4 (both ∪̲ – ∪ – – ∪ ∪ – ∪ –) and at *Ol.* 8, ep. v. 7 (– ∪ – – ∪ ∪ – ∪ ∪ –).

2. – ∪ ∪ – ∪ ∪ – ∪ ∪ – preceded by a sort of aeolic base: e.g. 8 and cf. 7 (the departure from the pattern both involve proper names). ∪ – – ∪ ∪ – ∪ ∪ – ∪ ∪ – seems comprehensible in context, with ∪ – – ∪ ∪ – ∪ – (cf. *Ol.* 1. 1) two cola earlier. Something very like – – – ∪ ∪ – ∪ ∪ – ∪ ∪ – – seems to be a motif in *Ol.* 5 (note str. v. 1, str. v. 2, ep. v. 1, and ep. v. 2 (2 + 3 Bowra).

3. Finally, with the end of 10–11, compare *Nem.* 6, str. v. 6 (καίπερ ἐφαμερίαν οὐκ εἰδότες οὐδὲ μετὰ νύκτας).‖

Let us now return to the central question; what was the *nomos* held to justify the conduct of Heracles? Many different theories have been put forward, but in essence they fall into two groups.

The first group takes *nomos* in this place to mean 'custom' or 'usage'. Before the publication of the papyrus, this kind of view was taken by Wilamowitz (*Platon*, ii ([1]1919 = [3]1962), 95; *Pindaros* (1922), 462), Ehrenberg (*Die Rechtsidee im frühen Griechentum* (1921), 119), Latte (*Kl. Schr.* 216), and Pohlenz (*Philol.* 97 (1948), 139 = *Kl. Schr.* ii. 337). The action, it is argued, was in itself unjust, but we are used to considering Heracles as good and Diomedes as bad. Since the publication of the papyrus, Theiler and Bowra have defended the view of Wilamowitz; Ostwald by a complicated argument has maintained that *nomos* denotes 'the attitude traditionally or conventionally taken to a norm' (p. 134), which seems to me much the same thing.

The other kind of view is that *nomos* here means the law of the universe, and particularly the law of Zeus. Schroeder, *Philologus*, 74 (1917), 202, argues that *nomos* here came close to the meaning of *moira* or *ananke*, meaning in effect 'a male goddess of fate'. A similar position was defended at greater length by H. E. Stier, *Philol.* 83 (1928), 227 f.; he thought *nomos* meant 'an inviolable order, which dominates the belief and will of gods and men'. In recent times this kind of view has been taken by Lesky in 1950 (*Österr. Zeitsch. f. öffentliches Recht*, 2 (1949),

588 f. = *Ges. Schr.* 493), by Gigante in 1956 and again in 1966, by Dodds in 1959, by Treu in 1963, and by Guthrie in 1969.

The holders of the first type of view, it seems to me, have made the mistake of not relating this passage to the attitudes found in Pindar's work in general. Pindar, like other believers in his religion, honoured Heracles as a great benefactor of mankind; in suppressing monsters like Geryones or Diomedes, he was not merely doing the bidding of Eurystheus or of Hera, but carrying out the will of Zeus. Gigante has abundantly illustrated this from Pindar's works; in particular, he stresses the prophecy of Teiresias in the First Nemean Ode, which tells how many monsters, ignorant of justice, Heracles by sea and land shall slay. Like the Cyclopes and Laestrygones in Homer, Geryones and Diomedes live outside the *themistes*; they are outlaws, monsters, whom any man valiant enough to challenge them can earn glory by killing.

What was abnormal, in the eyes of a believer in the Olympic religion such as Pindar, about these particular actions of Heracles? It was surely that in these two cases Heracles was the aggressor; neither Geryones nor Diomedes had provoked him. It was a commonplace of Greek ‖ thought that the aggressor, he who 'began unjust violence' (ἄρχειν ἀδίκων χειρῶν), was to blame. Yet in these two instances Heracles undoubtedly delivered an unprovoked attack. Yet did not Heracles act justly? Surely he did, Pindar answered, for in attacking these common enemies of gods and men, Heracles was carrying out the will of Zeus, and helping to enforce the order of the universe. Fragment 81—'I praise you in comparison with him, Geryones, but I am silent altogether about what Zeus does not approve'—seems to provide an exact parallel.

Why, it may be asked, did Pindar not say he was justified by Themis or Dike; why Nomos? Before the fifth century, Themis or Dike would doubtless have been preferred in such a context. But at this time the praises of Zeus' Law were often sung. Already in Hesiod, Eunomia with Dike and Eirene is one of the Horai, daughters of Zeus and Themis; as such she is prominent in the First Pythian Ode, where Pindar is praising the founder of the just city of Aitna. As the polis developed, law and law-abidingness acquired a special sanctity. But human law was only an extension of the divine law; Pindar, like Sophocles, would have agreed with Heraclitus that 'all human laws are nurtured by one divine law'. Some of Pindar's contemporaries, it is true, contrasted Nomos unfavourably with Physis, but Pindar took little notice of such people. In using Nomos with reference to the will of Zeus he was not acting under 'Orphic' influence, whatever that might be supposed to mean. Law for him was identical with the will of Zeus.

I add a note about an argument which Guthrie (*A History of Greek*

Philosophy, iii. 133) seems to think decisive. 'All these interpretations', he writes, speaking of the interpretations which, like my own, take *nomos* in the poem to mean the law of the universe, the law of Zeus, 'seem to ignore what Pindar plainly says: not that *nomos* is the will of Zeus, but that even Zeus is subject to *nomos*, which lords it over gods as well as men.' Pindar does not 'plainly say' that 'even Zeus is subject to *nomos*'; he gives *nomos* the title of 'king of gods and men'. That title is commonly given to Zeus, so that a hearer would very likely infer that Zeus' *nomos* was in question. Zeus is, in a sense, subject to his own *nomos*; in the *Iliad*, for example, he thinks of saving from death his son Sarpedon, but in the end declines to sacrifice to a sudden whim his own settled policy.[2] I cannot, therefore, share Guthrie's confidence in this argument.

[2] [Cf. Σ. *Hec.* 798–801, a passage often misinterpreted, but which is perfectly illustrated by Pindar's words.]

11(35)

Heracles at Eleusis

P. Oxy. 2622 and PSI 1391 [= *Pindar fr. 346 S.–M.*]

P. Oxy. 2622 (Part XXXII (1967), pp. 63 f., with plate v) contains a fragment of lyric poetry which in the opinion of its first editor, E. Lobel, 'may include part of Pindar's version of the meeting in Hades of Heracles and Meleager, Bacchylides' treatment of which has been recovered in an epinician for Hiero (v)'. Wilamowitz (*Pindaros*, 314 n. 1) suggested that Pindar described this meeting in the dithyramb whose opening is preserved in P. Oxy. 1604 (fr. 70b Snell = 86 Turyn = 61 Bowra). By a strange coincidence, PSI 1391 (Part XIV, pp. 62 f., with plate v: Pack[2], No. 1949) contains a commentary on the very passage part of which is in the Oxford manuscript. In reviewing Part XIV of PSI (*Gnomon*, 31 (1959), 111 f. [= *Greek Comedy, Hellenistic Literature, Greek Religion, and Miscellanea*, ch. 29, pp. 357 f.] I discussed the text of No. 1391, receiving invaluable aid from my friend and colleague Professor J. W. B. Barns; now that we have the Oxford fragment also, it is possible to advance further. The problems presented by the two texts cannot be considered in isolation from each other, and in studying the text and discussion of the Oxford fragment (O) which I shall print below, the reader will often need to refer to the text of the Florence fragment (F), different in one or two places from that which I published in *Gnomon*, loc. cit., which will be found later in my article. To begin with I shall reproduce Lobel's text of O; after discussing it I shall offer *exempli gratia* an attempt at restoration, some parts of which will be more speculative than others.

The hand is 'an unskilled upright uncial of medium size'; Lobel assigns it to the first century after Christ.||

Maia, NS 19 (1967), 206–29.

P. OXY. 2622

fr. 1

(*a*)]ρωικτ̣εαν[
]αμοσυνας̣[
]ϊαλατερπ̣ε̣ιφ̣ιλ. . [
]φερσεφοναιματ. . [
]ντελευ̣τ̣ανϊνεσεν[
]διδυμαις̣ε ʽδονευμο[
]. ραι
]π̣ορενηρακλειπρωτω[
]ντικελευθονεπισπησει. [
]. νιαδασαλοχος [
]αλλεγεμαν [
]τικαμινφθιμενων [
]τρεφεταικαι°σοʿνποντωι̣[
]μενος[ ] [
]α̣μ[. . .]. διοσυιον [
. . .

(*b*)]. . . [
]μιναντι̣α̣. [
]λεαγρονατερθ. [
]να(λε̣υ [

fr. 2

εν̣[
θα[
. .[
το̣[
φρ[
ει̣[

fr. 3

] [
]. ει. . ι[
] [
]υμνο̣[
]ι̣σερε[
] [

(*a*) (κρέccoν' ἐν) και]ρῶι κτ̣εάν[ων
(σοφὸν ἁγητῆρα)]αμοσύναϲ̣[
εὐνομ]ία⟨ι⟩ λατερπέι φ. λ̣ο̣. [
Ἐλευσινόθε] Φερσεφόναι ματρ̣ί̣ [τε χρυσοθρόνωι
θῆκεν ἀστοῖσι]ν τέλος ινεσεν[
]δ̣ιδύμαις εἶδον Εὐμο[λπ
]. ραι
]π̣ορεν Ἡρακλέι πρώτω[ι (ξένων)
]ντι κέλευθον επισπησει̣. [
Ἀμφιτρυ]ω̣νιάδας ἄλοχος
]αλλε γε μὰν
αὐ]τίκα νιν φθιμένων [
]τρέφεται καὶ ὅς' ἐν πόντωι [
]μενος[ ]
]α̣ μ[έγα]ν Διὸς υἱὸν [
. . .

(*b*)]. . :[
]μιν ἀντ̣ι̣α̣. [
Με]λέαγρον ἀτερθ. [
]νασλευ

(*a*) 1 (κρέσσον') vel (κρέσσονά τ') ex F Ll.-J. (ἐν) και]ρῶι Lobel 2 (σοφὸν ἁγητῆρα) ex F Ll.-J. 3 εὐνομ]ία⟨ι⟩ Ll.-J. φ̣ι̣λ̣ο̣. [(Lobel) vel φ̣υ̣λ̣ο̣. (Ll.-J.) 4 'Ἐλευσινόθε ex F Ll.-J. finem ex F Lobel 5 init. ex F Ll.-J. τέλος F (l. 25): τελευ̣τ̣αν O (τελετὰν Lobel) 6 Εὐμο[λπ Ll.-J. 8 (ξένων) Ll.-J. 10 suppl. Lobel 12 ad fin. (ψυχαὶ) Lobel 15 suppl. Lobel
(*b*) 3 suppl. Lobel

PSI 1391

Fr. B, col. i

. . κ]αλλιϲ̣[τ	
. .].ψεωϲ[	
. . . .]υντο[	
. . .]ν̣αιαιε̣[	
κ]ρέϲϲονα[.]ρ[	χ]ρ[ημάτων Bartoletti
ϲο]φὸν ἀγη[τ]ῆρα. [	λ̣[Bartoletti λ̣[έγει τὸν] κα- Merkelbach
.]α καιρὸν. [.]ϲ̣ϲ. [	τ]ὰ καιρὸν . . . . . γευό- Merkelbach
μενον τῶν κτ[	μενον τῶν κτ[ημάτων Merkelbach
ἐν καιρῶι δὲ καὶ[	[δόντα Merkelbach
κρείϲϲονα δὲ κατα[	κατὰ [φύϲιν Bartoletti
ὄντα καὶ ὑπεράν[ω τοῦ δι-	suppl. Bartoletti
αφόρου ὄντα πιθα[νῶϲ δὲ τὴν	suppl. Ll.-J. (1967)
εὐνο⟨μ⟩ίαν κατὰ ϲύν[θεϲιν	fin. suppl. Bartoletti εὐνο⟨μ⟩ίαν Ll-J. (1959)
εἴρηκεν λατερπ[έα διὰ	fin. suppl. Bartoletti
τὸ τοὺϲ λαοὺϲ τέρπ[ειν δηλον-	fin. suppl. Bartoletti
ότι. οἵ [δ' ἡν]ίκα ἂ⟨ν⟩	οἵ[δ' ἡν]ίκα ἂ⟨ν⟩ εὐνομ[ία Ll.-J. (1967)
εὐνομ̣[ία κα-	ὁπ[ην]ίκα εὐνοα̣[Bartoletti, vix recte
θεϲτήκῃ̣ κρείϲϲονα [καὶ	suppl. Bartoletti
ὑπερ]άνω τῶν χρημ[άτων	suppl. Bartoletti
ὄντ]α καὶ τοῦτ' ἐν καιρ[ῶι	suppl. Bartoletti
πρά]ττοντα, οὐκ εἰκῆι	suppl. Bartoletti
. .].ν. . ἐϲτι προϲεκτ̣ικόν.	legit Barns (1959)
. .]'Ελευϲινόθε Φερϲεφόνα̣	legit Barns: Ἐλευϲ{ε}ῖνοϲ Bartoletti
ματρί τε χρυϲοθρόνωι θῆ-	
κεν ἀϲτοῖϲι]ν τέλοϲ. φηϲὶν	suppl. Ll.-J. (1959)
. . . .]. . . ἐν̣ κρατίϲτω̣	
. . . . . . .τ]ῆϲ ἐν τῆι Ἀτ-	ex. gr. χωρίωι?
τι]κῆι Ἐλευϲῖνοϲ τοῖϲ αὑ-	suppl. Bartoletti
το]ῦ ἀϲτοῖϲ τελετὴν κατέ-	suppl. Bartoletti
ϲτη]ϲ̣ε̣ τ[ῆι] τε Φερϲεφό-	suppl. Bartoletti
νηι καὶ τ]ῆι Δήμητρι, του-	suppl. Bartoletti
τέϲτ]ι̣ κατέϲτηϲεν αὐτο̣ῖ̣ϲ̣	suppl. Ll.-J. (1959): -των̣ δ]ὲ̣ Bartoletti
]τ̣α̣ϲ̣ τῶν θεῶν	ἑορτὰϲ μεγίϲ]τ̣α̣ϲ̣ Bartoletti
]. που του	
].εν	
]ππο	

(traces of two lines more) ‖

Lobel supplies (ἐν) και]ρῶι: cf. F 6–7, 9, 19. (ἐν) και]ρῶι must have been preceded at no great distance by κρέϲϲον', or κρέϲϲονά τ':[1] cf. F 5, 10, 17.

Lobel suggests τλ]αμοϲύναϲ or μν]αμοϲύναϲ: see below. ϲοφὸν ἀγητῆρα comes from F 6: its placing is highly conjectural.

At F 13 εὔνοιαν can hardly be correct. εὔνοια is not found in lyric verse, whereas no word is likelier to have been given by Pindar the epithet λατερπήϲ than εὐνομία.[2] In the review quoted above [= op. cit., p. 358] I suggested reading εὐνομίαν at F 13; at F 16 the word read as ευνοα̣[is as likely to be ευνομ̣[. Lobel says that the omission of iota adscript is unexpected in such an early text, but this consideration hardly rules out my supplement. [W. J. Slater, *Lexicon to Pindar* (1969), takes no account of this supplement.]

Lobel reads φι̣λο̣.[but in dactylo-epitrite three successive short syllables would be surprising, and since only the bottom part of the supposed iota can be seen φυ̣λο̣.[seems possible. Something containing φῦλον might be correct.[3]

F 22 suggests that Ἐλευϲινόθε stood before Φερϲεφόναι. The form Ἐλευϲινόθε will probably mean not 'from Eleusis' but 'at Eleusis' in this place; see Kühner–Blass i. 294.

Barns pointed out (*Gnomon*, loc. cit. = op. cit., p. 358) that ll. 21–4 of F did not begin two letters' space further left than the rest, as the first editor, the lamented Vittorio Bartoletti, had supposed. The observation that the fragmentary col. ii seems to contain a similar ἔκθεϲιϲ does not invalidate Barns's observation, as Snell on Pindar fr. 346e, l. 2 (ii. 159) might be taken to imply, and Barns is now shown to have been right, for at l. 4 O reads not Δάματρι, supplied in this place by Bartoletti, but ματρί. After it Lobel has restored τε χρυϲοθρόνωι from F.[4] ||

Lobel accepts θῆκέν τε λαοῖϲι]ν, which Bartoletti had supplied at F 23–4. The observation of Barns mentioned in the previous note shows

[1] It is not surprising to find the expression κρέϲϲονα ... κτεάνων being explained in the commentary: apart from χρημάτων κρείϲϲων at Thuc. 2. 60, Dem. 58. 29, and Isocr. 1. 19, I have found no parallels. [But note Pindar, *Pyth.* 8. 92 κρέϲϲονα πλούτου μέριμναν and *Nem.* 9. 32 κτεάνων ψυχὰϲ ἔχοντεϲ κρέϲϲοναϲ ἄνδρεϲ.]

[2] See *Ol.* 9. 14 f. κλεινᾶϲ ... Ὀπόεντοϲ ... ἂν Θέμιϲ θυγατήρ τέ οἱ ϲώτειρα λέλογχεν μεγαλόδοξοϲ Εὐνομία; 13. 4 f. τὰν ὀλβίαν Κόρινθον ... ἐν τᾶι γὰρ Εὐνομία ναίει καϲιγνήτα τε, βάθρον πολίων ἀϲφαλέϲ, Δίκα καὶ ὁμότροφοϲ Εἰρήνα, τάμι' ἀνδράϲι πλούτου, χρύϲεαι παῖδεϲ εὐβούλου Θέμιτοϲ; *Pyth.* 5. 63 Apollo cures sickness and grants skill in music to whom he will, ἀπόλεμον ἀγαγὼν ἐϲ πραπίδαϲ Εὐνομίαν; *Paean* 1. 9–10 (of Apollo) τὰ]ν δὲ λαῶν γενεὰν / δαρὸν ἐρέπτοι / ϲώ]φρονοϲ ἄνθεϲιν Εὐνομίαϲ. See Andrewes, *CQ* 32 (1938), 89 f. Cf. Timotheus, *Persae*, *ad fin.*

[3] φῦλον occurs thrice in Pindar (*Pyth.* 3. 21; *Dith.* 2. 21; 3. 18).

[4] χρυϲόθρονοϲ serves in the epic as an epithet of Hera, Artemis, and || Eos. It occurs in *Nem.* 1. 37 (of Hera), *Pyth.* 4. 260 (of Cyrene), and Bacchyl. 14 B 1 (of Hestia). Aristophanes has it in a parody of choral lyric at *Av.* 950; and cf. Lyr. Adesp. *PMG* 953. 1 and 1023. 18 Page and *Hymn. Is.* (1028 Kaibel) 7.

that this supplement is too long for the space; θῆκεν ἀcτοῖcι]ν suggested by me in *Gnomon*, loc. cit., is preferable (note ἀcτοῖc at F 28).

The following word appears in O as τελευ̣τ̣αν: Lobel corrects to τελετάν. This must be what the scribe intended; but is it right? In the lemma at F 24 we find τέλοc, though τελετὴν is found in the explanation at F 28. τελετά might perfectly well occur in verse (see e.g. Pindar fr. 131a Snell). But the use of τέλοc as equivalent to τελετή is poetic: τελετή is actually used to explain τέλοc in F; and τελετὰν [seems] likelier to have displaced τέλοc than τέλοc to have displaced τελετάν.[5] Against this the word would appear to be followed in O by two short syllables; as Lobel says, the likeliest articulation of ινεcεν[seems to be ἵν' ἐc ἐν[, so that τελετάν should be preferred].

Why does the mention of a wise leader who ruled εὐνομίαι λατερπέι and founded the Eleusinian Mysteries come not long before an account of how Heracles in Hades saw the ghost of Meleager? The clue would appear to lie in the story of how Heracles before his descent into Hades to recover Cerberus was initiated at Eleusis. From about the end of the fifth century, this would seem to have been well known. We find it on an Attic pelike of the fourth century from the Crimea and on the red-figure crater in the British Museum known as the Pourtalès Vase;[6] and later, during the imperial period, we find it on the Lovatelli Urn and on the Terra-Nova ‖ sarcophagus.[7] It is mentioned in the pseudo-Platonic *Axiochus*,[8] a work not earlier than the fourth century, in Plutarch's life of Theseus and in the fourth book of Diodorus, who says that the Lesser Mysteries at Agrae were founded by Demeter in order to allow Heracles to be initiated. More significantly, it figures in the account of Heracles' descent into Hades given by Apollodorus in the

[5] Cf. Soph. *OC* 1050 (of Eleusis) οὗ πότνιαι cεμνὰ τιθηνοῦνται τέλη; fr. 837 Pearson: Aesch. fr. 387 Radt; Eur. *Hipp.* 25; Plato, *Rep.* 560 E (who is using poetical language). On τελετή, see C. Zijderveld, *Τελετή* (Diss. Utrecht, 1934); cf. p. 87.

[6] See G. Mylonas, *Eleusis* (Princeton, 1961), 210–13 (with illustrations); Nilsson, *Geschichte der gr. Religion*, i. 318 with Taf. 46; cf. Wilamowitz, *Der Glaube der Hellenen*, ii. 55 n. 1. On the pelike see E. Simon, *Antike Kunst*, 9 (1966), 72–86.

[7] See Mylonas, op. cit. 205 f. Cf. R. Eisler, *Orphisch-Dionysische Mysteriengedanken* (1925; repr. 1966), 159 with n. 4.

[8] 371 E καὶ τοὺc περὶ Ἡρακλέα τε καὶ Διόνυcον κατιόντας εἰc Ἅιδου πρότερον λόγος ἐνθάδε μυηθῆναι, καὶ τὸ θάρcoc τῆc ἐκεῖcε πορείαc παρὰ τῆc Ἐλευcινίαc ἐναύcαcθαι; Plut. *Thes.* 30; Diodorus 4. 25 *ad fin.* Diodorus 4. 14. 3 says that Demeter founded the Lesser Mysteries so that Heracles could be purified from the blood of the Centaurs; ΣAr. *Plut.* 845 that they were founded so that Heracles, although a foreigner, could be initiated. The same scholion describes the Lesser Mysteries as a προκάθαρcιc καὶ προάγνευcιc τῶν μεγάλων: ἐνθάδε in the *Axiochus* passage seems to refer to them. Apollodorus too says that Eumolpus purified Heracles from the blood of the Centaurs before initiating him. The story of Heracles' purification looks like an αἴτιον designed to explain the nature of the Lesser Mysteries. See Deubner, *Attische Feste*, p. 70. Stephanie West draws my attention to PRIM 20, col. i, 18 f. (1996 Pack^2), a speech supposed to be spoken by Heracles when refused initiation.

second book of his *Bibliotheca* (5. 12), where the capture of Cerberus figures as the last labour of the Dodecathlos. The *Bibliotheca* as we have it is scarcely earlier than the second century after Christ. But Norden in his great commentary on the sixth book of the *Aeneid* gives strong reasons for thinking that in that chapter Apollodorus is following an early epic account of the descent of Heracles.[9] The same account, in his opinion, lay behind the description of the descent of Dionysus in the *Frogs* of Aristophanes, the encounter of Heracles with Meleager in the fifth dithyramb of Bacchylides, and also several features of the descent of Aeneas in the sixth book of the *Aeneid* itself.

According to this Apollodoran narrative, the person who initiated Heracles at Eleusis was not Musaeus but Eumolpus. When therefore at the end of l. 6 we find the letters ευμο[, it seems natural to suggest that subject of this passage was the initiation of Heracles, and that the wise ruler praised by the poet was none other than Eumolpus.[10]

Originally there was doubtless only one Eumolpus (*HH*‖*Cer.* 474); but from an early date the Eumolpus who came from Thrace and fought for Eleusis against the Athenians under their king Erechtheus was distinguished from another person, sometimes from two other persons, of the same name.[11] The earliest author whom we find drawing the distinction is Andron of Halicarnassus, writing about 400 BC, who gives the following family tree:

Eumolpus I
Ceryx
Eumolpus II
Antiphemus
Musaeus
Eumolpus III

He makes Eumolpus III the founder of the mysteries and hierophant. This fits the received heroic chronology admirably, for in the time of Heracles the king of Athens was Theseus, who according to the list of kings current in the time of Andron was fourth in descent from Erechtheus. Jacoby in commenting on the fragment of Andron sug-

[9] See Norden, *Aeneis VI*, p. 5, and the passages of his commentary referred to in the second footnote on that page.

[10] Eumolpus founded the mysteries, according to Ister, *FGrH* 334 F 22; cf. Plut. *De exilio* 17 and Lucian, *Demonax* 34.

[11] For particulars of the Thracian Eumolpus who fought against Athens, see Colin Austin's commentary on the new fragments of Euripides' *Erechtheus*, *Recherches de Papyrologie de la Sorbonne*, 4 (1967), 11–67. Ister, *FGrH* 334 F 18 distinguishes two bearers of the name: Acestodorus (*FHG* ii. 464) made the second of them fifth in descent from the first; see Jacoby's commentary on the former passage. Andron (*FGrH* 10 F 13) is quoted in the same learned note in the scholia on Soph. *OC* 1053 (p. 48 De Marco); he is probably a 4th-c. writer (see Jacoby, and Schwartz in *RE* i. 2159, 11). On Eumolpus on Eleusinian vases see Simon, loc. cit. (n. 6), 89.

gested that his Eumolpid genealogy was invented in order to reconcile the story of Eumolpus' initiation of Heracles with the established legend of the war between Eumolpus and Erechtheus; he thinks Andron may have taken it over from an Atthis.

At this point we may recall the brief account of his descent to Hades given by Heracles to Amphitryon in the *Heracles* of Euripides.[12] He was fortunate, he says, in seeing the *ὄργια* of ‖ the initiates. Wilamowitz in his commentary took this to allude to the story of the initiation; the new text seems to prove that he was right. The plot of Euripides' play connects Heracles closely with Attica, for it makes him rescue Theseus from the underworld and makes Theseus in gratitude transfer to him the countless Theseia in Attica, which now became Heracleia.[13] We shall enquire presently how early this Attic connection of Heracles came into being.

Let us return to the text of l. 6. In connection with what two female persons or things may what persons have *seen* Eumolpus performing what action? I had rather not attempt to guess.[13a]

According to Apollodorus, loc. cit., Heracles was the first foreigner initiated at Eleusis; in order to qualify he had to be adopted by one Pylius.[14] So what Eumolpus gave to Heracles 'first' was probably the privilege of being initiated in spite of being a foreigner, and *πρώτω[ι* was probably followed, perhaps at some distance, by *ξένων*.

]ντι is probably the end of the dative singular of a participle referring to Heracles; the late epic word *ἀντικέλευθος* is not likely to have been used here. I can make nothing of *ἐπιcπηcε̣ι̣[* (see Lobel); is it part of an optative standing in a dependent clause?

Lobel suggests *Ἀμφιτρυ]ω̣νιάδαc*. The iota following the ny is oddly made; 'the dots are formed as dashes and placed beside, not above, the tip of iota, to which they impart the appearance of tau' (Lobel, p. 63). The only other iota in the piece formed in this way is the first letter visible in l. 3, where for reasons given above it seems reasonably likely that iota, and not tau, is intended. Here too iota is probably right, and

[12]
610 *Αμφ.* *ἦλθεc γὰρ ὄντωc δώματ' εἰc Ἅιδου, τέκνον;*
Ηρ. *καὶ θῆρά γ' ἐc φῶc τὸν τρίκρανον ἤγαγον.*
Αμφ. *μάχηι κρατήcαc ἢ θεᾶc δωρήμαcιν;*
613 *Ηρ.* *μάχηι· τὰ μυcτῶν δ' ὄργι' εὐτύχηc' ἰδών.*

Wilamowitz ad loc. recognizes the allusion to the story told by Diodorus 4. 26 that Persephone intervened in Heracles' favour, and notes that this must have been connected with the story of his initiation at Eleusis. [See G. W. Bond, *Euripides*, *Heracles* (1981), ad loc., quoting J. Boardman, *JHS* 95 (1975), 1 f.]

[13] See Philochorus, *FGrH* 328 F 18 with Jacoby's commentary.

[13a] e.g., the *daughters* of the Eleusinians may have seen Eumolpus within the twin *doors* of the temple.

[14] The same story is told by Plut. *Thes.* 30: Heracles and the Dioscuri were the first foreigners initiated (Xen. *Hell.* 6. 3. 6).

if it is Lobel's supplement is not easily avoided. *ἄλοχοc* presumably belonged to a different sentence or a different clause.

Lobel supplies *αὐ]τίκα*, and thinks ⟨*ψυχαί*⟩ followed *φθιμένων*.‖ The scene, he remarks, is that described by Bacchylides at 5. 62 f.,[15] where Heracles in Hades sees the countless spirits of the dead: we shall see presently that we already had reason to suppose that this comparison derived from the common epic descent of Heracles which Norden postulated. Bacchylides compares the souls to leaves; he and his source doubtless had in mind the comparison of the generations of leaves to those of men made by Glaucus in the sixth book of the *Iliad*. Virgil compares the souls waiting on the banks of Acheron for Charon's ferry first to leaves and then, with marvellous aptness, to birds assembling on the seashore before migrating.[16] Norden pointed out that Bacchylides and Virgil must have had a common model; Pindar, all probability, had the same model in mind. *τρέφεται* suggests that he, like Bacchylides and Virgil, first compared the ghosts with leaves.[17] If it was not leaves, it was some other object formed on earth, for *πόντωι* suggests that 'earth' stood in the preceding clause. What was the object formed at sea that formed the second object of comparison? Lobel thinks it was the waves, which are a type of infinite numerosity in the verses put into

[15]
ἔνθα δυcτάνων βροτῶν
ψυχὰc ἐδάη παρὰ Κωκυτοῦ ῥεέθροιc
οἷά τε φύλλ' ἄνεμοc
Ἴδαc ἀνὰ μηλοβότουc
πρῶναc ἀργηcτὰc δονεῖ. ‖

Cf. *Il.* 6. 146 f. (also 2. 468); there are other echoes in Mimnermus fr. 2. 1 and A.R. 4. 216 f. *ψυχαὶ* is hardly certain; the poet may have said, for instance, *φθιμένων ἀγέλαι.*

[16]
quam multa in silvis autumni frigore primo
lapsa cadunt folia, aut ad terram gurgite ab alto
quam multae glomerantur aves, ubi frigidus annus
trans pontum fugat et terris immittit apricis.
stabant orantes primi transmittere cursum,
tendebantque manus ripae ulterioris amore. (6. 309 f.)

Norden's view [cf. id., *Kl. Schr.* 508 n. 77] that the comparison of the souls to birds must have been suggested to Sophocles by the Heracles *Κατάβαcιc* seems to me improbable. Sophocles, describing the plague at Thebes, writes

ἄλλον δ' ἂν ἄλλωι προcίδοιc ἅπερ εὔπτερον ὄρνιν
κρεῖccον ἀμαιμακέτου πυρὸc ὄρμενον
ἀκτὰν πρὸc ἑcπέρου θεοῦ. (*OT* 175 f.);

the comparison of a spirit to a bird is natural enough. He is on safer ground when he suggests that the comparison of the advancing Trojans to migrating cranes at *Il.* 3. 2 f. may have been in Virgil's mind; I doubt, though, whether Von der Mühll is right when he makes the same suggestion regarding the comparison of the clamour of ghosts to that of birds at *Od.* 11. 605 (*Philol.* 93 (1938), 9 [= *Ausgewählte kleine Schriften*, 23–4]).

[17] For *τρέφειν* of the earth in a similar connection, cf. *Il.* 11. 740–1: Agamede knew as many herbs as the broad earth *τρέφει*: cf. Aesch. *Agam.* 1407 f. (*χθονοτρεφὲc ἐδανόν* in a similar disjunction). Cf. *Il.* 6. 146–7 *φύλλα . . . ὕλη / τηλεθόωcα φύεc*: Mimnermus fr. 2. 1 *οἷά τε φύλλα φύει . . . ὥρη / ἔαροc.*

the ‖ mouth of Phrynichus the tragedian by Plutarch;[18] his suggestion that *μενος* comes from a mention of the 'might' of the winds is likely to be right. The second object of comparison might conceivably have been grains of sand, as the not dissimilar comparison at *Pyth.* 9. 46 f.[19] might suggest; but I find Lobel's idea of a comparison with waves poetically more satisfying.

'There is no doubt', says Lobel, 'that (*b*) stands under (*a*), right-hand edge under right-hand edge. There is no external indication of the interval between them.' Presumably the interval was not very great; in Apollodorus the first spirits whom Heracles sees are the Gorgon and Meleager, and in Bacchylides he meets Meleager immediately after his arrival.

(*b*) 3 *ἄτερθ(εν)* presumably signifies 'apart' or 'by himself', as Lobel says; it may have been a preposition (cf. *Ol.* 9. 78 and *Paean*, 8. 77), and will then have governed something like *ἄλλων*. But it may have stood as an adverb, as at *Pyth.* 5. 96.

In conclusion, let us look back at (*a*) 2. In the praise of Eumolpus mention of Mnemosyne would not surprise; e.g. the poet may have written *cοφὸν ἁγητῆρα [Μν]αμοϲύνας [τε κόραιcι φίλον*. But there are many other possibilities.

From what poem does the fragment come? According to a scholion on *Il.* 21. 194, Heracles when he descended to the underworld to fetch Cerberus met Meleager, who asked him to marry his sister Deianeira. On his return to the upper world he hastened to Aetolia, found Deianeira being wooed by Achelous, fought him and delivered her.[20] This story is attributed to Pindar; it is clearly possible ‖ that he told it in the poem to which this piece belongs. The title of the dithyramb whose opening is preserved in P. Oxy. 1604 seems to have been *Κ]α̣τ̣ά̣[βαcιc] Ἡρακλέ̣ο̣υ̣[c] ἢ Κέρβεροc*;[21] Wilamowitz suggested (see above) that it contained the story told in the Homeric scholion. We cannot be sure that he was right. Pindar often mentioned or described the same

[18] For waves as a type of the innumerable Lobel quotes the verses put into the mouth of Phrynichus the tragic poet: *cχήματα δ' ὄρχηcιc τόcα μοι πόρεν ὅcc' ἐνὶ πόντωι / κύματα ποιεῖται χείματι νὺξ ὀλοή* (ap. Plut. *Quaest. symp.* 932 F) [cf. Aesch. *PV* 89–90 (H.-S.)]. Wilamowitz, *Griechische Verskunst*, 465 n. 1 observes that these look as if they came from an epigram of Hellenistic date, and compares the epigrams by Dioscorides on Thespis (*AP* 7. 410 = Gow–Page, *HE* 1585 f.) and Sositheus (*AP* 7. 707 = Gow–Page, *HE* 1607 f.). Lesky's statement (*Die tragische Dichtung der Hellenen*[3], 62) that Wilamowitz 'thought of Dioscorides or Sosibius' as their possible author rests on a confused recollection of his note.

[19] *οἶcθα . . . ὅcca τε χθὼν ἠρινὰ φύλλ' ἀναπέμπει, χὡπόcαι*
ἐν θαλάccαι καὶ ποταμοῖc ψάμαθοι
κύμαcιν ῥιπαῖc τ' ἀνέμων κλονέονται. Cf. *Il.* 2. 800; 9. 385.

[20] Pind. fr. 249a S.–M. (ii. 75) [=ii. 79 ll.]

[21] Ibid. 73 [=75].

incident in more than one work, as the publication of P. Oxy. 2442 with its new narrative of the infant Heracles' killing of the snakes has recently reminded us,[22] and the story of Heracles' battle with Achelous would be somewhat surprising to find in a dithyramb named after the descent to Hades. Turyn takes the Homer scholion to refer to the paean which contained the mention of Achelous found in fr. 70 + 249b S.–M. = 71 Turyn = 233 Bowra; that is one possibility.

Both the dithyramb *Cerberus* and the new fragment are in dactylo-epitrite. We know the metrical schema of the strophe of the former, but not that of the epode; with the material we possess, it is not possible to prove or to disprove that the new piece fits the pattern of the strophe. The fragment that mentions Achelous is also in dactylo-epitrite; here too the evidence of the metre yields no results. I think it quite likely that the new fragment is from the poem mentioned in the Homeric scholion, and also that it is from the dithyramb *Cerberus*; but I would not care to see it taken for granted that all three of these pieces, or even any two of them, come from the same poem.

I do, however, think it likelier than not that the new piece is the work of Pindar. Lobel has remarked that κτέανον occurs in no other lyric poet; Pindar's remains contain twelve instances. ἀγητήρ is found at *Pyth.* 1. 69.[23] Further, εὐνομία λατερπής (l. 3) is a markedly Pindaric expression. These are only straws in the wind, but taken in conjunction with the fact that Pindar is known to have told this story they amount to something.

Even if the poem is not by Pindar, it can hardly be later than the fifth century. We have noticed that it has two features—the meeting of Heracles with the numberless multitude of ghosts, and soon afterwards with Meleager—with the common source of Bacchylides and Virgil as Norden conceived it. That source was in ‖ his opinion summarized by Apollodorus, and the new text has in common with Apollodorus, in all probability, the initiation of Heracles at Eleusis by Eumolpus. This seems to warrant a re-examination of the problems raised by the Apollodoran narrative, in the hope that the new evidence, slight as it is, may prove to throw some light on them.

Let us examine the narrative of Apollodorus (2. 5. 12). First Heracles goes to Eleusis, where Eumolpus initiates him, after first purifying him of the blood shed during his slaying of the Centaurs. Being a foreigner, he had to be adopted by an Athenian called Pylius before he could be initiated. Heracles then descends by way of Taenarum. When the spirits of the dead see him, they flee except for the Gorgon Medusa and

[22] *Paean*, 20 (ibid., pp. 61 f. [=63 f.]); cf. *Nem.* 1. 33 f.

[23] Cf. P. Oxy. 2447, fr. 37. 2 (?) [and note Ibycus, *PMG* 282(*c*) 4. 5].

Meleager. He draws his sword on Medusa, but is warned by Hermes that she is nothing but an empty wraith: we remember that in the *Odyssey* Heracles tells Odysseus that when he descended to fetch Cerberus, he was helped by Hermes and Athene. Near the gates of Hades he sees Theseus and Peirithous, who are held prisoner there in consequence of their attempt to carry off Persephone. Heracles takes Theseus by the hand and awakens him, but when he tries to raise up Peirithous the earth quakes and he lets go. He sees also Ascalaphus, the person who betrayed to Pluto that Persephone had eaten a pomegranate-seed in Hades, and was therefore in his power; Demeter has punished him by placing a huge stone over him, which Heracles now rolls away. Wishing to give the spirits blood—obviously in order to make them talk, as Odysseus does in the Nekyia of the *Odyssey*—he kills one of the cattle of Hades, and has to wrestle with their herdsman, Menoetes. He breaks Menoetes' ribs, but lets him go at the request of Persephone. When he asks Pluto to lend him Cerberus, Pluto replies that he may take him if he can vanquish him without the aid of the weapons he is carrying. Cased in his cuirass and covered by his lionskin, he flings his arms around the brute's head and grips and chokes it till it yields. Then he ascends by way of Trozen. Demeter turns Ascalaphus into a short-eared owl; and Heracles, after showing Cerberus to Eurystheus, takes him back to Hades.

Let us now consider what light is thrown upon this narrative by the writers with whom Norden has connected it. It will be convenient to start with Aristophanes, who in the *Frogs* makes use of the by now not uncommon comic theme of a descent to Hades.‖ Dionysus wishes to visit the underworld and induce its rulers to release one of the great tragedians to return with him to earth. He therefore calls upon Heracles, who is known to have accomplished the same journey with success when sent to fetch up Cerberus. Heracles describes the stages of the journey (137 f.), and when the journey is actually performed, his account turns out to be in every particular correct. The last two editors of the *Frogs*[24] are surprisingly vague on the subject of the poet's sources; but when we consider how closely each experience of Dionysus seems to parallel an experience of Heracles, we cannot help suspecting, with Norden, that a standard account of the descent of Heracles was in his mind, and we shall find this suspicion reinforced by certain correspondences with the Apollodoran narrative and those others believed by Norden to depend upon it.

Dionysus crosses the lake of Acheron on Charon's ferry, as we shall presently find reason to believe Heracles had done in the earlier

[24] L. Radermacher in *SB der Öst. Akad. der Wissensch.*, ph.-hist. Kl. 198/4 (1921; repr. 1954); W. B. Stanford (London, 1958) [; also D. Del Corno (1985).].

account. Then he expects to meet snakes and monsters (278 f.), and Xanthias pretends to see Empusa; since Meleager was separate from the other ghosts, this stage may well be parallel to the stage at which, in the Apollodoran narrative, Heracles meets the Gorgon. They are near a great expanse of mud and dung (273), where the wicked are being punished for their crimes on earth; this seems to correspond with the stage at which in Apollodorus Heracles sees Theseus, Peirithous, and Ascalaphus. Then he hears beautiful music (313), sees light like the light of day (155 f.), and in a myrtle-grove finds happy parties of men and women. The myrtle is sacred to Demeter and Persephone, and these men and women are repeatedly referred to as 'initiates'. The attempts made by various scholars to show that the parodos of the *Frogs* is meant to represent a particular religious procession seem to me misplaced;[25] but I feel no doubt that the songs in which the ‖ initiates celebrate Iacchus and the Eleusinian goddesses are meant to suggest the atmosphere of the Eleusinian cult. They sing these songs not in Attica, nor in any part of the upper world, but in Hades; they are initiates, who have not only been initiated but have led a virtuous life on earth (457 f.). That in an Athenian comedy the blessed in the next world should be represented by persons who have been initiated at Eleusis is by no means surprising. But it is seen to be especially natural if it is the case that in the epic account of Heracles' *katabasis* which was known to the poet and probably to many of his audience the hero's initiation at Eleusis played an important part. It may well be that the blessed were represented by initiates, or at least had some initiates among their number, in the epic poem also.

It is the initiates who finally show Dionysus the way to Pluto's palace (435 f.), a point that we shall have occasion to remember later. When he gets there, he gets a hot reception from the janitor of Hades, who

[25] Radermacher rightly (p. 185) says, 'Das Lied... muß als eine freie Phantasieschöpfung gelten, zu der Eleusis ein Stück von Rahmen lieferte'; cf. C. Segal, *HSCP* 66 (1961), 240 n. 78. Most earlier editors had taken the view found already in the scholia that the parodos is meant to represent the procession at the Great Mysteries. Gerhard in *Philol.* 13 (1858), 210 f. and Tucker in *CR* 18 (1904), 416 f. and again in his school edn. of 1906 argued that the procession of the Lesser Mysteries was in the poet's mind; they are followed by G. T. W. Hooker, *JHS* 80 (1960), 112 f. Hooker seems ignorant of the earlier discussions of the topic, and gravely underestimates the difficulty of persuading others to adopt his view that each stage of the ‖ journey of Dionysus is somehow localized in Attica as well as in Hades. The frogs of the secondary chorus indeed speak of the *temenos* of Dionysus at Limnae as their own (219); but they are not now there but in the lake of Acheron, being presumably defunct. Hooker's attempt to prove that the meadow in which the initiates are dancing is the meadow near this *temenos*, and not the meadow of Elysium, appears to rest on the application to it of the epithet ἔλειον (352). This is not enough; on the λειμών of the blessed, see Dieterich, *Nekyia*, 19 f. The attempt of M. Tierney, *Proc. Royal Irish Acad.* 42 (1935), 199 f. to show that the parodos described the procession at the Lenaea has been refuted by Deubner, *Gnomon*, 12 (1936), 506, as has been noted by Wüst, *Bursian*, 263 (1939), 60 and Newiger, *Gnomon*, 32 (1960), 753–4.

remembers how he once 'darted off choking Cerberus and escaped with him' (467 f.). That harmonizes with the account of Heracles' battle with Cerberus given by Apollodorus. Then the supposed Heracles receives an invitation to a delicious dinner brought him by a handmaid of Persephone (503 f.). In Apollodorus it is Persephone who begs off the herdsman Menoetes; other authors too say that she gave Heracles a specially kind reception, and this is natural if we remember that before starting out he had secured initiation into her own mysteries.[26] Persephone is roasting a whole ox for Heracles, and the landladies who soon after arrive in ‖ pursuit of him complain that he has wolfed huge quantities of their food without payment (549 f.). In Apollodorus he roasted a whole ox, but with the laudable purpose of giving blood to the ghosts so that they could speak. The comic poet has seized an obvious chance to make a joke about the notorious gluttony of Heracles, but we see that the original put a different complexion on the incident. Late in the play, Pluto kindly allows Dionysus to return to earth with the tragedian of his choice, just as in the Apollodoran narrative he allowed Heracles to return with Cerberus.

Aristophanes did not religiously follow one 'source' in the manner of a historian. But the general correspondence between the Apollodoran and the Aristophanic *katabasis* remarked by Norden becomes more noteworthy now that we are able to lay a heavier stress on the Eleusinian connections of Heracles. Its significance will be increased by a consideration of the affinities between these two narratives and that contained in the sixth book of the *Aeneid*.

Apart from the Homeric Nekyia, and apart from the various theological and philosophical sources that contributed to the Virgilian *katabasis*, Norden recognized Virgil's use of a *katabasis* of Orpheus, doubtless the same used for the Aristaeus episode of the Fourth Book of the Georgics, and of the *katabasis* of Heracles summarized by Apollodorus. Let us examine the successive stages of the journey of Aeneas, noting how far they correspond with those of the journeys of Heracles and Dionysus.

Soon after Aeneas and his guide the Sibyl have entered Hades, they meet various personified abstractions of unpleasant nature (273 f.) and then a horde of monsters—Centaurs, Scyllas, Briareus, the Hydra, the Chimaera, and finally Gorgons and Harpies (285 f.). These seem to correspond with the single Gorgon in Apollodorus, and with the snakes and beasts in Aristophanes. Aeneas draws his sword and points it at

[26] Cf. Norden, *SBPAW* 1934, 659 f. = *Kl. Schr.* 503 f. For Heracles' connections with the Eleusinian goddesses, see Bayet, *Les Origines d'Hercule romain*, 390 f. D. M. Lewis draws my attention to an amusing instance of the notorious ψυχρότης of Timaeus which shows the celebrity of this connection, *FGrH* 566 F 102b.

them; he sheathes it only when the Sibyl assures him that they are only shadows (290–4). Two centuries ago the learned editor of both Virgil and Apollodorus, Christian Gottlob Heyne, observed that the Sibyl's warning to Aeneas echoed the warning of Hermes to Heracles in the Apollodoran narrative.[27] When Bacchylides makes Heracles aim an arrow at the shade of Meleager and be told by him not to shoot against the spirits of the dead, he seems to have the model of the *katabasis* of Heracles in mind.[27a] ||

Next Aeneas and the Sibyl are ferried across the lake of Acheron in the barque of Charon (245 f.). In Bacchylides and, as we now see, in Pindar, the comparison of the countless souls to leaves came right at the start of the journey through Hades; but Virgil holds it back in order to apply it to the countless spirits waiting on the shores of Acheron to be ferried to the other shore (305 f.). Charon accepts the living passengers with reluctance; it was against his will, he says, that he accepted earlier living passengers, Theseus and Peirithous, who came to abduct Persephone, and Heracles, who dragged Cerberus from the very throne of Hades to the upper world (388–97).[28] It has long been realized that we have here an echo of an earlier *katabasis*.[29] Servius in his note on the passage says that we read in Orpheus that when Heracles came down Charon in terror at once accepted him as a passenger, and later spent a year in fetters as a punishment for having done so.[30] This does not show that the *katabasis* of Heracles was ascribed to Orpheus; for this allusion to the earlier descent of Heracles may easily come from the *katabasis* of Orpheus himself, which we have good reason to think was also used by Virgil.

Between the lake of Acheron and the inmost part of Hades Virgil locates five classes of souls, whose description is for the most part based on the Nekyia of the *Odyssey*.[31] When the travellers reach the inmost

[27] On *Aen.* 6. 290–4.

[27a] [Professor Evelyn Harrison, *AJA* 71 (1967), 43–5 makes a strong case for identifying figure D on the east pediment of the Parthenon not with Dionysus but with Heracles, pointing out that it is not surprising to find him near the Eleusinian goddesses; cf. my note in *AJA* 74 (1970), 181 (no. 44).]

[28] This passage is no safe guide to the place where Cerberus may be found in Hades; see Norden^3, p. 466.

[29] The importance of Heracles' Eleusinian connections in the *Frogs* is stressed by E. Lapalus, *REG* 47 (1934), 1 f., in an article which is a strange mixture of silliness and shrewdness.

[30] 'Lectum est in Orpheo [= fr. 296 Kern] quod, quando Hercules ad inferos descendit, Charon territus eum statim suscepit; ob quam rem anno integro in compedibus fuit.' Abel and Lobeck, *Aglaophamus*, 812 inferred from this passage that there was an Orphic *katabasis* of Heracles, incautiously, for this could clearly come from a *katabasis* of Orpheus himself. See G. Ettig, *Acheruntica* (Leipziger Studien 13, 1891), 286 f. and Norden on 384–416 [cf. *Kl. Schr.* 508 n. 77].

[31] See Norden on 417–547 (B); for Virgil's use of Homer in Book VI in general, see Knauer, *Die Aeneis und Homer*, 107 f. Norden's suggestion that the mention of Tydeus, Parthenopaeus, and Adrastus derives from the Heracles *katabasis*, which itself took it from a Theban epic, is far from certain to be right.

part of Hades, they come first to Tartarus, where among other sinners they find Peirithous (601) and later Theseus (617 f.); Virgil has preferred the version of the story that makes him not escape. There is no mention of Ascalaphus, but this need not ‖ surprise us; Virgil did not need him. Next the travellers come to Elysium, where the blessed, illuminated by the purple light that shines from a wider aether and from a sun and stars all their own, entertain themselves by wrestling or by songs and dances (635–44). Their musical director is 'the Thracian priest', a figure usually identified with Orpheus (645 f.). The special light, the music, and the athletics all occur in a famous fragment of one of Pindar's dirges (*Threnos* 7 on S.–M. ii. 109 f. [= Maehler ii. 117 f.]); the two former features occur also in the *Frogs*. In the *Frogs* Dionysus has been told by Heracles that the blessed people will direct him to his destination, the palace of Pluto (161 f.), and so they do (435 f.). In the epic *katabasis* Heracles must have gone, like Dionysus, to Pluto's palace; but Aeneas is looking for Anchises. The Sibyl asks the inhabitants of Elysium where they will find Anchises (666 f.). Her question is answered by Musaeus,[32] a fact that may have some significance in the present inquiry.

Musaeus has been called 'a pale copy of Orpheus, a kind of Attic double of his'.[33] Like Eumolpus, he sometimes figures as a Thracian by birth; but again like Eumolpus he is connected with Eleusis from an early date. Most often, he is called an Athenian, first in the *Rhesus* (945).[34] Sometimes he is the father of Eumolpus, but more often he is his son, as he is called by Philochorus. We have seen how in the Eumolpid family tree given by Andron he is the great-great-grandson of the first Eumolpus, the grandson of the second, and the father of the third, who founds the Mysteries. Diodorus says that he was hierophant at Eleusis, and initiated Heracles; Pausanias says that the one poem assigned to him that is genuine is a cult-hymn written for the Eleusinian family of the Lycomidae; he was buried at Athens, both on the Hill of the Muses and at Phalerum.[35] From about the end of the fifth century, at least, his connection with Athens and Eleusis is well attested; he is also connected with Orpheus, and more than one poem ascribed ‖ to Orpheus was addressed to him. From the fifth century it was thought that Orpheus and Musaeus were famous poets, earlier

[32] On Musaeus see Rzach in *RE* xvi. 757 f. and Jacoby on Philochorus, *FGrH* 328 F 208 (full bibliography).

[33] 'Musaios ist sozusagen dessen (des Orpheus) abgeblaßte Kopie, eine Art attischer Orpheus': Ernst Maass, *Orpheus*, p. 138.

[34] Cf. Diog. Laert., prooem. 1. 3; the *Suda* s.v. says Ἐλευσίνιος ἐξ Ἀθηνῶν.

[35] [Jacoby on Philochorus] fr. 208; cf. the epigram in Diog. Laert., loc. cit. (= Preger, *Inscr. Gr. Metr.*, no. 238); Eusebius, *Chr.* 749 (test. 18 Kern). He wrote a poem called Εὐμολπία (fr. 11 D.–K. ap. Paus. 10. 5. 6). [For a different genealogy, tracing his descent from the Moon, see A. Henrichs, *ZPE* 58 (1985), 1–8.]

than Homer and Hesiod.[36] But from early times doubt attached to the attributions of their works; we have noted the scepticism of Pausanias regarding most of them. None of the surviving fragments ascribed to Musaeus need be very early. Clement quotes two fragments which were thought to have been copied by Homer, but no doubt it was the other way round,[37] and many fragments are shown by their subject-matter to have originated long after the end of the archaic period. Plato in his *Republic* (363 C–D) names Musaeus as a known authority on the afterlife; he blames him for describing the blessed as engaged in a perpetual symposium, with nothing to do but to get drunk. From this passage it has often been inferred that a poem about the underworld known to Plato was supposed to have been written by Musaeus. Now Norden thought with good reason that Virgil used a *katabasis* of Orpheus as well as one of Heracles, and it may well be that Musaeus owes his place of honour in Virgil's underworld to his well-attested Orphic associations. But we have to reckon with the possibility that he owes it to his equally well-known connections with Eleusis, or to both.

Norden was surely right in maintaining that Virgil must have known an epic *katabasis* of Heracles, known also to Aristophanes and Bacchylides and, as we may now add, Pindar. Much the most interesting consequence of our new knowledge that the episode of Heracles' initiation almost certainly goes back to that *katabasis* consists, it seems to me, in the light it throws upon the origins of the Virgilian Sibyl. The descent of Odysseus takes place under the auspices of Circe. That of Aeneas takes place under the auspices of the Sibyl of Cumae, who not only prescribes the proper ritual to be performed in order to win the favour of the rulers of the underworld, but herself accompanies Aeneas to guide him through Hades.[38] Radermacher need not have complained that the notion that a hero needed a guide for such an adventure was un-Greek, and have sought its origins in Jewish and Christian apocalyptic literature.[39] In the Apollodoran narrative Hermes is there to warn Heracles that the Gorgon is nothing but a ghost; Hermes in ‖ his capacity as Chthonios and Psychopompos is peculiarly suited to this function. Heracles before descending was first purified of the guilt of murder and then given a special claim on the friendship of the queen of the world below by the favour of an august personage, Eumolpus; he was then conducted to the palace of Hades by the divine messenger himself. Pindar's description of a 'wise leader', who rules by means of *eunomia* 'that makes the people to rejoice'

[36] Plato, *Apology*, 41 A couples them with Homer and Hesiod, Cf. Ar. *Ran.* 1032–4.

[37] Frr. 4–5 D.–K.

[38] Norden, pp. 43 f.

[39] *Das Jenseits im Mythos der Hellenen* (1903), 14 f.

suggests that Eumolpus was presented as a figure of great dignity, as we should expect in the founder of the mysteries and proto-hierophant. No doubt the favourable reception which Persephone gave to the descending hero was in large measure due to the impeccable credentials he had received from the head of an institution so dear to her and to her mother.

Virgil must have wished that it lay within his power to allow Aeneas to be advised and helped, at the start of his expedition, by a person who enjoyed the same kind of special relationship with the chthonic powers as had Eumolpus in the earlier *katabasis*, and to be guided through the nether regions by a person familiar with their topography and unlikely to be intimidated by their more disagreeable features. The Sibyl of Cumae was properly speaking a prophetess and a priestess of Apollo. In Virgil she does indeed prophesy (77 f.); but the main prophetic office must be reserved for Anchises, the counterpart of the Homeric Teiresias. Virgil had to contrive to give the Sibyl a connection with the underworld powers, and did so most adroitly. He made her also priestess of the cult of Hecate by the lake of Avernus;[40] since Hecate is either a *πρόπολος* of Artemis or is identical with Artemis in one of her aspects, this manœuvre does not lack plausibility. When he published the first edition of his great commentary, Norden believed that this was one of the traditional functions of the Cumaean Sibyl; later, Corssen convinced him that it was a Virgilian innovation.[41] We can understand more clearly the impulse towards such an innovation, now that we know that between the Homeric Circe and the Virgilian Sibyl there stood the august figure of the Eleusinian hierophant.

Odysseus had needed no guide through Hades, but now a guide was needed. Following a suggestion made to him by Wilamowitz, Norden (p. 154) suggested that the notion of a guide had derived ‖ from the mystery religions with their concept of the *μυσταγωγός*. In a sense this suggestion is confirmed, for in the Heracles *katabasis* the hero was accompanied by Hermes, and the *katabasis* had a definite connection with Eleusis.[42] One must admire Virgil's discretion in retaining the Sibyl, who has already discharged the function of Eumolpus, to perform that of Hermes also. Not only is Aeneas less able than Heracles to retain interest and admiration while having an Olympian god to guide him, but the presence of Mercury would have detracted from the majesty of the prophetic shade of Anchises, in the encounter with whom the descent of Aeneas reaches its climax.

[40] See *Aen.* 6. 35; 117–18; 564 f.

[41] Corssen, *Sokrates*, 1 (1913), 1 f.; cf. Gercke, *Die Entstehung der Aeneis*, 176 f. See Norden on 9–13 and on 103 f.

[42] The use of the Ascalaphus story, which looks as if it were of Eleusinian origin, may be held to supply further evidence of this.

Finally, is there anything we can say with confidence about the date and nature of the epic *katabasis* of Heracles? First, do we have to believe in it? It might be argued that both Pindar and Bacchylides could have derived their descents of Heracles from the *Cerberus* of Stesichorus, of which we know almost nothing,[43] or that Bacchylides copied Pindar and later Virgil did the same; there are good reasons for thinking that Virgil may have read some Pindar. Both possibilities may be discounted; we have seen that the author of the *katabasis* had a marked interest in Eleusis and in Athens, so that he must either have been Athenian or have wished to give pleasure to the Athenian public. The poem cannot have been written before the time at which Theseus as an Athenian national hero was being exalted as a friend and peer of Heracles, nor before doctrines of purification were widely known and the mysteries widely popular. It can scarcely have been earlier than the beginning of the sixth century, and may have been appreciably later; perhaps round about 550 BC. Its contents seem to have influenced Euripides, and very probably the writers of the Atthides.

Who can the author have been? The story of how Heracles fetched up Cerberus is mentioned both in the *Iliad* and in the Nekyia of the *Odyssey*. Von der Mühll has tried to resolve the awkward problem of the composition of the eleventh book of the *Odyssey*, in which a conjuration of Teiresias suddenly turns into a visit to Hades, by suggesting that the poet used, and partly incorporated in his work, an earlier *katabasis* of Heracles.[44] This guess seems ‖ to me impossible to verify, and indeed some indications make against it. It is not the most economical explanation of the difficulty it was meant to solve, for Von der Mühll is obliged to suggest that the catalogue of heroines (225–332) was derived from another source, the Hesiodic *Ehoiai*; the only part of the book which can be argued to point towards a *katabasis* of Heracles is that between 565 and the end, containing the mention of Minos, Orion, Tityus, Tantalus, Sisyphus, Heracles himself, and Theseus and Peirithous, and even this can be explained in other ways. Von der Mühll thinks that the idea of Odysseus drawing his sword upon the ghosts, which he does to stop them from drinking the blood before he is ready to speak with them, was suggested to the poet of the *Odyssey* by Heracles drawing his sword upon the Gorgon. Even if Von der Mühll's theory is correct, the *katabasis* of Heracles used by the poet of the *Odyssey* was probably not our *katabasis*. It is not in itself inconceivable that the poet who brought the *Odyssey* to its final form

[43] The speculations of Vürtheim, *Stesichoros' Fragmente und Biographie* (1919), 21–3, are well dealt with by Bowra, *Greek Lyric Poetry*², 94.

[44] *Philol.* 93 (1938), 1f. [=*AKS* 18f.]; cf. *RE* Suppl. vii (1939), 724f. He has ‖ convinced Merkelbach, *Untersuchungen zur Odyssee*, 191; but not Page, *The Homeric Odyssey*, 49 n. 7.

might have used a poem written as late as about 550, but it is not especially likely. Further, there is one indication that the author of the *katabasis* may have known the Nekyia of the *Odyssey*. At the start (43), Odysseus is frightened of the swarms of ghosts; at the end (633) he is frightened again by the thought that Persephone may send up the head of the Gorgon. Soon after the beginning of Heracles' descent, he sees the Gorgon; instead of being frightened, he threatens her with his sword. Does it not look as if the author of our *katabasis* wished his hero to improve on the performance of Odysseus?

We know a little about epic *katabaseis* that might be relevant.[45] Pausanias (9. 31. 5) mentions a Hesiodic *katabasis* of Theseus and Peirithous. He also implies (10. 28. 4) that there was another in the *Nostoi*, we do not know by whom. There was another in the *Minyas*; of the five fragments, one mentions Charon (fr. 1 Davies, p. 144), another mentions Amphion and Thamyris as paying for their misdeeds in Hades (4); another mentions the death of Meleager (3); and another was thought to have a mention of Heracles just after one of Orion [(5: fr. 6 Kinkel; but see R. Philippson, *Hermes*, 55 (1920), 269)]. These data have prompted several guesses. Karl Otfried Müller identified the Prodicus of Phocaea to whom ‖ Pausanias (4. 33. 7) ascribes the *Minyas* with the Prodicus of Samos or Perinthus said by Clement to have written a *katabasis* of Orpheus;[46] and concluded that the two poems were the same;[47] the forty-year-old Wilamowitz agreed with him. That seems unlikely; the identification of the two poets is precarious, and what is known of the *Minyas* looks more like epic than like Orphic material. Welcker thought the *Minyas* contained an account of the descent of Heracles and his rescue of Theseus.[47a] But how could this figure in a poem with such a name? Welcker's theory that it was somehow tacked on to an account of Heracles' war against Orchomenos under Erginus does not convince; nor is there positive evidence in favour of Rohde's guess that the descent in the *Minyas* was that of Theseus and Peirithous.[48] Whatever the *Minyas* contained, it is unlikely to have been our poem; the problem of how this work, with its obvious Athenian connection, came to bear such a title would take some resolving.[49]

[45] See Ganschinietz, *RE* x (1919), 2401 f.; Ettig, op. cit. (n. 30), 278.

[46] *Strom.* 1. 21. 131 (= test. 222 Kern); the *Suda* calls him Herodicus (test. 223 Kern).

[47] *Orchomenos*, 18, cf. 212; Wilamowitz, *Homerische Untersuchungen* (1888), 222; cf. Lobeck, *Aglaophamus*, 360; Robert, *Nekyia*, 79; Ganschinietz, *RE* x (1919), 2399 f.

[47a] [The same view was taken by M. Croiset in *Mélanges Henri Weil* (1898), 73 f. (oddly reprinted by W. M. Calder III and J. Stern, *Pindaros und Bakchylides* (Wege der Forschung 134, 1970), 405 f.), only he thought that Pindar and Bacchylides took it from Stesichorus.]

[48] *Der epische Cyclus*, i. 239 f.; Rohde, *Psyche*, ch. 7, n. 3. See Ettig, op. cit. (n. 30), 282 f.

[49] Polygnotus' picture of Hades showed Meleager, Thamyris, and Charon (Paus. 10. 28 f.). On the Apollodoran chapter on the *katabasis* of Heracles, see Ettig 391 f. He notes that the occurrence of the lion-skin as an attribute of Heracles indicates comparative lateness.

Its author was probably an Athenian or a person belonging to the orbit of Athenian culture. Who he was I would rather not guess; but though it must have been distinct from the *katabasis* of Orpheus, it is just conceivable that it may have been one of the poems circulating under the name of Musaeus, who as we have seen served as a kind of Attic Orpheus, a link between Orphism and the Eleusinian cult.[50] Again, it may have been one of the many poems about Heracles, or even part of one of the poems about Theseus.

Norden doubts whether Virgil read the *katabasis* in the original; he is inclined to think that he derived his knowledge of it from a mythographical handbook, not unlike what we know as Apollodorus, but fuller. He may be right, but we can hardly be very positive; much early epic literature existed as late as the time of ‖ Hadrian—we have only to remember how much seems to have been available to Pausanias—and Virgil was a man of most uncommon erudition. I wonder if he really found in a handbook, even if he possessed a very full one, the comparison of the ghosts to leaves or waves, which is not the kind of detail that a handbook normally contains. If Virgil did know the descent of Heracles from a handbook, perhaps he may have taken this particular detail from his reading of an author not unfamiliar to him—I mean Pindar.[51]

[50] Wilamowitz, *Der Glaube der Hellenen*, ii. 58–9.

[51] See A. Seidel, *De Vergili studiis Pindaricis* (Diss. Breslau, 1925).

I am obliged to R. Kassel, D. M. Lewis, and B. B. Shefton for useful suggestions, and to N. J. Richardson for a valuable discussion of the problem, as well as for several corrections.

[Later relevant literature: R. J. Clark, *Phoenix*, 24 (1970), 244 f.; G. Thaniel, ibid. 25 (1971), 237 f.; F. Graf, *Eleusis und die orphische Dichtung Athens in vorhellenistischer Zeit* (RGVV 23, 1974); R. G. Austin, *Aeneidos liber sextus* (1977), 130 (who patronizes Norden as well as me in an intolerably superficial note). On R. J. Clark, *Catabasis: Vergil and the Wisdom Tradition* (1979) see Graf, *Gnomon*, 53 (1981), 545–8; on R. J. Quiter, *Aeneas und die Sibylle: die rituellen Motive im sechsten Buch der Aeneis* (Beiträge zur kl. Philologie 162; 1984) see Graf, *Gnomon*, 58 (1986), 360–3.]

III. TRAGEDY

12(R36)

Review of von Fritz, *Antike und moderne Tragödie*

Kurt von Fritz, *Antike und moderne Tragödie: Neun Abhandlungen*. Berlin: de Gruyter, 1962. Pp. xxix + 511.

ONLY two of the nine papers printed in this book appear here for the first time; of the others, three came out in periodicals during the thirties and four during the fifties. But the book derives a real unity from the consistency of the author's aims and methods, whose nature is indicated in an introduction which is not the least valuable section of this important work.

The introduction deals with the old and still continuing controversy between the 'moralizing' and 'aesthetic' interpreters of Greek tragedy: von Fritz approaches it by the characteristic method of comparing the Greek attitude in this matter with the theories of later dramatists. First he discusses Corneille's theory that tragedy should warn against evil passions and encourage virtue, and shows that it not only starts from a wrong interpretation of Aristotle, but gives a wholly insufficient account of tragedy, even in its Cornelian form. There follows a fair and lucid discussion of Brecht's notion of a drama that should inspire its audience with the will to remedy social institutions that the poet finds deplorable. The adherents of the 'epic theatre' would do well to pay attention to von Fritz's discussion, which is full of interesting observations; for example, he points out that the famous 'Verfremdungseffekt' is not wholly alien to Greek tragedy, with its masks, elaborate costumes, and choruses. He draws a clear line of demarcation between tragedy properly so called and the kind of social drama which aims at remedying abuses, and which even when written with the skill of an Ibsen or a Shaw seems dated and superficial once the circumstances which have provoked it belong to the past.

Von Fritz is above all things an Aristotelian. Again and again he corrects a wrong interpretation of the text of the *Poetics*; indeed, the last paper in the book decisively refutes the thesis of F. Grayeff that the ninth chapter of the *Poetics* contains grave interpolations, a thesis in which von Fritz rightly sees an example of the danger of interpreting

Gnomon, 34 (1962), 738–47.

the words of a text without sufficient regard to the things they signify. But von Fritz's fidelity to Aristotelian principles is seen not only in what is *καθόλου* but in what is *καθ' ἕκαστον*. One cannot read his book without becoming conscious that in most respects Aristotle's general theory of tragedy is the most profound of those he analyses. Approaching the subject from an Aristotelian viewpoint, von Fritz offers both a theory of tragedy far more comprehensive than those of most contemporary aestheticians and an analysis of Greek tragedy far more penetrating than those offered by the popular handbooks of today.

Morality, according to von Fritz, is the stuff of tragedy; but he means by this something different to what would be meant by Corneille or Brecht, or to what would be meant by those naïve interpreters of the moralizing school who think a Greek tragedian meant his drama as a kind of sermon ‖ in which he could develop moral theories, often new and independent ones, for the edification of his audience. Von Fritz would not allow that the moral content of tragedy could be reduced to any set of precepts, let alone to any set of recipes for social action; he holds that its aim is not to communicate opinions but to exercise an effect, by transporting us for a moment from the cramping mental conditions of daily life into a higher emotional plane. Tragedy, he argues, must truly and genuinely represent the general conditions of the human situation, and it must do so in terms of true and genuine moral values (p. xxvi; cf. p. 14). He rightly argues that these moral values do not vary according to the poet's caprice, a truth that he particularly well illustrates in the paper on the Orestes myth, to be discussed presently.

Von Fritz effectively restates the importance of Aristotle, both for the theory of tragedy in general and for the analysis of Greek tragedy in particular. Workers in both fields should be grateful to him for suggesting a fresh start on Aristotelian principles: but they may well feel that Aristotle needs more revision and expansion than he is always willing to allow. As regards the general theory, a valuable indication of the lines along which such a revision might proceed is given by A. M. Quinton, *Proceedings of the Aristotelian Society*, Suppl. 34 (London, 1960), 145 f., whose discussion of the aesthetics of tragedy starts from a sound and appreciative treatment of Aristotle's theory. Von Fritz's book is more concerned with Aristotle's analysis of Greek tragic practice; and here his sympathetic treatment does not entirely resolve the doubts expressed by Wilamowitz in his famous chapter called 'Was ist eine attische Tragödie?' Von Fritz says (p. 90) that on the whole the basic principles of Aristotle's philosophy play no great part in the *Poetics* in general and in the sections on tragedy in particular. It is true that Aristotle's metaphysics do not dominate his aesthetics to the same extent as those of, say Hegel. But Aristotle's metaphysics derive from those of Plato; and though he allows for growth as Plato does not, his doctrine of form, like Plato's, inclines him to assume that each object of analysis has one proper form and one only, which once it has been attained remains constant. Too much of what Aristotle asserts of tragedy in general is true of the *Oedipus Tyrannus* and other plays of his own favourite pattern, but false of the majority. It is easy to show that Aeschylus, an author who flourished before tragedy had 'attained its nature', is not sufficiently accounted for by his analysis.

Von Fritz is right to reject the notion that tragic poets may deliberately seek to introduce new moral doctrines; the moral standpoint of tragedy does not vary according to the poet's whim. But it may well vary from one civilization to another, according to variations in the religious and ethical background against which tragedies are set. A thorough analysis of Greek tragedy must take scrupulous account of the differences between Greek and modern moral attitudes, and here von Fritz's analysis might sometimes be carried further, notably in his papers on the *Alcestis* and on tragic guilt.||

Von Fritz observes the close connection of the controversy between moralizers and aestheticians with the disputes over the consistency of characterization in Greek tragedy that began in 1917 with the posthumous publication of Tycho von Wilamowitz's memorable study of the dramatic technique of Sophocles, and by the dispute over dramatic unity in tragedy that arose from it. Von Fritz recognizes the value of Tycho's work in destroying the 'bourgeois-sentimental' approach to the characters of tragedy which as he says is typified in Ewald Bruhn's commentaries on Sophocles, but which could easily be exemplified from the work of far more celebrated scholars. Tycho's book, he allows, has helped to make it clear that the tragedians do not study psychological niceties for their own sake; but he thinks Tycho went too far in denying them any consistency in characterization. Ancient ἦθος is not the same as 'character' in modern literature; but its laws demanded that the characters should be consistent, as Aristotle and other ancient theorists assert and as the study of the texts confirms. In the same way, von Fritz argues, the reaction against the romantic interpretation of tragedy went too far in denying it the principle of dramatic unity. Aristotle condemned the tragedy that was 'episodic', and in von Fritz's view the evidence of the texts bears him out.

Von Fritz's belief in the consistency of tragic characterization is supported by the three papers on the Haemon scene in the *Antigone*, the 'Trugrede' in the *Ajax*, and on the Danaid trilogy (all first published in the thirties) and in the new paper on the figure of Eteocles in the *Seven against Thebes*. The long studies of the *Alcestis* and *Medea*, like the new paper which examines the use of the myth of Orestes by the three great tragedians, argue that these plays accord with the general conception of tragedy which von Fritz maintains. All these principles play a part in the detailed study of tragic guilt and poetic justice with which the book begins. Throughout the work, but especially in the three last-named papers, von Fritz uses his wide knowledge of modern European literature since the Renaissance to throw light upon Greek tragedy; and at the same time his analysis of Greek tragedy in the light of Aristotelian principles does much to increase our understanding of modern tragedy and tragic theory. A comparative study of many treatments of the themes of Alcestis and Medea after Euripides helps to throw light on Euripides' methods; and a survey of later conceptions of tragic guilt from Seneca onwards helps to elucidate the Greek conception.

Let me discuss first the four papers which I have described as illustrating von Fritz's notion of the unity of character.

That on the *Antigone* does so by contending that Haemon in his argument with Creon is actuated not by love for Antigone but, as he himself says, by genuine concern for Creon, a concern which the Chorus wrongly tries to minimize when at the close of the scene they sing of the power of Eros and imply that Haemon has acted under his influence (790 ὁ δ' ἔχων μέμηνεν). It is true that we have no reason to question the sincerity of Haemon's claim that he is trying to help his father for his own good; and it is true that by attributing his action to the power of love, which the Greeks, unlike ourselves, regarded as a kind of madness, the Chorus means to minimize the ‖ importance of his words; von Fritz's achievement in demonstrating that this is so is worthy of all praise. But he seems to me to go too far when he says that love plays no part whatever in determining Haemon's conduct in the play. It is true that the immediate motive of Haemon's suicide is anger with himself for having drawn his sword upon his father (1235). But would he have drawn his sword upon his father if he had not loved Antigone? Note Ismene's words at 570, which clearly show that affection existed between Antigone and Haemon.[1] That Haemon himself says nothing about it is altogether characteristic of Greek tragedy.

The article on the *Ajax* is directed against the use of the 'Trugrede' at 646 f. by Tycho von Wilamowitz to support his denial of the consistency of Sophocles' characterization; it modifies an attempt to refute Tycho previously offered by W. Schadewaldt, *Neue Wege zur Antike*, 8 (1929), 59 f. Von Fritz is wrong to accept Schadewaldt's claim that the 'Trugrede' is not a deliberate falsehood, but a λόγοc ἐcχηματιcμένοc which is really true but is intended to deceive; but he is right to reject

[1] [But see Lloyd-Jones and Wilson, *Sophoclea* (1990), 127.]

his contention that ll. 666 f. reveal 'mildness' and 'humility'. Clearly Ajax has to deceive his companions; he has to get out of the tent; and clearly the lines in question reveal not humility, but bitter irony. But is he right in holding that Schadewaldt has made an important discovery in finding an 'innere Entwicklung' in the character of Ajax? The delicate distinctions drawn in the last part of von Fritz's article do not suffice to show that Ajax's character and attitude have in any way altered. By Athene's action he has been placed in a situation in which he cannot live on true to his old self and without dishonour; he prefers to die, as could have been predicted.

The paper on the Danaid trilogy is also concerned with the consistency of characterization, for its most important feature is a discussion of the vexed question of whether the Danaids were opposed to marriage in general or to marriage with their cousins in particular. Von Fritz's suggestion that their loathing for the Aegyptiads aroused in the Danaids an aversion for marriage in general which they were persuaded to abandon only at the end of the trilogy offers the best way yet suggested of reconciling passages of the *Supplices* which at first sight seem inconsistent. It may be right; but even if it is, it does not go far towards telling us what happened in the play called by their name. For all we know, Aeschylus may even have followed the version of the legend in which all were killed except Hypermnestra; to assure the royal line of Argos, only she needed to survive. Almost everything concerning the last two plays of the trilogy remains uncertain.

Von Fritz's theory has been built upon by R. P. Winnington-Ingram, *JHS* 81 (1961), 141 f. [cf. now *Studies in Aeschylus* (1983) ch. 3]); few of his suggestions are impossible, but none is certain. He follows M. L. Cunningham, *RhM* 96 (1953), 223 f. ibid. 105 (1962), 189 f.) in reviving the conjecture of Nathansky, *WS* 32 (1910), 21 f. that the *Aegyptii*, in spite of its name, had a chorus consisting of the Danaids. If Miss Cunningham's identification of P. Oxy. 2251 (fr. 280 in my appendix to the Loeb Aeschylus = 451h Radt) is right, this is highly probable. Miss Cunningham's suggestion is not incompatible with B. Snell's probable supplement in l. 3 of this fragment, and has certainly more plausibility than I at first allowed (op. cit., p. 571). But it is still true that many hospitable houses and hospitable persons may have suffered destruction in lost plays of Aeschylus; and in default of further evidence it would be a pity if acceptance of this attractive possibility were to become an article of faith.

One of the most important contributions to the understanding of the *Seven against Thebes* was made by Friedrich Solmsen, *TAPA* 68 (1937), 197 f.) [= *Kl. Schr.* i (1968), 106 f.], who has demonstrated from the text that from l. 653 onwards Eteocles acts under the influence of the

Erinys brought down upon him by his father's curse. 'It is misleading', Solmsen there writes, 'to speak of his choice or to suggest that Aeschylus, when he wrote his play, was fascinated by the problem of free will, the relation of human will and fate, or anything of the kind.' The *Festschrift* presented to the late Werner Jaeger (*HSCP* 73, 1958) contains two attempts to go further in the same direction. Erwin Wolff (pp. 89 f.) contends that, since Eteocles has already left the stage to make his postings for the battle before the beginning of the first stasimon ‖ (pp. 282 f.) at l. 653 it is too late for him to send any other champion against his brother, for no other champion is available. In this fact Wolff discovers the working of the curse; it was inevitable that circumstances themselves should leave Eteocles with no alternative. Accepting Wolff's conclusions, H. Patzer (pp. 97 f.) tries to affirm the unity of the two parts of the play before and after l. 653 by showing that even in the first part Eteocles is partially blinded by the operation of the curse. Does not his treatment of the Chorus in the scene following the parodos, Patzer asks, show nervous irritability? and why does he not think of saving Thebes from the danger of destruction by renouncing his own title to the throne? Only at l. 653, Patzer argues, does Eteocles become aware that his own death in battle is inevitable; before that, he supposes that the brothers can divide their inheritance with the iron without both perishing.

In answer to Wolff, von Fritz points out that to insist that, because Eteocles has gone off to post his men before the beginning of the first stasimon, therefore the audience must think of the posting as having been completed during the singing of that stasimon is to credit Aeschylus with an unreasonable degree of naturalism; the words spoken by Eteocles at ll. 282 f. simply supply a motive for his leaving the stage at that moment. The tenses in which Eteocles, during the scene with the Messenger, refers to his posting of the Theban defenders could all perfectly well refer to the immediate future. Eteocles uses three futures (408, 621, 672), two perfects (448, 473), one present (553), and one aorist (505); and von Fritz rightly points out that all these, including the two perfects, which are instantaneous in sense, are most naturally taken as referring to the immediate future. He completes the refutation of Wolff's theory by asking what answer, if Wolff were right, Eteocles would have made to the plea of the Chorus that he should not risk an unbearable pollution by taking the field against his brother. 'I am sorry', he ought to have replied, on Wolff's view, 'but someone has to fight him; and at this stage of my preparations I have no one left to send against him except myself.' What Eteocles does say is very different, and makes for a far superior effect. Von Fritz's case may be confirmed by the reflection that we should not expect the curse to work simply by placing Eteocles in circumstances that leave him no choice. We should expect it also to work upon him through Ate, who takes away men's judgement at a critical moment; and that, as Solmsen has shown, is what the text shows to be the case.

Why is Eteocles visited by Ate? Is he a blameless victim of the curse of Oedipus? or is Patzer right in arguing that he is not as innocent as is usually assumed? Von Fritz shows that his conduct towards the panic-stricken women of the chorus reveals harshness, but not criminality; he is right to stop them from spreading alarm and despondency in the beleaguered city. Further, he refutes Patzer's contention that Eteocles might have saved the city by renouncing his title to the throne. The rule of Polyneices as he is depicted in this play would be as disastrous for Thebes as that of Clytemnestra and Aegisthus is for Argos in the *Oresteia*; the friendship with Tydeus and Capaneus and the condemnation of the virtuous Amphiaraus leave no doubt. Eteocles knows from the beginning of the play that his own death is certain; a man confident of survival would hardly include the Erinys among the gods to whom he prays at the beginning of the battle (69 f.). Von Fritz allows that Eteocles has no choice in the sense that he is able to decline to sacrifice his life for Thebes. But he is free, von Fritz argues, in the sense that he cannot help acting as he does because he is the man he is: a man fully conscious of his responsibility and unable to bear the shame that would result if he evaded it (p. 213).

Von Fritz has shown that Eteocles is, in a sense, innocent; when he goes to fight his brother, he is doing his duty. But it is still true that Eteocles is, in one sense, guilty; his action in going to fight his brother is not only a duty, but a crime. That is proved beyond dispute by the language of the Chorus when they implore him not to go. They warn him that the pollution that will result from such an act will never grow old (682); they exhort him not to let himself be carried away by θυμοπληθὴς δορίμαργος ἄτα (686 f.); they denounce the ὠμοδακὴς ἵμερος that is impelling him to reap a bitter harvest of blood that it is wrong for him to shed (692 f.). Eteocles cannot deny ‖ that what they say is true; he can reply only that the curse of Oedipus is come upon him.

The notion that an act may be at the same time a duty and a crime is to most ways of thinking paradoxical. Yet it is the obvious solution of the problem of Eteocles' decision; and it is supported by the best possible parallel. Agamemnon at Aulis has no choice but to sacrifice his daughter; without her death the fleet cannot sail to Troy, and Zeus has ordered that it shall sail. Yet the Chorus speaks of the sacrifice in language that unmistakably brands it as a crime (218 f.). Agamemnon has no choice but to perform this action which is both crime and duty; but his decision to perform it does not simply follow from the circumstances. Ate is sent to take a way his judgement (223 f.). Why is Agamemnon placed in this impossible position? I have argued (in an article called 'The Guilt of Agamemnon' [= ch. 21]) that it is because he is the son of the accursed Atreus. In the same way, I suggest, Eteocles is put in an impossible position by the working of the curse of Oedipus, itself in all probability a consequence of the earlier curse upon Laius.

Von Fritz's defence of the feeble line 472 against Eduard Fraenkel's athetesis (*SB der Bayer. Akad.* 1957, 29 f.) does not convince. He is right in taking the use of the perfect with *καὶ δή* to indicate the rapidity of the instantaneous decision, and it is strange that he does not recognize that this usage can hardly follow the hesitant utterance which Eteocles, if l. 472 is kept, begins with. Von Fritz prints l. 634 with the manuscript reading *κἀπικηρυχθεὶϲ*: but]*ηρυθειϲ* in P. Oxy. 2333 points strongly to *κἀπιγηρυθεὶϲ*: cf. *Suppl.* 460.[2]

In the other new paper von Fritz offers a careful examination of the treatment of the myth of Orestes in the three tragedies that deal with his revenge and in the *Orestes* of Euripides.

Von Fritz holds that Euripides was in a sense 'polemizing' against his predecessors. But he demonstrates that the 'polemic' did not, as is commonly asserted, take the form of a protest against the moral values which their plays exemplified; for Euripides represents the situation at Argos at the moment of Orestes' return as being substantially different from what it is in the plays of Aeschylus and Sophocles, and this difference in itself accounts for the different moral attitude to the action of Orestes which the two Euripidean plays express. This helps to support von Fritz's view that the moral values of a tragedy are determined by the situation itself, and do not vary according to the poet's caprice, an unpalatable truth to the modern advocates of a 'littérature engagée'. Euripides' 'polemic' against his predecessors, if that is the right word for it, indicates not a wish to substitute a new set of moral values for theirs, but an eagerness, natural in a competitor for tragic prizes at a time when poets constantly treated of the same legendary themes, to present a familiar story in a striking and unfamiliar way. The same eagerness will account for his disrespect to Aeschylus in the matter of Orestes' footprints and in that of the Messenger's speeches in the *Seven against Thebes* (*Phoen.* 751–2, a passage I unfortunately forgot to quote when I discussed the footprint scene in *CQ*, NS 11 (1961), 177 f. [= ch. 26]).

Von Fritz in his study of the *Alcestis* uses a comparative analysis of

[2] [Hutchinson ad. loc. defends *κἀπικηρυχθεὶϲ*, saying that 'a herald will announce Polynices the rightful king'; with the papyrus reading, the clause he thinks, 'could not be amplified by 635'. Why not? Polyneices mounts upon the wall and makes proclamation to the city, 'uttering a paean to its destruction'.]

many later treatments of the theme to bring out the significance of Euripides' own choice of method by contrast with the modifications which the taste of later ages found necessary. Wieland, Alfieri, Hofmannsthal, Browning's 'Balaustion's Adventure', and the 'Dream of Alcestis' of the American poet Theodore Morrison are all made use of; not, however, the recent *Alcestis* of Thornton Wilder, which appeared first in a German translation; but the piece most helpful to the understanding of Euripides is Goethe's satire on Wieland's production. Von Fritz agrees with Lessing that this gives no adequate interpretation of Euripides' play; but it points ‖ to certain significant factors which von Fritz's own analysis does not reckon with.

Von Fritz believes that Admetus is depicted as an unreflecting egoist, who is only gradually aroused by his wife's loss to the realization of his own selfishness. Alcestis' concentration upon her children, and not her husband, during her farewell scene seems to him evidence for a coldness between them; the scene with Pheres helps to expose Admetus' selfishness; and even his entertainment of Heracles at the time of his bereavement is discreditable, since it is done not so much from generosity as from an unworthy concern for public opinion. Von Fritz seems to have written this paper before seeing the excellent treatment of the play by A. M. Dale in the introduction to her valuable commentary (Oxford, 1954; see in particular pp. xxii f.). Miss Dale ascribes the prevalence of interpretations like that of von Fritz to 'our inveterate modern habit of regarding a drama almost exclusively in terms of its characters' (p. xxiv); she believes (like A. Lesky, quoted by von Fritz on p. 301) that Euripides has not the least interest in the kind of person Admetus was. She points out that Alcestis' preoccupation with her children during her last moments is no evidence for an estrangement from her husband; the words ἐγώ cε πρεcβεύουcα (282) are by ancient standards a sufficient indication of her feelings. The scene with Pheres is wholly understandable in terms of Euripides' delight in the rhetorical amplification of both sides of an argument; and the importance of Admetus' φιλοξενία as a central theme can hardly be disputed (note especially the prologue and the ode beginning at l. 568). Miss Dale well observes that when Admetus says ἄρτι μανθάνω (940), he means not 'I see now that I ought never to have accepted this sacrifice', but 'I see now that Alcestis is better off in dying than I in living, since I have lost my happiness and my reputation.' It is hard to believe that any ancient Greek, any more than any modern Chinese or Japanese, threatened with death but able to survive by means of a woman's sacrifice, would have dreamed of refusing.[3]

[3] [See the excellent criticism of von Fritz in *MH* 29 (1972), 124–40 + *MH* 30 (1973), 130–43 = A. Rivier, *Études de littérature grecque* (1975), 101–38.]

It is strange that von Fritz should have been deceived by the romantic misunderstanding of the character of Admetus, when in his paper on the *Medea* he has given a valuable warning against the romantic misunderstanding of the character of Jason. That paper begins with a useful résumé of what is known of the myth before Euripides, in which one or two points of detail call for comment.

Von Fritz is inclined to accept the statement of Dicaearchus that Neophron's *Medea* preceded that of Euripides; he might have referred to the article of E. A. Thompson, *CQ* 38 (1944), 10 f. He thinks it likely that Euripides' *Aegeus* was earlier than his *Medea*, but not for the reasons given by Wilamowitz in *Hermes*, 15 (1880), 482 f. = *Kl. Schr.* i. 18 f. Von Fritz puts forward the curious argument that Euripides is not likely to have written after his *Medea* a play in which 'the negative side of Medea's character appeared so exclusively as in the *Aegeus*'. It is more pertinent to recall that B. B. Shefton in a careful scrutiny of the artistic evidence (*AJA* 60 (1956), 159 f.) has shown that the vases that connect Medea with Theseus' encounter with the bull of Marathon are all later than about 430 BC. The Bologna cup by the Codrus Painter showing Aegeus with Aethra and Medea belongs to the mid-thirties (Shefton, op. cit. 163 n. 29); if we could feel sure that this reflected knowledge of the *Aegeus*, the *terminus ante quem* could be set back a little. These facts supply a slight indication that the *Aegeus* was the earlier of the two plays; but as Shefton observes (161 n. 15) Medea's flight to Athens and her marriage to Aegeus were Attic legend before Euripides.

Von Fritz goes on to give a careful analysis of the scene between Medea and Jason (446 f.) in which he protests with good reason against the current fashion of supposing that Jason is depicted as a mere hypocrite. In the circumstances in which he finds himself, it is understandable that he should accept Creon's offer; especially according to the way of thinking of the ancient world, which was very different from our own ‖ where women, love, and marriage are concerned. There is no reason to doubt him when he says that he is marrying the princess partly for his children's sake, nor to question the sincerity of his claim that he is well disposed towards Medea and eager to help her. It is also understandable that Medea should receive his overtures as she does. When he tells her that it is to Eros and Aphrodite, and not to her, that he owes his preservation in Colchis, there is, as von Fritz shows, a certain justification for his claim in terms of ancient heroic ethics; yet it is not surprising that Medea hears his claim with bitter indignation.

Using the same method that he has applied to the *Alcestis*, von Fritz now goes on to use a comparative analysis of successive treatments of the story to throw light on Euripides' reasons for handling the legend in the particular way he did. We may profitably compare the not dissimilar treatment of the Medea theme by W.-H. Friedrich, *Nachr. der Akad. der Wiss. in Göttingen*, phil.-hist Kl. (1960/4), 67 f. Von Fritz's study confirms the results of his detailed examination of Euripides' play itself by leading to the conclusion that no author of a play about Medea, except the eighteenth-century French poet Longepierre, has found it necessary to depict Jason as a 'ready-made villain', as a modern critic has absurdly called Euripides' Jason. His excellent treatment enables us to see the tragic conflict of the play in its full severity.

Perhaps the most important paper of all is the long study of 'Tragische Schuld und poetische Gerechtigkeit in der griechischen Tragödie', first printed in 1955.

Von Fritz begins by drawing attention to a misapprehension of the *Poetics* which has had serious consequences; most interpreters have confused the imperfection of character which fits a person to be a tragic hero (1452^{b}30 f.) with the ἁμαρτία μεγάλη which involves him in misfortune (1453^{a}7 f.). The English expression 'tragic flaw' fits only the former, the German expression 'tragische Schuld' only the latter; but both expressions have long been commonly applied to both concepts, with unfortunate results. Did Aristotle mean the ἁμαρτία μεγάλη to be a crime or an error? Using the results of the examination of the word and its cognates by P. W. Harsh, *TAPA* 76 (1945), 47 f., but checking them against the evidence of selected tragedies, von Fritz shows that 'error' is the right meaning, and that such an error is not necessarily criminal. The confusion has been exploited from an early date to father upon Aristotle a doctrine of 'poetic justice' altogether foreign to his way of thinking.

When tragedy came to life again at the end of the Middle Ages, it was influenced not only by Greek tragedy, but by the Stoic tragedy of Seneca and the medieval morality plays. Neither Stoic nor Christian drama demands the principle of poetic justice; for both, virtue should be its own reward and sin its own punishment; yet as von Fritz says the influence of the two was bound to recommend it.

Both in his plays and in his theoretical treatises, Corneille did homage to the principle, yet he was too honest to claim that the Greek tragedians had everywhere observed it; the signal counter-instance furnished by the *Oedipus Tyrannus* brought this home to him. Shakespeare fortunately knew nothing of Aristotle's rules as they were interpreted across the Channel, and did not observe poetic justice. Von Fritz rightly observes that Shakespeare's conception of tragedy stands nearer to that of the Greeks than any other modern writer's. It is a pity he diminishes the force of this observation by his continual reiteration of the now highly unfashionable doctrine of A. C. Bradley that the fate of every Shakespearian hero follows necessarily from his own character (he makes an exception of Hamlet). Is this any truer of Macbeth than it is, say, of Sophocles' Ajax? But Shakespeare was an isolated exception. On the Continent poets and scholars accepted the crude formulation of poetic justice given in André Dacier's ‖ commentary on the *Poetics* (1692), which in von Fritz's words makes the *Oedipus Tyrannus* into 'a kind of glorified *Struwelpeter* for adults'. Lessing unfortunately neglected to correct this misapprehension; and the German drama of Schiller's contemporaries was marred by a quasi-Calvinistic conception of predestined guilt that was ill fitted to help in the production of genuine tragedies.

The same miscontruction of Aristotle did not fail to affect Hegel's conception of tragedy as involving a collision of the right exemplified by the 'ethical institution' to which the hero is committed with some other kind of right, and the reaffirmation of eternal justice as a whole against the pretensions of its parts that results from his destruction. Von Fritz points out that this theory wholly fails to account for some of the most famous tragedies. An obvious counter-instance is provided by the *Oedipus Tyrannus*; but even Hegel's favourite example, the *Antigone*, provides another. Who but a bigoted nationalist, and one grossly deficient in aesthetic sensibility, would have argued that Creon and Antigone represented moral viewpoints of equal validity? But von Fritz gives Hegel credit for advancing a view closer to Aristotle's than any of the then prevailing heresies, and for insisting that tragedy must depend on the limitations of human nature. It is a pity he does not go on to discuss the views of Schopenhauer and Nietzsche regarding tragic guilt. But in the present state of letters he cannot be reproached for using space to castigate the neo-Lutheran absurdities of a modern critic of no significance, and to indicate in gentle terms that the statement that tragedy represents 'einen Fall aus dem Schein in die Wirklichkeit'—the only sense to be extracted from a wordy pronouncement by Heidegger—is, as far as it has meaning, false.

Von Fritz has done valuable service in correcting such an important misconception of Aristotle and in revealing the full extent of its effects. But Aristotle's treatise, even when its words are rightly understood, hardly gives a satisfactory account of the part played by guilt in Greek tragedy. Certainly the Greek tragedians possessed nothing like the childish notion of poetic justice entertained by many dramatists in modern times. But we may well hesitate to deny that poetic justice of a kind was observed by Aeschylus, an author in whom Aristotle seems to have felt comparatively little interest. A famous chorus of the *Agamemnon* denies that the gods punish the innocent and insists that they punish only the guilty, either themselves or their descendants after them. All that we know of Aeschylus suggests that his plays regularly exemplified this doctrine. In the technical sense of the word employed by Aristotle, the *ἁμαρτία* of Orestes was the killing of his mother. But how does Orestes come to be in the position of having either to kill his mother or to incur the awful penalties with which Apollo threatens him in case of his refusal? He is the son of Agamemnon. And how does Agamemnon come to be in the position of having to sacrifice his daughter, thus provoking his wife's enmity, or to disband his fleet? He is the son of Atreus, the perpetrator of the Thyestean feast. In the same way, in the Theban trilogy Eteocles is the victim of the curse of Oedipus; that curse in turn was in all probability a consequence of an earlier curse on Laius.

It seems hardly credible that the oracle given to Laius (*Sept.* 742 f.) was given to an innocent man. Can we guess at the nature of Laius' guilt? The account we have of his abduction of Chrysippus and the curse laid upon him by Chrysippus' father, Pelops, comes from a late source (Σ Eur. *Phoen.* 1760), and we cannot feel certain that the story in this form was ‖ known to tragedy before Euripides; although the confidence of L. Deubner, *SBPAW* 1942, 10 n. 5 [= *Kl. Schr.* (1982), 644], 'daß Aeschylus den erotisch-pathologischen Stoff behandelt hat, ist ganz unwahrscheinlich', will not be shared by those who remember what is known of the Achillean trilogy.[4]

It seems that in the case of the later generations of the Labdacidae and the Atreidae, the *ἁμαρτία* in Aristotle's sense must be distinguished from the *ἀρχὴ κακῶν*, the original source of guilt. This distinction does not exist in the case of Xerxes, whose guilt derives from the same action that involves him in catastrophe; the same may be true of the Danaids, whose guilt presumably derived from their treacherous murder of the Aegyptiads.

So also in Sophocles there are some heroes in whose case *ἁμαρτία* and *ἀρχὴ κακῶν* are identical. This is true of Heracles, whose fate is precipitated by his injustice towards the house of Eurytus; it is manifestly true of Creon in the *Antigone*. The doom of Ajax is brought upon him by his disrespect towards Athene. His guilt, like that of Cassandra in the Agamemnon, is of a special sort; the relations of the Greek gods to men are different from those of other gods, and mortals who offend them do not forfeit human sympathy; Prometheus is another tragic hero whose guilt is of this special kind. The two Sophoclean heroes whose guilt is the subject of dispute are Antigone and the Oedipus of the *Tyrannus*.

4 [See *The Justice of Zeus* (2nd edn. 1983), 120 f.]

Antigone's ἁμαρτία in Aristotle's sense is the act of burying her brother. In spite of Hegel, that act is represented by the poet not as a crime, but as a duty; and if Antigone, like Electra, is shown as having all the impulsiveness and obstinacy that go with her heroic qualities, these qualities do not constitute a cause of guilt. Antigone commits no crime, and yet she comes to a sad end. Is Sophocles concerned to explain how this can happen? or is he indifferent to the doctrine proclaimed with so much emphasis by the Chorus of the *Agamemnon*? Although the guilt which is fatal to Antigone plays no part in the action of Sophocles' play, he has given clear indication of its nature. The inherited curse upon the Labdacids forms the main theme of the second stasimon (582 f.), one of the most impressive odes in all tragedy; and when the Chorus explicitly points to this curse as the cause of Antigone's doom, she herself does not deny it (856 f.).

There remains the curse of Oedipus. The more honest and perceptive critics in all ages have seen that Oedipus' ἁμαρτία in the Aristotelian sense—the killing of his father and marriage with his mother—does not involve guilt in any subjective sense. The attempt to show that he is punished for impetuosity and irascibility is just as unconvincing in his case as in Antigone's. Is Oedipus then a wholly innocent victim? Sophocles felt no obligation to mention the cause of Oedipus' guilt; it lay for him outside the limits of the play. But the Oedipus of the second play accounts for it in Aeschylean terms, as the consequence of a family curse:

θεοῖϲ γὰρ ἦν οὕτω φίλον
τάχ' ἄν τι μηνίουϲιν ἐϲ γένοϲ πάλαι (964–5).

In the *Oedipus Tyrannus* also it is likely that this traditional cause of Oedipus' guilt is taken for granted; yet Sophocles felt no need to mention it. Nor, in all likelihood, would he have mentioned the part played by the curse in the undoing of Antigone, had it not offered him an opportunity for poetic explanation, as did the curse upon the Pelopids at *Electra* 504 f. Sophocles is not always concerned to employ guilt as a topic; but there is no reason for denying that his view of human guilt and divine justice resembled that of Aeschylus.

Does even Euripides ever explicitly represent a person free of guilt of any kind as meeting with disaster? The question is a complicated one, and can hardly be gone into within the limits of this already long review. But it might prove harder than at first sight most people would imagine to prove that the answer to this question should be Yes.[5]

Von Fritz's clear, honest, and intelligent book makes a really substantial contribution to the understanding of tragedy, ancient and modern. It deserves and will repay most careful study.

[5] [Recent discussion of the problem is surveyed by Suzanne Saïd, *La faute tragique* (1978).]

13(R47)

Review of Jones, *On Aristotle and Greek Tragedy*

John Jones, *On Aristotle and Greek Tragedy*. London: Chatto and Windus, 1962. Pp. 284.

HOUSMAN's low estimate of his classical colleagues as literary critics is strongly borne out by the character of the most fashionable handbooks upon Greek tragedy, which approach the subject from a point of view very like that represented in Shakespearian criticism by Andrew Bradley. A sharp corrective is administered by this remarkable book, whose author taught law before he began teaching English, but has never been a professional classical scholar. The work is imagined by the author as being 'the first part of a trilogy dealing with the presentation of the self in western literature'. It is by no means easy reading, for its style and vocabulary are harsh, technical, and unappealing; and though in many places this style helps the author to express concisely complicated thoughts, there are others where its artificiality introduces unnecessary difficulties. The author's ruthless and lawyer-like demonstration of his thesis seems to me a little over-rigid, as I shall try to show. But he has produced by far the most important critical study of the subject to appear for many years, and any reader with a serious interest in Greek tragedy as literature has no choice but to make the effort which the reading of his book demands. Most of my space must be used to show where I differ from Mr Jones; but the differences between us are far less numerous than the agreements.

The opening section starts by correcting two misapprehensions of the *Poetics* that have led to grave misunderstanding of Aristotle's whole approach to tragedy. Max Kommerell had already shown that the belief that Aristotle thought that every tragedy must have a hero rests on a mistranslation in the thirteenth chapter of the *Poetics* of which even Bywater was guilty (*Lessing und Aristoteles* (Frankfurt, 1940), 160; cf. pp. 189, 228, 293). Mr Jones shows how this error has been used to obscure the implications of the statements in the seventh chapter that 'tragedy is an imitation not of human beings but of action and life' and that 'the stage-figures do not act in order to represent their characters; they include their characters for the sake of their actions'. In defiance

Review of English Studies, NS 15, no. 58 (1964), 221–4 [= *Classical Survivals* (1982), 111–15.].

of these doctrines, the imported concept of the hero has been used to interpret tragedy in terms of an interest in individual character that is altogether modern. Having done much to elucidate Aristotle's theory, Mr Jones goes on to examine the extant plays in the light of his results.

At the start he is encouraged by the observation that the lack of individuation in the characters of the *Suppliant Women* of Aeschylus helped to maintain the belief in its early date, almost universal from the beginning of this century until the publication of new evidence in 1952 showed that it was almost certainly written after the *Persians* and the *Seven Against Thebes*. Aeschylus' art was thought to have developed from naïve beginnings represented by the *Suppliant Women* to the comparatively advanced characterization which the *Oresteia*, performed two years before his death, was supposed to show. This view of the *Oresteia* is strongly disputed by Mr Jones, who believes it to be concerned hardly at all with the individual members of the House of Atreus, almost entirely with the House itself and the re-establishment of its normal existence which has been ‖ disturbed by the usurpation of Agamemnon's murderers. Mr Jones allows even the characters of the *Oresteia* the smallest possible degree of individuality; he shows convincingly how modern critics' interest in individuals has blinded them to much of the effect of the play as a whole, and he justly complains that both those who make out Agamemnon as a hybristic tyrant and those who regard him as a perfect gentleman 'operate at the same level of individualistic humanism'.

Mr Jones finds fault with those who have applied the word 'primitive' to Aeschylus' thought; but he himself has been carried too far by his anthropological approach to Aeschylus' characters, a few of whom do reveal the beginnings of individual portrayal. The basic principle of ἦθος, the word which in ancient literary criticism roughly corresponds to our word 'character', is that the persons in a story should have the qualities which their actions presuppose. A few of Aeschylus' characters act in a way that seems to presuppose both good and bad qualities; and this mixed nature is clearly indicated in their presentation. Two such characters are Agamemnon in the *Oresteia* and Eteocles in the *Seven Against Thebes*. Agamemnon is both the brave king and warrior and the accursed son of Atreus, doomed from birth; Eteocles is both the brave king and warrior and the accursed son of Oedipus. Both have to be made comparatively sympathetic; and the disastrous errors of judgement which they make are explained in Homeric terms by saying that in virtue of the curse they have inherited Zeus has sent Ate, the spirit of destruction, to take away their wits. Agamemnon is neither wholly good nor wholly bad, but has elements of both, as he has in Homer, of whose all-pervasive influence on early tragedy Mr Jones seems not sufficiently aware. In exalting the importance of the House of Atreus, he underrates that of the individual heroes who are its members, which is as real for Aeschylus as it was for Homer. The double character of these two kings seems to me to have obliged the poet to give them a more distinct individuality than his other characters (for Agamemnon, see *CQ*, NS 12 (1962, 187 [= ch. 21, pp. 283 f.] and for Eteocles, *Gnomon*, 34 (1962), 740 [= ch. 12, pp. 195 f.]). Clytemnestra, says Mr Jones, 'stands outside the norm'; her abnormal qualities have been given her not for their own sake, but because her actions require that she possess them. This is both true and important and it is true of other characters beside Clytemnestra; but because Clytemnestra's striking characteristics are not there for their own sake, but for the sake of the action, it does not follow that the poet is not free to exploit them to secure subsidiary effects; and when in Mr Jones's words 'a modern sensibility snatches hungrily at the touches of reckless moral autonomy . . . which Aeschylus bestows on her', it has rather more justification in doing so than Mr Jones allows.

'Sophocles's achievement', says Mr Jones, speaking of his treatment of individual character, 'was to realise privacy'; 'in the place of family solidarity', he writes, 'we have individuation and (in a vague provisional sense) personalising of consciousness'. Sophocles' work certainly shows a development in this direction, but a rather less abrupt development than Mr Jones allows; for example, his comparison of Sophocles' treatment of the revenge of Orestes in his *Electra* with Aeschylus' handling of the same story in *The Libation-Bearers* || sometimes lays a wrong emphasis on the differences between them. Aeschylus' play is part of a trilogy, but Sophocles' an independent drama in which he chose to concentrate attention on the figure of Electra, whose heroic nature is revealed in her clashes with her sister and her mother. Like those of Aeschylus, the characters of Sophocles reveal no idiosyncrasy, but the more important ones are given a distinct individuality, always in a manner suitable to the requirements of the action. It is indeed wrong to say (or to think that Aristotle said) that every play must have a hero in the sense of a central figure around whom everything revolves. But every complete play of Sophocles does contain a figure of heroic stature, whose character is deliberately contrasted with those of ordinary mortals. Sometimes the hero stands out in contrast with his enemies, like Ajax, Philoctetes, or Electra in relation to Clytemnestra; sometimes an obvious foil for him is created, as Ismene for Antigone, Chrysothemis for Electra; Creon's relation to Oedipus in *Oedipus Tyrannus* has something in common with both these types. Yet always character is subordinate to plot and action, and this important fact Mr Jones admirably demonstrates. His section about

Sophocles is full of acute observations and profound insights. Here, in particular, one regrets the brevity of the book, for with more space available it could have contained more of the analysis of particular sections of the texts that is best exemplified by his excellent treatment of an episode in the *Philoctetes* (pp. 162 f.); best of all is the chapter called 'Men and Mutability'. I believe that he underrates the *Ajax* through a failure to understand it; that he mistakenly neglects the important part played in the second Oedipus play by what is going on at Thebes; and that he mistakes the nature of Antigone, to which the family curse (see 582 f., and cf. *El.* 515 f.) is relevant in a way recalling Agamemnon and Eteocles in Aeschylus. But these differences of opinion must be elaborated elsewhere.

The section about Euripides is on a smaller scale than the rest, and is less satisfying. If Sophocles attained privacy, Euripides according to Mr Jones attained inwardness; he instances some of the many passages in which a character is shown debating with himself about the merits of a proposed course of action and inclining now one way and now the other. His favourite example is the perplexity of Agamemnon at the opening of the *Iphigeneia in Aulis* as he debates whether or not he shall entice his daughter to Aulis by a trick so that she may be sacrificed. Euripides, like Aeschylus, liked to exploit dilemmas like that of Agamemnon at Aulis or that of Orestes when he had to decide whether or not to kill his mother. Aeschylus, as was natural for a writer of his date, gave a theological explanation for the decisions of his characters; he could say that Zeus sent Ate to take away Agamemnon's wits so that he sacrificed his daughter, or that Apollo told Orestes that he must kill his mother. Euripides was not concerned to deny the existence of the gods (a fact clearly realized by Mr Jones, who has no sympathy with the popular notion of this poet as a propagandist); but he preferred to get his effects by showing different passions at work within the human mind and struggling with each other to influence its decision. Whether these passions are or are not implanted by the gods is of no special consequence, though it is worth observing that in some plays, in the *Hippolytus* for instance, the || poet tells us that they are. Euripides certainly shows an access of interest in the human passions, and not only in their effects, but in their detailed workings. But this does not justify the somewhat vague statement that he shows an access of interest in human beings, or in human consciousness. When he presents the consciousness at work, it is always either registering impressions that conform to certain standard types (even in Alcestis' death scene, which is here misrepresented through failure to recognize the convention) or debating a course of action by presenting arguments on both sides. Nowhere does the poet show the slightest interest in human idiosyncrasy; certainly not in connection with the farmer to whom Aegisthus marries Electra in the *Electra*. Characters of humble status but of heroic virtue were familiar to the fifth-century audience, which knew well the part played in the *Odyssey* by Eumaeus and Philoetius; although poor, the farmer is of good birth; and as the trial scene in the *Orestes* indicates, Euripides had good reasons for showing his partiality for honest farmers. The debates of Euripidean characters with themselves, like their innumerable debates with one another, are never far removed from rhetoric. Does their author show more interest in human beings than Sophocles, whose characters, for all their lack of idiosyncrasy, are more vivid and better able to arouse our sympathies? Sophocles' form is perfectly adapted to his matter, so that his characters seem to speak with the straightforward utterance of passion.

The main thesis of this book, that Greek tragedy must be understood not in terms of character but in terms of plot and action, seems to me to have been demonstrated with triumphant success; a major factor in that success is the author's understanding of the religious outlook of early Greece and his freedom from the usual tendency to interpolate modern religious notions. Most modern presentations of the subject err in exaggerating the accessibility of Greek tragedy to our understanding. Mr Jones supplies a valuable corrective, but has gone a little too far in the opposite direction. 'It turns out to be our bad luck', he writes on the last page but one, 'that Greek tragedy is superficially intelligible in a modern way.' That sentence might almost imply that the fifth-century Greeks inhabited a world of thought mysteriously different from our own, like that of primitive savages whose cast of mind is strangely remote from that of normal civilized beings. If instead of crudely assimilating the early Greek world outlook to one or other of those familiar to ourselves we are willing to start again from scratch and make a careful and dispassionate investigation of the data bearing on it that we possess, there is no reason why we should approach with undue despondency the effort needed to enter in imagination into the state of mind necessary in order to understand Greek tragedy. Mr Jones's book has done a good deal to increase our prospect of success.

14(R56)*

Review of Snell, *Scenes from Greek Drama*

Bruno Snell, *Scenes from Greek Drama.* (Sather Class. Lectures 34.) Berkeley and Los Angeles: University of California Press, 1964. Pp. x + 147.

THE Sather Classical Lectures have lately been the subject of an eloquent eulogy by Sterling Dow (*Fifty Years of Sathers*, 1965), and the series certainly includes some books of great value. But the invitation to give a series of lectures at a specific place and time is not always an unmixed benefit even to a distinguished scholar. The author of the book before me has rendered services to classical studies whose importance would not be easy to exaggerate; and these, no less than the personal qualities which have so endeared him to colleagues, pupils, and audiences in many countries, make it highly distasteful to criticize adversely anything that he has written. Yet the truth must be stated that this book is deeply disappointing. From the beginning of his notable career Professor Snell has been a passionate apostle of a particular kind of *Geistesgeschichte*. If this is uncongenial to many scholars of a younger generation, that is in the nature of *Geistesgeschichte*, and these must remember that our own point of view will be equally unattractive to their successors. But even a reviewer who was more sympathetic than I am to Snell's outlook in such matters might well take exception to the way in which, in the greater part of this book, *Geistesgeschichte* is forcibly extracted from words and sentiments not really suitable for this use. In this review I shall not often criticize ‖ Snell's point of view, much as it differs from my own; I shall be more concerned to criticize his method, and his conclusions.

The first lecture deals with the speech of Achilles in the Florentine papyrus which most scholars have assigned to the *Myrmidones* of Aeschylus (fr. 132c Radt). Definite proof of its Aeschylean origin is lacking, and Page and I in our Loeb volumes were right to remind the public of this fact; but there is much to be said for the generally accepted view, which Snell thinks certain.

His argument depends on ll. 9–14, which he prints as follows (see

* [Cf. *CR*, NS 23 (1973), 192 f. = ch. 15.]

Gnomon, 37 (1965), 12–17.

pp. 139 f. for a text and apparatus criticus of the fragment based on fresh study of the papyrus):—

ἄγ' εἴ]περ εἷc ὢν, ὡc λέγουcι cύμμαχοι,
ἥccα]ν τοιαύτην ἔκτιc' οὐ παρὼν μάχηι,
κείροι]μ' ἐγὼ τὰ πάντ' Ἀχαϊκῶι cτρατῶι.
τοῖον] δ̣' ἀφεῖναι τοὖποc οὐκ αἰδώc μ' ἔχει·
τίc γὰρ] τ̣οιούτ[ο]υ̣ς̣ εὐγενεcτέρουc ἐμοῦ
ἀρχοὺc ἄ]ν [εἴ]π̣οι καὶ cτρατοῦ ταγ̣ [ε]ύ̣ματα;

Snell translates (3) as follows: 'If, as my confederates assert, I alone am the cause of such a devastating defeat, then I will [spoil] everything for the Achaean army. For who can maintain that such leaders and such military detachments are nobler than I?'

The supplements in the first three of these lines are Snell's own, and they are far from certain. Neither *ἄγε* nor *ἄγετε* is as far as I know used with an optative of wish following, at any rate in tragedy; Croenert's revision of Passow (pp. 74 f.) has no instance; and though *ἡccάομαι* occurs in tragedy, *ἧccα* seems not to be found in verse. The oddity of the optative of wish *κείροι]μ'* is obscured in the translation by the rendering 'I will spoil'; what the words mean is 'may I spoil'. A wish is different from a statement of intention, and a statement of intention is what Snell's interpretation requires, as his translation clearly shows. Nor is the verb *κείρειν* specially likely to have stood in this place. In l. 12 *τοιόνδ'* is preferable to *τοῖον δ'*: and in l. 14 *ταγ[ε]ύματα* must mean not military detachments', but 'commanders', being a synonym of *ἀρχούc*.

I shall return presently to the weaknesses of this text; but first let me examine the conclusions Snell draws from it. He observes that the threat of stoning is not in Homer, and thinks it implies that Achilles is a traitor to the state according to a way of thinking foreign to the epic, but characteristic of the fifth century. In replying to a threat by declaring that he feels no *αἰδώc*, Achilles, in his view, is 'throwing shame and decency overboard in pursuit of an aim of his own which he considers right' (8). 'On the one hand', Snell writes (p. 11), 'the state demands the unconditional observance of its legal system; on the other man develops an awareness that whenever right is on his side, he must stand up for it unconditionally.'

If l. 12 is to be assigned such far-reaching importance, it is imperative to determine if possible what precisely is the saying which *αἰδώc* does not restrain Achilles from uttering. This saying is contained in ll. 9–11; unfortunately the beginnings of these lines are lost, and as I have already shown the new supplements which Snell offers are anything but certain. If we could believe that they, or something like them, were correct, then Achilles would be asserting that he was not ashamed to express the wish that he might '[spoil] everything' for the Achaean army. Most editors (for instance Page, Mette, and I) have adopted a different type of supplement, according to which Achilles would be asserting that he was not ashamed to boast that he was all in all to the Achaean army. Which sense is likelier? Luckily the two following ll. 13 and 14 offer a means of checking the relative probability of the two supplements; for the highly probable *γὰρ* in l. 13, accepted by Snell as by all recent editors, would imply that they supplied a justification for what the speaker has said immediately before. Now if one has threatened to '[spoil] everything' for the Achaean army and declared that *αἰδώϲ* will not restrain one, it is very odd indeed to justify oneself by saying, 'For who can maintain that generals like these are nobler than I?' Often people's being ‖ nobler than oneself does not restrain one from fighting them. But if one has declared that one is not ashamed to boast that one is all in all to the Achaean army, then it is natural to justify one's claim by insisting that one is nobler than its generals. *αἰδώϲ* is regularly felt by a man for those whose birth is nobler than his own; thus in the Doloneia (*Il.* 10. 237 f.) Agamemnon tells Diomedes not to choose a partner for the dangerous night-raid simply out of deference for his nobility:

> *μηδὲ cύ γ' αἰδόμενοc cῇϲι φρεϲὶ τὸν μὲν ἀρείω*
> *καλλείπειν, cὺ δὲ χείρον' ὀπάϲϲεαι αἰδοῖ εἴκων,*
> *ἐϲ γενεὴν ὁρόων, μηδ' εἰ βαϲιλεύτερόϲ ἐϲτιν.*

The reason given by Achilles to justify his not being restrained by *αἰδώϲ* seems very like that given by Agamemnon for Diomedes' possible unfortunate choice; the *αἰδώϲ* would be no more than the respect a man may feel for those who are *βαϲιλεύτεροι* than he is. That is why even if something on the lines of Snell's not very probable supplements in 9–11 could be accepted, this passage is wholly unable to bear the weight of the considerable implications that Snell wishes to attach to it.

The second lecture is devoted to the lost *Ἱππόλυτοϲ Καλυπτόμενοϲ* of Euripides and the third to his surviving *Ἱππόλυτοϲ Cτεφανηφόροϲ*. Snell shares the optimistic view taken by many scholars, but argued in most detail in the careful dissertation of C. Zintzen (*Analytisches Hypomnema zu Senecas Phaedra*, 1960) that much of the content of the lost play can be reconstructed with the aid of Seneca's play on the same subject. Barrett (*Euripides, Hippolytos*, pp. 15 f.) has given warning that this procedure is by no means safe; I must explain why I agree with him, as I have done already at *JHS* 85 (1965), 164 [= ch. 36, p. 419]).

Snell points out that the scene in which Phaedra expresses the wish to hunt (Sen. *Phaedra* 358–403) is generally thought to be based on Eur. *Hipp.* 170 f. This scene, he argues, fits badly into Seneca's play, so that we can assume that the other more coherent parts come from a single Attic tragedy. This assumption is clearly not safe, and Barrett's warning that Sophocles' *Phaedra* may have been made use of must not be ignored.[1] Snell offers one new argument from a point of detail, which is based on fr. 430 Nauck^2 = fr. C Barrett:

> ἔχω δὲ τόλμης καὶ θράσους διδάσκαλον
> ἐν τοῖς ἀμηχάνοισιν εὐπορώτατον
> Ἔρωτα, πάντων δυσμαχώτατον θεόν.

Snell imagines that the words ἐν τοῖς ἀμηχάνοισιν denote a very special kind of moral helplessness by means of an expression 'typical of the early lyric poets' (it occurs at Archilochus 128 W. and at Sappho fr. 130 L.–P.). 'Helplessness', he writes (34), 'begins when in the vicissitudes of the emotions one loses contact with other men and stands alone; and the portrayal of that condition links Euripides with the archaic poets.' In fact in any period of Greek literature any person who finds it impossible to get what he desires can be said to be reduced to ἀμηχανία, and Phaedra in any version of the story must have been in that position. Eros in particular reduces his victims to it (cf. Sappho, loc. cit.); for the type of expression exemplified in l. 2 of the fragment, compare Aesch. *PV* 59, Arist. *Eq*. 758–9.

I see no reason to suppose that the use of ἐν τοῖς ἀμηχάνοισιν here is as refined and specialized as Snell believes, nor has the passage any bearing on the relation of Seneca's *Phaedra* (in which nothing corresponding to it is found) to Euripides' first *Hippolytus*. As with αἰδώς in the Achilles fragment, so with ἐν τοῖς ἀμηχάνοισιν here; Snell has pounced on the occurrence of a word that in his special kind of *Geistesgeschichte* has a particular significance, and has read into it implications which the context does not warrant.‖

This flimsy argument plays an important part not only in Snell's attempt to show that Seneca can be safely used to reconstruct the first *Hippolytus*, but in his attempt to determine the significance of this play in the development of moral thinking during the fifth century. 'The essential point', he writes on p. 46, 'is that he extends the old ἀμηχανία of passion to moral frailty and that he lets Phaedra become conscious of the consequences of this disposition.' Do we really know enough about the first *Hippolytus* to feel sure of this? 'Seneca's *Phaedra* according to this development', Snell continues, 'fits in exactly between the Admetus and the Medea of Euripides, and so it does not seem hazardous to identify her in broad outline with the Phaedra of the first *Hippolytus*, which in all probability was written between the *Alcestis* and the *Medea* (p. 48). To me it seems extremely hazardous.

[1] [See now O. Zwierlein, 'Senecas Phaedra und ihre Vorbilder', *Abh. der Mainzer Akademie*, 1987, no. 5.]

In the third lecture Snell reiterates the view previously expressed by him and other scholars that, when the Phaedra of the extant *Hippolytus* protests that she knows that she is right but is still prevented by her passions from acting on her knowledge, the poet is consciously 'polemizing' against the moral intellectualism of the Socratics. Barrett (op. cit., pp. 229 f.) has rightly protested against this opinion; I have given reasons for agreeing with him [ch. 36, p. 419 f.], but I must now add further reasons. The view of the nature of moral action taken in Phaedra's speech, and also in other places in Euripides, has more in common with the traditional view adopted in Greece before that time than is generally recognized. When an Aeschylean character makes a disastrous decision, Zeus has sent Ate to take away his wits. But even in Aeschylus, Ate acts upon such a character through the medium of passions that we can understand; parallel with the external motivation coming from the gods, we find an internal motivation coming from the passions. Euripides omits or prefers not to stress the external motivation; yet the domination of the actor's intelligence by his passions, the triumph of *θυμόϲ* over *βουλεύματα*, is simply the developed form of an internal motivation present already in the older poet. The irrationalism of Euripides has far more in common with the irrationalism of Aeschylus than with the rationalism of the Socratics; and it may have come into being as a development of the former rather than as a protest against the latter. Can we be certain that Euripides was adopting a new and original position, meant as a criticism of new and original positions occupied by contemporary thinkers who were not poets, and not rather voicing an attitude which he shared with most of his contemporaries?

The fourth lecture, on the *Antiope*, calls for little comment; the most valuable part (pp. 80 f.) is that in which Snell argues that Antiope's reflections about knowledge and whether it can help us in misfortune can be harmonized easily enough with the anti-intellectualism expressed in the utterances of Medea and of Phaedra. A long discussion of the debate between Zethus and Amphion on the active and contemplative lives brings little that is new, and is protracted by the inclusion of remarks about Orphic and mystical notions of the happy life that are interesting in themselves but ‖ not especially to the point. Snell traces back the idea of changing one's whole mode of life to the satyrs in the *Isthmiastae*, who are weary of their life with Dionysus, and to Phaedra, who longs to go hunting because she will then be near Hippolytus; but these cases are so different that I see little point in bringing them together.

The fifth and sixth lectures are devoted to the *Agen*, the satyric drama satirizing Harpalus which was ascribed by some to Alexander the Great and by others to the orator Python of Byzantium, but was more probably by the otherwise unknown Python of Catane, doubtless a contemporary of Alexander.

Athenaeus (595 F f.) makes two apparently inconsistent statements about its production: first he says that it was produced at a Dionysiac festival on the Hydaspes and then that it was produced after the famous flight of Harpalus. If the Hydaspes is to be identified with the Jhelum, in the Punjab, the play must have been produced while Alexander was there in the summer or autumn of 326 BC. But Harpalus fled to Athens 'about February 324' (see E. Badian's excellent article about him, *JHS* 81 (1961), 24).

The accepted view of the play's date was established by Beloch, *Griechische Geschichte*, iv/2. 434 f., who argued strongly for placing it after the flight of Harpalus. He therefore held that the Hydaspes meant by Athenaeus was not the Jhelum, but another Hydaspes in Media. Virgil, *Georg*. 4. 211 mentions a 'Medus Hydaspes', and so does Petronius (123. 239). Petronius may be echoing Virgil, but the statement that Pompey discovered the Hydaspes may rest on real knowledge.

The argument of Beloch that in 326 Alexander was too busy to organize theatrical performances is rightly brushed aside by Snell; troops always need diversion, and Alexander might well have made use of this one. Further, Beloch's contention that the 'flight' mentioned in fr. 1. 4, must be the famous flight of Harpalus is over-positive; Snell is right in arguing (pp. 114 f.) that we cannot be sure that the famous flight is meant, not that his own explanation of what it means is certain.

But there is another consideration that makes strongly against Snell's dating. In 326 Harpalus was one of the most powerful men in Alexander's empire, high in the royal favour; in 324 he was a friendless exile. In which of these years is he likelier to have been mocked in a performance given at the court of Alexander? Snell maintains (118 f.) that the earlier date is likelier, because only in the days of his power would there have been any point in satirizing Harpalus; his argument shows a strange ignorance of the character of courts and despotisms. He rightly remarks (pp. 120 f.) on the similarity between the fragments of the Agen and those of Theopompus' *Letter from Chios* addressed to Alexander (*FGrH* 115 F 251–4); that too must belong after the flight of Harpalus.

This argument may be strengthened if we consider the situation at Athens that is represented in the dialogue contained in fr. 1 of the *Agen*. One speaker says that while the Athenians were, in their own opinion, slaves, they had enough to eat, but that now they have to live on pea mash and fennel; the other answers that Harpalus has sent them wheat.

This fails altogether to fit any situation before or during 326; but it may very well have been the case in late 325 or early 324. The unsettled conditions of that year are powerfully depicted by Badian (op. cit. 25 f.). A great body of disaffected mercenaries, hostile to Alexander, was concentrated at Taenarum; Leosthenes was known to be in touch with them, and in all probability other Athenian politicians maintained varying degrees of contact. The dramatic date of the *Agen* clearly preceded Harpalus' flight, for he is represented as being still in Asia Minor. While he was still there, he had made presents to the Athenians; cf. Diodorus 17. 108. 3 *εἰϲ δὲ τὰ παράλογα τῆϲ τύχηϲ καταφυγὰϲ πορίζόμενοϲ εὐεργετεῖ τὸν Ἀθηναίων δῆμον*, a statement that may well refer to the gifts of wheat mentioned in the *Agen* (fr. 1. 14–15). Harpalus probably reached Athens in the second or third week of July 324 (Badian, op. cit. 41–3), and the composition and per‖formance of the *Agen* must be placed after the news of this event had reached Alexander. Beloch pointed out that we have evidence that in the autumn of 324 Alexander celebrated a Dionysiac feast at Ecbatana, in Media (Ephippus ap. Ath. 538 A, Arrian 8. 14. 1, Diodorus 17. 110. 7, Plut. *Alex*. 72, all quoted by Beloch, p. 435). That date would be perfectly suitable for the production of the *Agen*. As Beloch (p. 436) pointed out, we know little of the geography of ancient Media, and the name Hydaspes may well have been applied to the modern Karatschi, the river that flows through the valley of Hamadan in which Ecbatana is situated.

The reason for Snell's special interest in the *Agen* is that he believes that Harpalus represented for its author 'a burning political question, indeed a fundamental problem of Alexander's kingdom'. He thinks that Python, and also Theopompus in the Letter from Chios, was concerned to make a serious protest against the blasphemy committed by Harpalus in demanding divine honours for his successive mistresses, Pythionice and Glycera. He is taking both men's utterances more seriously than they deserve. The author of a dramatic entertainment satirizing a fallen minister may be expected to make fun of, and in all probability to exaggerate, the follies of his victim's private life. But such mockery cannot necessarily be assumed to convey a serious protest; if a comedian made in 1964 a number of jokes about Miss Christine Keeler, he does not necessarily deserve to be respected as the author of a solemn remonstrance against immorality in high places. Nor do the ponderous lucubrations of Theopompus deserve to be taken much more seriously; had Harpalus only retained the royal favour, he might have made his pig a god before Theopompus would have dared to breathe a word against him. When Theopompus (*FGrH* 115 F 253) speaks of 'those who have fallen for Alexander's kingdom and the freedom of the Greeks', Snell thinks that he 'believed (and demanded) that the Macedonian dominion could be reconciled with the freedom of the Greeks' and that he 'represents the traditional ideological opinion of the Greeks in as much as they were prepared to make the best of the Macedonians'. Snell has not considered that on the lips of the despot and his partisans the word 'freedom' did not mean quite what it does to him. He is right in regretting that the interesting fragments of the *Agen* have not received more attention; but his attempt to redate the work is unconvincing, and his attempt to prove that it is an important document of *Geistesgeschichte* fails completely.

His treatment makes some good points of detail. The word *ἄορνον* (fr. 1. 2) certainly suggests the scene of a *νεκυομαντεῖον*, and the suggestion that the satyrs took the parts of the *μάγοι* who called up the spirit of the dead Pythionice is attractive. In 1. 10, Snell strangely puts in the text Steffen's poor conjecture *κραίνουϲιν* and omits to mention Schweighaeuser's very probable *κατέχουϲιν*. The frequency of resolutions in the fragments is indeed surprising, but the suggestion that Python may have treated iota before a vowel after a long syllable as consonantal lacks a satisfying parallel.

15(R99)

Review of Snell, *Szenen aus griechischen Dramen*

Bruno Snell, *Szenen aus griechischen Dramen.* Berlin: de Gruyter, 1971. Pp. vii + 210.

THE first four chapters of this book consist of a German version of Snell's 'Scenes From Greek Drama' (Sather Classical Lectures 34, Berkeley, 1964), with text and footnotes somewhat amplified to refute the criticisms of reviewers, particularly Mrs Anne Burnett (*CP* 62 (1967), 288 f.) and myself (*Gnomon*, 38 (1966), 12 f. [= ch. 14]); the last three chapters consist of three lectures on late tragedy delivered at the London Institute of Classical Studies, the last of which, about the *Exagoge* of Ezechiel, appeared first in *Antike und Abendland*, 13 (1967), 150 f.

The lecture on the Florence papyrus attributed to Aeschylus' *Myrmidons* appears with some additions; but an unwary reader of the new version who did not study the text of the papyrus carefully might still not realize that Snell's attempt to make this text into an important document of *Geistesgeschichte* depends entirely on his own improbable supplement of l. 11. Mrs Burnett correctly stated (op. cit. 289) that 'the thing, mention of which *aidos* does not disallow, has been defined not by the papyrus but by Snell'. Snell asserts (p. 2 n. 2) that the new fragment published by Bartoletti in *American Studies in Papyrology*, 1 (1966), 121 f. definitely proves the piece in PSI 1211 to come from the *Myrmidons*; it is written, he claims, in the same hand as P. Oxy. 2163 frs. 1–6. PSI 1208–10 are indeed in the same hand as these Oxyrhynchus fragments (see Lobel, P. Oxy. XVIII (1941), 23), but not PSI 1211, which is quite different, as anyone who troubles to compare the photographs may see for himself. True, there is more to be said for assigning PSI 1211 to this play than I allowed when I published it (as fr. 286) in my supplement to the Loeb Aeschylus, and J. R. Rea, *ZPE* 7 (1971), 93–4 has strengthened the case by acutely reading the end of l. 14 as τὰ β̣ἑ̣λ̣τ̣ατα.

Snell's attempt to use a speculative reconstruction of Euripides' first *Hippolytus* to make *Geistesgeschichte* survives with little change; anyone

Classical Review, NS 23 (1973), 192–4.

who cares to look up the instances of ἀμήχανοϲ in early lyric quoted by him at p. 35 n. 19 may judge for himself whether they support the very special sense of the word which he finds in *Hippolytus* I fr. C Barrett (cited on p. 28). In the face of the ἀμηχανία that becomes common in poetry during the lyric age, he says, 'hilft nur das Dulden und Ausharren' (p. 43). Is the notion expressed by the verb τλῆναι insignificant in, say, the *Odyssey*? Virtually unchanged, also, is the attempt to show that Euripides in his *Medea* and *Hippolytus* II was engaged in a debate with Socrates. The chapter on the *Antiope*, also, survives almost unaltered: the delicate distinction drawn (p. 87), to meet Mrs Burnett's criticism, between Antiope's remark that ignorance is almost a pleasure (fr. 205 Nauck^2) and the Simonidean Danae's envy of her infant's ignorance is too subtle for me. The wholly unconvincing redating of Python's *Agen* and the surprising attempt to represent it as a serious criticism of 'die neuen Formen, wie sich staatliche Autorität etabliert' (p. 133 n. 64) reappear with few changes; energetic self-defence over minor details does little to conceal the fundamental weakness of the whole elaborate structure.

Nobody finds Snell's general theory of 'the discovery of the mind' more unconvincing than I do, and in *The Justice of Zeus* (Sather Classical Lec||tures 41, 1971 [2nd edn. 1983]) I have now explained my reasons. But it would be unfair to a distinguished scholar to judge the theory solely on the basis of its presentation in these four chapters, which do not show him at his best.

Ch. 5 is about Astydamas' *Hector*, to which Snell assigns the two papyri published by Page, *GLP* as no. 29 (p. 160) [= *TrGF* 60 frs. 1i and 2b] and P. Hibeh 174 (II, pp. 9 f.) [= *TrGF* fr. 1h], but not the piece published by R. A. Coles at *BICS* 15 (1968), 110 f. [= *TrGF*, Adesp. fr. 649]. The words ἐγώ τ' ἐμαυτοῦ χειρον[which follow l. 10 Page [= *TrGF* 60 fr. 1i. 11] are made to yield somewhat far-reaching conclusions about the development of mental concepts. Lines 22 f. [= ibid. 18 f.] read:

ἀϲπὶϲ δ' οὐ διῆκ' εἴϲ[ω ξίφοϲ
ἀλλ' ἴϲχεν αὐτοῦ, δεϲπ[ότην δ' ὁπλιϲμάτων
τὸν καινὸν οὐ προ⟨ὔ⟩δωκε.

I doubt whether either Astydamas or Astydamas' Hector knew of the idea that a shield might have magical power (p. 152); cf. Men. *Aspis* 14–16.

Ch. 6 contains a little about Agathon, more about Chaeremon. Aelian, *VH* 2. 21 says that Agathon often quarrelled with his lover in order to give him the pleasure of becoming reconciled; Snell wonders (p. 157) whether a fifth-century poet can have done this. When

Aristotle (*Rhet.* 3. 1413^{b}13) calls Chaeremon a ποιητὴς ἀναγνωστικός, that does not mean, Snell rightly says, that he did not write for the stage; he might have quoted Otto Zwierlein's demonstration (*Die Rezitationsdramen Senecas* (1966), 128 f.) that ἀναγνωστικός is a stylistic term; yet as Snell points out (p. 159) the acrostich of the poet's name discovered by R. Kannicht in P. Hibeh 224 (II. pp. 148–9) [= *TrGF* 71 fr. 14b] certainly shows that some of the verses of the polymetric *Centaur*, at any rate, were designed to appeal to the reader rather than the hearer. Snell illustrates this 'literary' quality from the famous fragment of the *Oeneus* (*TrGF* 71 fr. 14) and from other pieces. *TrGF* 71 fr. 2 is now earliest attested not in the 'Alexander Brief-Roman' but at (see p. 168 n. 27) Menander, *Aspis* 411. The list of the fragments on papyrus given by C. Collard, *JHS* 90 (1970), 22 f. may now be augmented by that at ll. 425–6 of the same play, thanks to E. W. Handley's brilliant emendation at the end of l. 427.

The most interesting chapter of the book is that devoted to the *Exagoge*, a drama in which the Alexandrian Jew Ezechiel, under the influence of Euripides, described the exodus of the children of Israel, 269 lines of which have been preserved. A scene in Act III presents Moses hearing the voice from the burning bush; I am surprised that Snell, even after reading Zwierlein's discussion of the question (op. cit., pp. 138 f.) thinks it not out of the question that the work may have been performed (p. 186). In order to bring out its specially Jewish elements, he compares it with the *Persae*. The comparison is interesting, though I wonder if the differences are quite as great as Snell thinks; even in the *Iliad*, the βουλή of Zeus determines the course of history, and in Aeschylus Zeus appears to assign the various nations their share of power (see e.g. 92, 762). Snell makes the interesting observation that at roughly the same time Naevius was writing his epic on the Punic Wars, with a not dissimilar conception of a special preference given by the great god to a particular favoured nation. Clearly this element in Ezechiel owed little to Greek sources. About Naevius and Ennius one cannot be so sure; in their work also it strikes us as profoundly Roman, yet if we knew more of the numerous Hellenistic epics about the foundations of cities and the deeds of kings and generals—see Snell, p. 186, and ‖ cf. K. Ziegler, *Das hellenistische Epos* (2nd edn. 1966)—things might appear quite differently.

An appendix puts the monologue of Neophron's *Medea* (*TrGF* 15 fr. 2) in its place in the development of *Geistesgeschichte*; Snell is inclined, with von Fritz, to trust the didaskalia and think Neophron's play earlier than that of Euripides (p. 203) [; see above, p. 200].

16(R82)

Review of de Romilly, *Time in Greek Tragedy*

Jacqueline de Romilly, *Time in Greek Tragedy*. (Messenger Lectures, 1967.) Ithaca, NY: Cornell University Press (London: Oxford University Press), 1968. Pp. viii + 180.

THE concept of time in various early civilizations has been the subject of a vast literature, many of whose authors have been led by an exact study of terminology to conclude that the people they have studied had a notion of time which was in some way peculiar or abnormal. Some of the pitfalls into which such an approach may lead have been well indicated by J. Barr, *Biblical Words for Time* (London, 1962). The most influential contribution to the topic by a classical scholar has been Herman Fränkel's paper 'Die Zeitauffassung in der griechischen Literatur', first published in 1931 and reprinted with changes in *Wege und Formen frühgriechischen Denkens* (Munich, 1960). Fränkel rightly points out that Homer is indifferent to exact chronology and to temporal sequences in general; when Homer needs to express the passage of time in general fashion, he uses the more concrete notion of 'the day'. In Archilochus 'das Ich konstituiert sich eigentlich erst jetzt, in seiner Gefährdung durch den Tag, als eine besondere Welt, die mit der Außenwelt, die nun erst zum Gegenpart geworden ist, in Beziehung tritt' (p. 8); during the lyric age the ego becomes conscious of itself as trying to escape being carried along by the stream of time. Time is first accorded 'den Rang und strahlenden Glanz der erfüllenden, vollendenden Zukünftigkeit' (p. 9) by Solon, who paves the way for the coming greatness of Athens. Here first we find an abstract conception of time; Fränkel is over-sceptical about the Chronos of Pherecydes (see Kirk in Kirk–Raven 56).[1] Pindar also entertains such a conception, but remains an archaic author, but Aeschylus ceases to be archaic and becomes classical; in his earlier works χρόνοϲ means only the future or the interim, but in the *Oresteia* it is used also of the past (p. 12).

This approach involves grave dangers, as has nowhere been better pointed out than by Momigliano in his paper 'Time in Ancient Historiography' (in *History and Theory: Studies in the Philosophy of History*,

[1] [=Kirk–Raven–Schofield 57.]

Classical Review, NS 20 (1970), 302–4.

Wesleyan University Press, 1966), 8 f. [= *Quarto contributo di storia degli studi classici e del mondo antico* (1969), 22 f.]. It is one thing to point out that the early epic deals with time and the passage of events in terms of certain well-marked literary conventions; it is another to conclude that its authors were unaware of time as ordinary people know it, but held some mysterious conception of its nature peculiar to themselves. In so far as Fränkel's paper describes the manner of referring to time and temporal sequences in Greek authors from Homer to Aeschylus, it is a valuable analysis; in so far as it draws from literary conventions, and what may be accidents of verbal usage, the conclusion that Solon, Pindar, and Aeschylus progressively invented the concept of time as we know it—and ‖ whether Fränkel meant that or not, that is how he has been interpreted by his followers—its method and results are open to objection.

Mme de Romilly begins by invoking Fränkel; he ended and she begins with Aeschylus. If the consciousness of time was acute and ripe during the fifth century, she says (p. 4), then each of the great tragedians must have had views about it, both spoken and unspoken. Tragedy deals with change, and therefore with time (p. 5); it suggests a perpetual reflection about the events which it describes in connection with the past and future, and it discusses and meditates about intricate causes and responsibilities (p. 11). 'Therefore', Mme de Romilly continues, 'it always presents us with a more or less conscious philosophy of time.' But does it?

Aeschylus often—not 'always' (p. 7)—wrote in continuous trilogies and, as Mme de Romilly shows in the chapter she devotes to him and as others have shown before her, moves between past, present, and future in a way greatly facilitated by the presence of the chorus; in his works time tends to appear as the power that will eventually bring justice. Sophocles writes single plays, and tends to concentrate on the catastrophe of their central figures; time in Sophocles, Mme de Romilly says (p. 88), is 'not, as in Aeschylus, the means of justice, but the cause of unsteadiness and lability in human life'. In the little of his works which we possess Aeschylus indeed tends to consider time as the bringer of justice; but it follows from his view of human life that human beings may see time as something inscrutable and unreckonable, for time will reveal what is the will of Zeus. Sophocles tends to stress this second aspect of time's nature, but it still follows from his view of life that time may be seen as the bringer of justice, as it is seen, to take an obvious example, in his *Electra* (see e.g. 179; cf. *OC* 614 etc.). The works of Euripides are, as Mme de Romilly acknowledges, so numerous and various that it is not easy to single out a definite attitude towards time. It would be easier if he had written less.

The book's general conclusions are summed up (p. 141) in these words:

Homer had known but a fragmented and disorderly kind of time, where as Fränkel says 'the day' was the main notion. Then arose the idea of a continuous time, including a whole sequel of events. This idea culminated in Aeschylus' tragic time. Through Sophocles' alternate changes, we have seen time dissolving into an uncertain drift so that in Euripides 'the day' again becomes all that we know. Yet there is a difference (time demands that there should be one!). This new 'day' has now become tragic, precisely because it is felt as an isolated fragment of a broken *chronos*, it is not that one does not think of *chronos* as a whole, but that this *chronos* finally turns out to be irrational, and to evade all human calculations. For we are left with our emotions, which implies both the feeling of something amiss and the growth of new interests.

The first sentence echoes Fränkel, whose theory as we have seen gives rise to doubts. Did the idea of a continuous time, 'including a whole sequel'—I think the writer means 'sequence'—arise only after Homer? Did Sophocles and Euripides not describe a whole series of events set in continuous time? Did Aeschylus and Euripides not connect time with mutability of human fortune? Were Aeschylus and Sophocles unaware that mortals cannot know the future? The truth is that a certain attitude to time—attitude to, not conception of—was bound to be expressed by poets holding the general view of the universe, man, and the gods which was common to Aeschylus and Sophocles and, to a large extent, to Euripides also. That obvious fact is only obscured by the || excessive preoccupation with the *development* of Greek thought that is so characteristic of Fränkel and others of his generation.

This book contains much that is true and useful about the poetical and metaphorical use of time in the tragedians and about the dramatic exploitation of its passage. This is particularly true of the second chapter, on the personification of time, and the sixth, on 'Young and Old in Greek Tragedy', though both chapters sometimes suffer from a tendency to generalize about the poet's attitudes on the basis of limited material; may not, for example, the absence from extant Aeschylus of any complaint about old age (note *Agam.* 72!) or any suggestion that old men are not always reasonable (see pp. 151 and 154) be fortuitous? But when Mme de Romilly tries to tell us something significant about the different 'philosophies' of time supposed to have been held by the three great tragedians, it seems to me that she is not successful.

Although the friends employed to check Mme de Romilly's English might have served her better—there are many small mistakes in idiom, and sentences like 'The past can be mentioned as being of similar features to the present' or 'Hippolytus ... is a sweet, idealistic youngster' should never have been allowed to stand—much of the speaker's charm comes through even in a foreign language. P. 11: did Medea kill more than one brother? P. 18, l. 4: for 'Teiresias' read 'Calchas'.

17(29)

Problems of Early Greek Tragedy

Pratinas and Phrynichus

No chapter in the history of Greek literature is more obscure than the origins of tragedy, and indeed its whole history before the second quarter of the fifth century. The beginnings are veiled in darkness; even the statement of Aristotle that tragedy originated from the dithyramb is not universally accepted; and those who do accept it disagree about the nature of the early dithyramb and its influence on tragedy. Recent accessions to our knowledge have only made this darkness darker. Till 1952 it was widely believed that we possessed a play written by Aeschylus early in his career, perhaps as early as about 500 BC, when he was some twenty-five years old. In that year the publication of a scrap of papyrus[1] put paid to that delusion. We must now recognize that the *Suppliants* of Aeschylus was in all probability produced in 464 or 463, and is therefore later than the *Persians*, produced in 472, and the *Seven against Thebes*, produced in 467. For the half-century between the date given for the first performance of Thespis at the Dionysia and that of the production of the *Persians*, we have no specimen of a tragedy.||

This deep and long-lasting obscurity is suddenly lit up, like the night sky of Argos by the beacon, by the bright star of Aeschylus. How natural that historians of tragedy should have tended to assume that Aeschylus was not only one of its supreme exponents, but also its real inventor! Gilbert Murray's book on Aeschylus, published in 1940, has for its sub-title *The Creator of Tragedy*; this reflects the influence of Wilamowitz, who in both his introduction to Greek tragedy of 1889 and his book on Aeschylus published twenty-five years later took this view.[2] All the ingredients of tragedy existed, says Wilamowitz, before

[1] P. Oxy. 2256, fr. 3, published as fr. 288 in Smyth–Lloyd-Jones, *Aeschylus*, ii (LCL, London, 1957), 595–8, with discussion and bibliography; see also Lloyd-Jones, *The 'Supplices' of Aeschylus: The New Date and Old Problems* (*AC* 33 (1964), 356–74 [= ch. 19].

[2] Wilamowitz, *Einleitung in die griechische Tragödie* (Berlin, 1921), 44–120; *Aischylos: Interpretationen* (Berlin, 1914), 240.

Estudios sobre la tragedia griega (Cuadernos de la Fundación Pastor 13, Madrid, 1966), 11–33.

Aeschylus; what was lacking was what is specifically dramatic, and this Aeschylus supplied. Aeschylus, according to Wilamowitz, invented tragic dialogue, and so discovered the dramatic element; Aeschylus gave the primitive goat-song as its content the heroic saga that was the essential subject-matter of tragedy.[3]

For us Aeschylus is indeed the first tragedian of real importance; but does it follow that we can be sure that none of his predecessors made an important contribution to the development of tragedy? Very little, to be sure, is known of them; and before this question can be answered, that little must be examined with some care.

Thespis passes for the 'inventor' of the genre; he is said to have inaugurated the tragic performances at the Dionysia in 534 BC. Can we know anything about his works? The *Suda* gives four titles: *The Funeral Games of Pelias* or *Phorbas*, *The Priests*, *The Young Men*, *Pentheus*. Aristoxenus, the well-known writer on music who was a pupil of Aristotle, accuses another fourth-century philosopher, Heraclides Ponticus, of having written tragedies which he passed off as the work of Thespis. Some scholars have claimed that no certain know||ledge about the works of such an early writer can have come down to the compiler of the *Suda*, and have therefore assumed that the four titles, and also the four fragments attributed to Thespis,[4] must have come from the tragedies forged in his name by Heraclides. This assumption is not altogether safe. Even if it is allowed to be unlikely that any complete tragedy by Thespis can have survived until the Alexandrian period, we cannot be sure that the titles of some of his works, and perhaps even extracts from them, cannot have been preserved. Even before the time of Aristotle there existed a not inconsiderable literature about tragedy, part of it written as early as the fifth century. Several works that formed part of it are quoted by Aristotle in the *Poetics*; the book of Glaucus of Rhegium, *On the Earliest Poets and Musicians*, appeared probably not much later than 400 BC.[5] While it is not certain that treatises like this contained reliable information about Thespis, we cannot rule out the possibility.

The titles credited to Thespis by the *Suda* are therefore worth examining, if only with caution. Two of them, *The Priests* and *The Young Men*, tell us nothing; but the other two refer to episodes of heroic saga of just the kind that furnished subjects to the tragedians of the fifth century. *The Funeral Games of Pelias* had been the subject of a

[3] Wilamowitz, *Einleitung*, 92–3.

[4] See Nauck, *Tragicorum Graecorum Fragmenta* (2nd edn., Leipzig, 1889), pp. 832–3 [and now *TrGF* i^2, p. 65].

[5] Kranz, *Stasimon: Untersuchungen zu Form und Gehalt der griechischen Tragödie* (Berlin, 1933), 4–6, with notes on p. 268. [G. Lanata, *Poetica pre-platonica: testimonianze e frammenti* (1963), 270 f.; K. Ziegler, *Der kleine Pauly*, ii (1979), 812.]

narrative lyric poem by Stesichorus; so had the story of Orestes, later to be made famous by so many tragic treatments; as to the Dionysiac legend of Pentheus, no subject could be better suited to a tragedy. We cannot rule out the genuineness of these titles simply on the ground that Aristotle (*Poet.* 1449^{a}19) tells us that tragedy took a long time to throw off the satyric element and become dignified. For all we know, Aristotle had in mind the character not so much of early tragedies as of rustic || performances originating from the dithyramb that had preceded their appearance; from his point of view any early tragedy, even one dealing with Pentheus, may well have seemed lacking in dignity. There is, then, a possibility, though it falls short of certainty, that these titles are genuine. What of the four fragments assigned to Thespis? The fourth may be at once ruled out, and the third seems unlike work of the sixth century. But the first two might, for all we know, be genuine, and if they are, the plays that contained them may well have had some dialogue.

With this inconclusive result in mind, we may pass on to the other early tragedians of whom something is recorded. Choerilus is a writer of uncertain date. The ancient life of Sophocles makes him compete against Sophocles as early as 468, and Eusebius' date for his *floruit*, 482, would make him about contemporary with Aeschylus; but according to the life in the *Suda* he first took part in the tragic competition between 523 and 520. The two known fragments ascribed to him certainly seem archaic, though not more so than some phrases to be found in Aeschylus; they comprise the kennings 'bones of earth' for 'stones' and 'veins of earth' for 'rivers'.[6] His only known title points to a subject highly suitable for tragedy. Alope, daughter of the Megarian king Cercyon, was killed by her father for having become pregnant by Poseidon. Cercyon is well known to Attic legend as one of the bad men killed by Theseus; Alope's son Hippothoon gave his name to one of the Athenian tribes. This story counts as a local Attic legend; Aeschylus wrote a satyr-play called *Cercyon* and Euripides a tragedy called *Alope*. Still, Choerilus' date is uncertain, and these data, few as they are, are of little value for our present purpose.

A much more promising source of information about early tragedy seems to be offered by Pratinas. He is said to have || competed against Aeschylus and Choerilus between 499 and 496; he was presumably dead by 467, when a tetralogy of his was exhibited by his son Aristias.[7] Pratinas and Aristias are said to have written the most famous satyr-plays apart from those of Aeschylus. The *Suda* says that Pratinas was

[6] Nauck, op. cit. 719–20 [; *TrGF* i^2, 67–8.]

[7] Hypothesis to Aeschylus, *Seven against Thebes*; cf. P. Oxy. 2256, fr. 2 and see Kakridis, *Ποικίλα Ἑλληνικά* (*Ἑλληνικά* 13 (1954), 165–74 [= *Μελέτες καὶ Ἄρθρα* (1971), 261–2.

the first to write satyr-plays, and it is usual to reconcile this with the derivation of tragedy from satyric performances by supposing with Pohlenz[8] that by about 500 tragedy had grown so remote from its satyric origins that Pratinas, a native of Phlius, reintroduced the satyr-drama from its original Peloponnesian home, where it had survived unchanged. This is possible, but much less certain than has often been supposed; how can we be sure that the statement in the *Suda* that Pratinas invented satyr-plays is anything more than an instance of the common tendency for the earliest author known to have been eminent in any genre to be credited with its invention?[9]

Several fragments of Pratinas have been preserved by Athenaeus; one of them is seventeen lines long, and is of much interest.[10] It is a lyric piece in which a chorus of satyrs protests energetically against a performance in which the aulos accompaniment drowns the words it is supposed to accompany. Athenaeus calls the piece a hyporchema, a name given to lyrics meant to accompany a vigorous dance. The names which the Greeks applied to different kinds of lyric composition are often vaguely and inconsistently used,[11] and the word 'hyporchema' ‖ might possibly have been applied to a lyric from a drama; but on the whole it is likely that by using this term Athenaeus indicates that he or his source took the piece to be not a dramatic but a choral lyric composition.

Against what kind of performance may we suppose that the protest of the chorus of satyrs was directed? Hardly, I think, against the musical accompaniment of tragedies by Phrynichus. Since Dalecampius published his translation of Athenaeus in 1583, supporters of this view have argued that the poet actually puns on Phrynichus' name. 'Beat him, with his breath of a variegated toad,' the satyrs cry (ll. 10 f.), 'scorch him with his spittle-wasting reed, with his chattering deep voice, his gait without tune or rhythm, him whose body was shaped beneath the gimlet.' The Greek word for 'toad' is φρῦνος or φρύνη and the mention of a toad is taken to allude to Phrynichus. But Wilamowitz rightly said that the comparison of the aulos to a toad 'would be appreciated only by one who has disturbed a common fat toad from its sleep amid the strawberries';[12] the likeness of a clarinet or oboe in the hands of an unskilful player to a spitting toad is so close that there is no ground for imagining that this or any other name was punned on. Nor is it likely that an early tragedian would have allowed

[8] Pohlenz, 'Das Satyrspiel und Pratinas von Phleius', *NGG* (1926), 298–321.

[9] See Kleinguenther, *Πρῶτος εὑρετής. Untersuchungen zur Geschichte einer Fragestellung* (*Philol.* Suppl. 26/1, Leipzig, 1933).

[10] [*PMG* fr. 708 = *LGS* fr. 407 = *TrGF* 4 fr. 3.]

[11] See Harvey, 'The Classification of Greek Lyric Poetry' *CQ*, NS 5 (1955), 157–75)

[12] Wilamowitz, *Sappho und Simonides* (Berlin, 1913), 134.

that drowning of words by music that was later to be one of the most scandalous features of the new dithyramb that was brought in late in the fifth century by poets like Timotheus and Philoxenus. The case for thinking the piece to be a protest against the innovations of Simonides' rival, the dithyrambic poet Lasus, is not much stronger; there is rather more to be said for the possibility that the later innovations of Melanippides are in question; but there is no compelling reason for supposing any particular innovations to be attacked at all. All that is necessary is to suppose that the song of satyrs followed a performance by the aulos-player,‖ which may not have been accompanied until they themselves began to sing.[13]

At this point we may as well take note of the words used by Athenaeus in introducing the quotation, words which have often been too hastily brushed aside. He says that 'at a time when the dancing-floors were occupied by hired aulos-players and choreutai, people complained because the aulos-players did not accompany the choruses, as had been the custom, but the choruses accompanied the aulos-players.' During the fifth and fourth centuries before Christ, tragic and dithyrambic choruses at Athens had to consist of citizens, and the mention of hired choreutai caused Wilamowitz to suggest that Athenaeus had wrongly supposed that conditions during the fifth century were like those of his own time, when such people would be hired professionals. Still, even during the fifth century the choreutai had their expenses discharged by the choregos; and after the introduction of the new music of Timotheus and others, considerable expertise must have been required from both musicians and choreutai, so that there may well have been some competition for the services of the best performers. Something like the state of affairs described by Athenaeus—or his source, probably Aristoxenus—may well have existed towards the end of the fifth century.

Bearing in mind that Athenaeus or his source seems to have thought the piece to be a choral lyric composition, let us consider other fragments attributed to Pratinas. Like the five fragments preserved by Athenaeus, the three testimonies preserved by Pseudo-Plutarch, *On Music* probably come from a writer on music, most likely Aristoxenus. Had we no other information about the author than what they provide, we should suppose him to have been a choral lyric poet, markedly interested in the history of poetry and music. Pratinas is quo‖ted as saying that a younger Olympus, and not the famous aulos-player, wrote the famous 'nome of many heads'; that Xenodamus wrote hyporchemata and not paeans; that the lyric poet Thaletas saved

[13] See A. M. Dale, 'Stasimon and Hyporcheme', *Eranos*, 48 (1950), 14–20; *Words, Music and Dance* (Inaugural Lecture at Birkbeck College, London, 1960), 11 f. [= *Collected Papers*, 167 f.].

Sparta from a plague (frr. 7–9). He praised the Spartans for excellence in forming choruses (fr. 4); he specially recommended the Aeolic harmony (fr. 6); he claimed originality in terms recalling those employed by other lyric poets (fr. 5). [Moreover, the language and metre of the remains of Pratinas are much more like such late fifth-century dithyrambic poets as Timotheus than they are like what we should expect about 500 BC.] And yet the life of Pratinas in the *Suda* makes no mention of any lyric poetry, but only of tragedies and satyr-plays. It is usual to suppose that Pratinas wrote choral lyrics besides these, but that the author of the life omitted to mention this. We should account for the known facts more satisfactorily without having advanced much further on the path of speculation were we to guess that there were in fact two men called Pratinas, one a tragedian and writer of satyr-plays who flourished about 500 and another a lyric poet, active towards the end of the fifth century. This is no more than a speculation; but the arguments that have prompted it may serve as a warning of the extreme uncertainty of our information about Pratinas, and of the danger of using even the single long fragment as evidence for the state of early tragedy or satyric drama. The one extant fragment which I would assign with much confidence to the early Pratinas is fr. 1, containing the questionable statement that the quail has a sweet voice.[14]

So far our examination of what is recorded of the earliest tragedians has brought almost no results. True, some things that are asserted of these writers might be held to cast doubt on the claim of Wilamowitz that it was Aeschylus who made heroic saga the subject of tragedy; but this can hardly be taken for granted. The titles assigned to Thespis, while not ‖ certainly spurious, are not certainly genuine; Choerilus may have produced his *Alope* after Aeschylus had become active, and the same is true of the *Persians* and *Tantalus* of Pratinas. It is another matter when we come to Phrynichus.

Phrynichus was at least five years older than Aeschylus, perhaps a good deal older. His first victory is placed between 511 and 508, at about the time of the fall of the Pisistratids; he must therefore have been born before 530. His career spanned a long period, for he won one of his victories as late as 476. Aeschylus first competed in the tragic contests in 500, so that the careers of the two poets overlapped over a period of at least twenty-four years. It follows that each may have taken over certain practices from the other, and we must guard against

[14] This is quoted as coming from the *Dymanai* or *Karyatides*; the former name should certainly be emended with Toup to *Dymainai*, as Latte in his Hesychius, s.v., saw and as the new Alcman in P. Oxy. XXIV has confirmed. [This dating of the long fragment has been attacked by R. Seaford, *Maia*, NS 30 (1978), 81 f. (cf. M. L. West, *Greek Metre* (1982), 101 n. 69 and R. Kannicht, *TrGF* i² (1986), 345) and defended by B. Zimmermann, *MH* 43 (1986), 145 f.]

assuming that the older of the two must have been in every way the more old-fashioned. Still, Phrynichus was the older, and in two places Aristophanes makes it clear that at the start of Aeschylus' career Phrynichus was his chief rival (*Frogs* 908 f.; 1296 f.). Any feature which he and Aeschylus had in common may have been taken over by the older from the younger poet; but we cannot assume that this has been the case without definite evidence.

Aeschylus, according to Wilamowitz, was the first to make heroic saga the subject of tragedy. We have already discovered certain indications that heroic saga may have been the subject-matter of tragedy before Aeschylus; there is no doubt that it was the subject-matter of Phrynichus, and it seems most unlikely that this was due to imitation of Aeschylus' practice. In his *Egyptians* and his *Danaides* he handled a heroic subject known from Hesiod's *Catalogue*, to say nothing of an early epic called the *Danais*, and later to be treated in an Aeschylean trilogy. The mention of two plays on the same subject may indicate that Phrynichus, like Aeschylus, sometimes wrote a trilogy on a continuous theme; if so, one wonders who introduced the practice. Hesiod's *Catalogue* may also have been Phrynichus' source for the story of *Alcestis*. Because the *Alces*||*tis* of Euripides was presented in lieu of a satyr-play and has certain burlesque elements, it is often held that the *Alcestis* of Phrynichus must have been a satyr-play; but this is hardly a safe inference. It is true that in Phrynichus' play Apollo persuaded the Fates to allow Admetus to escape death, provided another victim were offered in his place, by entertaining them and making them very drunk. This touch is primitive, but not necessarily burlesque; Aeschylus thought it proper to mention it in his *Eumenides* (723–8) and the other detail of the plot that is known to us, the appearance of Death at the bedside of Alcestis to cut from her head a lock in token of her dedication to the infernal gods, does not necessarily suggest a semi-comic treatment. We cannot be sure that the *Alcestis* of Phrynichus was satyric; and even if it was, its plot had a strong element of heroic legend.

Phrynichus wrote a play about Actaeon, another undoubtedly heroic subject. The fate of Actaeon, son of Cadmus' daughter of Autonoe by the god or hero Aristaeus, was in all probability treated in Hesiod's *Catalogue*; the daughters of Cadmus must have figured in that poem.[15] One of them, Semele, was the mother of Dionysus, and the

[15] P. Oxy. 2509 contains a hexameter text dealing with Actaeon that Lobel thinks may come from the *Catalogue* (XXX, p. 5); Athene comes to the cave of Chiron on Mount Pelion and prophesies that one day there shall be born Dionysus, who shall hunt upon the mountain with Actaeon's dogs, and what looks like a description of Actaeon's death follows. The story that Actaeon was punished for seeing Artemis naked is not attested before the Hellenistic age; the old legend is that he was eliminated as being Zeus' rival for the love of Semele. [See M. L. West, *The*

story may well have counted as a Dionysiac legend; Aeschylus later handled it in his tragedy *Toxotides*, and two minor tragedians wrote plays called *Actaeon*.

Phrynichus' *Women of Pleuron* seems to have told how Meleager's life was ended when his mother Althaea hurled into the fire the burning brand whose consumption by the flames || was fated to be followed by his death. Again the story is heroic; we find it in a papyrus fragment of an early epic, probably Hesiodic,[16] which may have served as source to Phrynichus and also to Bacchylides in his fifth dithyramb. Phrynichus' play about the Libyan giant Antaeus, who wrestled with Heracles, may have been satyric, like Euripides' play about a similar Libyan character, *Busiris*; so may even his *Tantalus*, though this seems unlikely. Still, we have more than enough evidence to show that most of Phrynichus' subjects came from heroic saga. Can we really be sure that he chose such subjects only because Aeschylus had set him the example? I think not.

But early epic and choral lyric were not the only sources of Phrynichus' plots. So far as we know, he was the first tragedian to dignify by tragic treatment themes from recent history that seemed to possess the necessary weight and dignity. We do not know which of the tragedians instituted this practice, but Phrynichus is the first whom we know to have adopted it. It is in no way surprising that it should have happened, for the fifth-century Greeks drew no hard and fast line between history and myth. Herodotus might ridicule his predecessor Hecataeus for tracing back his own descent to a divine ancestor, but in his own work the frontier between myth and fact is never clearly defined, and even the sceptical Thucydides in his *Archaeologia* in book 1 treats the Trojan War as a real event and those who fought in it as real people.[17]

The first contemporary event dramatized by Phrynichus was the sack of the great Ionian city of Miletus in 494, which formed the disastrous climax of the ill-fated revolt of the Ionian cities against Persian domination. Despite the danger then threatening her from her near neighbours in Aegina, Athens sent twenty triremes to help the rebels, but long before the end of the revolt she had withdrawn her aid. At the start the || rebels won important victories, but in the midsummer of 495 they lost a decisive sea-battle at Lade near Miletus, and the fall of the great city followed. The horrors of the event were notorious, and seem to have been unsparingly depicted in Phrynichus' play. In

Hesiodic Catalogue of Women (1985), 88 with n. 129, who thinks that 'its plan and style are sub-Homeric rather than pseudo-Hesiodic'; and *Greek Comedy, Hellenistic Literature, Greek Religion, and Miscellanea*, 326 n. 84.]

[16] [Fr. 280 M.–W.]

[17] 1. 3, 9, 11–12.

consequence, he was fined a thousand drachmas, according to Herodotus (6. 21), for 'having reminded the citizens of their own grief'. The reason alleged sounds suspiciously aesthetic. The doctrine that poetry causes the soul to suffer experiences of its own through contemplating the good or ill fortune of others, in extant literature first found in the writings of the sophist Gorgias, may well have existed earlier in the fifth century; still, one wonders whether the prosecutor would have made use of it in framing his indictment. Was it simply that the audience found the subject too distressing for their comfort? Or had the choice of a subject a political purpose, so that the poet was punished by political antagonists? The second possibility cannot be ruled out, particularly in view of an event connecting Phrynichus with Themistocles, which we shall come to presently; for even at this time Themistocles was urging Athens to arm herself against the Persian menace. Whatever the cause, it is well to remember that the fine was not a very heavy one.[18]

Of the treatment of the subject in Phrynichus' play we know nothing, nor has any fragment been preserved. Wilamowitz thought it must have been a kind of oratorio, containing lyrics and narration but no action;[19] but as we shall see presently this view rests on dubious inferences. Perhaps the methods used by Aeschylus to describe the Argive onslaught upon Thebes may give some clue as to the means adopted by Phrynichus to describe the Persian onslaught upon Miletus; in any case, it is hard to see how a play on this subject can have been entirely without action.||

The ancient hypothesis of the *Persians* of Aeschylus contains a valuable testimony taken from a book *On the Plots of Aeschylus* by that Glaucus of Rhegium whom I mentioned earlier. He said that Aeschylus' play was modelled on the *Phoenician Women* of Phrynichus, which also described the defeat of Xerxes in the campaigns of 480–479. We happen to know that in 476 Phrynichus won a victory in the tragic competition with a trilogy for which Themistocles was the choregos; Plutarch, from whom we learn this fact, quotes an inscribed tablet set up to commemorate the victory by Themistocles himself. Bentley's guess that the *Phoenician Women* formed part of this particular trilogy is not certain to be right, but it is exceedingly attractive.

According to the hypothesis Phrynichus' play began with a prologue speech in trimeters by an eunuch, who was engaged in putting out seats for the royal counsellors of Persia, and in so doing announced the news of the defeat. That might lead one to suppose that the royal counsellors formed the chorus of this play, as they did that of Aeschylus' *Persians*;

[18] See Freymuth, 'Zur *Μιλήτου ἅλωϲιϲ* des Phrynichos', *Philol.* 99 (1955), 51–69.

[19] Wilamowitz, *Einleitung*, 91–2; *Aischylos: Interpretationen*, 240.

but the title indicates that the chorus consisted of Phoenician women, and this is made probable by two fragments in which the chorus seems to be speaking of its home in Sidon (9. 10). The chief naval strength of Xerxes lay in the Phoenician fleet, which suffered heavy casualties at Salamis, so that the wives and mothers of its sailors would form an appropriate chorus for a play on this topic. The presence of the royal council suggests a setting in Persia, and the words of the chorus 'leaving the town of Sidon' may indicate that they too are in Persia for the time being.

Perhaps the royal councillors were mutes, like the jurors in the *Eumenides*; perhaps they formed a secondary chorus, like the handmaids of the *Danaids* at the end of the *Suppliants*; but there is another possibility. The life of Phrynichus in the *Suda* says that he wrote a play called *The Just Ones*, or *The Persians*, or *The Councillors*. Was this the same as *The Phoeni||cian Women*? The possibility cannot be ruled out; but four different titles is a great many for one play, and more probably the tragedy mentioned in the *Suda* is a separate work. If so, it may well come from the same trilogy as *The Phoenician Women*. We have seen that there is one piece of evidence (p. 231 above) that Phrynichus sometimes composed a trilogy on a continuous theme. Ancient authors not infrequently confuse one play of a trilogy with another, and not the least plausible explanation of the evidence would be that the play that began with the setting out of seats for the council was really *The Just Ones*, or *The Persians*, or *The Councillors*. In that case, *The Phoenician Women* will have been another play belonging to the same trilogy. Of its actions we know nothing, and the various attempts that have been made to 'reconstruct' it have got nowhere. It has been contended that the announcement of the defeat right at the start, in contrast to *The Persians*, where it is held back until the queen of Persia and the chorus have had time to express their deep anxiety, shows that Phrynichus' treatment of the subject must have been markedly less dramatic than that of Aeschylus. Even that inference is obviously most unsafe.[20-6]

The indications are that he differed greatly in style and method from his younger contemporary, Aeschylus. Can we form any notion of the difference? All that we know of him suggests that his poetry was typical of the artistic atmosphere of Athens during what must have been his formative years, when under the patronage of the cultivated Hipparchus the city was becoming the chief centre of art and culture in the Ionian world. Lasus and Simonides were active there, but perhaps more characteristic of the time were Ibycus and Anacreon. The ornate

[20-6] [I omit a passage concerning the Gyges drama, *TrGF* Adesp. fr. 664, on which my views have changed: the metre, which is the only solid basis for conjecture, points to the later dating. See N. Holzberg, *Živa antika*, 23 (1973), 273–86; Edith Hall, *Inventing the Barbarian* (1989), 65 n. 37.]

amplitude of the Italiote differs markedly from the elegant simplicity of the Ionian; yet both have in common a mannered elegance, a refined sensuousness. In the *Thesmophoriazousai* of Aristophanes the effeminate poet Agathon appeals to the example of these poets, together with Alcaeus, to justify his principle that in poetry 'handsome does as handsome is'. 'And besides it is inelegant', says he (159 f.), 'that a poet should appear rough and shaggy. Remember the great Ibycus and Anacreon of Teos and Alcaeus, who gave harmony its flavour, how they wore turbans and enjoyed Ionian luxury. Phrynichus too—you will have heard of him—was hand‖some and wore handsome clothes; that is why his plays were handsome too; for what one writes is bound to be like one's character.'

The impression of Phrynichus given by this passage is strongly confirmed by all that is known to us about his lyrics. The elderly Athenians afflicted with a passion for jury-service who fifty-four years after the production of the *Phoenician Women* come in the middle of the night to wake their friend Philocleon arrive before his house 'carrying lamps and humming ancient melodies, honeyed, Sidonian, Phrynichan, delightful' (*Wasps* 219–20). Eight years later Aristophanes' chorus of birds addresses with these words the muse of bird-song: 'Muse of the thickets, Muse various, with whose aid in gardens and on mountain-tops I perch upon the leafy ash-tree and with songful throat reveal the sacred strains of melodies for Pan and songs to be danced to the Mountain Mother. From these Phrynichus like a bee sucked the fruit of melodies immortal, ever carrying away sweet song' (*Birds* 737 f.). Does this exquisite stanza echo an actual lyric by Phrynichus? We cannot prove it, but we are fully justified in thinking it possible.[27] The influence of cult poetry on the lyrics both of tragedy and comedy was of course profound; but these words do not mean that Phrynichus in his melodies was influenced by the cult-hymns which we know to have been written in honour of Pan and of the Mother of the Gods. The thought is simpler and more poetic, that the poet borrows his melodies from the song of birds; the same claim was made for Walther von der Vogelweide and by Milton on behalf of Shakespeare.[28]

Lamentably little of Phrynichus' lyrics has survived, but even that little seems to testify to an affinity with the eastern half of the Greek world. One fragment (6) has two and per‖haps three successive examples of the greater asclepiad; this metre, known to us from the monodic lyric of the Lesbians, does not occur elsewhere in what we have of tragedy. Ionics are of course used by Aeschylus and other poets; but in view of their supposed associations with the east, their

[27] Fraenkel, *Beobachtungen zu Aristophanes* (Rome, 1962), 209.

[28] I side with Wilamowitz, *Griechische Verskunst* (Berlin, 1921), 436 n. 2 against Fraenkel, loc. cit.

occurrence in one of the very few lyric fragments (14) may possibly be significant. Ion of Chios told how Sophocles after a dinner-party was moved by the beauty of a boy who brought him wine to quote from Phrynichus the line 'and on his crimson cheeks there shines the light of love': *λάμπει δ' ἐπὶ πορφυρέαις παρῆισι φῶς ἔρωτος* (fr. 13). It has a languorous richness that to us seems more Ionian than Athenian.

With music went the dance, and Phrynichus was famous for his dances. One of his dance movements figured in the parody of a dance performed by the three sons of Carcinus with which the *Wasps* concluded (1490–2). Plutarch quotes a beautiful elegiac epigram, probably of Hellenistic date, about his dances: 'and my dances gave as many figures as at sea, when the billows are stormy, the cruel night makes waves'.[29] According to one writer of late Hellenistic date, Phrynichus and other early tragedians actually gave dancing-lessons.[30]

Did these celebrated lyrics make up the chief part of Phrynichus' plays? A Peripatetic writer says that he and the tragedians of his time 'were principally musicians'.[31] Aristotle's remark that Aeschylus reduced the part of the chorus and gave the chief place to the spoken element[32] is often taken to mean that in the plays of all tragedians older than Aeschylus the lyric part must have greatly predominated, and together with his statement that Aeschylus brought in the second actor, led Wilamowitz to argue that Aeschylus invented dialogue. If ‖ Aeschylus invented dialogue, Phrynichus was quick to copy him; we have some eight lines of his in dialogue metre, none of them in what Aristotle says was the earliest dialogue metre, the tetrameter, but all in trimeters.[33] The life in the *Suda* says that Phrynichus invented the tetrameter; the statement is absurd, for Archilochus and others were writing tetrameters long before Phrynichus, but it may reflect the fact that Phrynichus was writing dialogue not much later than or indeed actually before Aeschylus.

Only in one other respect is Phrynichus stated to have been an innovator; he is credited in the life in the *Suda* with the introduction of the first female character; but we cannot be sure that this was his only innovation. Plutarch couples him with Aeschylus as having brought tragedy on so that it dealt with myths and sad events,[34] and all the evidence we have is in accordance with his statement. The high praise given him by so good a judge as Aristophanes is in no way discounted

[29] Plut. *Quaest. symp.* 732 F; see Wilamowitz, *Griechische Verskunst*, 465 n. 1.

[30] Aristocles, quoted by Ath. 22 A.

[31] Ps.-Arist. *Probl.* 920 ᵃ11.

[32] *Poetics*, 1449ᵃ15 f.

[33] The tetrameters quoted in the Ammonius scholia on the *Iliad* in P. Oxy. 221 have nothing to do with Phrynichus. [I wonder; see *TrGF* 3 fr. 10a.]

[34] *Quaest. symp.* 615 A.

by the nature of the few remains. Can we safely take for granted that his work consisted almost entirely of lyrics, that it had little dramatic power, or that it made no important contribution to the development of tragedy? Surely not; today we know even less about the first half-century of tragedy than Wilamowitz thought we knew.[35]

[35] I am grateful to D. Antonio Pastor for many kindnesses.

18(6)

Zeus in Aeschylus[1]

AFTER a hundred and thirty years of controversy, the interpretation of the *Prometheus Bound* is still the subject of debate. To the Romantic poets of the revolutionary era, the Titan tortured by Zeus for his services to mankind appeared as a symbol of the human spirit in its struggle to throw off the chains which priests and kings had forged for it.[2] But to the distinguished Hellenists who after the fall of Napoleon laid the foundations of the great century of German scholarship, no such naïve and one-sided view of the *Prometheus* seemed tolerable. It was partly, perhaps, that the political atmosphere discouraged an interpretation adverse to authority; but the other writings of Aeschylus himself seemed to offer strong evidence against this view. Elsewhere in Aeschylus, they could argue, Zeus was treated with profound respect. In the *Supplices*, he is continually appealed to by the chorus of Danaids, who miss no opportunity to extol the supremacy of his power. Zeus' omnipotence is the burden of a celebrated section of the first chorus of the *Agamemnon*; although scholars have differed widely in the details of their interpretation of this passage, most are agreed that it expresses theological doctrines that are at once subtle and sublime, and many have discovered profound significance in its puzzling allusions to 'learning by suffering' and to a *χάρις* that comes to men from the gods. Chiefly upon evidence derived from these two plays, many scholars have credited their author with the invention of a peculiar personal religion, tending to exalt Zeus at the expense of the other members of the Olympic pantheon, and crediting him with the purpose of perfecting men in goodness through the discipline of suffering. Some have gone so far as to detect tendencies to monotheism in Aeschylus.[3] A fair

[1] This paper was read to the Oxford branch of the Classical Association in Nov. 1953. I gratefully acknowledge the kind help of Professor D. S. Robertson, Professor E. R. Dodds, Mr G. S. Kirk, and in particular Professor D. L. Page. [Dodds allowed me to read his paper on 'The *Prometheus Vinctus* and the Progress of Scholarship', written in 1946, but published only in *The Ancient Concept of Progress* (1973), 26–44.]

[2] See K. Heinemann, *Die tragischen Gestalten der Griechen in der Weltliteratur*, pp. 12 f.; O. Walzel, *Das Prometheussymbol von Shaftesbury zu Goethe*; L. Séchan, *Le Mythe de Prométhée*, 15 f., 95 f.

[3] First Haas in *Archiv für Religionswissenschaft*, 3 (1900), 163 f.; see the list of writers taking this view given by Wm. Nestle, *Griechische Studien*, 66 n. 14. So also H. J. Rose, *A Handbook of Greek Literature*, 159: 'Practically, Aeschylus was a monotheist.'

Journal of Hellenic Studies, 76 (1956), 55–67.

specimen of the usual kind of view is that of Nilsson,[4] who begins by hesitating to pronounce Aeschylus a monotheist; Aeschylus, he warns us, is not a religious innovator preaching a new form of religion, but a profoundly pious poet;[5] but later, he comes dangerously near to this view. The power of Zeus, he argues, is so much magnified that at one point he seems more a principle than a personal god.[6] In dealing with the famous fragment of the *Heliades* (fr. 70):

> Ζεύϲ ἐϲτιν αἰθήρ, Ζεὺϲ δὲ γῆ, Ζεὺϲ δ' οὐρανόϲ,
> Ζεύϲ τοι τὰ πάντα χὤτι τῶνδ' ὑπέρτερον.

Nilsson leaves open the possibility that the poet 'went yet further', and embraced a pantheistic view of nature.[7] Twice he speaks of 'Aeschylus' Zeus-religion', and finally he remarks with some surprise that it had no influence on posterity.[8]

Here, then, is the problem which perplexed the contemporaries of Hermann and Welcker and which still perplexes us: how is the presentation of Zeus given in the *Prometheus Vinctus* to be reconciled with the picture of him found in the other plays of Aeschylus?

This disagreement is in large measure due to our ignorance regarding the trilogy of which the surviving Prometheus play formed part. Of the *Προμηθεὺϲ Λυόμενοϲ* we possess a number of fragments, and it is clear that the release of Prometheus by Heracles, predicted in the existing play, occurred in this work, as its name implies. Of the *Προμηθεὺϲ Πυρφόροϲ*, on the other hand, we know scarcely anything. Welcker's opinion that it was the first play of the trilogy[9] has been generally given up since Westphal pointed out that the narrative of past events in the *Prometheus Vinctus* makes it improbable that another play had gone before.[10] But the extreme rarity of quotations lends

[4] *Griechische Religionsgeschichte*, i. 707 f. The list of books and articles whose authors hold this kind of view about religion in Aeschylus could be made very long indeed. See, for example, Schmid–Stählin, *Griechische Literaturgeschichte*, i/2. 265 f.; Wilamowitz, *Der Glaube der Hellenen*, ii. 133 f.; M. Pohlenz, *Die griechische Tragödie*, i^2. 141 f.; Wm. Nestle, *Neues Jahrbuch*, 1907, 225–46, 305–33 = *Griechische Studien*, 61 f.; G. Murray, *Aeschylus: The Creator of Tragedy*; W. Jaeger, *Paideia*, i. 263–4 (English version); W. K. C. Guthrie, *The Greeks and their Gods*, 65, 258; F. M. Cornford, *Plato's Cosmology*, 361 f.

[5] P. 708, top: 'er ist kein Systembauer, sondern ein von einem hohen Glauben an seine Religion ergriffener Dichter'.

[6] P. 711, near top: 'seine Macht wird so gesteigert, daß er an einer Stelle mehr als ein Prinzip denn als ein persönlicher Gott erscheint'. The 'one place' is, of course, *Agam.* 160 f., on which see below.

[7] Other evidence for pantheism in Aeschylus is lacking: and without a knowledge of the context of this fragment it is unsafe to argue from it. In default of any evidence to support the pantheistic interpretation, I suspect that it means much the same as the concluding line of the *Trachiniae*:

> κοὐδὲν τούτων ὅ τι μὴ Ζεύϲ.

[8] P. 711: 'auf die Religion der Nachwelt hat seine Zeusreligion kaum Einfluß gehabt'.

[9] *Die Aeschyleische Trilogie Prometheus* (1824), 7 f.

[10] *Prolegomena zu Aeschylus' Tragoedien* (1869), 207 f. [But see now M. L. West's surprisingly dogmatic article, *JHS* 99 (1979), 130–48.]

colour to the suspicion that *Προμηθεὺς Πυρφόρος* may be simply another way of referring to the satyr-play ‖ produced with the *Persae*, *Glaucus Potnieus*, and *Phineus* in 472 BC and called, in all probability, the *Προμηθεὺς Πυρκαεύς*. Professor D. S. Robertson has suggested[11] that the author may have died without putting the finishing touches to the trilogy and that some of the odes may have been supplied by one of his sons, Euphorion and Euaion, or by his nephew, Philocles. Without new evidence this theory cannot be proved right or wrong; but the convenient explanation which it offers of the abnormalities of style and language[12] and the yet more striking abnormalities of metre[13] presented by the *PV* make it a possibility which must be taken seriously. In another place I hope to discuss in detail the limited evidence which bears on the reconstruction of the trilogy; here, I wish to concentrate upon the central problem of the conflict between Zeus and Prometheus.

Some scholars have contended, against all appearances, that the sympathy of the audience is meant to lie, not with Prometheus, but with Zeus. Others have argued that the remaining play or plays of the trilogy must have shown a development in Zeus' character; from the savage tyrant of the *PV* must have emerged the beneficent monarch of the *Supplices* and *Agamemnon*. Others, unwilling to accept either view, have denied that the surviving Prometheus play can be by Aeschylus, a view to which the abnormalities referred to certainly lend some colour.[14] The first of these three opinions was originated by Schoemann,[15] was taken over by Wecklein,[16] and is familiar to English readers from the school edition of Sikes and Willson (Macmillan, 1898), which after fifty years is still the most useful commentary to the play in our language.[16a] Prometheus, Schoemann argued, did mankind no real good; all he did was to invent science and technology, mixed blessings. Holders of this view describe the conflict in the language of Christian or Jewish theology. Prometheus has sinned and is receiving just punishment, and in the last play he will repent and acknowledge his transgression. Some of them even argue that the conduct of Zeus

[11] *PCPS* 169–71 (1938), 9–10.

[12] See esp. J. Wackernagel, *Verhandlungen der 46. Versammlung der Philologen* (1902), 65 f.; id., *Studien zum griechischen Perfektum*, 11 f. [= *Kl. Schr.* ii. 1008 f.]; J. D. Denniston, *The Greek Particles*, pp. lxvii f.; bibliography in W. Schmid, *Untersuchungen über den Gefesselten Prometheus*, 1–4 and cf. ibid. 41 ff.

[13] See E. C. Yorke, *CQ* 30 (1936), 116 f., 153 f.; J. D. Denniston, ibid. 73 f., 192 f.; W. Kranz, *Stasimon*, 226–8.

[14] Wackernagel seems to have arrived at this opinion for purely linguistic reasons: see above, n. 12.

[15] *Opuscula*, iii. 81 f., 95 f., 120 f.; see also introd. to his edn. of the play.

[16] In the introd. to his edn. of the play (2nd edn. 1878; English translation by F. D. Allen, Boston, 1891).

[16a] [See now M. Griffith, *Aeschylus, Prometheus Bound* (Cambridge, 1983).]

described in the *PV* was only part of a higher design for the destiny of mankind,[17] a design oddly reflected in Zeus' intention, frustrated only by Prometheus, of exterminating them (see ll. 231–4). This opinion has gone out of fashion during the present century, and no wonder.[18]

Dissen[19] was the first to suggest that the trilogy showed an evolution in the character of Zeus, an evolution which by the beginning of the last play of the trilogy must have transformed the youthful tyrant of the *PV* into a beneficent ruler, who of his own free will releases his noble enemy. This is a view which during the last century remained comparatively unpopular, but which during the present one has become accepted by most scholars. It was adopted by Wilamowitz;[20] and among living scholars it is held by Nilsson (loc. cit.), Festugière,[21] and Professor E. R. Dodds (in an unpublished paper read last year to the Leeds branch of the Classical Association, which he has with the greatest kindness allowed me to make use of). Wilamowitz (*Aischylos: Interpretationen*, p. 150) quoted in illustration of it Friedrich Theodor Vischer's explanation of the savage cruelties perpetrated in the name of the Jehovah of the Old Testament: 'In those days the good God himself was young'; and he drew attention to the constant stressing, in the *PV*, of the youth of Zeus and the newness of his power. By way of parallel, the holders of this opinion point to the change of character they say is experienced by the Eumenides in the play named after them.

Is the notion of a god thus changing in character one that Aeschylus can possibly have entertained? Schmid (op. cit. 89) flatly asserted that it was impossible; the Greek notion of divine growth, he contended, is expressed in Callimachus' address to Zeus:

ἀλλ' ἔτι παιδνὸς ἐὼν ἐφράσσαο πάντα τέλεια (*Hy.* 1. 57).

And since Schmid wrote, two eminent authorities on Greek religion, Farnell and Karl Reinhardt,[22] have expressed agreement.||

Believers in the development theory have certainly failed to support

[17] See J. A. K. Thomson, *HSCP* 31 (1920), 1–37.

[18] It has been revived in an extreme form by E. Vandvik, *The Prometheus of Hesiod and Aeschylus* (Skrifter utgitt av Det norske Videnskaps-Akademi i Oslo, hist.-filos. klasse, 1942/2); the experiment does not seem to me successful.

[19] Ap. Welcker, op. cit. 40 ff.

[20] Aischylos: Interpretationen (1914), 149–50; *Der Glaube der Hellenen*, ii. 133–4.

[21] *Rev. phil.* 22 (1948), 156 f. Cf. E. Myers in E. Abbott (ed.), *Hellenica* (1880), 20; L. Campbell, edn. of *PV* (1890), introd.; Janet Case, *CR* 18 (1904), 99 ff.; H. Weil, *Études sur le drame antique*, 82 f.; S. H. Butcher, *Harvard Lectures on Greek Subjects*, 17 f.; Wm. Nestle, loc. cit. (n. 4); J. T. Sheppard, *Greek Tragedy*, 62 f.; P. Mazon, *Eschyle*, i. 153 f.; M. Croiset, *Eschyle*, 163; J. A. K. Thomson, op. cit. 34; O. J. Todd, *CQ* 19 (1925), 61 ff.; M. Pohlenz, *Die griechische Tragödie*, i². 76 f.; G. Thomson, edn. of *PV* 12 f.; A. Bonnard, *Rev. de théol. et phil.* 1933, 206 f.; F. Vian, *REG* 55 (1942), 190 f.; F. Solmsen, *Hesiod and Aeschylus*, 147 f.; H. D. F. Kitto, *JHS* 54 (1953), 14 f.; id., *Greek Tragedy*, 64 n. 1; L. Séchan, *Le Mythe de Prométhée*, 55 f.

[22] L. R. Farnell, *JHS* 53 (1933), 40–50; K. Reinhardt, *Aischylos als Regisseur und Theologe* (1949).

it by any convincing parallel. That of the Eumenides was exploded by Farnell, who points out (op. cit. 47) that they do not change in character. A handsome offer from Athene, backed by the threat of Zeus' thunder (*Eum.* 826–8), induces them to abandon their design of avenging upon Athens the acquittal of Orestes, and to take up their quarters on the Acropolis. They do not repent of their earlier persecution of Orestes, any more than they hand over their duty of pursuing the slayers of their own kin to the court of the Areopagus or to the Athenian state, as so many scholars have implied they do. And without a parallel, the development theory is not easy to defend. The most ingenious argument for it known to me is that of Professor Dodds, who argues that its opponents have confused the character of the Aeschylean Zeus with that of the eternal and unchanging god whose first appearance in Greek thought is no earlier than Plato and Aristotle. The Zeus of the fifth-century poets, he points out, differed from this god in being within time. Zeus had not always been the monarch of the universe, but Ouranos and Kronos had gone before him; Zeus had been a child, and his birthplace could be visited.[23] No one can deny that the Zeus of Aeschylus' age was within time; but one may question whether it follows from this that his character could develop. Professor Dodds draws attention to the constant stressing of the youth of Zeus, and argues that this may indicate that when Zeus was older he will have become wiser.[24] But Reinhardt has pointed out (op. cit. 68) that this can easily be explained by the poet's desire to contrast the older generation of gods, to which Prometheus belonged, with the younger generation of Zeus and his brothers; for a similar contrast between older and younger gods, see *Eum.* 150, 731, 778–9, 808–9, 882–3. The advocates of the development theory fail to support it by adducing any certain example of such a change of character in any god. And I shall

[23] See Wilamowitz, *Der Glaube der Hellenen*, i. 127–32.

[24] Professor Dodds supports his argument by a reference to Moschion [*TrGF* 97 fr. 6]. This fragment describes the early time when men lived like beasts without houses, crops, or tools:

> ἦν δ' ὁ μὲν νόμος
> ταπεινός, ἡ βία δὲ cύνθρονος Διί.

At last Time, the mother and nurse of all things, educated them:

> εἴτ' οὖν μέριμναν τὴν Προμηθέωc πάρα
> εἴτ' οὖν ἀνάγκην εἴτε τῆι μακρᾶι τριβῆι
> αὐτὴν παραcχὼν τὴν φύcιν διδάcκαλον.

and they tilled the soil, built cities, and buried their dead. Professor Dodds suggests that Moschion wrote with the Prometheus-trilogy in mind; he thinks the first of the two passages I have quoted may allude to the replacement, at the end of the trilogy, of Bia by Dike as the *πάρεδρος* of Zeus. Now in l. 16 it is not certain that *Διί* is the right reading; both the MSS of Stobaeus originally read *νηί*, which has been corrected to *Διί* in P. The correction may be right, but Canter's *δίκηι* has slightly more palaeographical probability. But even if *Διί* were unquestionably right, the inference regarding the Prometheus-trilogy would not be certain.

argue later (see p. 259) that in view of what we are told about Prometheus' knowledge of the danger that threatens Zeus, the development theory is not necessary to the working out of the plot.

I have given reasons for hesitating to accept either the view that Zeus throughout the trilogy must have been seen in a more sympathetic light than that in which the surviving play reveals him or the view that the later play or plays showed a development in Zeus' character. Schmid, who rejected both these opinions, felt himself obliged to deny the authenticity of the single surviving play;[25] the linguistic and metrical evidence which he adduced, while by no means lightly to be brushed aside, would not, I suspect, have led him to so extreme a step had not the content of the play conflicted with his whole conception of Aeschylus. He held that the presentation of Zeus in the other plays of Aeschylus was one so elevated, so advanced, and so profound that its author could not possibly have written the *PV*, with its very different picture. No, this play must be by someone else; and Schmid invented a sophist of the last quarter of the fifth century, probably an Ionian and certainly a blasphemer and a degenerate, to serve as whipping-boy.

Now if the development theory is regarded with suspicion, as I think it must be, how can Schmid's view or one like it be ruled out? Can this be done by questioning any of the assumptions common, so far, to all investigators of the problem? All alike have maintained the purity, nobility, and refinement of the theology of the other plays of Aeschylus, so utterly at variance with the theology of the *PV*. It may be worth while to re-examine the evidence on which this assumption rests.

I will begin with the *Supplices*, which has more about Zeus than any other play of Aeschylus but the *PV*, and which has supplied much of the evidence alleged in favour of an Aeschylean Zeus-religion. Apart from occasional invocations of Zeus as ξένιος (627, 672 cf. *Agam.* 61, 362, 748), cωτὴρ τρίτος (26 cf. *Eum.* 759–60, fr. 55. 4) or ἀλεξητήριος (8), the chorus of Danaids appeals to him in three capacities; as their ancestor through Epaphos, his son by Io, as Hikesios, the protector of suppliants, and as the special defender of the rights of Dike, Justice. There are also a number of invocations which lay special stress on the supremacy of Zeus' power. The first two of these four aspects of Zeus do not here require investigation. Any other Olympian might have been the divine ancestor of the chorus; and any other Olympian might have had the protection of suppliants as a special charge. The two aspects that remain must be examined; I will take first the last-mentioned. The power of Zeus is celebrated above all in two well-

[25] Op. cit. With ludicrous over-confidence, he treats this conjecture of his own, in his history of Greek literature, as an established fact: Schmid–Stählin, op. cit. i/2. 193, 281 f.

known passages, one in the parodos (86–103) and the other at the end of the second stasimon (590–9).‖

†εἴθ᾽ εἴη Διὸc εὖ παναληθῶc
Διὸc ἵμεροc.† οὐκ εὐθήρατοc ἐτύχθη;
παντᾶι τοι φλεγέθει
κἀν cκότωι μελαίναι cὺν τύχαι
μερόπεccι λαοῖc.

πίπτει δ᾽ ἀcφαλὲc οὐδ᾽ ἐπὶ νώτωι
κορυφᾶι Διὸc εἰ κρανθῆι πρᾶγμα τέλειον.
δαυλοὶ γὰρ πραπίδων
δάcκιοί τε τείνουcιν πόροι
κατιδεῖν ἄφραcτοι.

ἰάπτει δ᾽ ἐλπίδων
ἀφ᾽ ὑψιπύργων πανώλειc
βροτούc, βίαν δ᾽ οὔτιν᾽ ἐξοπλίζει.
πᾶν ἄπονον δαιμονίων.
ἥμενοc ὃν φρόνημά πωc
αὐτόθεν ἐξέπραξεν ἔμ-
παc ἑδράνων ἀφ᾽ ἁγνῶν.

100 em. Wellauer: *τὰν ἄποινον δαιμονίων* **M** 101 em. Paley: *ἥμενον ἄν͂* (= *ἄνω*) **M**

The opening of the first is probably corrupt, and has not been convincingly emended; 'would that the desire of Zeus may in all truth be of Zeus' is very curious sense. The chorus continues: 'Is not it (he?) easy to track down? See, everywhere it (he?) blazes, even in the darkness, with a black fate for mortal peoples. And it falls safely and not on its back, any act that the nod of Zeus ordains shall be accomplished. For thick and bushy do the paths of his mind extend, inscrutable to the eye. And from their high towers of hope he hurls mortals to utter destruction, yet he arms no violence against them. Effortless are all things for the gods: seated he somehow accomplishes his purpose, come what may, from his sacred seat.' The passage from the second stasimon runs as follows:

τίν᾽ ἂν θεῶν ἐνδικωτέροιcιν
κεκλοίμαν εὐλόγωc ἐπ᾽ ἔργοιc;
αὐτὸc ὁ πατὴρ φυτουργὸc αὐτόχειρ ἄναξ,
γένουc παλαιόφρων μέγαc
τέκτων, τὸ πᾶν
μῆχαρ, οὔριοc Ζεύc.

ὑπ᾽ ἀρχᾶc δ᾽ οὔτινοc θοάζων
τὸ μεῖον κρεῖccον ὧν κρατύνει.
οὔτινοc ἄνωθεν ἡμένου cέβει κάτω.
πάρεcτι δ᾽ ἔργον ὡc ἔποc

σπεῦσαί τι τῶν
βούλιος φέρει φρήν.

'Which among the gods might I invoke with reason on the strength of juster actions? The Father himself is the King whose very touch begat us, the mighty architect of our race of old is he who can cure all ills, the kindly Zeus. And seated beneath no other's power, he that is greater rules over what is less. He does not pay reverence below while another sits above; his is the power to accomplish a deed as easily as a word, any deed his counselling mind brings forth.'

It is important to recognize that this exaltation of Zeus' power is not new. These passages recall the nod with which Zeus assures Thetis of his help at *Il.* 542–30 and Zeus' boasting that he could draw up gods and earth and sea upon a golden chain at *Il.* 8. 18–27. Zeus in Homer may seem a less remote and dignified figure than Zeus in Aeschylus outside the *PV*. But this may be simply because he is shown at closer range. In the *Iliad* just as much as in Aeschylus, Zeus effectively controls the action: *Διὸς δ' ἐτελείετο βουλή*. This may apply to the Zeus of the *PV* also.

The Aeschylean Zeus, like the Homeric, has power to accomplish any action, or almost any action, by his mere fiat; he is, in one sense, omnipotent, and is called so three times in Aeschylus (*Sept.* 255, *Supp.* 816, *Eum.* 918; cf. *Supp.* 1048–9, *Eum.* 650–1, *PV* 61). But at this point Professor Dodds's warning against confusion between the fifth-century Zeus and the later *θεὸς αἰώνιος* of the philosophers must be borne in mind, though the context is different from that in which it was originally delivered. Zeus' omnipotence is limited, and so is different from the kind of omnipotence that is familiar to modern thinking. Some of its limitations are ascribed to an unwillingness to transgress the decrees of fate; as in the *Iliad* Zeus regretfully decides not to save Sarpedon, so in the *Agamemnon* (1022–6) and the *Eumenides* (644–51) he is said to have forbidden others to raise men from the dead, and we infer that he would not do this himself. One limitation of his power does not seem voluntary; in the *PV*, he cannot kill the immortal Prometheus (1053; cf. 933). Moreover, Zeus is ‖ not safe against the remoter consequences of his own actions; just as Kronos begat a son more powerful than himself, so will Zeus also, unless Prometheus warns him first. The will of the later philosophic god is identical with the will of destiny; that of the Aeschylean Zeus is not. In the *PV* (518), we learn that Zeus cannot escape Fate; and the last words of the *Eumenides* imply a similar distinction between their powers.

Just as Zeus is omnipotent, but not omnipotent in the modern sense, so is he eternal, but not eternal as we use the word. The rule of Zeus, says the chorus of the *Supplices*, will never end (574, *δι' αἰῶνος κρέων ἀπαύςτου*); but unlike that of the Platonic god, it had a beginning. Another epithet of the philosophic god that has been misleadingly applied to Zeus is 'perfect'. At the opening of the second stasimon of the *Supplices* (524), Zeus is invoked as follows:

> ἄναξ ἀνάκτων, μακάρων
> μακάρτατε καὶ τελέων
> τελειότατον κράτος, ὄλβιε Ζεῦ.

'King of Kings, most blessed of the blessed ones, strength mightiest to accomplish of all that have this power.' Professor Fraenkel's note on *Agam.* 973 well brings out the sense of the epithet *τέλειος*. The common rendering 'most perfect of the perfect' is as inaccurate as it is vague, and carries with it a most misleading suggestion of metaphysical perfection.

I come now to the conception of Zeus as the special champion of the rights of Dike. At *Supp.* 143 f., the chorus appeals to Dike as the daughter of Zeus; cf. *Cho.* 948 f., where her name is probably etymologized as *Διὸς κόρα*. Zeus holds the balance and dispenses injustice to the wicked and holiness to the law-abiding (403 f.); the power of Zeus is just (437); Zeus is implored to look with hostile eye, 'in accordance with justice', upon the violence of the pursuers (811 f.). This connection also receives special stress in the *Seven against Thebes*; there we are repeatedly told that Zeus and Dike are on the side of Eteocles and the defenders; this is implied at 443–6, 565–7, and 630, and is clearly stated at ll. 662–71, where Eteocles calls Dike the maiden daughter of Zeus and affirms that Polyneices, from his earliest years, has had no part in her. There is more about the relation of Zeus and Dike in a fragment of great interest published in Volume XX of the Oxyrhynchus Papyri (P. Oxy. 2256, fr. 9(A) [= fr. 281a Radt]), of which I reproduce the text with Mr Lobel's supplements. The fragment has not been assigned to any particular play,[26] but is probably from a satyr-play. The speaker of

[26] Professor D. S. Robertson has suggested (*CR*, NS 3 (1953), 79–80) that the person whose mischievous behaviour is described in the concluding lines is Ares, and that Dike is leading up to an account of his trial for the murder of Halirrhothius. Ares, however, is nowhere else credited with such a persecution of travellers as that described in ll. 34 f.; and one recalls that this feature would suit Cycnus, not the son of Poseidon killed by Achilles, but the son of Ares killed by Heracles. This Cycnus used to harry pilgrims on the way to Delphi ([Hes.] *Sc.* 478–80, Eur. *HF* 389–94).

Wecklein and Nauck credit Aeschylus with a play *Κύκνος*, but only tentatively, because of Ar. *Ran.* 963:

> Κύκνους ποιῶν καὶ Μέμνονας κωδωνοφαλαροκόμπους.

Professor Robertson has warned me that in this passage the association with Memnon suggests that the other Cycnus is meant (cf. Pindar, *Ol.* 2. 82 and *Isth.* 5. 39). Still, the pseudo-Hesiodic Cycnus is a familiar figure in literature and art (cf. [Hes.] *Sc. passim*; Pindar, *Ol.* 10. 15; Eur. *HF*,

the *ῥῆcιc* with which the fragment begins is clearly Dike herself; she is explaining the manner in which she came by her historic privileges.

μακαρων.[
αυτηθεῶν[
.]*α*'. . . *λ*^{λ}..*πε*^{π}*ε*.*δ*.[
.].[. . . . . .].*ν*.[
ἵζει δ' ἐν αὐτῶι.[.]..[.]..[.].[
δίκηι κρατήcαc τῶ̣ιδε.[
πατὴρ γὰρ ἦρξεν, ἀνταμ[
ἐκ τοῦ δέ τοί με Ζεὺc ἐτίμ[*ηcεν*
ὁτιὴ παθὼν ἡμ[.].[
ἵζω Διὸc θρόνοιcιν[. . .]*ιcμένη*[
πέμπει δέ μ' αὐτὸc οἷcιν εὐμεν[
Ζ̣[*ε*]*ύ̣ς̣, ὅcπερ ἐc γῆν τήνδ' ἔπεμψέ μ'*. . [
. [. .]*εcθε δ' ὑμεῖc εἴ τι μὴ μά*[*την*] *λέγω*.
— . [. .]*οὖ*[*ν προ*]*cεννέποντεc εὐ*. [.]*ήcομε*[*ν*;
— *Δίκην μι*[. . .]*ον πρεcβο*. *η*. *ε*. . . *ρο*. [.
— *ποιᾶc δὲ τ*[*ιμ*]*ῆc ἀρχ*. . . . . . . *ειcα*[;
— *το*]*ῖc μὲν δ*[*ι*]*καίοιc ἔνδικον τειν*. . *ο*.°[,
—].*cα θέ*[*c*]*μ*[*ι*]*ον τόδ' ἐν βρ*[*ο*]*το*[*ῖc*.
— *τοῖc δ' αὖ μα*]*ταίοιc*.[.].[.]. . . .[. .]. . . .*φ*. [.
— *πότερον ἐ*]*πωιδαῖc ἢ κατ' ἰcχύοc τρόπο*[*ν*]; ‖
— *γράφουcα*] *τἀμπλακήματ' ἐν δέλτωι Διό*[*c*.
—].*ωι δε πίνακ' ἀναπτύccει*[*c*] *κακ*[;
—]*ηι cφιν ἡμέρα τὸ κύριον*.
—]*εκτέα cτρατῶι*.
—]*έχοιτό μ' εὐφρ*[*όν*]*ωc*.
].[. . . . . . .]*ηcâτα*[. .] *ἔχω*[ ] [
]*ν*[. . . . . .]*ο ἐπιcπέ*. . .[. . . .]. .[.] [
πό]*λιc τιc οὔτε δῆμοc οὔτ' ἔτηc ἀνήρ*
τοιάνδε μοῖραν π[*αρ*]*ὰ θεῶν καρπουμένη*[.
τέκμαρ δὲ λέξω τῶι τόδ' εὐδερκὲ[*c*] *φέρε*[.
ἔθρε[*ψ*.] *παῖδα μάργον ὃν τίκτει*[
Ἥρα μιγεῖcα Ζηνὶ θυμοιδ[
δύcαρκτ[*ο*]*ν, αἰδὼc δ' οὐκ ἐνῆ*[*ν*] *φρ*[*ον*]*ήματι*,
].*υκτα τῶν ὁδοιπόρων βέλη*
].*ιδωc ἀγκύλαιcιν ἀρταμῶν*
] *̓ν ἔχ*[*αι*]*ρε κἀγέλα κακόν*
]*ν ὄζοι φόνοc*
]*μουμένη*
].*ιπρ*[. . . . .
] *νον χέρα*.

loc. cit.; *Alc.* 503; see Engelmann in Roscher's lexicon, ii/1, 1690; he occurs ten times in Dr Jacobsthal's mythological index to Sir John Beazley's *Attic Red-Figure Vase-Painters*, the other Cycnus never). But it is hard to think of a supplement that will accord with this suggestion. For another suggestion about this fragment, see Ed. Fraenkel in *Eranos*, 52 (1954), 61 f. [See Radt, p. 383, on l. 31.]

]οὖν ἐνδίκωϲ κικλήϲκεται
]νιν ἔνδικ' [. . . . .]. οϲ

In l. 5, it seems probable that *αὐτῶι* refers to the throne once occupied by Kronos, but now by Zeus, and that Zeus is here described as taking his seat upon it. Lines 6 and 7 seem to have said that Zeus gained his victory justly, because his father began the fight.[27] From this time, the goddess continues, Zeus did her honour; the reason seems to be that Zeus, having been attacked first (*παθών*, l. 9), took counter-measures against Kronos, and thus had Justice on his side. Dike now sits (l. 10) beside the throne of Zeus, and he sends her to those mortals to whom he is well disposed. If this is untrue, the goddess's hearers will know it. How, then, is she to be addressed? Dike is her name. And what is her *τιμή*, her particular prerogative? For just men, she protracts their righteous existence; but the wanton or rebellious she chastises. How? By writing down their offences in the books of Zeus, so that when the day of reckoning comes, they will be brought to judgment.

The conception of justice suggested by this passage is not, it must be conceded, a very advanced one. In particular, the account of how Dike acquired her privileges seems oddly primitive. Zeus' reason for bestowing them is strangely personal for the sublime Zeus of Aeschylus; it would hardly be surprising if a god who came to grant Dike her privileges in such an accidental way should be unwilling to allow his own practice in private matters to be regulated strictly by her laws. And the sort of justice implied by this passage, and also by those from the *Supplices* and the *Septem* which I have quoted, seems to be a rough, eye-for-an-eye justice, best expressed in the words of that fragment from the *Μεγάλα Ἔργα* which was the favourite quotation of the emperor Claudius:

εἴ κε πάθοι τά τ' ἔρεξε, δίκη κ' ἰθεῖα γένοιτο.[28]

It may, of course, be argued that the mythology of a single play, and that very probably a satyr-play, is no reliable guide to the general conception of Zeus and Dike entertained by Aeschylus. I can only answer that I have found in Aeschylus nothing that is inconsistent with this notion, but much that is consistent with it; and I shall presently try to support this opinion by an examination of the *Oresteia*. But first I must point out the close correspondence of the fragment with the *Works and Days* of Hesiod, a writer whose great importance for the understanding of Aeschylus has lately been given increased and necessary recognition. Zeus, says Hesiod (252 ff.), sends to earth thirty

[27] This passage helps to make clearer what should be clear from Eur. *HF* 1169 and id. fr. 825, that there is no reason for altering ἦρξεν at *Agam.* 1529 to ἔρξεν.

[28] [Hesiod, fr. 286, 2 M.–W.]

thousand watchers to observe the doings of men. The maiden Dike is Zeus' daughter, honoured by the Olympian gods, and when any man does her violence, she takes her seat by Zeus and proclaims the injustice of men, that he may punish them.

> τρὶϲ γὰρ μυρίοι εἰϲὶν ἐπὶ χθονὶ πουλυβοτείρηι
> ἀθάνατοι Ζηνὸϲ φύλακεϲ θνητῶν ἀνθρώπων,
> οἵ ῥα φυλάϲϲουϲίν τε δίκαϲ καὶ ϲχέτλια ἔργα,
> ἠέρα ἑϲϲάμενοι, πᾶϲαν φοιτῶντεϲ ἐπ' αἶαν.
> ἡ δέ τε παρθένοϲ ἐϲτὶ Δίκη, Διὸϲ ἐκγεγαυῖα,
> κυδρή τ' αἰδοίη τε θεοῖϲ, οἳ Ὄλυμπον ἔχουϲιν,
> καί ῥ' ὁπότ' ἄν τίϲ μιν βλάπτηι ϲκολιῶϲ ὀνοτάζων,
> αὐτίκα πὰρ Διὶ πατρὶ καθεζομένη Κρονίωνι,
> γηρύετ' ἀνθρώπων ἄδικον νόον, ὄφρ' ἀποτείϲηι.

Fishes, beasts, and birds, we are told (274 ff.), eat one another, as there is no Dike among them. To men, on the other hand, Zeus has given Dike, which is far the best of all things; for to the man who recognizes what is just and is willing to speak it out, Zeus gives prosperity; but as for the man who bears false witness upon oath and does outrage against Dike, his race after him becomes feeble.

> τόνδε γὰρ ἀνθρώποιϲι νόμον διέταξε Κρονίων.
> ἰχθύϲι μὲν καὶ θηρϲὶ καὶ οἰωνοῖϲ πετεηνοῖϲ
> ἐϲθέμεν ἀλλήλουϲ, ἐπεὶ οὐ δίκη ἐϲτὶ μετ' αὐτοῖϲ.
> ἀνθρώποιϲι δ' ἔδωκε δίκην, ἣ πολλὸν ἀρίϲτη
> γίνεται. εἰ γάρ τίϲ κ' ἐθέληι τὰ δίκαι' ἀγορεῦϲαι
> γινώϲκων, τῶι μέν τ' ὄλβον διδοῖ εὐρύοπα Ζεύϲ.
> ὃϲ δέ κε μαρτυρίηιϲι ἑκὼν ἐπίορκον ὀμόϲϲαϲ
> ψεύϲεται, ἐν δὲ δίκην βλάψαϲ νήκεϲτον ἀάϲθη,
> τοῦ δέ τ' ἀμαυροτέρη γενεὴ μετόπιϲθε λέλειπται.
> ἀνδρὸϲ δ' εὐόρκου γενεὴ μετόπιϲθεν ἀμείνων.

With this doctrine the Oxyrhynchus fragment, and also the *Supplices* and the *Septem*, is in no way inconsistent. It remains to enquire whether the *Oresteia* provides any evidence for an exalted and sophisticated conception of Zeus.

The chorus of the *Agamemnon* in its opening anapaests declares that it is Zeus, Zeus Xenios, who has sent the Atreidae against Troy; the purpose of Zeus is being worked out, and no attempt to appease him can succeed (60 f.). Of this we are reminded at the beginning of the first choral ode. The portent of the eagles comes from Zeus, and it carries with it Zeus' promise that the expedition shall achieve its purpose. But the portent 'though favourable, is fraught with blame':

> τούτων δ' αἰτεῖ ξύμβολα κρᾶναι
> δεξιὰ μέν, κατάμομφα δὲ φάϲματα (144–5)

The subject of the sentence is Artemis; and whom is Artemis asking to fulfil the portent? The common use of κραίνειν to describe Zeus' making valid of his ordinances, as well as the general sense of the passage, suggests that it is Zeus. To supply Δία as the subject of κραίνειν is certainly not easy. But in this context the likelihood of a reference to Zeus is considerable; and there is a passage of the *Prometheus* which seems to me to supply a parallel. At *PV* 99–100 Prometheus says:

> φεῦ φεῦ, τὸ παρὸν τό τ' ἐπερχόμενον
> πῆμα στενάχω, ποῖ ποτε μόχθων
> χρὴ τέρματα τῶνδ' ἐπιτεῖλαι.

Most editors think that τέρματα is the subject of the *oratio obliqua* clause here; and they take ἐπιτεῖλαι as a rare instance of the active of this verb being used, as the passive so often is, of the rising of stars or planets or of things that are compared to stars or planets. So they arrive at the translation 'when is an end of my labours fated to arise'. Another way of explaining this passage would be to take ἐπιτεῖλαι in its common sense of 'to command' and to supply Δία from the preceding sentence as its subject: then one would render it 'when is he destined to ordain the end of these labours?' Why then is ποῖ used where we should expect πότε? ποῖ occurs with a verb meaning 'to end' to denote the point in time at which something is to end at *Pers.* 735 etc.; see Index Aeschyleus s.v. ποῖ II; and here τέρματα . . . ἐπιτεῖλαι has the force of such a verb. This seems to me open to rather less objection than the usual way of dealing with the problem.[28a]

The chorus, then, is faced with a dilemma; Zeus has sent the Atreidae against the Trojans, but Zeus will concede to Artemis her demand for vengeance against the tearers of the hare. True, Zeus' responsibility is most obscurely indicated; but it may fairly be argued that the reluctance of both Kalchas and the chorus to admit it openly supplies a good reason why it should be darkly suggested instead of clearly stated. In his warning against the dangers that will follow, Kalchas does not only prophesy the sacrifice of Iphigenia; he hints at still more grievous consequences of that sacrifice:

> μίμνει γὰρ φοβερὰ παλίνορτος
> οἰκονόμος δολία, μνάμων μῆνις τεκνόποινος. (154–5)

The chorus' dilemma explains something which I have not seen otherwise accounted for: it explains why the chorus chooses this point to break off its narrative of the events at Aulis to appeal to Zeus. Zeus only has the power to stave off the danger of which the prophet has given warning. This is the context in which is set the famous passage which has so often been appealed to, seldom with any reminder of the nature of that context, by the believers in a 'Zeus-religion'.

[28a] [M. Griffith in his commentary of 1983 also takes ἐπιτεῖλαι as transitive, but translates 'where (one) must set the limit . . .'.]

'Whoever Zeus is, if this name is acceptable to him, by this do I address him.' This invocation is indeed singular; but its singularity has long since been explained in terms of a type of ritual ‖ address not uncommon in ancient religion. In order to invoke a god successfully, you must address him by the right name; accordingly, you either enumerate a whole string of titles or, to save time and to ensure not leaving out the title which the god really does approve, you adopt the form we find here. Many scholars have acknowledged this, but have none the less argued that Aeschylus here makes use of a relic of primitive religion in order to serve advanced convictions of his own.[29] If any convincing evidence of an advanced or spiritual conception of Zeus can be adduced from Aeschylus, this possibility will receive confirmation; at this stage of the investigation, we can only observe that, taken by itself, the passage is wholly explicable as an echo of a primitive manner of invoking gods.

'Measuring all things against him, I can compare to him none but Zeus, if I am in truth to cast from my reflecting mind the vain burden.' What is 'the vain burden'? *μάταιοϲ* and its cognates are often used of the 'vain' or 'frivolous' conduct of those who deny or defy the power of Zeus; and Professor Fraenkel (op. cit. 102–3) takes 'the vain burden' to mean the burden borne by those who think such thoughts. I cannot understand how this meaning can have been clear to the audience. The 'vain burden' is presumably a burden which the chorus is seeking to cast from its own mind. Perhaps this is simply the burden of worry about the fortunes of the house of Atreus. Perhaps it is the burden of trying to think of someone, other than Zeus, with whom Zeus can be compared, a burden which the chorus do, at this minute, jettison. Whatever it is, I doubt if this expression has any religious implication.

What follows shows little trace of an advanced conception; rather it recalls the crudest myths of Hesiodic cosmogony. Ouranos ruled first till Kronos overthrew him; then Kronos ruled till he, in turn, was overthrown by Zeus; sing to Zeus a hymn of victory and you will have managed to attain good sense.

'Good sense' in this connection means the recognition of the feebleness of men in comparison with the gods; in the next stanza it is echoed by *φρονεῖν* at l. 176 and *ϲωφρονεῖν* at l. 181 (see Fraenkel on l. 176 (p. 105)). What is the meaning of the first three lines of the next stanza, from which many scholars have extracted quasi-Tolstoyan theories of 'redemption through suffering'? Professor Fraenkel translates it: 'it is Zeus who has put men on the way to wisdom by establishing as a valid law "By suffering they shall win understanding" '. The use here of a familiar proverbial expression cannot be

[29] See Fraenkel, ad loc. (Aeschylus, *Agamemnon*, ii. 100).

dissociated from a famous passage of the *Works and Days*, in which Hesiod solemnly warns his deplorable brother to choose Dike in preference to Hybris. In the end, he assures him, Dike will be triumphant:

δίκη δ' ὑπὲρ ὕβριος ἴςχει
ἐς τέλος ἐξελθοῦςα. παθὼν δέ τε νήπιος ἔγνω. (218–19)

Here we have the earliest application to the dealings of the gods with men of a proverbial expression which appears earlier in the Homeric phrase ῥεχθὲν δέ τε νήπιος ἔγνω (*Il.* 17. 32, 20. 198). How far has Aeschylus moralized or theologized this piece of popular wisdom? Many have used it to contend that he thought sufferings were sent by the gods to refine or purify the character of mortals. But the actual victims of this law in the *Oresteia*, Agamemnon, Clytemnestra, and Aegisthus, scarcely bear out this contention. They are not purified or ennobled; they are simply killed; and the after life to which they are consigned is not one that offers scope for moral evolution. Other people, it may be argued, will learn from their example. But it is not easy to see what those people will learn, if not the uselessness of defying Zeus' law; and this the victims themselves, in their last moments, may be said to learn also. Now though this doctrine is in a sense ethical, it cannot be called advanced ethics.

The stanza continues: 'And before the heart there drips, instead of sleep, pain that reminds of injury; and even to the unwilling discretion comes.' This clearly reinforces, in other words, what has been said immediately before. What comes next is very difficult. To brush aside the easy correction βίαιος as one dictated by subjective considerations is, I think, a mistake. The strongest argument for this change is a grammatical one; a perfect, as Professor Fraenkel has shown in his note on l. 407, describes a state and not an action, and therefore cannot be modified by an adverb of the sort that normally describes an action, such as βιαίως. Not that this argument in favour of βίαιος is decisive. ἡμένων, unlike βέβακεν at l. 407, is a participle; and it is arguable that since the words cέλμα cεμνὸν ἡμένων are in effect equivalent to ἀρχόντων they can be modified by just such an adverb as ἀρχόντων might be.[30] In any case, the difference to the sense is not a great one. Keep βιαίως and you get the sense, 'And it is, I think, a kindness coming from the gods who by violence sit on their dread bench.' Read βίαιος and you get: 'And it is, I think, a kindness done by violence that comes from the

[30] 'Mr. Conington observes with truth that βιαίως means "violently" rather than "powerfully"; and on the strength of this he prefers βίαιος, and translates "strange as it may seem (που), the free gift of the gods is forced on men". But we may fairly reply that the poet merely meant βιαίως ἀρχόντων, "ruling by the law of constraint, and not allowing mortals to follow their own headstrong wills with impunity" ' (Paley, ad loc.).

gods who sit on their dread bench.' Neither version leaves any excuse for sentimentalizing this χάρις, which is clearly of the sort that would normally be described ‖ as a χάρις ἄχαρις. The word χάρις is being used ironically; what is described as a χάρις is the forcing upon men by the gods of the kind of σωφρονεῖν that comes to them against their will. The notion that the maintenance by the gods of a definite scheme of punishment for offences against their laws lets us 'know where we are', and is therefore on that account called a χάρις, seems to me far too hard to supply from the context, as well as far too sophisticated, for any audience to have understood it.[31]

To sum up, the so-called 'Zeus-hymn' seems to me to yield no evidence whatsoever in favour of 'advanced conceptions', let alone an 'Aeschylean Zeus-religion'. On the contrary, it is set entirely within the primitive framework of the theology of the *Works and Days* of Hesiod.

So is the part played by Zeus throughout the trilogy. At *Agam.* 1485 f., the Chorus finally confirm the fear already present in their minds during the parodos, and roundly affirm that Zeus is responsible for all that has passed:

> ἰὼ ἰὴ διαὶ Διὸς
> παναιτίου πανεργέτα.
> τί γὰρ βροτοῖς ἄνευ Διὸς τελεῖται;
> τί τῶνδ' οὐ θεόκραντόν ἐστιν;

I do not think it possible to doubt that this is one of those passages in which the chorus pronounces a judgment which the poet wishes us to accept. One thinks at once of that passage of the *Niobe* which Plato singled out for special reprobation, the passage whose speaker asserts that the god first manufactures a pretext, when he is bent on the destruction of a mortal.[32] The surrounding context of these lines, which is known to us from a papyrus fragment published by Vitelli and Medea Norsa in 1933, makes it easier to recognize what should in any case have been surmised, that the speaker of these lines does not assert that Zeus punishes the innocent; a view which, if it could safely be regarded as the poet's own, would be at variance with a famous

[31] W. Headlam (*Agamemnon*, 187, followed by F. M. Cornford, *Plato's Cosmology*, 361) suggests that ποῦ might be written instead of που, and the sentence taken as a question; he would then translate, 'Where is there any joy of deities who sit upon their awful seat violently?' He may be right in taking the sentence as a question, but his translation scarcely suits the sense. If the sentence were a question, it would be better rendered, 'And where is the favour shown by the gods who sit on their dread bench through violence?' A close parallel is found in the desperate appeal of Croesus on the funeral pyre (Bacchyl. 3. 37–8): ὑπέρβιε δαῖμον, [πο]ῦ θεῶν ἐστιν χάρις;, where χάρις means 'grace', 'favour', 'gratitude'. In this context, the Chorus may well ask where is the χάρις of the gods; the implication would, of course, be that it is hard to see any. Cf. P. Oxy. 2251. 4–5 (fr. 451h Radt). [Headlam's suggestion, which may well be right, has now been revived by M. Pope, *JHS* 94 (1974), 100–13.]

[32] Fr. 154a. 15–16 Radt.

passage of the *Agamemnon* (750 ff.) in which the Chorus certainly seems to speak for Aeschylus in asserting the opposite opinion. But the *Niobe* passage says nothing about *why* Zeus decides utterly to destroy a house; it only says that, once he *has* decided to, he manufactures some immediate cause for trouble.

The Chorus of the Agamemnon recognizes that Zeus has ordained the king's death; but it warns Clytemnestra that the law of Dike requires that she and Aegisthus in their turn shall pay the penalty; and in spite of her talk of a compact with the daimon, she knows that they speak the truth (1560–77). In the next play, Orestes is the instrument of this law; Apollo is not lying when he tells the Erinyes (*Eum.* 616 ff.) that every utterance of his shrine has behind it the authority of Zeus (cf. 17–19, 713, 797). The speech in the *Choephori* (264 ff.) in which Orestes implores Zeus to help him to his vengeance is one of great interest, though it has never been a favourite with the advocates of Zeus-religion. Note especially ll. 255 ff.:

> *καὶ τοῦ θυτῆρος καί cε τιμῶντος μέγα*
> *πατρὸς νεοccοὺc τούcδ' ἀποφθείρας, πόθεν*
> *ἕξειc ὁμοίαc χειρὸc εὔθοινον γέραc;*
> *οὔτ' αἰετοῦ γένεθλ' ἀποφθείραc, πάλιν*
> *πέμπειν ἔχοιc ἂν cήματ' εὐπειθῆ βροτοῖc.*
> *οὔτ' ἀρχικόc coι πᾶc ὅδ' αὐανθεὶc πυθμὴν*
> *βωμοῖc ἀρήξει βουθύτοιc ἐν ἤμαcιν.*

'If you destroy the nestlings of the eagle (Electra and myself), how shall you have from such a hand as ours, the banquets that are your due? If you cut off the race of the eagle, you will never more be able to convey to mortals the signals that they must obey; and if this whole royal stock wither away, never more will it do duty to your altars on the days of the sacrifice of oxen.' Neither here nor anywhere in the great Kommos which follows this scene does the alleged advanced conception of Zeus make its appearance; we find only an appeal (382 f.) to Zeus as the sender of *ὑcτερόποινον ἄταν* to the evil among mortals. The later utterances of the Chorus conform to the pattern we have come to expect. The sword of Dike will strike down those who have transgressed against the majesty of Zeus (639–45); Zeus is prayed to grant Orestes victory at 783 ff.; after the killings, the Chorus declares that Dike is their author (947 ff.).‖

In the *Eumenides* the issue lies between the law of Zeus, who as the champion of Dike demands through his *προφήτηc* Apollo that the doer shall suffer, and the ancient *τιμή* of the Erinyes, who pursue the slayers of their own kin. At the trial, Orestes calls upon Apollo to prove that his action was just (609 f.): he does not deny the action, but let Apollo say whether it was just or not. Apollo replies that it was just; this assertion he confirms by his authority as *μάντιc*, and this in turn he justifies by declaring that all his utterances as *μάντιc* are dictated by the commands of Zeus. What Zeus orders must be just; and he calls upon the jury to follow Zeus' will (614 f.). The Chorus asks whether Zeus really can have commanded the oracle to order Orestes to kill his mother; and Apollo replies by explaining that the murder of a woman is a much less serious matter than the treacherous killing of a great prince and warrior. 'As you make it out', reply the Chorus, 'Zeus reckons more of a father's death than of a mother's; but Zeus threw his own aged father into chains.' The remark gets Apollo on the raw. With extreme rudeness he points out that prisoners can always be released, but that once a man is dead, he cannot be recalled to life.

ΧΟ. *πατρὸc προτιμᾶι Ζεὺc μόρον τῶι cῶι λόγωι.*
αὐτὸc δ' ἔδηcε πατέρα πρεcβύτην Κρόνον.
πῶc ταῦτα τούτοιc οὐκ ἐναντί', ὡc λέγειc;
ΑΠ. *ὦ παντομιcῆ κνώδαλα, cτύγη θεῶν,*
πέδαc μὲν ἂν λύcειεν, ἔcτι τοῦδ' ἄκοc,
καὶ κάρτα πολλὴ μηχανὴ λυτήριοc.
ἀνδρὸc δ' ἐπειδὰν αἷμ' ἀναcπάcηι κόνιc
ἅπαξ θανόντοc, οὔτιc ἔcτ' ἀνάcταcιc.

When the votes are found to be equal and Athena has to decide the issue, she justifies her acquittal of Orestes on what has always been felt by most people to be a trivial ground; she herself has no mother, and is therefore more sympathetic to the father's rights. In fact, one may well wonder why the Erinyes ever agreed to her appointment as arbiter; for as daughter of Zeus and sister of Apollo she could hardly appear likely to support the ancient powers against the new ones. When the verdict is given, they scarcely accept it with a good grace. Their first impulse is to blast the land of Athens with their power, and only in the face of a direct threat of Zeus' thunder do they at length accept Athene's tempting offer (826–8),

κἀγὼ πέποιθα Ζηνί, καὶ τί δεῖ λέγειν;
καὶ κλῇδαc οἶδα δώματοc μόνη θεῶν
ἐν ὧι κεραυνόc ἐcτιν ἐcφραγιcμένοc.

We are commonly told that one of the main purposes of the trilogy is to show how the primitive vengeance of the Erinyes was replaced by the rule of law; but nowhere does the text say that the Eumenides are now handing over their functions to the Areopagus. They nowhere promise to give up their present pursuits; and nothing in the play justifies the belief that the transition from the blood-feud to the trial by jury is the main theme of the trilogy.[32a]

An examination of the surviving plays apart from the *Prometheus*

[32a] [See now my contribution to '*Owls to Athena*': *Essays on Classical Culture presented to Sir Kenneth Dover* (1990), 203–11 = *REG* 102 (1989), 1–9.]

seems to lead to what many people will find the startling conclusion that Aeschylus' conception of Zeus contains nothing that is new, nothing that is sophisticated, and nothing that is profound. But is this conclusion, after all, surprising? No advanced theology of the kind Aeschylus is so often credited with is found in any Greek author before Plato, a fact which Nilsson (see n. 8) acknowledges with some surprise. If Aeschylus was in truth a solitary pioneer of an advanced theology, one would expect to find Plato acknowledging the fact with gratitude and admiration; but this is not the case. In Books II and III of the *Republic*, Plato delivers a fierce attack upon the famous poets of Greece for the corrupting character of the myths which they relate, and from this attack Aeschylus is far from being exempt. The first charge brought against the poets, that of representing the gods as doing violence to one another, is not specifically levelled against Aeschylus, though the *Prometheus* itself could have furnished an excellent example. But as an instance of the second crime of the poets, that of representing the gods as being responsible for evil, Plato quotes the passage of the *Niobe* which I have mentioned (*Rep*. 380 A); as an instance of the third crime, that of making out the gods as capable of deceit, he quotes the disguise assumed by Hera in the *Semele* (ibid. 381 D); and at the conclusion of the second book he cites, in summing up his remarks upon this head, Thetis' famous denunciation of Apollo's treachery (fr. 350, quoted at 383 A–B). Of the crimes of representing death as terrible and of crediting gods and heroes with unbecoming behaviour or with unbecoming sentiments, Aeschylus is not specifically accused, but it is clear that he would not have been acquitted. Now whatever Aeschylus might have said about the other gods, it is highly probable that if he had put forward a conception of Zeus that verged in places on the severe and abstract monotheism which Plato would have approved, Plato would not have included him in his general condemnation of ‖ the poets. Plato, it is true, had little sense of historical accuracy and was capable of unfairness to his predecessors; but I find it hard to believe he would have been so unfair and unintelligent as this.[33]

It seems possible, and indeed probable, that Plato lumped together Aeschylus with the rest simply because Aeschylus shared the common beliefs and assumptions about the gods. Aeschylus certainly makes Zeus supreme; so is he supreme in Homer. Numerous passages could be cited to show that Aeschylus, just as much as Homer, believed in the efficacy of gods other than Zeus, of Apollo and Athene, Artemis and Ares, Hera and Poseidon.[34] His presentation of the gods could not be

[33] I am grateful to Mr D. A. Rees for urging upon me the usefulness of this argument from Plato.

[34] All mentions of each god and daimon in Aeschylus are conveniently listed by W. Kausche,

more patently anthropomorphic. They have mouths, feet, arms, and eyes;[35] they have special means of transport and use special weapons.[36] Not only are the other gods often introduced upon the stage, but in the *Psychostasia* Zeus himself figured as a character. The gods constantly pursue mortal women, especially Zeus.[37] If Aeschylus had ever heard of Heracleitus or Xenophanes and their attempts to inculcate a more refined notion of divinity, there is nothing in his works to prove it.[38] Further, there can be little doubt that the view of Zeus and Dike that we find in Aeschylus is not materially different from that of Hesiod. Any advance we do find has appeared well over a hundred years before Aeschylus in the work of another Athenian poet, Solon, who presents the notion of Zeus as the champion of Dike and also the doctrine of *κόρος* followed by *ἄτη* or *ὕβρις* in a form in no important way different from what we find at *Agam.* 750 ff. (see fr. 13 West). Scholars have expended much learning and ingenuity in tracing a gradual development in ethical profundity from Homer to Hesiod, from Hesiod to the lyric poets and early elegists, from lyric to Aeschylus, from Aeschylus to Sophocles. Much of their work has no relation to what is in the texts, but is simply a product of the nineteenth century's insistence upon progress in all matters and its conviction that the power of poetry must reside chiefly, if not wholly, in the ideas, and especially the ethical ideas, which it expresses. In fact, as Plato naturally recognized, the similarity of general outlook between all these poets is far more striking than the differences between them. Almost the only scrap of Aeschylus' talk that has come down to us is the remark (in Ath. 8. 347 E) that his dramas were 'slices from the great banquets of Homer'. I suggest that this may be a more valuable indication of his poetic views than is commonly supposed.[38a]

Mythologumena Aeschylea (Diss. Philol. Halenses, 9 (1888), 129 f. See ch. 2 of W. Kranz's *Stasimon* (pp. 34 f.), entitled 'Die göttliche und die menschliche Welt in der alten Tragödie'.

[35] Mouths, *PV* 1032, fr. 350, 5; feet, *Eum.* 294; arms or hands, *Supp.* 313, 1066, *PV* 849, *Cho.* 395, fr. 327; eyes, *Supp.* 812–13, *PV* 654, 903.

[36] Transport, *PV* 287–8, 394–6, *Eum.* 403–5. Weapons, *PV* 405, 924–5, *Sept.* 131; Apollo and Artemis must have used arrows in the *Niobe*. Zeus' thunder is mentioned at *Sept.* 255, 429–30, 444–6, 513, 629–30, *PV* 358 ff., 667–8, 915 f., 1016, 1043–4, 1061–2, 1082–3, *Eum.* 826–8.

[37] Apollo pursues Cassandra in the *Agamemnon*; Poseidon pursues Amymone in the *Amymone*; Boreas doubtless pursued Oreithyia in the *Oreithyia*. Zeus pursues Io in the *Supplices* and *PV*, Europe in the *Kares*, Semele in the Semele-trilogy, Danae in the Perseus-trilogy, and no doubt Callisto in the *Callisto*.

[38] I do not trouble to show how feeble is the evidence adduced by those who have tried to prove him acquainted with the works of these writers. Wm. Nestle's remarks, in the article already quoted (*Griechische Studien*, 122–3), provide a fair specimen. Neither shall I deal with the attempts which some have made, encouraged by a light-hearted remark of Cicero ('Aeschylus, non solum poeta sed etiam Pythagoreus . . .', *Tusc.* 2. 23), to prove Aeschylus a Pythagorean or an Orphic. On the Lycurgus trilogy and its connection with Orpheus, see the sensible remarks of I. M. Linforth (*The Arts of Orpheus*, p. 10).

[38a] [The remark is unlikely to be authentic, but is an excellent critical judgement.]

It is time to return to the problem of the Prometheus trilogy: how are we to reconcile the presentation of Zeus in the surviving play as a harsh and ruthless tyrant, the torturer of the great champion of mankind, with the 'advanced' Zeus we find elsewhere in Aeschylus? Does the rejection of the view that Zeus' character 'developed' in the course of the trilogy oblige us to follow Schmid in believing that the surviving play is not by Aeschylus? If the view of Zeus in Aeschylus for which I have argued is correct, there is no reason to suppose the Zeus of the *PV* to be un-Aeschylean. Zeus may well honour Dike because she was on his side against his father, and uphold her τιμή among men. But for Zeus himself to feel scruples in dealing with a challenge to his own authority would be quite another thing. Any government maintains justice among its subjects, but few governments refer to the arbitration of abstract justice disputes with powers which have challenged their own authority. Further, one must be on one's guard against supposing that the god who is for mankind is necessarily to be thought of as 'good' and the god who is against mankind is necessarily 'bad'. The fifth-century Zeus was not a democratic god, who could never, for fear of losing his job, do anything not in the best interests of the human race. Hesiod, in whose works Aeschylus' mind was soaked, records without expressing disapproval Zeus' annihilation of at least one of the four successive races of mankind. Nor does he feel the fifth race, to which he himself belongs, to have received particularly favourable treatment at Zeus' hands. The gods, he says (*Op.* 42 f.), have concealed from men the means of subsistence (βίος); a man could easily do enough work in a day to keep himself for a year, had Zeus not concealed subsistence from us in his anger against Prometheus. No resentment against Zeus is expressed; that would be presumption, and would invite punishment; but neither is his benevolence stressed. What is emphasized, in ‖ Hesiod and also in tragedy, is the supremacy of his power and the uselessness of trying to deceive him or in any way resist his will. The whole of Greek poetry from Homer to Sophocles abounds with illustrations of this attitude. The gods by their laws encourage righteousness among men. But they themselves are not obliged to obey those laws, nor should we be reasonable to expect it. The poets talk not of the righteousness of the gods, but of their power, and of their insistence that we be righteous; they insist upon the excellence of their laws, but still more upon the foolishness of trying to oppose their will. There is a sentence in the *Magna Moralia* which well illustrates one facet of this difference: 'It would be odd if anyone were to say he loved Zeus' (*Mgn. Mor.* 1208ᵇ30 ἄτοπον εἴ τις φαίη φιλεῖν Δία).[39]

[39] See E. R. Dodds, *The Greeks and The Irrational*, 35, with n. Arist. *Rhet.* 1391ᵇ1–3 says that εὐτυχία makes men φιλόθεοι. But the context indicates that this means little more than that they are disposed to accept the decisions of the rulers of the universe in an uncomplaining spirit. 'Of

There is no reason why a god whom his worshippers regard with this degree of respectful detachment should not be portrayed by one of them as acting ruthlessly, particularly since in this instance Zeus was defending his own supremacy against a challenge. Zeus has a far closer and more personal relation to the action of this play than he has to that of any other surviving work of Aeschylus. Prometheus is a personal enemy, who has defied Zeus' authority; why should Zeus show him mercy, any more than he showed mercy to any other of his defeated rivals? Of course, it is unreasonable to argue that the Zeus of the *PV* is the object of the actual hostility of the poet. Aeschylus is not 'for' or 'against' either Zeus or Prometheus. His immediate purpose is not educative, but dramatic; and he must so hold the balance between the protagonists in the central conflict as to exploit to the full the dramatic possibilities which the story offers. Prometheus is indeed shown with much sympathy; but it does not follow that the audience is meant to regard Zeus with the indignation that we naturally feel. The warnings of the Chorus against excessive *αὐθαδία* have been seized upon by those who have regarded Prometheus himself as being the guilty party. This is ridiculous; but the warnings do serve to remind us that if Prometheus' attitude is magnificent, it is also unrealistic. And unrealism of this kind, the refusal to recognize one's own helplessness in the face of superior power, is a fault that Aeschylus, like other Greek poets, constantly reprobates. The point is made still clearer by the warning of Oceanus, a character whom some scholars have absurdly misunderstood. He offers to intercede with Zeus, at his own risk (325 f., 337 f.); and Prometheus himself acknowledges that some risk could be involved (340–1). What more he could have done, I cannot see. As his offer is rejected, there is nothing left for him to do but to warn Prometheus to moderate the violence of his utterances, a warning which Prometheus would have done well to heed. For all he gets by his rash defiance is an intensification of his punishment.[40]

With this in mind we return to such positive indications as we possess of how Zeus and Prometheus became reconciled.[41] Among these, the

the major Olympians', Dodds writes, 'perhaps only Athena inspired an emotion that could reasonably be described as love.' It is probable that in other cities beside Athens the local tutelary god inspired a similar feeling (Hera at Argos, Apollo at Delphi and Delos, etc.).

[40] Much misunderstanding has been caused by ll. 342–3:

μάτην γὰρ οὐδὲν ὠφελῶν
ἐμοὶ πονήϲειϲ, εἴ τι καὶ πονεῖν θέλειϲ.

Denniston (*GP* 303) correctly explains the last phrase as meaning 'if you *want* to labour'; no doubt of the sincerity of Oceanus' offer is implied.

[41] If there was a god to whom the Greeks assigned a not dissimilar relation of partiality, protection, and patronage towards the whole of humanity, it was Prometheus. And they did not conceive of Prometheus as having any share in the government of the universe. This is the prerogative of Zeus, *βιαίωϲ ϲέλμα ϲεμνὸν ἥμενοϲ*.

prophecies uttered by Prometheus himself in the *Desmotes* have a peculiar authority; and if we examine them, we find that again and again Prometheus declares that by his possession of the secret about Thetis he will force Zeus to release him. 'I swear', he says (167 f.), 'that the chief of the blessed ones shall yet have need of me, tortured though I am in mighty fetters, to reveal the new design through which he stands to lose his sceptre and his prerogatives. And never shall he coax me by the honey-tongued incantations of Persuasion, never shall I cower before his harsh threats and reveal this secret, till he release me from those cruel bonds and be willing to render me atonement for this outrage.' 'I know', he says later (185 f.), 'I know that Zeus is hard and keeps Justice by his side; but none the less, soft shall his resolution be when by this means he is crushed; and making smooth his rugged temper, he shall come to union and to friendship with me no less eagerly than I.' Observe that in this early scene with the Chorus, Prometheus says that his knowledge of the secret will force Zeus to release him; this statement he repeats at 524–5. In the final scene (908 f.) he talks of the disaster Zeus will suffer if he makes the fated marriage as though it is bound to overtake him. That, no doubt, is intended as a way of threatening Zeus, who as he must know, will hear his words. For though Prometheus knows that Zeus will eventually release him, Zeus can be made to do this only if he knows of the threat which overhangs him. We know, in any case, that Zeus did not suffer destruction; and we may therefore easily infer that it was by revealing the secret that Prometheus purchased his release. Now if Zeus has in the meantime reformed in character, it is odd that he should need the threat of impending disaster to lead him to pardon his noble adversary; this argument, which I have not before seen used, surely confirms arguments against believing that Prometheus owed his release to a change of heart on the part of Zeus which I set out earlier. The change of attitude by the *Eumenides* is ‖ indeed a parallel, but one which indicates the opposite of what it has often been supposed to. The Eumenides do not change their character, but they do a deal with Athene, and in consequence their attitude changes. So, I suggest, must Zeus have done a deal with Prometheus, and changed his attitude in consequence.

The deal once concluded, it is likely that the trilogy ended in an atmosphere of mutual concession and reconciliation, such as Heracles was a peculiarly fit person to promote. It may well have been indicated that Zeus would not renew the attempt to destroy the human race which Prometheus (231 f.) has claimed to have frustrated, we do not know how. But in view of the explicit words of Prometheus' prophecy, it seems certain that Prometheus' possession of the secret regarding

Zeus' dangerous marriage must have played an indispensable part in making this reconciliation possible. Why was this motive necessary to the poet? Firstly, no doubt, because it was part of the myth; secondly, because no Greek of Aeschylus' time or earlier thought of Zeus as having that special relation of patronage and protection towards mankind which the Jews and Christians have assigned to God.[42] Zeus is indeed the champion of Dike, a rough retributive justice; he insists that men, like gods, shall keep his law. This is an important function, and it plays a central part in the action of Aeschylus' plays. That the moral background it provides has great importance, I am very far from wishing to deny. What I do deny is that they contain an *advanced* morality involving an ascription to Zeus of greater power than earlier poets had assigned him and a belief in his regeneration of mankind through suffering. This rude morality serves as an indispensable background to the action. But there is no reason for thinking that the dramatist's main purpose was to commend it to his audience. There was nothing new in it, and among most Athenians there will have been little disposition to dispute it. Aeschylus probably never thought of doubting it, and to that extent his plays may certainly be said to recommend it. I will conclude by quoting some words of Hermann's, set originally in the context of a by no means wholly acceptable polemic against Welcker's reconstruction of the trilogy, but applicable to other opinions besides Welcker's. 'Sed aliud longe est fabulae rationem exponere, aliud explicare consilium poetae, praesertim scaenici, qui non docere, sed placere populo, aut si docere, non docere per aenigmata abstrusae cuiusdam sapientiae, sed per viva constantiae, fortitudinis, animi magnitudinis exempla velit. Ab eo trilogiam factam credere, ut istiusmodi doctrinae philosophicae satisfaceret, alienissimum a sana ratione est, neque utilius, quam quaerere (et quaesiverunt quidam) quomodo in Iovis persona crudelissimi tyranni exemplum proponere potuerit. Neque habuerunt ista apud Graecos offensionem, nec potuerunt habere, ut in religionibus, quae totae ex huiuscemodi fabulis essent compositae' (*Opusc.* viii. 255–6).

[42] Since Aeschylus everywhere works against the background of the system of moral order I have tried to describe, and therefore in a sense may be said to 'teach' that it is true, this statement may be thought to go too far. But the 'teaching' which poets are said by the Aristophanic Aeschylus (*Ran.* 1054–5) to provide was probably held to consist chiefly in the depiction of 'viva constantiae, fortitudinis, animi magnitudinis exempla' (cf. ibid. 1019–22) and not in the propagation of new religious or philosophical doctrines.

[I reprint this article with reluctance, simply because it played a part in the discussions of the time, and people may find it convenient to have a copy. I still agree with a good deal of its contents; but I have grown to dislike its tone, and I am not pleased when people who are apparently unacquainted with my later writings on related topics quote it as though it represented my most recent views. If Aeschylus' theology is comparatively simple, that does not mean that it is not profound; see *The Justice of Zeus* (2nd edn. 1983), 209 n. 100).]

19(23)

The *Supplices* of Aeschylus
*The New Date and Old Problems**

No papyrus discovery of modern times has given a ruder shock to orthodoxy than the fragment of a didascalic notice published in 1952 in Part XX of the Oxyrhynchus series.[1] Most living scholars had been brought up to take it for granted that the *Suppliant Women* was the earliest play of Aeschylus, produced perhaps as early as the beginning of the fifth century. This fragment indicates that the Danaid trilogy, of which the *Suppliant Women* is the first play, won first prize in a year in which the runner-up was Sophocles; and Sophocles competed first in 470, 469, or 468. The indication of date in the notice is mutilated; we read only the letters ΕΠΙΑΡ[. ΑΡ might possibly be the beginning of the word *ἄρχοντοϲ*, but it was more probably the beginning of the name of Archedemides, archon in 464/3; in either case, the name of Sophocles supplies a *terminus post quem*. Desperate attempts were made to save the orthodox belief. The production of the late sixties might, it was argued, have been a revival; or the play might have been written forty years earlier and produced only in the sixties; the three leading advocates of these suggestions averaged about eighty years of age. But Albin Lesky (*Hermes*, 82 (1954), 1 f. [= *Ges. Schr.* 220 f.]) pointed out that in the present state of the evidence it would be perverse to deny that a first production in the sixties is virtually certain, and this view is now generally accepted. Without wishing to abuse the advantage given us by the new evidence in order to patronize our predecessors, we may find it useful to investigate the causes of such a persistent and such an || important error. Such an inquiry may help us to avoid similar mistakes in future; it may also throw light upon the *Suppliant Women* in particular and upon the art of Aeschylus in general.

* This article represents the text of a paper read at the Fourth International Congress of Classical Studies at Philadelphia, Pa., in Aug. 1964. I am grateful to those friends who have discussed with me the problems here treated, and especially to Richmond Lattimore and Rudolf Kassel.

[1] P. Oxy. 2256, fr. 3 (Part XX, p. 30). [For a full bibliography and a detailed treatment of this question, see now A. F. Garvie, *Aeschylus*' Supplices*: Play and Trilogy* (1969).]

L'Antiquité classique, 33 (1964), 356–74.

Most of the scholars who had assigned the Danaid trilogy to a late date had argued from supposed allusions to contemporary political events; most of those who put it early had argued from an analysis of its style, diction, metre, and dramatic technique. The former method was employed first at the beginning of the nineteenth century, when the school of Boeckh, Otfried Müller, and Welcker began to apply the data of history and archaeology to the interpretation of Greek literature. Otfried Müller,[2] following Boeckh, felt little doubt that the trilogy must have been produced while the Athenian alliance with Argos made in 461 was in effect. In the *Oresteia*, produced in 458, Agamemnon rules not in Mycenae, as in Homer, nor in Sparta or Amyclae, as in the lyric poets, but in Argos; and Orestes promises that Argos will requite his acquittal by an Athenian court by helping Athens in time of war. That is generally held to allude to the alliance. Now in the *Suppliant Women*, after Danaus has announced the ratification by the Argive demos of the king's decision to protect his daughters against their pursuing cousins, the Aegyptiads, the Chorus expresses its gratitude by a solemn prayer to all the gods that Argos may enjoy prosperity and be free from war and pestilence (625 f.). This ode conforms to a standard pattern for invoking blessings on a city that has its origins in early epic:[3] in this it resembles the ode of benediction upon Athens sung by the chorus of the *Eumenides* (916 f.). Müller connected this ode with the friendship between Athens and Argos. He believed that at this time Argos must have had a democratic constitution; and he saw allusions to this fact in several passages in the play which insist upon the Argive king's incompetence to make a decision that may involve war without the consent of the popular assembly.

Other scholars protested that the alleged political allusions || gave no clear evidence for the play's date.[4] The ode of blessing upon Argos, they objected, is wholly suited to its context in the play: it is natural for the Danaids to return thanks, and to do so in terms of a known pattern of benediction. The force of this objection must be acknowledged: without denying that the ode may be intended to allude to the alliance, we can hardly accept it as a certain indication of the play's date. But what if it is taken together with the alleged allusion to democracy at Argos? During the great scene in which the Danaids supplicate the Argive king Pelasgus, the king warns them that he cannot promise them support without the consent of the whole citizen body (368–9).

[2] *Eumenides* (Göttingen, 1833), 123; cf. *Griechische Literaturgeschichte*, ii^3. 88; cf. Boeckh, *Tragoediae Graecae principum reliquiae* (1808), 54.

[3] See G. Müller, *De Aeschyli Supplicum tempore atque indole* (Diss. Halle, 1908), 67 f.

[4] Notably G. Müller in the dissertation just quoted. Although Müller reached a wrong conclusion, his work has great merits; his refutation of the historical arguments for a late date is particularly good.

The Danaids insist that he, the king, is the sole ruler, dwelling on the notion of monarchical supremacy with an intensity characteristic of their oriental origin (370 f.). Later in that scene the king fears that if he acts without the consent of the *demos* the people may say of him, in case of a disaster, 'In doing honour to strangers you have ruined the city' (397 f.). Finally he summons Danaus before the assembly, complaining as he does so of the people's readiness to make complaint against their ruler (480–8). After the king has persuaded the multitude Danaus returns, and vividly describes how the sky was thick with right hands raised in favour of the decision (607–8). Later, in the hymn of blessing, there is mention of 'the people, that rules the city' (699).

'The praise of democratic Argos in these lines' it was claimed as recently as 1960,[5] 'is totally irrelevant in any mythological situation.' That statement is a little sweeping. Even an epic kingdom had its popular assembly, as we know from several places in the *Iliad*. 'The Greeks at Troy', writes Andrewes,[6] 'will feel and will be safer if they have a king to tell them what to do, but—and that is the point of the meetings and the speeches—they must also feel sure about their king, confident that ‖ he can cope with their situation and with them, and that he will continue to fulfil their own obscurely felt purposes.' Aeschylus' allusions to the powers of the assembly are perfectly consonant with this state of affairs. Even the king's fear of being found fault with by his subjects could well be found in Homer: at the crisis of his fate Hector is prevented from taking refuge in Troy by the thought that people will reproach him for not having taken the advice of Polydamas to order a retreat earlier.[7] It may well be that the poet has been led by the added importance of assemblies in his own times to lay special emphasis on the powers of the epic *demos*, as Euripides does in his *Orestes*; but in doing this he is not guilty of an anachronism. The possibility that he had a contemporary Argive democracy in mind must certainly be reckoned with; but we cannot feel certain that it was not the importance of the assembly in his own city that was in his mind.

The need for caution at this point becomes even greater when we consider how little we know about the Argive constitution during the fifth century. Herodotus[8] says that after the disastrous defeat by Sparta in 494 Argos was ruled for a time by slaves. This statement is usually interpreted in the light of the testimony of Aristotle and Plutarch that after this event certain of the perioikoi—'the best' of them, in

[5] By W. G. Forrest, *CQ*, NS 10 (1960), 240 l. 3. My disagreement with Mr Forrest over the *Suppliants* does not mean that I fail to appreciate his ingenious and learned article.

[6] A. Andrewes, *Probouleusis* (Oxford, 1954), 7.

[7] *Il.* 22. 99 f.

[8] 6. 83.

Plutarch's words—were admitted to the citizenship.[9] Did their admission make the constitution more democratic? And if so, how much more? We can hardly hope to know. The slaves, according to Herodotus, were eventually expelled from Argos and seized Tiryns, from which they were ejected about the middle of the sixties. After that the Argolid was again controlled by the descendants of its former rulers. Did the constitution become less democratic? Again, we do not really know. Themistocles, who took refuge in Argos about 470/69, was eventually forced to leave in obedience to Spartan pressure. Was that because an aristocratic, and therefore pro-Spartan government was in power? or was it simply due to Argive weakness? The fact that Argos made offerings at Delphi at about this time has been thought to strengthen the || former supposition; but Delphi was a panhellenic shrine, and the argument is not a safe one. Again, it has often been inferred that when Argos made the alliance with Athens in 461 a democratic government must have been in power. But this inference is not valid; the ancient rivalry of Argos and Sparta must have been strong before democracy was thought of in either place. To sum up, our evidence for the nature of the Argive constitution during this period is almost negligible, though such indications as there are make somewhat against the likelihood that it was democratic. After all, where in mainland Greece outside Athens was there a real democracy before the Athenian Empire had advertised this form of government?

We must conclude that it is possible, but far from certain, that the play contained 'praise of Argive democracy'. If it did, then it is likely that as Otfried Müller supposed Argos was democratic at the time of its production. That runs counter to the views of other historians,[10] who believed that a democratic Argos, containing enfranchized perioikoi, had taken in Themistocles about 470, but that an aristocratic Argos ruled by the descendants of those defeated at Sepeia in 494 had expelled him a few years later. They suppose that the supplication of the Danaids at Argos was meant to remind the Athenian audience of 463 of the supplication of Themistocles in about 470, and that the praise of Argive democracy was meant not for the present prospective allies of Athens but for an Argive democracy suppressed by them some years earlier. Both these suppositions are unlikely. If the play praises the Argive democracy, it probably praises a contemporary democracy, as Müller thought. But we have seen that independent evidence for the existence of democracy at Argos in 463 is not strong. Its weakness does

[9] Arist. *Pol.* 1303^{a}6; Plut. *De mul. vit.*, 245 F. See the discussion by Forrest, op. cit. 221 f. [cf. M. Wörrle, *Untersuchungen zur Verfassungsgeschichte von Argos im 5. Jahrhundert vor Christus* (Diss. Munich, 1964), reviewed by G. L. Cawkwell, *JHS* 88 (1968), 219.]

[10] See Forrest, op. cit. 239, who quotes E. Cavaignac, *Rev. phil.*, 45 (1921), 102–6.

not favour the possibility—and it is no more—that the play does praise Argive democracy at all. We must conclude that even if Boeckh and Müller were right in thinking the play referred to the political situation of the late sixties, the grounds adduced by them for dating at that time were far from sufficient to prove their case. Of other attempts to date the trilogy by features of its content—the explanation of the Nile flood, the treatment of the myth of Io and its ‖ relation to that in the Prometheus, the supposed allusion to a particular building in the mysterious passage about the goddess who occupies the *cεμνὰ ἐνώπια*—the less said the better.[11]

Those who have argued from the play's style, diction, metre, and technique have argued with few exceptions, for an early date. Let me first deal briefly with the linguistic and stylistic arguments. It has commonly been maintained that these strongly favour an early date; and yet the upholders of this view have not so far offered a close and detailed examination of the relevant material.[12] That would need considerable space; at present it will be enough to point out that the assertions made by believers in an early date have not so far been adequately grounded. For the least incomplete statement that any of them has offered we must go to such a gravely inadequate production as the late F. R. Earp's study of the style of Aeschylus.[13] The *Suppliant Women* has not more compounds than other plays, nor more *hapax legomena*, nor more similes or metaphors; the number of epic words is higher, but not significantly higher, than in the other plays. There are thirty words which Earp considers to be rare, nine more than in the *Persae*, which comes next in this respect; but of these thirty ten occur in two brief passages, amounting together to less than sixty lines, in which the poet is clearly making a special effort to convey the effect of outlandish speech and which are, moreover, deeply corrupt.

So much for the evidence from diction. Earp finds the style markedly different from that of Aeschylus' other works. 'In parts of it', he writes, (p. 63) 'the various forms of ongkos are not laid on with the brush but spread with a trowel, and very little of the surface escapes. We find here all the qualities paro‖died by Aristophanes . . .'. In fact we find these qualities everywhere in Aeschylus. Do we find in the *Suppliant Women* anything too extreme for the poet who called fishes 'children of the undefiled' (*Pers.* 578), or dust 'mud's thirsty sister' (*Agam.* 494–5) or who wrote the first stasimon of the *Seven* (287 f.) with its astonishing description of the horrors that accompany the sack of cities?

[11] On the Nile flood, see G. Müller, op. cit. 28 f.; on the Io myth, 31 f.; on the goddess (*Supp.* 144 f.), 14 f.

[12] Authority is no substitute for argument, nor will the isolated treatment of details suffice.

[13] F. R. Earp, *The Style of Aeschylus* (Cambridge, 1948). Studied in his way, according to this author, the tragedians 'become three old friends whose little ways we know' (p. 5).

Metrical arguments advanced in favour of the early date are not substantial. The strict synapheia observed in the simple dochmiacs of the supplication scene may be a singularity, but is scarcely a safe indication of earliness.[14] Advocates of the late dating were able to point to one archaic feature of the *Persae* which is wanting in the *Suppliant Women*. Aristotle says that in early times the regular dialogue metre of tragedy was the tetrameter, and in the *Persae* it is used for two whole scenes. The other party retorted that the *Agamemnon* also contains two tetrameter scenes. But there the scenes are shorter, and the tetrameters seem to be used just as they are in later tragedy, to convey the notion of speed and excitement. The use of tetrameters in the *Persae* is different, and the difference may well be significant.

The most noteworthy metrical argument deployed on either side of the controversy rests on the observation of E. C. Yorke in 1936[15] that the incidence of resolutions would indicate for the *Suppliants*, with one resolution in eleven trimeters, a date between the *Persae*, with one in nine, and the *Seven*, with one in twelve; in all the other plays, resolution is considerably more frequent. As Yorke recognized, this test is not one that can be applied rigidly or by itself. But Zieliński[16] had shown that in the case of Euripides, where our material is far ampler than it is for Aeschylus, the order indicated by the incidence of resolution closely corresponds with the order so far as it is known from external evidence. That warrants us in treating this indication with special respect, especially now that it is confirmed by the evidence of the papyrus.

We now come to arguments which have been far more influential in creating belief in an early date, those based on dra||matic technique. Aristotle (*Poet.* 1449^{a}17) says that the choral element in early tragedy was very great, and that Aeschylus reduced it. The choral element in the *Suppliants* is greater in proportion to the whole than in any other surviving play of Aeschylus, and hence it is inferred that it must be earlier than the others.[17] This argument has had great influence, and yet its weaknesses are obvious. First, it is not safe to assume that the choral part was reduced at a uniform rate, so that each successive play had less chorus than the one before. Next, the subject-matter of the *Suppliants* clearly required that the choral element should be very great; can we be sure that a poet writing after the choral element in most plays had been considerably reduced would not have made an exception for a play dealing with this subject? In this play the chorus is protagonist, but that does not help to prove that the play is early. Even

[14] G. Müller, op. cit. 61 f.

[15] *CQ* 30 (1936), 116 f.

[16] *Tragodoumenon libri tres* (Cracow, 1925), 133 f.; cf. E. B. Ceadel, *CQ* 35 (1941), 66 f. [and A. M. Dale, *Euripides, Helen*, p. xxv].

[17] See G. Müller, op. cit. 52 f.

in the *Eumenides* it is at least deuteragonist, and that play was produced in 458; even as late as Euripides' *Bacchae*, we find a chorus deeply implicated in a play's action.

Closely connected with the argument from the large amount of choral utterance is the argument from the poet's restricted use of the second actor.[18] Like the *Persae* and the *Seven*, the *Suppliants* is a play for two actors, but there are only two passages, neither of them very long, where two actors take part in a dialogue together. Throughout the great supplication scene which is the kernel of the play, the chorus sings lyrics and the coryphaeus speaks iambics in the effort to persuade the king while Danaus *πατὴρ καὶ βούλαρχος καὶ ϲταϲίαρχοϲ* (11–12), *πρόνοοϲ καὶ βούλαρχοϲ* (969–70), stands silent on the stage. Only when the supplication is over and the king has consented does Danaus give tongue,[19] and then it is only to ask the king for an escort for himself and his daughters as they enter the city. Later in the play there is another dialogue between actors, between the king and the Egyptian herald (911 f.) Aeschylus, it is argued, was not used to the luxury of a second actor, and therefore failed to put it to the effective use he might have done.

It must be acknowledged that in one scene, at least, the scene between Pelasgus and the Egyptian herald, Aeschylus does put ‖ the second actor to an effective use. Compare the scene with other tragic episodes of the same nature—one thinks of the dialogues of Prometheus with Hermes (*PV* 944 f.) and of Demophon with the herald of Eurystheus in Euripides' *Heraclidae* (120 f.)—and the scene in the *Suppliants* emerges well from the comparison. Was the writer of that scene really incapable of using a second actor skilfully in other places? To answer the question we must ask in what other places he might have used one. We can only answer, in the supplication scene; the other scenes are only brief interludes that give little occasion for a dialogue. Would the supplication scene be more effective if Danaus shared with the chorus and the coryphaeus the task of supplication? From the standpoint of modern dramatic naturalism the scene would have greater verisimilitude; but whether the emotional effect at which the poet aims would have been more successfully conveyed is a different question. The assignation of the main part to the chorus is not inconsistent with the techniques used by Aeschylus in his other works. When in the *Persae* the Messenger first announces the defeat at Salamis (249 f.) an epirrhematic scene of thirty-five lines between messenger and chorus takes place before the Queen of Persia says a word. When she does speak, (290 f.) the Queen explains her silence as being due to

[18] Ibid. 46 f.

[19] At l. 490; the speech must be given to Danaus (cf. ll. 500, 504).

the shock she has received; this is not the only case in which a silence imposed on Aeschylus by a technical reason is motivated by a dramatic one. After that epirrhematic scene comes a dialogue between Queen and Messenger in which both speak long rheseis; out of 242 lines (290–531), only two are spoken by the chorus. The next scene in this play to contain an epirrhematic element is that of the invocation of Darius' ghost. After the initial stasimon, Darius opens with a speech of thirteen trimeters (681 f.). The chorus in a brief stanza in ionics protest that they are too terrified to speak with him; the king in three tetrameters urges them to speak; and the chorus in an antistrophe at all points symmetrical to the strophe repeat their refusal. At this point the King instructs the Queen to replace the chorus as his interlocutor, which she continues to be throughout the tetrameter scene that follows; it contains a stichomythia, but also several long rheseis. In the *Seven* there are two epirrhematic scenes between Eteocles and the chorus; in the first, (181 f.) the king pleads with the Theban women not to spread despondency by excessive prayer and la‖mentation, in the second (677 f.), the women plead with the king not to take the field against his brother. Both scenes are unlike that in the *Persae* and like that in the *Suppliants* in the sense that each comprises an agon, an attempt by one party to win over the other to his point of view. Each of these scenes consists of a succession of brief dochmiac stanzas by the chorus punctuated by trimeters, never more than three together, spoken by the actor; each continues with a one–one stichomythia; after the first comes a long rhesis by the actor; after the second, a stasimon. Had he wished the poet could have introduced an actor, Antigone, perhaps, or Iocaste, to plead with Eteocles; yet all trimeters are spoken by the coryphaeus.

Aeschylus' practice seems to be that when an epirrhematic episode verges into one containing long speeches, an actor takes over from the chorus; but so long as not more than three trimeters at a time are being spoken on the side the chorus represents, these are given to the coryphaeus. The reason for this arrangement seems clear enough; in a scene of this kind the trimeter utterances on the side of the chorus are so closely bound up with the lyrics that they cannot be given to an actor [or not at any rate to an actor who does not sing]. If the chorus takes the main part throughout the supplication scene, that does not mean that they and not Danaus take the lead in the family's counsels; we are explicitly told more than once that this is not so. The predominance of the chorus in this scene is due to a technical reason, inconsistent with the principles of modern naturalism but fully comprehensible in the context of an Aeschylean tragedy.

The belief that in the *Suppliants* the chorus numbered not twelve, as in the rest of Aeschylus, but fifty originated in consequence of the belief that the *Suppliants* was a very early play. But once in being it helped to maintain the belief in an early date, and its credentials must certainly be re-examined now. Otfried Müller in 1833 first combined the statement of Pollux that the tragic chorus originally numbered fifty with the assumption that the fifty Danaids must be represented by fifty choreutae; despite the sound sense of Pickard-Cambridge,[20] the belief still flourishes.‖ The number of the Danaids is never actually mentioned in the text; but the Danaids say that Aegyptus has fifty sons, and the number appears in most known versions of the story. Wilamowitz was a strong believer in the theory,[21] and indeed enlarged it with a remarkable extension. He suggested (first in 1923)[22] that in the scene in which the Danaids are threatened by the Egyptians' herald, the lyrics customarily assigned to the herald should really be given to a secondary chorus of the herald's fifty Egyptian henchmen. There are good reasons for thinking that the play had one such subsidiary chorus. The lyrical dialogue in the exodos is generally thought to take place between the main chorus of Danaids and a subsidiary chorus of their handmaidens.[22a] If the handmaidens have no special function, there is no reason why they should be explicitly introduced at ll. 975 f.; and it is hardly possible, after the passionate unanimity of the Danaids against marriage that has been maintained throughout the play, for half of them suddenly and temporarily to defend Aphrodite's functions. A satisfactory parallel is supplied by the subsidiary chorus of Propompoi in the exodos of the *Eumenides*. No such evidence for the chorus of Egyptians is furnished by the text. Yet Wilamowitz's theory was believed by Paul Maas and Walther Kranz.[23] Five years after the publication of the new evidence for the date of the trilogy, Denys Page[24] wrote of the *Suppliants*: 'It is a play of movements and multitudes. The scene is filled, and vacated, by very large throngs of persons, for the most part, of dusky complexion and unfamiliar costume. There is a moment when we stop and ask, how many persons are in sight? On a mound or platform are Danaus and his fifty daughters, each (unless the poet misleads us) with an attendant girl; in the scene below are a herald and his Egyptian escort, presumably equal in number to the screaming women whom they would carry away; to

[20] *Dithyramb, Tragedy and Comedy* (1st edn.), 87 f.; *Theatre of Dionysus in Athens*, 31 f.; *Dramatic Festivals of Athens*, 241 f.

[21] *Aischylos: Interpretationen* (1914), 4.

[22] Ap. Maas, *Griechische Metrik*, § 76; cf. *Hermes*, 64 (1929), 461 f. = *Kl. Schr.* iv. 479.

[22a] [H. Friis Johansen and E. W. Whittle in their learned commentary (*Aeschylus: The Suppliants* (1980), iii. 306 f.) have not persuaded me that this is not the case.]

[23] Maas, loc. cit. (on pp. 53–4 of my English translation); Kranz, *Stasimon* (1933), 272.

[24] In his introduction to Denniston and Page, *Aeschylus, Agamemnon* (1957), p. xxx.

them enters the king of Argos with a bodyguard presumably ‖ equal in number to the Egyptians whom they repulse. Can the total be less than 203? The multitude, the outlandish costumes, the black faces, the threats of the one party and the screams of the other, compose a scene without parallel in the remains of Greek tragedy.' Again, W. S. Barrett, in his great edition of the *Hippolytus* published in 1964 writes (p. 368) that in this passage a secondary chorus of Egyptians, 'announced at 825 f., sings in alternation with the main chorus'. But look at 825 f. and at every other part of the text, and you find no plurals but singulars; the text contains not one scrap of support for Wilamowitz's theory. It is true that the members of a chorus might be referred to in the singular; but the onus of proof is on those who believe in a plurality.

Let us examine the reasons for Wilamowitz's belief. 'So long as it is uttering its threats', he writes,[25] 'the secondary chorus can be singing, but when its members rush to the attack and try to lay hands on individual girls, this is no longer possible, so the herald speaks . . .'. In fact there is nothing whatever in the text to suggest that the herald or any one person 'rushes to the attack' or 'tries to lay hands on individual girls'. Behind this naturalistic argument lies the assumption that a single herald could not throw fifty Danaids into the state of panic that is apparent from their utterances. Since we are dealing with a medium so utterly remote from modern stage naturalism as Aeschylean tragedy, it is hardly safe to assume that one herald could not have terrified fifty Danaids. But if it turns out that there is no need to imagine that there were any more than twelve choreutae, the main support for Wilamowitz's conjecture will collapse. Maas (loc. cit.) advocated the theory on the ground that it harmonizes with his rule that characters of low social standing do not sing in tragedy.[26] Whether a herald is of low social standing is debatable; and though Maas in defence of his rule deprived the Nurse in Sophocles' *Trachiniae* of her lyrics, he still had to admit one somewhat considerable exception, the Phrygian slave's aria in the *Orestes* of Euripides. This is not the kind of rarity that can be ruled out simply because it is rare. A further objection against Wilamowitz's theory has been impressed upon me by Richmond ‖ Lattimore. He points out that it is hardly consonant with Greek practice for a herald to arrive at a foreign city with a *démarche* accompanied by a military force. What we should expect is what happens in the analogous scene in Euripides' *Heraclidae*: the Herald arrives alone, presents his demand, and when it is refused goes off to tell his master, who will now bring up his army (214 f.). If we follow the natural indications of the text, that is what happens here.

[25] *Hermes*, 64, loc. cit.

[26] See *Greek Metre*, § 76 [cf. *Kl. Schr.* 47 f.].

We must now return to the arguments for thinking there were fifty members of the chorus. It rests on a combination of the statement of Pollux (1. 110) that the early tragic chorus numbered fifty with the naturalistic assumption that fifty Danaids could be represented only by fifty choreutae. I wonder how many of the subscribers to this theory have been aware that in the passage to which they assign such weight Pollux says that the number continued to be fifty until the production of the *Eumenides*, when the fifty Erinyes so much scared the Athenians that they reduced the numbers. Anyone who has more than a superficial acquaintance with the *Onomasticon* is aware that this statement is only too characteristic of its author, whose authority in this matter we should be wise not to take too seriously.[27] Wilamowitz's statement that it would be 'absurd' for twelve choreutae to represent fifty Danaids (*Aischylos: Interpretationen*, 4) is the mere assertion of a naturalistic prejudice. In a play about this subject, the Danaids would have to be the chorus. It would hardly be putting too great a strain on the imagination of the audience to ask them to take twelve to represent fifty; the poet made it a little easier for them by making only one reference, and that oblique, to the number of the sisters. The decisive parallel is supplied by the *Suppliants* of Euripides; there the chorus numbers fifteen, but we are told four times in the text that it consists of the mothers of the seven chiefs who have perished before Thebes (12, 100–2, 598, 963–5). 'The number seven', writes Wilamowitz in the introduction to his translation of that play,[28] 'only signifies a ‖ conventional number, which is practically equivalent to a proper name.' Exactly; so, in the *Suppliants* of Aeschylus, does the number fifty.[28a]

The question of the number of the sisters affords a convenient transition from arguments based on technique to arguments based on content. It has been argued that the legend, with its strong element of the folktale, could have been used only in a very early tragedy. Such an argument need not detain us; one has only to remember, say, the legend of Oedipus. The feature that has struck some people as particularly primitive is of course the number of Danaids and Aegyptiads. Naturalistic prejudices die hard when the number of children people have is in question, yet this number is a matter of fashion, and fashion varies. According to the manuscripts of the Agamemnon

[27] The author of the most objectionable statement of this view known to me (*CR*, NS 7 (1957), 1) remarks that 'the whole question of Pollux's sources and credibility deserves a monograph'. If he had troubled to examine any part of the considerable literature of the subject, or even the *Onomasticon* itself, even his confidence might have been shaken.

[28] *Griechische Tragödien*, i⁴. 221.

[28a] [Another parallel is supplied by the *Eumenides*: the canonical number of Erinyes, at least during the classical period, was three.]

(1605) twelve children of Thyestes were served up to him by Atreus; Page follows Emperius in altering that number to two, remarking that twelve is 'a ludicrous multitude in this context'. Ludicrous, maybe, to most of us; but not, I think, to the author of the Prometheus plays, the *Sphinx*, the *Phorcides*, the *Psychostasia*. The theme clearly lends itself to the composition of an Aeschylean trilogy, similar in many respects to the *Oresteia*. The issue between Danaids and Aegyptiads supplies a conflict, not unlike that between the two branches of the house of Pleisthenes; and by means of the device of treating the antipathy of the Danaids towards their cousins as though it were an antipathy towards marriage in general, Aeschylus contrived to make the clash of individuals symbolize a clash of cosmic principles.

The persons offered by the legend are suited to Aeschylus' purpose; but the poet's characterization of them has seemed to many to prove the trilogy to be an early work. Danaus in particular has been severely criticized. He has been found dull, prosy, irrelevant; he has been called a mere appendage of the daughters who call him the leader of their company. The question of Danaus' character is clearly bound up with that of the handling of the second actor. If in the supplication scene the Chorus does the talking, that does not mean that Danaus simply obeys his daughters' will. They tell us that he is their leader, and there is no reason why we should not believe them. Throughout the play identity of purpose between father and daughters is complete, and when they speak together, Danaus is always in com||mand. That is the point of his warnings to his daughters to behave discreetly; if we consider them without first clearing our minds of naturalistic prejudices, it is easy to mistake their purpose. A recent writer on the style of Aeschylus[29] has said that Danaus 'is portrayed as given to proverbs and maxims and admonitions', and takes this as an instance of the weakness in characterization of the early Aeschylus. If the word 'portrayed' means 'characterized as an individual', Danaus is not 'portrayed' at all. He is simply the πατήρ and βούλαρχος who must show wisdom and resource in defending the interests of his family. For him to warn his daughters against the desires their beauty may excite is highly relevant to the theme of the nature of marriage and the power of Aphrodite which must have been central to the trilogy, as we can see from the last scene of the *Suppliants*, the speech of Aphrodite from the *Danaides*, and the subject of the accompanying satyr-play, the *Amymone*. The didactic manner of Danaus' utterance, far from being meant to characterize him as a somewhat fussy and sententious old gentleman, simply expresses his function after the austere and formal fashion of Aeschylean tragedy. In Danaus as a person Aeschylus takes no interest

[29] W. B. Stanford, *Aeschylus in his Style* (Dublin, 1942), 112.

whatsoever, any more than he takes an interest in the Queen of Persia, or in Orestes—whose lack of colour is such a grievous disappointment to such members of the individualizing school as have remarked it—or in Oceanus, whose static dignity has been so grotesquely misrepresented as the false civility of a disloyal friend.

Neither is Aeschylus interested in the personality of Pelasgus, which some modern critics have analysed with great care. Some have pronounced him a weak character because, like Agamemnon or the Homeric Hector, he shows respect for public opinion; others have praised the psychological skill shown by the poet in depicting the mental turmoil into which he is thrown by the need to make his agonizing decision. The dramatic possibilities of that decision are indeed exploited with great power; but the poet is not concerned to show the effect upon Pelasgus as a person having a particular character of the situation in which he finds himself. Pelasgus is simply a good and conscientious king, con||fronted with a grim dilemma: either he must receive the suppliants, thus risking war with all its horrors, or he must bring down on himself and his people the wrath of Zeus, protector of suppliants. The moment the Danaids play their trump card, the threat of suicide, the king throws in his hand; in the Aeschylean world the hazards of war, grievous as they may be, are nothing compared with the pollution to which the suicide at the altars of rejected suppliants will expose the *polis*. Like Eteocles, like Agamemnon at Aulis, Pelasgus can make only one choice.

What the ancients called ἦθος required that the characters in a work of literary art should be depicted as being the kind of people likely to perform the actions that the story gives them. The persons of Aeschylus have enough character to satisfy that requirement and no more. One factor making it hard for us to see this has been the existence of three of them who strike us as having, in a sense, marked individuality. Clytemnestra is a peculiar case; the actions given her by the story are of such a special kind that in order to seem capable of them she must seem abnormal, and to explain them the poet suggests that she, like her sister Helen, is a peculiar kind of being sent into the world by Zeus in order to accomplish a destructive purpose. Eteocles and Agamemnon have acquired peculiarity in a different way, from the duality of the function assigned them by the plot of the trilogy. Each is the good king, the shepherd of his people, and yet each has laboured from birth under a curse; their actions must reflect this double destiny, and hence their character acquires a double aspect. It is by an accident that these three characters acquire individuality, an accident that may be called the beginning of character-study in European drama.[30]

[30] I have developed these reflections somewhat further in a review of John Jones's remarkable

We may now look back and survey the arguments used to try to date the trilogy before the publication of the papyrus. Before that time, the materials for determining the date did not exist. Walter Nestle's writings during the thirties[31] certainly ‖ ought to have warned scholars that their confidence in the early date was much exaggerated, and he will always be remembered with admiration for his stand against dogmatic orthodoxy. But even if Nestle had been spared to offer the exhaustive treatment of the subject that he promised, the severe limitations of our knowledge of Aeschylus would hardly have allowed him to put the case for a late date beyond reasonable doubt. Seven complete plays, with fragments, is simply not enough; after a hundred and fifty years of doctoral dissertations, people who confidently claim to *know* the date of Sophocles' *Electra* or *Trachiniae* are living in a private world. But now that we have external evidence for the date of the Danaid trilogy, we can see that the internal evidence in no way makes against its reliability; and it may be instructive to examine the reasons for the old dogmatism.

At the root of it lies a prejudice, in essence metaphysical, which originated in antiquity; the belief that an art form develops in the same way as an animal or a plant, and that its development can be studied by the same kind of method. That prejudice can be seen at work in the earliest and in many ways the best study of Greek tragedy known to us. Aristotle had taken over from his master, Plato, the belief that each individual and each abstract entity is what it is by virtue of its conformity to a fixed type, its participation in a given form. He adapted Plato's theory of forms in the light of his own favourite scientific discipline, that of biology: each species had its proper nature, to whose full realization its whole development must tend. Tragedy 'attained its nature' (*Poet.* 1449^{a}15) apparently when Sophocles wrote the pattern drama, the *Oedipus Tyrannus*; earlier efforts were seen by Aristotle as marking stages in a movement, irregular, perhaps, but none the less continuous, that culminated at that point in time. Aristotle must have thought of Aeschylus, when he thought of him at all, as an imperfect struggler towards the ideal, the Sophoclean pattern. That way of looking at the growth of tragedy as a kind of linear progression has been taken for granted as the only way by the great majority of modern scholars. Wilamowitz did not question it; how much less most of his immediate successors, obsessed with the notion of development to such a point that in a series of ponderous

book *Aristotle and Greek Tragedy* (London, 1962) in *Review of English Studies*, NS 15/58 (1964), 221 f. [= ch. 13]. Cf. also *Gnomon*, 34 (1962), 740 [= ch. 12, p. 191]; *CQ*, NS 12 (1962), 187 f. [= ch. 21].

[31] *Die Struktur des Eingangs in der attischen Tragödie* (Tübinger Beiträge, 10, 1930), p. 1; *Gnomon*, 10 (1934), 413; also in the preface to the reprint of Droysen's translation of Aeschylus (Króners Taschenausgaben 152).

tomes they remorselessly exhibit the growth of early Greek literature as a process culminating in Plato, smearing the sublime simplicity of the early Greeks || with a thick, clogging treacle of Hegelian metaphysics! Yet common sense reminds us that the analogy between a plant or animal on the one hand and an art form on the other is not necessarily a close one; even in the case of a plant or an animal, the question as to what exactly is its proper nature and when it has attained it is one that different people may answer in different ways. The dangers involved in the attempt to confine within the straitjacket of Platonic or Aristotelian idealism the untidy, wasteful, infinite pullulation of the forms of life is particularly well illustrated by the result of treating Aeschylean tragedy as a mere imperfect stage on the way to Sophoclean perfection, or to the ideal tragedy as it is conceived by a modern taste nurtured on Shakespearian individualism and Ibsenian naturalism.

Suppose we try the experiment of treating Aeschylean tragedy, at least for a moment, as though it had a form and principles of its own, ceasing for the time being to regard it as an imperfect groping towards the perfections realized by later dramatists. Our material is so scanty that we shall run the risk of making some mistakes; but we know enough about it for the attempt to be worth making, and by making it we may become alive to features of tragic technique that would otherwise escape us.

In an Aeschylean tragedy the chorus is as a rule deeply involved, at least emotionally, in the action; it may be deuteragonist or even protagonist. The choral part is greater in proportion to the whole than in later tragedy, especially when the chorus is most deeply implicated in the action. The actors are there in order to enact the plot, not in order to display their individual characters; in so far as they have character, they have such character as their acts require. Stichomythia is frequent, and is of a stiffer and more formal kind than that of Sophocles; free dialogue is rare, and in the *Suppliants* and the *Seven* scarcely occurs at all. An agon may take the form of stichomythia or, when a special effect of intensity is aimed at, of an epirrhematic scene. Long rheseis by actors may follow one another, with or without the interposition of brief choral comments, but no agon consisting of long rheseis is found in what we have of Aeschylus. The chorica do not neatly divide the dialogue portion into scenes of approximately equal length; the space between two chorica may be occupied by a single rhesis. Actors as well as choruses speak in a manner that is grand, formal, ceremonious. Neither the lan||guage nor the style has any close relation to that of ordinary life; humble figures, the herald in the *Agamemnon* and the nurse in the *Choephori*, are the licensed exceptions that prove the rule. The style and diction, with all their richness and with all their beauty, preserve an element of archaic roughness, distasteful to certain persons and to certain ages, yet giving to the verse a special simplicity and a special strength.

Bernard Ashmole in a memorable paper (*PBA* 48 (1962), 217) has lately reminded us of what he calls 'the pained surprise with which our grandfathers first beheld the sculptures of Olympia'. Those sculptures and the tragedies of Aeschylus have many qualities in common. Even in periods, like the present, when it is not fashionable to dislike these qualities, many people, some of them by no means without taste, view them with aversion. No wonder, when we consider the low esteem enjoyed by Aeschylus between his own time and the nineteenth century; in the violence with which the very people who most insist on the excellence of all genuine Greek tragedy declaim against any passage which they have persuaded themselves is an interpolation, we can recognize a reaction that is only human. It has been easier to patronize the *Suppliants* as a monument of primitive drama than to understand it; yet the qualities which it exhibits, whether they deserve praise or not, are those that are typical of Aeschylean tragedy. The force and passion of the opening anapaests make them almost the finest beginning to any play of Aeschylus; the lyrics, though damaged by corruption, are of superb quality; the scene with the herald, though the most corrupt of all, is dramatic to a high degree; and above all the supplication scene is hardly much inferior to the great kommos of the *Choephori*. What has told most against the play in modern estimation is the stilted character of much of the trimeter part. That is a feature of any play of Aeschylus, but most of all of this one. Those who dislike it have a perfect right to; but some will find it less displeasing once they have got used to the undoubted facts that Aeschylus wrote in a convention even more remote from modern naturalism than either of the other great tragedians, and that in character for its own sake he took no interest whatsoever.

20(R93)

Review of Lebeck, *The* Oresteia*: A Study in Language and Structure*

A. Lebeck, *The* Oresteia*: A Study in Language and Structure*. Harvard: The University Press; London: Oxford University Press, 1971. Pp. x + 222.

AMERICAN scholars, on the whole, still know less Greek and Latin than their European counterparts; the Rousseauite sentimentality in education which wastes the years when the memory is at its best, which is only now beginning to ruin European schools, has been at work in the United States for many years, powerfully supported by the disastrous influence of John Dewey. But American scholars are often more sensitive to literature as literature than scholars here; they often know more of literature outside the classics; and they very often—for good or ill—know more of modern literary criticism. The example of books like Caroline Spurgeon's well-known study of the images of Shakespeare may explain the frequency with which Americans have analysed the imagery of Greek poets; the best specimen of the genre so far is R. F. Goheen's interesting work on *The Imagery of Sophocles'* Antigone (1951). At its worst this kind of study merely provides a formula which enables people to write articles about the ancient poetry without a proper knowledge of the language. At its best it may result in valuable work; but it requires taste and intelligence of a high order. This Miss Lebeck has.

Many scholars, Headlam and Fraenkel, for example, have been aware that the recurrence and variation of certain significant images, as well as the repetition of certain significant words, forms an important instrument of the technique of Aeschylus. Wagner, who during his Paris period studied Greek with F. S. Lehrs and covered the margin of his copy of Droysen's translation of Aeschylus with notes, consciously borrowed his musical technique of the leitmotiv from the verbal technique of the Greek poet.[1] O. Hiltbrunner's *Wiederholungs- und Motivtechnik bei Aischylos*, a Göttingen dissertation by a pupil of Deichgräber (Berne, 1950) is a useful study of the topic. But Miss Lebeck

[1] [See my article in *Blood for the Ghosts* (1982), 126–42.]

Journal of Hellenic Studies, 92 (1972), 193–5.

seems to be the first author to have devoted a monograph in English to the detailed study of the imagery of part of Aeschylus' work.

'The significance of a recurrent image', she writes soon after the start of an introductory section (pp. 1–5) that needs to be read carefully, 'unfolds in successive stages.... The form which repetition or recurrence takes in the *Oresteia* is that of proleptic introduction and gradual development'. Parts I, III, and IV of her book take us through the *Agamemnon*, *Choephori*, and *Eumenides*, analysing the 'prolepsis' and development of the various images; Part II covers the entire trilogy, showing how 'images are interwoven by associative repetition'. Part Two consists of sections with the following headings: Action and Suffering—Sacrifice—The Hunting Net—The Robe—The Net—The Bond of Fate—The *Telos* of Marriage and the *Telos* of Death—The Carpet Scene—The Endless Flow of Blood.

To give a notion of the method, let us consider some of these sections. Under that of 'Sacrifice' the author remarks that this theme is touched on as early as the anapaestic prelude to the parodos; that in the parodos the eagles are said to 'sacrifice' the hare; that the actual sacrifice of Iphigeneia is described soon after; and that both in the parodos, obscurely, and later in the play, openly (on pp. 61–3 Miss Lebeck quotes 1407–9 and 1432–3, and might have quoted 1501–4) it is applied to the killing of Agamemnon. Under the heading of 'The Hunting Net', Miss Lebeck connects the metaphor of the net said in the first stasimon to have been cast over Troy with the thing often called or likened to a net used to entangle Agamemnon. In her last section of this part Miss Lebeck connects the notion that blood once spilt cannot be recalled, often in the past recognized to be an '*Oresteia* motif', with other images of blood.

In the three parts dealing with the separate plays,|| Miss Lebeck deals, for the most part, with images of more restricted application. Thus she shows how the comparison of the Atridae to eagles in the anapaestic prelude to the parodos of the *Agamemnon* foreshadows the eagles of the prophecy of Calchas; and she suggests that the expressions *ἐμπαίοις τύχαιςι ςυμπνέων* (*Agam.* 184) and *φρενὸς πνέων δυςςεβῆ τροπαίαν* (ibid. 219) take colour from the memory of the winds that blew from the Strymon to detain the Achaean fleet—'the storm without was mirrored by the storm within'. The book contains many more observations of this kind, always interesting, sometimes illuminating, but at other times unconvincing.

The sensitivity and intelligence of the writer make the reading of the book a pleasure. But he must read it with his wits about him, for her method has some dangers. First, she is too prone to take it for granted that a striking, or even not particularly striking, word or image *must* be interpreted in the light of the occurrence of an identical or similar word or image elsewhere in the trilogy. The three plays were originally written for *performance*, *in a theatre*, and I doubt whether the most acute and refined audience could have been expected to pick up some of the cross-references which Miss Lebeck asks us to believe in. Who, for example, hearing the Erinyes called μέτοικοι at *Eum.* 1011, would remember τῶνδε μετοίκων at *Agam.* 57? Are all the uses of μανθάνειν listed at pp. 26 f., or of τέλος at pp. 71 f., or of λάχος at pp. 158 f., really relevant to one another? Are all the uses of μανθάνειν, πάσχειν, and their cognates said to be relevant to πάθει μάθος really so, and are all the instances of δρᾶν quoted on p. 60 f. relevant to the δρᾶν/παθεῖν antithesis? I can just stomach the notion of a pun on Πάλλας and πάλος at *Eum.* 753–4 (p. 159), but I recoil from that of a pun on λέων and λεώς, at *Agam.* 825–7 (p. 181 n. 6). It is true that the lion-cub in the second stasimon of the *Agamemnon* and the infant Orestes of the Nurse's memories in the *Choephori* have in common an inability to control their natural functions; but I am not persuaded that the latter is meant to remind us of the former, nor that 'the recollections of Orestes' nurse contrast the innocence of his childhood with the cruel act which he must now commit'. A connection between the fable of the lion and the Nurse's words, if it could possibly occur to the mind of a member of the audience at this point, could only distract attention from the real function of the Nurse's speech, which is to contrast Clytemnestra's reaction to the report of her son's death with one of unmixed grief, and incidentally to show, with powerful dramatic irony, how Aegisthus is lured into walking into the trap without protection.

Miss Lebeck's principal fault is over-ingenuity, and this is especially in evidence in her treatment of ambiguity. Empson is not altogether a safe author for the interpreter of Greek poetry. 'Pursuing the customary methods of classical scholarship', Miss Lebeck writes, 'one is sometimes tempted to treat ambiguity as if the author were at fault, as if the clarity of normal diction were beyond his grasp. Yet the ambiguity of Aeschylus is not easy to achieve; it comes about neither by accident nor inability, but by design.' No doubt some of the ambiguities which people find in Aeschylus are of this kind. But when Miss Lebeck goes on to pronounce that 'when argument arises over meaning, the statement that claims to be exclusively right is categorically wrong', she is uttering a dangerously sweeping generalization. Besides ambiguities of the kind she means, there are others which exist for modern scholars simply because they have not understood the poet's meaning. There is a good deal in Aeschylus which none of us understand, either because it is corrupt or because it is difficult, and we lack material for comparison; and for one reason or another some scholars find more ambiguities there than others.

Take, for example, Miss Lebeck's treatment of the account of the sacrifice of Iphigeneia (*Agam.* 228 f.). She agrees (p. 81) that πέπλοιcι περιπετῆ means 'falling about Agamemnon's robes'; but she argues that the mention of Iphigeneia's robes flowing downwards in l. 239 shows that it *also* means 'with her robes falling about her'. Those who have argued for the first of these two meanings and against the latter have stressed the hard grammatical fact that πίπτω has no passive, and have shown that Soph. *Aj.* 907 provides no safe evidence for thinking that περιπετήc can have a passive sense. Miss Lebeck complains that this argument makes out the language to be normal, but that in fact the thought and language of this passage are not normal, but lyrical. Do these two predicates necessarily exclude each other?

Much of Miss Lebeck's general interpretation is admirable. She believes that a main function of the Cassandra scene is to supply the facts about Atreus and Thyestes which give the clue to the understanding of all the mysterious hints given earlier; and she deals firmly with N. G. L. Hammond's unfortunate attempt to undermine this view (see p. 177 n. 25; this is the main point made in my article in *CQ,* NS 12 (1962), 187 f. [= ch. 21]). In an excellent note (pp. 173–4) she says that I was both right and wrong to deny novelty and profundity to Aeschylus' doctrine of learning by suffering. Unlike Zeus, I have developed; see now *The Justice of Zeus* (2nd edn.), 209 n. 100 and the addendum on pp. 234–5. Does Artemis really punish the Atridae because she is a goddess of birth and fruitfulness (p. 35, with n. 23)? Such crimes as theirs are punished by Zeus. Much of the dramatic significance of the mentions of Helen lies, surely, in the fact that Clytemnestra is her sister. I wish people would not talk about 'the carpet scene': a carpet is meant to be trodden on, and the most important thing about Clytemnestra's priceless tapestries is that they were not.

In the matter of the central problem of the great kommos of the *Choephori,* Miss Lebeck inclines to side with Lesky rather than with Schadewaldt. I do not think she quite does justice to the latter's argu||ment. Certainly Orestes has been confronted, like his father at Aulis, with two alternatives each of which is intolerable, and there can be no easy solution. But does that mean that Orestes' mind is not made up before the beginning of the kommos, or indeed the play? The kommos is directed not at Orestes, who needs no incitement, but to the ghost; and Schadewaldt has given good reason for rejecting Schütz's transposition of ll. 434–8 to follow l. 455. νοcφίcαc ὀλοίμαν (438) does not express a genuine death-wish; from Homer on 'may I die, if only I can kill X' is simply a way of saying, 'I would give anything to kill X'. Do the Erinyes really change their nature (see p. 145)?

Miss Lebeck is well read in the literature of the subject. But at *Agam.* 361 the net can be shown to be a fishing-net (*HSCP* 73 (1968), 98 [= ch. 24, p. 311]); and the correct interpretation of *Eum.* 490–1 is due to K. J. Dover, *JHS* 77 (1957), 230 f.[2] On παῖδεϲ ἄπαιδεϲ at *Eum.* 1034 (p. 133), see D. Fehling, *Hermes*, 96 (1968), 142 f.[3] The helmsman's bench is not (p. 165) a 'rowing-bench'; and in n. 5 on pp. 213–14, Verrall is taken more seriously than he deserves to be.

The book is not suitable for beginners, for it must be read with caution. But scholars seriously interested in Aeschylus will find that it well repays a careful study.

[2] [= *Greek and the Greeks* (1987), 161 f.]

[3] [And *Die Wiederholungsfiguren vor Gorgias* (1969), 287.]

21(15)

The Guilt of Agamemnon[1]

In recent years the general view of the theology and morality of Aeschylus which we still find expressed in the most popular handbooks of Greek tragedy has come under fire;[2] fire which its defenders have so far been unwilling or unable to return. That Aeschylus was a bold religious innovator propounding advanced doctrines can no longer be assumed without argument; neither can one take for granted that his outlook on morality in general and on justice in particular was as advanced as it was once usual to maintain. Aeschylean justice, it is now beginning to be realized, had more in common with the ancient Hebrew justice that demanded eye for eye and tooth for tooth than with the exalted conceptions attributed to the poet by modern theorists. But whatever view we take of Aeschylus' notion of justice, we are not likely to dispute the paramount importance of justice in his work, and especially in the *Oresteia*. If I begin, then, with the assumption that the *Oresteia* is concerned with justice, human and divine, I shall be on safe ground.

The first and greatest of its three plays shows how the leader of the Greek expedition against Troy, the chosen instrument of Zeus' chastisement of the Trojans, comes to a miserable end. The train of events that leads to this conclusion has been set in motion long before the play begins, when the Greek fleet is assembled at Aulis on its way to Troy. The goddess Artemis becomes incensed with Agamemnon, and sends unfavourable winds that prevent the fleet from sailing. Either the great expedition, ordered by Zeus, must be abandoned, or the king must sacrifice his own daughter to appease the goddess. He consents to the sacrifice. This action earns him the bitter enmity of his wife, who at home in Argos plans his murder. She has at hand an instrument ready to her purpose. Agamemnon's father, Atreus, has long ago massacred the children of his brother, and has served him at a banquet with their

[1] This paper formed the first of my J. H. Gray Lectures given at Cambridge in 1961; it has also been given at other places. I am grateful to those who have helped to improve it, and particularly to Professor E. R. Dodds and Mr G. E. M. de Ste. Croix.

[2] See D. L. Page's preface to *Aeschylus, Agamemnon*, ed. J. D. Denniston and D. L. Page (1957); and my article 'Zeus in Aeschylus' in *JHS* 76 (1956), 55 f. [= ch. 18, p. 238].

Classical Quarterly, NS 12 (1962), 187–99.

roasted flesh. One survivor has escaped, and he is now a grown man waiting for his revenge.

The constant preoccupation of the poet with guilt and retribution creates a strong impression in the hearer's mind that the exact assessment of Agamemnon's guilt must be important for the understanding of the play. And yet there is no agreement among scholars as to the nature of that guilt. Agamemnon has been sent against Troy by Zeus himself; and yet Zeus allows him to perish miserably. Why? Is it for having consented to his daughter's sacrifice? If so, how far is his punishment the work of Zeus, and how far is it the consequence of the wrath of Artemis? The motive for that wrath is itself a subject of acute controversy. Or is Agamemnon punished for his remorseless extirpation of the Trojans, and the destruction of their city together with its temples and its altars? What part is played in his destruction by the curse brought down upon his family by the monstrous action of his father, Atreus? Or is he punished for his own pride and arrogance? Most modern scholars, with the notable excep||tion of Eduard Fraenkel, have seen him, if not as an arrogant and cruel despot, at least as something not far removed from one. Or can it be that several, or all of, these factors contribute somehow to his ruin? If so, how far can we hope to assign to each its proper degree of importance in working to this end?

All these questions are controversial. The most learned of Aeschylean scholars, to whom every serious student of the play must acknowledge a large debt, has even warned us that 'it would be absurd to attempt an exact calculation as to the degree of efficacy in each of the different elements that work together towards Agamemnon's fatal end'.[3] It is indeed important to guard against attacking the complicated task of unravelling these twisted strands with any excessive confidence that we shall reach a clear-cut answer. Yet it is agreed that the trilogy is concerned with justice, guilt, and retribution; and that seems to me to justify a fresh attempt to discover how the poet meant us to suppose these notions are exemplified in his work. Whether the results which are arrived at are absurd will be for the reader to judge.

The Chorus in its opening anapaests (60 f.) strongly asserts that the cause of the Atreidae against the Trojans is a just cause. They have been sent against Troy by Zeus, the guardian of the law of host and guest: Zeus, who has been outraged by Paris' crime against the sacred laws of hospitality. At the beginning of the first stasimon (367 f.), the point is further reinforced. The Chorus has just been told that Troy has fallen. 'They can speak of a stroke from Zeus', they begin; 'this, at least, one can make out.' Later in the play the same truth is strongly

[3] Ed. Fraenkel, *Aeschylus, Agamemnon* (1950), iii. 625.

insisted on both by the Herald and by the King himself. And yet it is by the will of Zeus, as the loyal elders themselves finally acknowledge, that Agamemnon comes to his miserable end.

The reasons begin to emerge in the parodos, in that great choral ode which describes what has happened ten years earlier, when the Greek fleet lay encamped at Aulis on its way to Troy. The portent of the eagles that tore and devoured a pregnant hare has taught Calchas, the prophet of the Greek army, that Troy is destined to fall to the expedition; it has taught him also that Artemis is incensed against its leaders, the Atreidae. In the whole play nothing is more controversial than the reasons for Artemis' anger, but in an investigation of the guilt of Agamemnon the problem of her motive is not one that we can avoid.

'In time', says Calchas (126 ff.), 'this expedition captures Priam's city; and all the plentiful herds of the people before the walls shall Fate violently ravage. Only may no envious grudge from the gods strike beforehand and cast into darkness the great bit for Troy's mouth that is the host encamped. For in pity Artemis bears a grudge against the winged hounds of her father that slaughter the poor trembling hare with all her young before the birth; and she loathes the feast of the eagles. . . . The Fair One, kindly as she is towards the helpless offspring of ravening lions and pleasant to the suckling young of all creatures that roam the wild, demands fulfilment of what these things portend; favourable is the portent, yet fraught with blame. And I invoke the blessed Healer, that she prepare not against the Danaoi lengthy delays in port caused by adverse winds that hold fast the ships, striving to bring about another sacrifice, one without song or banquet, a worker of quarrels born in the house and fearless of the husband. For there abides a terrible, ever re-arising, treacherous keeper of the house, unforgetting Wrath, child-avenging.' ||

The ancient epic called the *Cypria* accounted for the wrath of Artemis by means of a story not mentioned by Aeschylus. According to Proclus' summary of the plot of this lost work (OCT of Homer, v. 104),[3a] Agamemnon had shot a stag, and in his triumph boasted that as an archer he surpassed even Artemis. A similar story is told by Sophocles in his *Electra* (563 f.). That story is not mentioned here; but can we rule out the possibility that it was, none the less, the reason for the wrath of Artemis that Aeschylus had in mind? If that is so, it follows that he has set down Artemis' anger to an obscure and arbitrary grievance, a grievance so trivial that it is not worth mentioning in the play at all. It would certainly be unsafe to deny *a priori* that this could be the case; but the conclusion is such a strange one that it seems hardly reasonable to adopt it without further examination of the evidence. Does the portent give us any clue to the reason for the goddess's anger?

[3a] [= Davies, *EGF*, p. 32, ll. 55 f.]

Calchas says that she is angry because she loathes the feast of the eagles; and the eagles, he says, stand for the Atreidae. Here, say some scholars, we have the explanation of her anger: she hates the eagles, and the eagles stand for the Atreidae; therefore, she conceives a hatred for the Atreidae. This interpretation seems to me to rest on an intolerable confusion between the world of the portent and the world of the reality it happens in order to symbolize. The eagles and the hare belong to the world of the portent; that portent symbolizes an event which is to happen in the real world. The eagles stand for the Atreidae; so it is natural to infer that the hare must stand for some other figure or figures belonging to the real world. We can hardly avoid supposing that it stands for the Trojans and their city. So when Calchas says (137) Artemis abhors the eagles' feast, he must mean that Artemis abhors the coming destruction of Troy, which the Atreidae are destined to accomplish.

I believe that this conclusion is confirmed by the words of Calchas' explanation of the portent. But the point is not to be grasped immediately, for like most Greek prophets Calchas casts his interpretation in riddling language. 'In time', he begins (126 f.),[4] 'this expedition captures Priam's city; and all the plentiful herds of the people before the walls shall fate violently ravage.' This is strange language. We should have expected that the tearing of the pregnant hare would stand for the annihilation of the Trojans, not only men, women, and children, but even the unborn; we can scarcely help remembering the speech of Agamemnon to Menelaus in the sixth book of the *Iliad* (57 f.), in which he declares that not even the unborn children of the Trojans shall escape his vengeance. Yet when it comes to the explanation of the portent, we are told that the Achaeans will destroy the Trojan . . . cattle!

That seems incredible; and I have suggested that the explanation lies in the habit Greek prophets had of referring to people by the names of animals.[5] If so, 'the abundant herds of the people' will mean 'the abundant herds *that are* the people'. This is confirmed by the presence of the words 'before the walls'; for the Trojan cattle did not perish before the walls, but the Trojan men did perish 'in front of the city'.

If it is correct, the close correspondence of the portent with the future reality ‖ must be taken as established. The eagles stand for the Atreidae; the hare and its young stand for Troy and its inhabitants. What reason does Calchas give for the pity felt by Artemis for the

[4] *χρόνωι μὲν ἀγρεῖ Πριάμου πόλιν ἅδε κέλευθος,*
πάντα δὲ πύργων
κτήνη πρόϲθε τὰ δημιοπληθέα
μοῖρα λαπάξει πρὸϲ τὸ βίαιον.

[5] *RhM* 103 (1960), 76 f. [= ch. 23, pp. 305 f.].

hare? He says that Artemis is the patroness of the young of all wild animals; and according to many modern interpreters this fact in itself is enough to explain her anger against the Atreidae. But this is out of the question. Just as both eagles and hare correspond to figures of the real world, so must the motive assigned to the goddess for championing the hare represent a motive for championing what the hare represents in the world of reality. We have seen that in the real world the hare represents the Trojans. Has Artemis a special motive for championing the Trojans that may correspond in the world of reality to the motive assigned her in the world of the portent for championing the hare?

She has, in fact, an excellent motive; for in the *Iliad* and in the whole poetical tradition Artemis together with her brother Apollo appears as a loyal partisan of Troy against the invaders. This supplies a motive for her hostility to the Atreidae that is fully sufficient to explain her action.[6] The last scholar to put forward a view at all similar to this weakened his case by regarding the sacrifice of Iphigeneia as 'an atonement payable in advance for the destruction of Troy'.[7] This language is too legalistic: it is a mistake to talk of 'sin' and 'atonement' in this connection. In Aeschylus, as in Homer, the lesser gods have a position in no way comparable with that of Zeus; they may range themselves on either side in the Trojan conflict, but Zeus for the time holds the balance and will in the end decide the issue. Artemis' blow against Agamemnon is one move in the struggle; it is the attempt of a pro-Trojan goddess to strike at the invaders before the invasion: Artemis must be seen not as a judge punishing a sin, but as a powerful enemy striking at an enemy. Zeus will not prevent Artemis from bringing about the sacrifice; and Calchas hints that this may have consequences beyond itself. Why may it have these consequences? 'There abides', he says (152 f.),[8] 'a terrible, ever rearising, treacherous keeper of the house, unforgetting wrath, child-avenging.' That is usually taken as an allusion to Clytemnestra; indeed, some scholars have thought that it identifies the Wrath with her. But if I am right in translating *παλίνορτος* by 'ever rearising', the reference cannot be limited to her. There is a possibility (see Denniston–Page, ad loc.) that the word may mean 'arising in the future'; and in view of that I do not press the point. But it is worth noticing that if *παλίνορτος* here could bear its natural meaning, the reference would be to a child-avenging wrath that is 'ever again arising'. And that could only be the ancient wrath of the House of Atreus.

[6] [On the wrath of Artemis, see now *JHS* 103 (1983), 87–102 = *Greek Comedy, Hellenistic Literature, Greek Religion, and Miscellanea*, ch. 24.]

[7] D. Daube, *Zu den Rechtsproblemen in Aischylos' Agamemnon* (Zürich, 1939), pp. 147 f.

[8] *μίμνει γὰρ φοβερὰ παλίνορτος οἰκονόμος δολία, μνάμων μῆνις τεκνόποινος.*

After the narrative of Calchas' prophecy, the Chorus enters upon the famous invocation that is often called the 'hymn to Zeus' (160 f.). Why does the Chorus choose this moment for the invocation? The question is not one which every editor of the play has tried to answer; but the choral lyrics of Aeschylus are not normally irrelevant to the dramatic situation, and there is no reason why this one should form an exception to the rule. What is the situation at this point? Zeus has sent the Atreidae against Troy; but Artemis has confronted them with the intolerable choice between abandoning the expedition Zeus has ordered or consenting to Iphigeneia's killing. Where might Agamemnon have looked ‖ for help? And where might the elders of Argos appeal in the face of the anxiety that even now, ten years later, still torments them in consequence of what happened at Aulis? Only Zeus could have helped him, and them, to cast from their minds 'the burden of futile worry' (165). Zeus' power is over all, and he teaches men, by means of bitter experience, to obey his stern law of reciprocal justice. Artemis has faced Agamemnon with a terrible alternative. Zeus has sent him against Troy; surely he can hope for aid from Zeus.

Yet the Chorus does not appear at all confident that such aid will be forthcoming. 'Why not?', the audience may wonder. The Chorus gives no indication of the reason for its fears; at this point, the audience can only ponder on the riddling final words of the prophecy of Calchas. But, in the light of a full knowledge of the play, the reader may well wonder, 'Will aid from Zeus be forthcoming for the son of Atreus?'

From the invocation the Chorus returns abruptly to the scene at Aulis, and Agamemnon's grim dilemma. Should he have given up his expedition and gone home? Many scholars have been of this opinion. But in his brilliant introduction to the play D. L. Page has argued that Agamemnon has no choice. Zeus, he has pointed out, has ordered the expedition; it is his will that Troy shall fall. Hear the words attributed to Agamemnon (214 f.): 'That they should desire with passion exceeding passion a sacrifice to still the winds, a sacrifice of maiden's blood, is right in the sight of heaven'.[9] It is no use trying to water down the final word *θέμις*, whose emphatic position no less than its solemn associations lends it great weight in this place. Yet we must notice that Agamemnon's action is described by the Chorus in words that leave no doubt

παυςανέμου γὰρ θυςίας
παρθενίου θ' αἵματος ὀρ-
γᾶι περιόργωι cφ' ἐπιθυ-
μεῖν θέμις· εὖ γὰρ εἴη.

I follow Page in accepting Bamberger's emendation of *περιόργως* to *περιόργωι cφ'*. At *PV* 944 (cited by Fraenkel as a parallel supporting the manuscript reading), I would render *τὸν πικρῶς ὑπέρπικρον* by 'you whose excess of bitterness shall taste bitter in your own mouth', supposing the adverb to involve the sense of the word *πικρός* illustrated by Fraenkel, op. cit. ii. 301 n. i.

that it is considered as a crime (218 f.): 'And when he had taken upon him the bridle of compulsion, and the wind of his purpose had veered and blew impious, impure, unholy, from that moment he reversed his mind to a course of utter recklessness. For men are made bold by evil-counselling shameless infatuation, the beginning of woe. So he brought himself to sacrifice his daughter, in aid of a war to avenge a woman's loss and as advance payment for his ships.'

We are faced with an apparently glaring contradiction. We must agree with Page that Agamemnon has no choice but to sacrifice his daughter; the expedition had to sail. Yet E. R. Dodds[10] is equally right in insisting that his action was, and is meant to be regarded as, a crime. The text is explicit on this point. Can it be that both are right? Can Zeus have forced Agamemnon to choose between two crimes, either of which was certain to result in his destruction? My answer to this question would be, Yes.

The words just now quoted which describe how Agamemnon made his decision imply that he is mentally deranged: τάλαινα παρακοπὰ πρωτοπήμων (222). These words recall the famous passage in the nineteenth book of the *Iliad* in which Agamemnon tries to account for his reckless behaviour in provoking Achilles. 'I was not responsible', he exclaims, 'but Zeus and my portion ‖ and the Erinys that walks in darkness, who while I spoke put cruel Ate in my mind' (86–8). Infatuation, παρακοπή, in the *Agamemnon* is hardly distinct from Ate in the *Iliad*; and Ate is commonly an instrument of Zeus.

Zeus is indeed determined that the fleet must sail; Agamemnon has indeed no choice. But how has Zeus chosen to enforce his will? Not by charging Calchas or some other accredited mouthpiece to inform the king of his decision; but by sending Ate to take away his judgement so that he cannot do otherwise. Does it follow that Agamemnon is not held responsible for his action? Certainly not. In Homer Agamemnon excuses his behaviour by pointing to the action of Ate on his mind; but it does not occur to him to deny his responsibility, or to shuffle out of paying the enormous compensation which he has promised to Achilles. It is the same in Aeschylus. Dodds (*PCPS*, NS 6 (1960, 27) [= *The Ancient Concept of Progress*, 56–7] is right in saying that where Plato said αἰτία ἑλομένου· θεὸϲ ἀναίτιοϲ, Aeschylus might have said αἰτία ἑλομένου· θεὸϲ παναίτιοϲ: it is curious that in spite of this he still thinks Agamemnon could have chosen not to kill his daughter. Zeus has taken away Agamemnon's judgement; but that does not absolve Agamemnon from the guilt his error will incur. Nothing could better illustrate the saying

[10] 'Morals and Politics in the *Oresteia*', *PCPS*, NS 6 (1960), 19 f.; on this point see pp. 27–8 [= *The Ancient Concept of Progress* (1973), 57].

of Aeschylus' Niobe that Zeus makes a fault in men, when he is determined utterly to destroy a house (fr. 154a, 15–16 R.).

But what leads Zeus to determine to destroy a house? A famous chorus of the *Agamemnon* (750 f.) gives a definite answer to this question; and it stands in such a context that we can hardly doubt that the belief which it expresses is meant to be regarded as a true one. Prosperity in itself, the Chorus insists, is not sufficient to arouse the anger of the gods; only crime brings down punishment on a man or on his descendants after him. Despite the Chorus's claim of originality, this doctrine is not, of course, peculiar to Aeschylus; Page (D.–P., p. 136) has reminded us that it is found in two places in the fragments of Solon, a writer not unfamiliar to Aeschylus' audience. It is likely to represent Aeschylus' own belief. If so, it is unlikely that Zeus' decision to destroy Agamemnon is without a motive.

Zeus has faced Agamemnon with an impossible alternative. Also, he has taken away his judgement, so that he takes a fatal course; not that the other choice would not have been equally fatal. Why has he done this? Why, in using Agamemnon to punish Troy, has he chosen a course which must lead inevitably to the ruin of Agamemnon? Do we know of any guilt previously attaching to the King himself? No. But do we know of any guilt attaching to his ancestors? More than half the play has elapsed before we hear anything of such guilt. But let me continue with the commentary on the play's successive scenes that I have begun, resuming from the scene that follows the parodos.

Running right through the play we find a deliberate parallel between the fate of the house of Priam and the fate of the house of Atreus; equally pervasive, only less important, is the parallel between the fate of Helen and the fate of Clytemnestra. Again and again we find this sequence repeated; first, pious moralizings as the working of Zeus' law is traced in the just punishment of Troy; next, gradually increasing realization, both by the audience and by the Chorus, that what is true of Troy may prove true also of Troy's conquerors; lastly, agonized apprehension. This is the pattern of scene after scene and chorus after chorus. It was the pattern of the Chorus's initial anapaests together with the parodos; it is the pattern of the scene between Clytemnestra and the Chorus that follows.||

In the first of her two great speeches in that scene (281 ff.), Clytemnestra describes the rapid journey from Troy to Argos of 'the light lineally descended from the fire of Ida' (311). Some people see nothing in the Beacon Speech but an irrelevant, if magnificent, geographical excursus. No one could be more reluctant than I to attribute to ancient authors anything like what is generally meant by the modern term 'symbolism'. But I cannot doubt that in Clytemnestra's mind the fire from Ida stands for the avenging fire of Zeus; nor that the Beacon Speech is highly relevant to the parallel between the fates of the Priamidae and that of the Atreidae which I have just mentioned. In the second of her speeches in this scene (320 f.) Clytemnestra paints for the Chorus a vivid picture of what she imagines to be happening in the captured city. If the conquerors show piety, she says, towards the gods of the conquered land and towards their shrines, then they may escape being conquered in their turn. But if they commit sacrilege, they may provoke revenge; and even if they avoid sacrilege, they may arouse the vengeance of the spirits of the dead. Clytemnestra's pretended fears are obviously her secret hopes. This speech looks forward to the later scene in which the Chorus gradually extracts from the innocently optimistic Herald the news of the storm that has scattered the returning ships. This disaster was directly provoked by the sacrilege Clytemnestra had anticipated, and its occurrence greatly facilitated the accomplishment of her plan; for it was owing to the storm that Agamemnon returned in a single ship and without his brother. The adventures of Menelaus after the storm formed the subject of the satyr-play that accompanied the trilogy, the *Proteus*; this, too, may help to explain the importance assigned by the poet to the brothers' separation.

The first stasimon begins on a note of triumph and ends on one of disaster. From the theme of the just punishment of Troy, the Chorus passes to that of Helen and of the lives sacrificed for her sake, and ends on a note of anxious foreboding (459 f.) 'My anxious thought waits to hear something yet shrouded in darkness. For the gods are not unwatchful of the killers of many; and in time the black Erinyes consign to darkness him who is fortunate without justice, reversing his fortune and ruining his life; and he has no protection once he is among the vanished. To be praised exceedingly is dangerous. . . . My choice is the prosperity that comes without envy. May I not be a sacker of cities, nor yet be made captive by others and consume my life.'[11] It is

[11] μένει δ' ἀκοῦϲαί τί μου
μέριμνα νυκτηρεφές.
τῶν πολυκτόνων γὰρ οὐκ
ἄσκοποι θεοί. κελαι-
ναὶ δ' Ἐρινύες χρόνωι
τυχηρὸν ὄντ' ἄνευ δίκας
παλιντυχεῖ τριβᾶι βίου
τιθεῖϲ' ἀμαυρόν, ἐν δ' ἀΐ-
στοις τελέθοντος οὔτις ἀλ-
κά· τὸ δ' ὑπερκόπως κλύειν
εὖ βαρύ· . . . .
κρίνωι δ' ἄφθονον ὄλβον.
μήτ' εἴην πτολιπόρθης,

remarkable that Agamemnon's own loyal councillors can seem to imply that he is 'fortunate without justice'. If he has killed many, is it not because he is the minister of Zeus' vengeance? If he has made war and sacrificed his daughter for the sake of Helen, has it not been at Zeus' order?‖

The scene of the Herald repeats the now familiar sequence of hope and triumph followed by the slow realization that all is not well; it ends with the Chorus forcing the Herald, much against his will, to describe the disaster of the storm. Then the second stasimon takes up once more the theme of Helen, and illustrates her nature by the fable of the lion cub. It shows Helen to be in a sense a daemonic being, one sent into the world for the express purpose of causing havoc and destruction. We are meant to remember that Clytemnestra is her sister; later in the play, the Chorus itself will observe the similarity of their careers (1468 f.). From the theme of Helen, the Chorus goes on to speak of guilt and divine justice (750 f.). Prosperity does not of itself provoke the anger of the gods; evil deeds alone bring down divine justice either on their doer or on his descendants after him.

Immediately after this famous passage the King enters the stage; we can hardly doubt that the words the Chorus has lately uttered somehow apply to his case. The elders welcome him. In the past they have criticized his conduct in making war to recover Helen; but now that his plan has been successfully accomplished, they are glad to greet him with enthusiasm and to warn him against secret enemies. Perhaps the presentation of the King himself may furnish some clue to the problem of his guilt. But the character assigned him by the poet has been, and is, the subject of acute controversy. 'It is a common view', wrote Fraenkel in 1942,[12] '. . . that king Agamemnon is either the villain of the piece or, at any rate, a reckless, overbearing and impious tyrant.' His own view is very different. For him Agamemnon is 'in everything . . . a great gentleman, possessed of moderation and self-control';[13] he is '"every inch a king"; his every word and gesture is expressive of a powerful sincerity'.[14] Page takes a view of Agamemnon's character not widely removed from that against which Fraenkel

μήτ' οὖν αὐτὸϲ ἁλοὺϲ ὑπ' ἄλ-
λων βίον κατέδοιμι.

κατέδοιμι is Valckenaer's emendation of the manuscript reading *κατίδοιμι*. It is usually supported by quoting *Il.* 6. 202 *ὃν θυμὸν κατέδων*; but a better parallel is furnished by Q.S. 3. 456 *βίοτον κατέδων ὀδύνῃϲιν*. To hold on to the manuscript reading and suppose that *βίον κατίδοιμι* can be construed with *ἁλούϲ* as though it were a verb of knowing (as Murray, Fraenkel, and Page [and now West] all do) seems to me impossible.

[12] *PBA* 28 (1942), 22.

[13] *Aeschylus, Agamemnon*, ii. 441.

[14] *Der Agamemnon des Aeschylus* (Zürich and Stuttgart, 1957), 23 [= *Kl. Beitr.* i. 341].

has so energetically protested. 'His first address does not endear him,' he writes, 'he is ready with pious phrases, he greets success with gratitude, but without surprise. . . . He neither mentions his wife nor expresses pleasure in his home-coming . . .' (D.–P., pp. xxxiii f.). When he gives in to Clytemnestra and fatally consents to make a triumphal entry into the palace, treading underfoot the purple tapestries normally reserved as offerings to the gods, that happens, according to Page, 'simply because he is at the mercy of his own vanity and arrogance, instantly ready to do this scandalous act the moment his personal fears of divine retribution and human censure are, by whatever sophistry, allayed' (D.–P., p. 151).

Let us investigate the reasons for this singular disagreement. Fraenkel seems to me to have established that his calling the gods 'jointly responsible' for his victory does not immediately convict the King of hybris; such language was for a Greek perfectly consistent with a properly respectful attitude.[15] But it cannot be denied that in his opening speech Agamemnon looks back upon his ruthless extirpation of his enemies with a fierce satisfaction. 'The blasts of destruction still have life; and the embers as they die with the dead city waft upwards the rich incense of its wealth' (819–20). 'There is no sentimental lamentation in this fine sentence,' writes Fraenkel (edn. ii. 378), 'but a true note of profound sympathy.' A few lines later Agamemnon says, 'The ravening lion leaped over the wall, and lapped his fill of the blood of kings' (827–8). I find ‖ no sympathy, profound or otherwise, in that sentence or in anything that Agamemnon says about the Trojans; and I find it difficult to deny that the complaisance with which he views the extermination of his enemies must bode ill for him. Clytemnestra has, we know, been hoping that the Greeks will commit some act of sacrilege and provoke the anger of the gods; and the Herald has told us, in a line most unconvincingly obelized by Fraenkel (527), that the altars and shrines of the gods are to be seen no more. Agamemnon now boasts of the city's total destruction; are we to suppose that he has somehow managed it in such a way as to leave the shrines intact?[16] It is true that in his vengeance Agamemnon has acted as the minister of Zeus. But it is no less true that it is dangerous to be a sacker of cities, and that the destruction of the Trojan temples must provoke divine resentment.

We must agree with Page that the grimness and harshness of Agamemnon make an unfavourable impression; but we cannot deny that there is much in the situation that makes this understandable. It is hardly reasonable to reproach him with his coldness to his wife; it seems

[15] *Aeschylus, Agamemnon*, ii. 371 f.; cf. *PBA* 18 (1942), 22–3.

[16] See Denniston–Page, p. 120.

clear that rumours of what is going on at home have found their way to him. Nor is his behaviour at any point undignified; here we must contrast him with Aegisthus, whom the poet has portrayed in a most unsympathetic fashion. Both recent editors have remarked on the meanness of his conduct and the vulgarity of his language: what purpose had the poet in depicting him in such a way if not that of showing his enemy in a comparatively sympathetic light? Further, we must note the trust and affection of the humble Watchman who speaks the prologue; he looks forward to clasping in his own his master's well-loved hand. Notice, too, the attitude of the Chorus. They acknowledge to the King himself that they have criticized his conduct in the past. But they are glad to welcome him back from Troy with a friendly greeting; and their sincerity is proved by their lamentations at the miserable end of him whom they call their 'kindly guardian' (1452). Fraenkel and Page are both right; we have here a character of light and shade. This conclusion is confirmed by a comparison of Aeschylus' Agamemnon with that of Homer; the two are remarkably alike. Homer's Agamemnon is not, on the whole, an agreeable character. He is proud and irascible, to such an extent that he becomes involved in quarrels with his allies that have disastrous consequences. He is utterly determined to exterminate the enemy, declaring to Menelaus that even the unborn children in the womb shall perish (6. 57 f.). He is ready to proclaim in open council that he prefers the captive concubine Chryseis to his wife Clytemnestra (1. 113 ff.). But these defects cannot blind the reader to his magnificent heroic qualities. He is a good fighter, at his best in a difficult situation; his management of affairs is, as Apollo says in the *Eumenides* (631–2), on the whole successful. Like many hot-tempered men he is capable of behaving with dignity and nobility, as his reconciliation with Achilles plainly shows.

Let us now examine the crucial scene in which Clytemnestra induces her husband to tread upon the purple tapestries (932 ff.). Why does Agamemnon end by succumbing to his wife's persuasion? Fraenkel (edn. ii. 441) argues that he yields partly out of chivalry towards a lady, partly because after long years of struggle he is weary and his nerve finally gives way. This is not convincing. Chivalry of such a kind seems to be a medieval and a modern rather than an ancient concept; and the psychological explanation that the || King sees through his wife but is too weary to oppose her has a decidedly modern ring.[16a] It is a far cry from Aeschylus' Agamemnon to Mann's Aschenbach; nor is such a notion firmly grounded in the text. Must we then believe with Page that Agamemnon secretly longs to make a triumphal entry, and eagerly grasps at the sophistical excuses offered by the Queen? Or should we

[16a] [See now p. 404 n. 8 below.]

rather accept a third explanation lately offered us by Hermann Gundert,[17] who has argued that Agamemnon surrenders because he has been outwitted, and that he has been outwitted because Zeus has taken away his wits?

With these three theories in mind let us turn to the text. 'In a moment of fear', says Clytemnestra, 'might you not have vowed to the gods that you would do this?' This is no argument; an offering made to discharge a vow would have been in honour of the gods, but what Clytemnestra is proposing would be in honour of the King himself. Agamemnon knows this, and might have said it; what he does say is that, on the advice of an accredited exegete, he would have done so. 'What do you think Priam would have done?', the Queen asks. This again is a sophism; Priam was not only a barbarian, but a man under a curse. This too Agamemnon knows and might have said; instead he is content with the dry answer, 'Yes, *he* certainly would have done it.' 'Have no scruple, then,' says Clytemnestra, 'for the reproach of men.' Agamemnon could have answered that the reproach of men did not worry him, but that what he dreaded was the anger of the gods. Instead, he lamely replies, 'Yes, but public opinion is a great power.' Considered in terms of what we know of Aeschylean morality, this answer surely indicates a moral blindness. 'But the man who arouses no jealousy is not enviable', says the Queen. Agamemnon knows that to incur φθόνος is dangerous; yet he can counter only with the feeble complaint that a woman ought not to desire contention. 'But for the fortunate', his wife answers, 'it is becoming to yield the victory.' 'Do you think victory in this contest so important?' 'Be persuaded; if you give in to me, you are the winner.' The King has no answer to this; and after removing his boots in a futile gesture of appeasement, he enters the palace.

Agamemnon's answers to the last two questions give a definite indication that he has provoked divine φθόνος: the more closely we consider them, the harder it becomes to accept Fraenkel's explanation of Agamemnon's conduct. Must we agree with Page that he gives in 'simply because he is at the mercy of his own vanity and arrogance'? Here we are troubled by the empirical fact that during a performance of the play we find ourselves at this point regarding Agamemnon not with contempt, but with compassion. Note in particular the lines that immediately precede the stichomythia (926 f.). The king has replied to his wife's long and effusive speech of welcome with a curt and almost brutal refusal to accept her praise. But the conclusion of his speech, summing up his attitude, makes him, almost for the first time, sympathetic. 'Apart from foot-wipers and embroideries sounds the

[17] In Θεωρία (*Festschrift für W. H. Schuchhardt*) (Baden-Baden, 1960), 69 f.

voice of fame; and good sense is the god's greatest gift. Men should call him happy who has ended his life in the prosperity that we desire. And if in all things I can act thus, I lack not confidence.' These do not seem the accents of hypocrisy. Yet in the scene that follows, Clytemnestra twists her husband round her little finger; he is as helpless as Thrasymachus before Socrates against her devastating dialectic.

How can we account for Agamemnon's rapid collapse? Page's view that under temptation he reveals his secret moral weakness is not a wholly convin‖cing explanation of the change in him. Here we must carefully consider the explanation offered by Gundert, that Zeus and his portion and the Erinys have put Ate into his mind, to use the words put into Agamemnon's mouth in the nineteenth book of the *Iliad* (ll. 86–7). A parallel which seems to me to lend strong support to Gundert's view is furnished by that scene in the *Seven against Thebes* in which the Messenger describes to Eteocles the seven champions who are arrayed against the seven gates. Against the first six Eteocles dispatches champions from his own command. At the seventh gate stands Eteocles' own brother Polyneices. Like Agamemnon, Eteocles is a harsh and grim character who is yet not unsympathetically portrayed. He knows that if he fights his brother he will not survive; he knows that pollution of the most hideous sort is caused by the shedding of a brother's blood; and yet he cannot bring himself to do as the Chorus wishes and send another in his place. The reason for this is clearly indicated in the text, as Friedrich Solmsen has shown in an important article;[18] Eteocles is in the power of the Erinys. In Agamemnon's case the evidence of the text is less positive; but I have little doubt that Gundert is right in thinking that the reason for his behaviour is the same.

Not that Gundert's explanation seems to me entirely sufficient; in a curious way I believe that he and Page are both partly right. Gundert goes too far in arguing that Agamemnon reveals no ὕβρις, but mere stupidity; for when Zeus takes away a man's wits, he sends upon him a moral blindness, *τάλαινα παρακοπὰ πρωτοπήμων*. Zeus' action in sending Ate upon Agamemnon causes Agamemnon to commit a crime; so far Page is right; but in so far as the crime is the result of Zeus' action, Gundert has supplied an element of the truth which Page's explanation has ignored. It is clear that we have come upon an anomaly similar to that which so much perplexed us in the matter of Agamemnon's fatal decision at Aulis. There Page argued that Agamemnon could not be held responsible; Dodds argued that his action was a crime, and was called a crime by the Chorus; both views, I have argued, contain an

[18] *TAPA* 68 (1937), 197 f.; see also my review of K. von Fritz, *Antike und moderne Tragödie* (Berlin, 1962) [*Gnomon*, 34 (1962), 741 f. (= ch. 12, p. 191)].

element of truth. Here too it is the same. In one sense Agamemnon is guilty; Page has shown that he utters words that are bound to bring down on him divine envy, and we know that he will presently pay the penalty. Yet in a certain sense Agamemnon is innocent; he acts as he does because Zeus has taken away his wits. But why has Zeus done so? For the same reason as at Aulis; because of the curse. As Agamemnon succumbs, vanquished by the irresistible persuasion of Helen's sister, the destined instrument of his destruction, we look upon him not with scorn, but with compassion. Guilty as he is, he is not, like Aegisthus, mean and contemptible; destined as he is to ruin, at once guilty and innocent, he is a truly tragic figure.

The King disappears into the palace; the Chorus sings the third stasimon, full of ominous foreboding; and we are already waiting for Agamemnon's death-cry. But we are kept waiting till the end of the Cassandra scene. That scene occupies nearly 300 lines, not much less than one-fifth of the entire play. The power and beauty of that scene are so overwhelming that it is easy to forget to inquire what is its function in the unfolding of the plot. What is that function? Cassandra makes a desperate effort to get across to the uncomprehending Chorus a warning of Agamemnon's mortal danger which it is in‖evitably bound not to grasp. This provides a wonderful opportunity for the working up of an uncanny atmosphere and for the gradual building up of suspense. But this is not all. Since the narrative of the prophecy of Calchas, the audience has felt that there is some dark factor in the situation which has only been hinted at; something which if known would do more to explain the sinister forebodings of the Chorus than any vague talk of murmurs in the city against the princes. What that something is is instantly known to the foreigner Cassandra, who Clytemnestra has supposed may be ignorant of Greek. No sooner does she begin to move in the direction of the door than she sees in a vision (1096) the murdered children of Thyestes. Soon after she exclaims that even now a mighty evil is being plotted in the house (1102); and she describes in confused and agitated utterance a vision of the approaching murder. During the first part of the scene Cassandra speaks in lyrics; that part concludes with her calling to mind the fate of her own family and nation, and recalling once more to the audience the parallel, so often suggested during the first four great odes, between the fate of the Priamidae and the fate of the Atreidae. Then by a last great effort she collects herself, and in trimeters instead of lyrics, in speech instead of song, she openly declares to the Chorus (1178 f.) that the house of Atreus is beset by the Erinyes; that it is haunted by the spirits of the murdered children; that she and Agamemnon are presently to die an awful death; and that they will not go unavenged. And just before her final exit, she returns once more (1287 f.) to the fate of Troy and the not dissimilar fate of its conquerors.

We cannot regard the Cassandra scene as a mere episode, one whose presence may be amply justified by its effect but which is not essential to the development of the plot. Cassandra supplies us, first obscurely and later at the climax explicitly, with the vital piece of information that gives the missing clue for which we have so long been seeking. One main contribution of the scene to the unfolding of the plot is Cassandra's futile warning; but a more important one is her bringing into the open, for the first time in the play, the origin and nature of the curse.

There follows the scene in which Clytemnestra, standing over the dead bodies of the murdered pair, boldly confronts the Chorus and exults in her revenge. Returning to the theme so often played on in the early lyrics of the play, the Chorus cries out against Helen; now her deadly work has achieved its final triumph. 'O mad Helen,' they exclaim (1455 f.), 'you who alone destroyed those many, all those many lives beneath Troy, now have you crowned yourself with the last, the perfect garland, not to be forgotten, by means of the blood not washed away.' Clytemnestra forbids the Chorus to blame Helen. Next the old men address the daemon of the house (1468 f.): 'Daemon, you who fall upon the house and the two Tantalids, and exercise through women an evil sway . . .'.[19] Now you have set right your utterance', the queen replies, 'by calling on the daemon of this race, thrice glutted.' 'Great is the daemon of whom you speak,' says the Chorus, 'evil is his wrath, insatiate of baneful fortune. Woe, woe, through the will of Zeus, the cause of all, the doer of all. For what is fulfilled for men without Zeus? Which of these things is not god-ordained?' ‖

These words of the Chorus are not spoken idly. We can now trace, from the *πρώταρχοϲ ἄτη* of Thyestes, the grand design of Zeus. The action of the Theban trilogy, almost the only other of which we have a reasonable knowledge, is determined from the start by the curse upon Laius; so, I feel certain, is the action of the *Oresteia* by the curse upon Atreus. From his birth Agamemnon's fate, like that of Oedipus or Eteocles, has been determined; he is the son of the accursed Atreus. Zeus uses him as the instrument of his vengeance upon Troy; but he uses him in such a fashion that his own destruction must inevitably follow. At the outset of the expedition, Artemis, a partisan of Agamemnon's enemies, demands of him blood for blood. Agamemnon cannot refuse, for it is Zeus' will that the fleet sail; and Zeus sends Ate to take away his judgement and force him to consent. The King bows to the

[19] My translation assumes that in l. 1470 one may restore responsion and at the same time remove the very real difficulty of *ἰϲόψυχον* by accepting Rauchenstein's conjecture *κακόψυχον*.

goddess's demand: and his consent brings down upon him the vengeance of his wife, who shares her sister's uncanny and daemonic nature, serving like her as an instrument of Zeus' destructive purpose. Even his righteous revenge upon the Trojans involves Agamemnon in yet further guilt. In one sense, it is a triumph of divine justice; in another, an atrocious crime; the instrument of Zeus' punishment of Troy must himself be punished. But such guilt as the King contracts from the sacrifice of his daughter and from the annihilation of Troy with its people and its temples is only a consequence of the original guilt inherited from Atreus; the curse comes first, and determines everything that follows. Zeus brings about the ruin of Priam; Zeus brings about the ruin of Agamemnon. The Chorus of the *Agamemnon*, like Sophocles' Hyllus,[20] can justly echo Homer's words at the beginning of the *Iliad* and say that all that has happened has been in accordance with the will of Zeus.

[20] *Agam.* 1485–6; Soph. *Trach.* 1278.

22(1)

The Robes of Iphigeneia

Aeschylus, *Agamemnon* 228–43[1]

λιτὰϲ δὲ καὶ κληδόναϲ πατρώιουϲ
παρ᾽ οὐδὲν αἰῶ τε παρθένειον
ἔθεντο φιλόμαχοι βραβῆϲ.
φράϲεν δ᾽ ἀόζοιϲ πατὴρ μετ᾽ εὐχὰν
δίκαν χιμαίραϲ ὕπερθε βωμοῦ
πέπλοιϲι περιπετῆ παντὶ θυμῶι
προνωπῆ λαβεῖν ἀέρδην,
ϲτόματόϲ τε καλλιπρώιρου
φυλακᾶι καταϲχεῖν
φθόγγον ἀραῖον οἴκοιϲ
βίαι χαλινῶν τ᾽ ἀναύδωι μένει.
κρόκου βαφὰϲ δ᾽ ἐϲ πέδον χέουϲα
ἔβαλλ᾽ ἕκαϲτον θυτήρων ἀπ᾽ ὄμ-
ματοϲ βέλει φιλοίκτωι,
πρέπουϲά θ᾽ ὡϲ ἐν γραφαῖϲ, προϲεννέπειν
θέλουϲ᾽, κτλ.

περιπετῆ in l. 233 has been taken in three different ways. Firstly, it has been thought to come from περιπίπτω and to bear an active sense. Except in one passage, which I will deal with presently, περιπίπτω and περιπετήϲ seem always to be active. The verb means first 'to fall around' or 'to fall over'; then 'to fall in with' or 'to meet' persons; and lastly and most commonly 'to fall in with' or 'to run into' circumstances, sometimes good, but more usually adverse. And the adjective is used as the usage of the verb would lead one to expect.

Although this active meaning is so much the commonest, it seems here to give poor sense. The only meaning it will yield is 'falling round (or 'falling into') her robes': and though some have argued that this can be used loosely to mean 'with her robes wrapped around her', they have been unable to support their view by any evidence.

[1] This note originally formed part of a paper read to the Cambridge Philological Society in Oct. 1951. I am obliged to Professor D. L. Page, Professor A. J. Beattie, and Mr G. S. Kirk for their assistance: and to Sir John Beazley for allowing me to quote from two very helpful letters in which he discussed the black-figure amphora to which I allude.

Classical Review, NS 2 (1952), 132–5.

Secondly, περιπετῆ has been derived, again, from περιπίπτω, but taken to ‖ have a passive meaning. The sense would then be 'with her robes falling about her', which is possible in this context. But whether περιπετῆ can indeed be passive is a doubtful question. The only alleged parallel that seems worth considering is Soph. *Aj.* 906–8. 'By whose hand', asks the Chorus, 'did he fall?' 'By his own,' replies Teucer, 'it is plain; this sword, which he planted in the ground, and on which he fell, proves it.'

> ἐν γάρ οἱ χθονί
> πηκτὸν τόδ' ἔγχος περιπετὲς κατηγορεῖ.

Jebb in his *Ajax* (1896) prints this text; but in his note on *Ant.* 1223 (1888: 3rd edn., 1900) he confesses to a liking for Musgrave's περιπετοῦς. And there are strong grounds for suspecting that this may be right. It is not simply that in all other tragic passages in which κατηγορεῖν is similarly used (e.g. Aesch. *Agam.* 271) the object is expressed, in the genitive case, and is not left to be understood. It is that an examination of the compounds of πίπτω ending in -πετής listed in Buck–Petersen[2] reveals no instance of such a compound being *passive*. Nor do I know an instance of a passive adjective in -ής being formed from a verb whose passive does not occur.

I conclude that for περιπετής to be passive is not impossible, but at the same time not probable. So long as it is derived from περιπίπτω, whether the sense be active or passive, I do not see how the words will yield the meaning 'wrapped round in her robes'; yet this appears in a scholium in M (περιεσκεπασμένην), in Stanley's Latin version (*vestibus involutam*), and in many editions up to the most recent date.

The only explanation which would justify this rendering is that suggested by Professor Maas, who has hinted (*CQ*, NS 1 (1951), 94 [= *Kl. Schr.* 42]) that περιπετῆ may come, not from περιπίπτω but from περιπετάννυμι. The chief objection to this lies in the absence of any parallel formation from literature earlier than the fourth century. The earliest real parallels, ἀναπετής and διαπετής, occur in Hippocratic treatises, the περὶ ἀδένων and the περὶ καρδίης.[3] The former is ascribed by Galen (xviiiA. 379 K.) to τις τῶν νεωτέρων Ἱπποκρατείων, and the author of the latter, with his excellent knowledge of anatomy, is unlikely to have lived earlier than Diocles of Carystus.[4]

A fourth way of taking περιπετής is possible. We know from ll. 228–30 that Iphigeneia has prayed for mercy and has called upon her father. She has done so, presumably, in the characteristic posture of a

[2] *A Reverse Index of Greek Nouns and Adjectives* (Chicago, 1944), 730–1.

[3] ἀναπετής at Περὶ ἀδένων 9: διαπετής at Περὶ καρδίης 10.

[4] See Gossen in *RE* viii/2 (1943), 1837 [; and now C. R. S. Harris, *The Heart and the Vascular System in Ancient Greek Medicine from Alcmaeon to Galen* (1973), 83 f.].

suppliant, kneeling, with her arms clasped about her father's knees. I suggest that she is still in this position when the king's henchmen seize her, and that πέπλοιϲι περιπετῆ means 'with her arms flung about his robes'.

πίπτω, προϲπίπτω, and ἀμφιπίπτω are all used to describe the position of suppliants: cf. Eur. *Supp.* 279 ἄντομαι ἀμφιπίτνουϲα τὸ ϲὸν γόνυ . . ., and Eur. *Hec.* 787 ἀμφὶ ϲὸν πίπτω γόνυ. περιπετήϲ is used of having one's arms flung about a person at Soph. *Ant.* 1223, where Haemon with the dead Antigone is described in the words τὸν δ' ἀμφὶ μέϲϲηι περιπετῆ προϲκείμενον. Jebb translates, 'while he was embracing her with arms thrown round her waist'. If my suggestion is right, περιπετῆ in the *Agamemnon* will have been used in an exactly similar way.

Most commonly suppliants are said to grasp the knees of those they supplicate: but in order to do this they must also clasp their robes. When Hecuba || asks Odysseus whether he did not once clasp her knees in supplication, he replies (Eur. *Hec.* 246):

ὥϲτ' ἐνθανεῖν γε ϲοῖϲ πέπλοιϲι χεῖρ' ἐμήν

It may be objected that the word πέπλοι suggests female garments; and I suspect that the natural inclination of modern readers to make this assumption may account for no one's having drawn attention to this possibility before. The articles on πέπλοϲ in Dindorf's *Thesaurus* and in LSJ do not mention πέπλοι in the plural as being worn by men, except by Xerxes in the *Persae*, who is an Oriental, and by Orestes in the *Orestes*, whose πέπλοι are not clothes but bedclothes. But in fact this usage is a common one; see *IT* 312, *Ion* 1208, *Hcld.* 49, 130, *HF* 627, 629, 1198. The first two passages of *HF* are worth noting. First, the children clasp the robes of their father, who tells them to let go: μέθεϲθ' ἐμῶν πέπλων. Two lines later he complains that they have not done so:

οὐ μὴν ἀφιᾶϲ', ἀλλ' ἀνάπτονται πέπλων
τοϲῶιδε μᾶλλον.[5]

The phrase παντὶ θυμῶι I would take as going with περιπετῆ: this is possible, in view of the idea of supplication conveyed by that expression. This will enable us to adopt the colometry of this passage suggested by Miss Dale,[6] with end-stop after πρωτοπήμων (223) and after παντὶ θυμῶι here. Professor Ed. Fraenkel takes παντὶ θυμῶι with λαβεῖν ἀέρδην: but he has not observed that in the only relevant parallels he quotes (Pind. *Nem.* 5. 31 and Aesch. *Eum.* 735 f.: to which add Pind. *Pyth.* 9. 96) the phrase παντὶ θυμῶι describes an effort that is

[5] Mr G. S. Kirk points out that a long robe was the usual costume of a sacrificer.

[6] *The Lyric Metres of Greek Drama* (Cambridge, 1948), 81 [; 2nd. edn. (1968), 82].

mental, and not physical. *προνωπῆ*, 'bent forward', will be predicative with *λαβεῖν ἀέρδην*[7] and the whole sentence will mean: 'And her father gave orders to his henchmen to take her as she clasped his robes with all her heart and to raise her high, bending forward, like a goat, above the altar'.

The main difficulty of the next sentence lies in l. 239. What is meant by *κρόκου βαφάς*? Let me deal first with the suggestion of Professor Maas (loc. cit.) that it may refer to blood, as *κροκοβαφὴς σταγών* at l. 1121 probably does. I can find no positive evidence that *κρόκος* or any of its derivatives was used to describe anything coloured red before the Roman poets of the Augustan age; and it would certainly be odd if Iphigeneia still wanted to address each of the Achaean chiefs by name, as she does at 240 ff., some moments after the severance of her carotid artery. This makes strongly against the relevance of the black-figure amphora illustrated by Maas, of which Sir John Beazley writes: 'It illustrates *δίκαν χιμαίρας, ὕπερθε βωμοῦ* and no more. It is a hundred years earlier than the Oresteia, and a very rude work for its time.'[8]

Most modern commentators have taken the same view as Ed. Fraenkel, who ‖ translates, 'She let fall to the ground her saffron-dyed raiment'. It is likely enough that *κρόκου βαφάς* refers to the *κροκωτός*, but the rest of this translation gives rise to two strong objections. Firstly, no one has yet produced a convincing motive for what seems an unnecessary piece of exhibitionism, characteristic neither of Iphigeneia nor of Aeschylus; no one has answered the question asked in 1895 by Wilamowitz (*Ind. schol. hib.* (Göttingen, 1895–6), 5–6 n. 2 = *Griechische Verskunst*, 182 n. 1): 'cur vestem demisit virgo?' And secondly, for Iphigeneia to throw off her robe *is not possible*. For Agamemnon has ordered his men to seize her and to hold her high above the altar; and we know that his order has been carried out. For when, at 240 ff., Iphigeneia wants to call upon her friends, she cannot do so. Why can she not do so? Clearly because the order to gag her has been carried out. And if she is already gagged, it follows that she is already firmly gripped.

Since Iphigeneia is already in position for the sacrifice, *κ. β. χέουσα* cannot mean 'casting off her robe'. Professor Beazley has suggested that the verb refers, not to a deliberate action of Iphigeneia's, but to an

[7] Ed. Fraenkel takes *προνωπῆ* to refer to Iphigeneia's position *before* she is seized. Since he also takes *παντὶ θυμῶι* with *λαβεῖν*, he is led to assume a very peculiar word-order, despite the warning of W. Ferrari (*ASNP*, 2nd ser., 7 (1938), 391).

[8] *πρέπουσά θ' ὡς ἐν γραφαῖς* need not mean that Aeschylus had in mind any particular painting. Professor Beazley writes: 'I suppose that the heart of the comparison is the speaking look combined with the inability to speak. Aeschylus is thinking of the great new art of his time, with its new power of expressing *πάθος*, not of the simple works of early periods.' Cf. *Isthmiastae* fr. 78a. 6–7 R: *εἴδωλον εἶναι τοῦτ' ἐμῆι μορφῆι πλέον, / τὸ Δαιδάλου μίμημα· φωνῆς δεῖ μόνον.*

accidental consequence of the violence of Agamemnon's followers. Accepting this, I would translate: 'with her robe of saffron hanging down towards the ground'. The Greek idiom by virtue of which ἀποβάλλω means not only 'I throw away' but 'I drop' or 'I lose' suggests that χέω may mean not only 'I pour', 'I shed', but also 'I have pouring (or 'hanging') down from me'. [Cf. *Il.* 22. 468, *Od.* 11. 588; Pearson on Soph. fr. 378. 2–3.] And though I can discover no exact parallel for this use of χέω, this view seems to me strongly confirmed by the fact that χέουcα is not an aorist, but a present. Scholars have been accustomed to prove that χέω may describe the shedding of a robe by quoting *Il.* 5. 734: πέπλον μὲν κατέχευεν ἑανὸν πατρὸc ἐπ' οὔδει. But here the verb is aorist, and the presence of the preverb helps to show that the action is thought of as completed. In the *Agamemnon* passage commentators have been able to maintain their view of χέουcα only by tacitly conspiring to treat this verb as though it were an aorist. This is shown by the general acquiescence in Maas's substitution of τὼc for the manuscript reading θ' ὡc in line 242. Fraenkel justifies this change by writing (ii. 139), 'θ' will not do, for πρέπουcα, which is purely descriptive, cannot be put on the same level as χέουcα (239), which carries on the narrative'. If only we are willing to treat χέουcα as a present, which it is, then there is no reason why it should not be on the same level with πρέπουcα; and to write τὼc instead of θ' ὡc becomes unnecessary.

Had scholars not assumed that ἐc πέδον must mean 'to the ground', I suspect that this interpretation might have been current long ago. In fact it need mean no more than 'towards the ground'. In two passages of the *Antigone* (269–70, 441) a person is said to incline his or her head ἐc πέδον. Obviously these heads were hanging, not 'to the ground', but simply 'downwards'. So Iphigeneia's robe hangs down, but it does not come off.

[For different views, see J. T. Hooker, *Ἀγών*, 1/2 (1968), 59 f.; C. Sourvinou(-Inwood), *CQ*, NS 2 (1971), 339 f.; S. Radt, *Mnemosyne*, 4th ser., 26 (1973), 121–2; D. Armstrong and E. A. Ratchford, *BICS* 32 (1985), 1–12.]

23(12)

Three Notes on Aeschylus' *Agamemnon*

I. 128–30

χρόνωι μὲν ἀγρεῖ Πριάμου πόλιν ἅδε κέλευθος,
πάντα δὲ πύργων
κτήνη πρόcθε τὰ δημιοπληθῆ
μοῖρα λαπάξει πρὸc τὸ βίαιον . . .

129 *προcθετὰ* M: *πρόcθε τὰ* VFTr

SHOULD we read *πρόcθε τὰ* or *πρόcθετα* (Pauw, adopted by Hermann)? Unlike all other modern editors, Denniston–Page adopt the latter reading; but they admit that it is very difficult. 'The sense "additional" ', they observe, 'is out of place, and the only other possibility is "given up", "made over" (by debtor to creditor), a sense attested in an inscription of the mid-fourth century from Mylasa (*SIG* 167. 12 = Schwyzer 746 A 12).' It is enormously improbable that so prosaic an expression should occur in early poetry, and the sense it gives would be exceptionally feeble. They also point out that on this view *πύργων* would have to mean the fortress as a whole without reference to its fortifications, which is abnormal, and *κτήνη* would have to mean not 'cattle', as it normally does, but 'possessions'. The second of these objections is, I think, graver than has been realized. It is true that Hesychius explains *κτήνη* as meaning *χρήματα*: but no passage in any extant author gives definite support to his allegation (not even Hesiod fr. 200. 9 M.–W., where we can by no means exclude the possibility that ‖ *κτήνεccι* means 'cattle'), and it may well be mistaken. Denniston–Page argue further in support of *πρόcθετα* that 'the interruption of *πύργων πρόcθε* by the subject of the sentence is very harsh'. This is true; but I do not agree that the different nature of the words inserted between genitive and postposition in Fraenkel's two Homeric parallels (*Il.* 4. 54 and 12. 445 f.) robs them of all their force, especially since it is generally agreed that Calchas' prophecy is expressed in harsh and crabbed oracular language.

In sum, I come very near to agreeing with Fraenkel that *πρόcθετα* is

Rheinisches Museum für Philologie, 103 (1960), 76–80.

'impossible'. But Denniston–Page give one other reason for accepting it that at first sight seems formidable. This is that if we read *πρόcθε τὰ* we must assume that *κτήνη* means 'cattle'; and in such a context the mention of cattle seems absurdly trivial. Further, the cattle are to be destroyed *outside the walls*. Everyone who has read the *Iliad* knows that the Trojans did not do anything so foolish as to keep their cattle in such a place; they kept them on the slopes of Ida, and when the Greeks managed to carry off any, they did not indiscriminately slaughter them.

There is one way of keeping the reading *πρόcθε τὰ* and at the same time getting out of the text the reference to the slaughter of the human inhabitants of Troy which all likelihood demands. This is to assume that we have here one of the instances, so common in oracular language, in which human beings are referred to by the names of animals. We all know that an individual person can be referred to as, say, a cow or a lion, as at ll. 1127, 1223 f., and 1258 f. of this play (see Fraenkel's notes on these passages). Similarly, collectivities may be so described. One recalls, for instance, the oracle of Amphilytus at Hdt. 1. 62. 4:

> *ἔρριπται δ' ὁ βόλος, τὸ δὲ δίκτυον ἐκπεπέταcται,*
> *θύννοι δ' οἰμήcουcι cεληναίηc διὰ νυκτόc*

(verses which strikingly recall the language of *Agam.* 355–61). Even in non-oracular language, such words as *ἀγέλη* or *ἑcμόc* are not infrequently used of humans. The word *δημιοπληθῆ* has been rightly explained since Hermann as equivalent to *πολλὰ δήμια*, that is to say, as meaning 'many and of the people'. The first half of this compound may just as well correspond to a defining as to a possessive genitive; in other words, the phrase might mean 'the many herds of the people' in the sense that the herds are the people's property; but it might equally well || mean it in the sense that the many herds *are* the people. I can see no linguistic objection to this way of understanding the compound.

This view has the advantage of enabling us to take *κτήνη* in its normal sense, of getting rid of the objectionable *πρόcθετα*, and of allowing us to suppose that by means of an oracular licence highly consonant with the character of the context, *κτήνη* is being used to make the expected reference to the Trojan people. It also enables us to understand why the 'cattle' are to be slaughtered *before the walls*. The men of Troy took their stand against the invader *πυλάων . . . πρόcθε* (*Il.* 12. 445, where *εἱcτήκει*, as Fraenkel points out, is inserted between these two words); it is their fate *ὀλέcθαι ἐϋκλειῶc πρὸ πόληοc* (*Il.* 22. 110, a passage that may have been in Aeschylus' mind when he wrote *Agam.* 1304). This interpretation seems to me to have considerable advantages over any other that has been suggested.

II. 1055–7

οὔτοι θυραίαι τῆιδ' ἐμοὶ ϲχολὴ πάρα
τρίβειν· τὰ μὲν γὰρ ἑϲτίαϲ μεϲομφάλου
ἕϲτηκεν ἤδη μῆλα πρὸϲ ϲφαγὰϲ πυρόϲ.

'All is darkness here', say Denniston–Page of ll. 1056–7; both they and Fraenkel clearly summarize the difficulties. Both they and he agree that the notion that *ἑϲτίαϲ μεϲομφάλου* is what Sidgwick calls 'a loose local genitive . . . vaguely indicating the region' is highly unconvincing. This is not a construction that we expect to meet with in such a context. Moreover Fraenkel is right when he says 'the obvious way of constructing the passage is to take *τὰ μὲν γὰρ . . . μῆλα* together'; but how can we effect this without leaving in the air the intervening words *ἑϲτίαϲ μεϲομφάλου*? There is a second difficulty. Fraenkel has given good reason for treating the words *πρὸϲ ϲφαγὰϲ πυρόϲ* as a crux; and Denniston–Page agree with him.

I suggest that the clue to the understanding of *ἑϲτίαϲ μεϲομφάλου* is given by the well-known fact that the Greeks when sacrificing began and ended with an offering to the goddess Hestia. The evidence is well summarized by F. Schwenn, *Gebet und Opfer* (Heidelberg, 1927), 120 f.; cf. A. Preuner, *Hestia-Vesta* (1864), and in Roscher, *Myth. Lex.* i. 2614 f.; W. Süss in *RE* viii/1 (1912), 1272 f.; P. Stengel, *Die griechischen Kultusaltertümer*, 114 n. 15; A. C. Pearson on Sophocles fr. 726 (vol. II, p. 329). *ἀφ' Ἑϲτίαϲ ἄρχεϲθαι* was a common proverb (see Leutsch–Schneidewin on Zenobius 1. 40 in *Paroem. Gr.* i. 14); in the present connection, it is particularly relevant to mention ‖ that at Olympia they sacrificed to Hestia before sacrificing to Olympian Zeus (Pausanias 5. 14. 6).

Suppose we write *Ἑϲτίαϲ* with a capital *E*, and take it as referring not to the hearth but to the goddess: the genitive will then be a simple possessive, and the difficulty of taking *τὰ μὲν γὰρ* with *μῆλα* will vanish. The sense resulting will be good; for since it was generally known that one sacrificed to Hestia first, to say 'Hestia's oxen are ready' will have been a natural way, at least in poetry, of saying, 'The sacrifice is ready to begin'.

What is the meaning of *μεϲομφάλου*? In the only other place where this epithet is attached to the noun *ἑϲτία* the reference is to the sacred hearth near the *ὀμφαλόϲ* at Delphi (Eur. *Ion* 462); but I see no reason why it should not also be an epithet of the goddess. It could conceivably have reference to her special position at Delphi; cf. *Hymn. Hom.* 24. 1–2, Aristonous 2, p. 164 Powell (Diehl, 2nd edn., ii. 301) (*Ἑϲτίαν . . . ἃ καὶ Ὀλύμπου καὶ μυχὸν γαίαϲ μεϲόμφαλον ἀεὶ Πυθίαν τε δάφναν κατέχουϲα ναὸν ἀν' ὑψίπυλον Φοίβου χορεύειϲ*). But she may well have the epithet simply because of the central position which her shrine occupies in each household: cf. *Hymn. Hom.* 5. 30 *καί τε μέσωι οἴκωι κατ' ἄρ' ἕζετο*, Simias fr. 9 Powell = fr. 5 Diehl *Ἱστία ἁγνά, ἀπ' εὐξείνων μέσα τοίχων*, *Hymn. Orph.* 84. 1 f. Quandt[2] *Ἑστία, . . . ἣ μέσον οἶκον ἔχεις πυρὸς ἀενάοιο μεγίστου*. In this context the adjective would function both as a solemn epithet of the goddess and as an indication of the spot at which the sacrifice was to take place. We certainly ought to print *Ἑστίας* with a capital *E*; but the truth is that in such a case the Greeks did not distinguish sharply between the god and the sacred object; compare *Ἥφαιστος* as the first word of the Beacon Speech at 281.

What of the crux at the end of l. 1057? Plutarch (*Mor.* 703 D) says that Hesiod (*Op.* 748) is right to warn people against eating or washing from vessels over which sacrifice has not been offered; they should eat, says he, *ἀπαρχὰς τῶι πυρὶ καὶ γέρα τῆς διακονίας ἀποδιδόντας*. This notion of rewarding the fire for its services may very well lie at the root of the custom of sacrificing first to Hestia; and this consideration seems to me somewhat to strengthen the case for the conjecture *προσφαγαὶ πυρός*, which Denniston–Page tentatively put forward (p. 163).

III. 1652

ἀλλὰ κἀγὼ μὴν πρόκωπος οὐκ ἀναίνομαι θανεῖν.

This is what the manuscripts offer, and what most editors have printed. They have taken it to mean, in Paley's words,‖ 'But I, too, be assured, do not object to die with my hand on my sword'; but we require the sense, 'But I am ready with sword advanced, and do not shrink from death'. E. Lobel (ap. Fraenkel, p. 788) obtains this by changing *οὐκ* to *οὐδ'*: Denniston–Page follow him: Fraenkel has preferred *κοὐκ*. Either of these suggestions might be right; but I had rather be more economical and simply insert a colon after *πρόκωπος*, leaving *οὐκ* alone.

There is no reason to be afraid of the asyndeton. A similar case is that of *PV* 54, where all editors print a single sentence:

καὶ μὴν πρόχειρα ψάλια δέρκεσθαι πάρα.

This is not impossible, but I should much prefer to insert a colon after *ψάλια*. Ellipse of the verb 'to be' is especially common with *ἕτοιμος* and its synonyms (such as *πρόχειρα* at *PV* 54) and virtual synonyms (such as *πρόκωπος* at *Agam.* 1652;[1] see A. C. Pearson on Eur. *Hel.* 1523, J. D. Denniston on Eur. *El.* 37). With *δέρκεσθαι πάρα* as an independent

[1] *πρόκωπος* may well have found its way in from the preceding line, displacing some other adjective. Herwerden suggested *ξιφήρης*, A. S. F. Gow (in *CQ* 8 (1914), 6) *ξιφουλκός*. I should not dare to change the text; but I think it possible that Aeschylus may have written *πρόχειρος*.

sentence, compare *ὁρᾶν πάρεστι* at *Agam.* 1354 and Soph. *Ant.* 1293 and *ὁρᾶν πάρα* at PSI 1211 (= Aesch. fr. 132c R.), l. 28. After it had occurred to me that it would be better to place a colon after *ψάλια*, I looked up Rostagno's facsimile of M, where a colon is plainly visible in this place [still not observed by M. Griffith or West in their editions].

Another place in Aeschylus where editors have been insufficiently alive to the possibility of asyndeton is *Supp.* 926. M offers

ἤκουσα τοὔπος οὐδαμῶς φιλόξενον,

which Wecklein, Mazon, and Murray in his first edition followed the older editors in keeping unchanged. Headlam made a step in the right direction when he wrote

ἤκουσα, τοὔπος ⟨δ'⟩ οὐδαμῶς φιλόξενον;

Wilamowitz, Weir Smyth, Würtheim, Murray in his second edition and Kraus all followed suit. But we need only to place a colon after *ἤκουσα*; then there is no need to insert δ'.

24(42)

Agamemnonea

I. AESCHYLUS, *AGAMEMNON* 72–4

ἡμεῖϲ δ' ἀτίται ϲαρκὶ παλαιᾶι
τῆϲ τότ' ἀρωγῆϲ ὑπολειφθέντεϲ
μίμνομεν . . .

FRAENKEL, following Ahrens, seems to me to have demonstrated that the word ἀτίτηϲ must mean 'one who does not, or cannot, pay'. Yet Denniston and Page are right when they say that '"unhonoured" seems the only possible sense here' and when they argue that the suggestion that 'not paying' could stand for 'unable to pay the debt of military service to the state' is not attractive.

We could reconcile the natural meaning of ἀτίται with the sense demanded by the context if we could suppose that the meaning was 'we who have not paid to our aged flesh the debt we owe it', i.e., the debt of death. The notion of death as a debt which all men owe is very frequent in sepulchral epigrams from the Hellenistic period onwards; see Lattimore, *Themes in Greek and Latin Epitaphs*, 170 f., and Gow on Callim., *epigr*. 37. 4 (Gow–Page, *HE* ii. 194) and on Theodoridas, *epigr*. 12 (ii. 544). The epigrammatists took over the notion from earlier poetry; even if the ascription to Simonides of *AP* 10. 105 [= Page, *FGE* 79] (θανάτωι πάντεϲ ὀφειλόμεθα, l. 2) is dubious, the notion occurs at Soph. *El*. 1173, Eur. *Alc*. 419, 782, *Andr*. 1271, fr. 10, Ar. fr. 468 K.–A., all quoted by Gow). But in all these places death is said to be owed to us. In the epigrams, we are said to owe death to Hades, or to Nature, or to the gods. I cannot find a place where an old man is said to owe death to his aged flesh, and in default of such a passage I hesitate to press home my suggestion. But the idea seems to me a natural one, and I do not think it lies open to the reproach of being too obscure that ‖ Denniston and Page have levelled against the suggestion that the debt of military service owed to the state is intended here.

Harvard Studies in Classical Philology, 73 (1969), 97–104.

II. 355–61

ὦ Ζεῦ βαϲιλεῦ καὶ νὺξ φιλία,
μεγάλων κόϲμων κτεάτειρα,
ἥτ' ἐπὶ Τροίαϲ πύργοιϲ ἔβαλεϲ
ϲτεγανὸν δίκτυον, ὡϲ μήτε μέγαν
μήτ' οὖν νεαρῶν τιν' ὑπερτελέϲαι
μέγα δουλείαϲ
γάγγαμον, ἄτηϲ παναλώτου.

Fraenkel in his note on *ὑπερτελέϲαι* (ii. 189 ff.) discusses the question of whether the poet is comparing the Trojans to fish or to game. Although he notes that the word *γάγγαμον* seems to favour fish, comparison with ll. 358 ff. and ll. 1375 ff. seems to him to favour game; 'besides', he remarks, 'the distinction made in *μήτε μέγαν μήτε νεαρῶν τινα* seems more appropriate if we are thinking not of fish but of game.' *νεαρῶν*, he says in his note on that word, suggests a comparison with young wild animals. He arrives at no definite conclusion, but is inclined to favour game.

Now though in two other passages of the *Agamemnon* Aeschylus uses the image of game being caught in a net, that is no reason why he should not use the image of fish here. Further, I do not see why *νεαρῶν* should not suggest young fish as well as it can suggest young animals. Fraenkel is therefore right in not considering the arguments that seem to favour game as being decisive, and I have a new argument that favours fish.

Babrius, *fab*. 4 (on p. 8 of B. E. Perry's very useful Loeb edition of 1965) reads as follows:

Ἁλιεὺϲ ϲαγήνην ἣν νεωϲτὶ βεβλήκει
ἀνείλετ', ὄψου δ' ἔτυχε ποικίλου πλήρηϲ.
τῶν δ' ἰχθύων ὁ λεπτὸϲ εἰϲ βυθὸν φεύγων
ὑπεξέδυνε δικτύου πολυτρήτου,
ὁ μέγαϲ δ' ἀγρευθεὶϲ εἰϲ τὸ πλοῖον ἡπλώθη.
Ϲωτήριόν πώϲ ἐϲτι καὶ κακῶν ἔξω
τὸ μικρὸν εἶναι· τὸν μέγαν δὲ τῆι δόξηι
ϲπανίωϲ ἴδοιϲ ἂν ἐκφυγόντα κινδύνουϲ.

The relevance of this fable will, I think, strike most readers as obvious. What most recommends it is the explanation which it offers of the singular fact that the older Trojans or beasts are denoted not by any word ‖ signifying age, but by *μέγαν* (358). In the fable of the fish it was their size that prevented the older fish from escaping from the net, so it is natural that they should be called not 'the old ones' but 'the big ones'. 'Loquitur tanquam de piscibus irretitis', says Blaydes (*Aeschyli Agamemnon* (Halle, 1898), p. 199).

I cannot find this fable in any collection older than that of Babrius [nor is any cited in the new Teubner edition of Babrius by M. J. Luzzatto and A. La Penna (1986), p. 6]; but that is no reason why it should not have existed in the time of Aeschylus. On Babrius' sources, see Perry, lix ff.

III. 738–49

πάραυτα δ' ἐλθεῖν ἐϲ Ἰλίου πόλιν
λέγοιμ' ἂν φρόνημα μὲν
νηνέμου γαλάναϲ,
ἀκαϲκαῖον ⟨δ'⟩ ἄγαλμα πλούτου,
μαλθακὸν ὀμμάτων βέλοϲ,
δηξίθυμον ἔρωτοϲ ἄνθοϲ.
παρακλίναϲ' ἐπέκρανεν
δὲ γάμου πικρὰϲ τελευτάϲ
δύϲεδροϲ καὶ δυϲόμιλοϲ
ϲυμένα Πριαμίδαιϲιν
πομπᾶι Διὸϲ ξενίου
νυμφόκλαυτοϲ Ἐρινύϲ.

The fourth syllable of l. 742 must be long, and clearly either *τ'* (Hermann) or *δ'* (Porson) must be supplied. *μὲν* . . . *τε* is of course not an uncommon combination, but it is far from being as common as *μὲν* . . . *δέ*. If many editors, of whom the latest is Fraenkel, have preferred Hermann's *τ'*, it is because the sense has seemed to them to indicate that *δ'* will not do. It has seemed so because, to quote Fraenkel, 'almost all translators render (unintentionally, we may well think) the four components of the list asyndetically'. They have been so impressed by what Fraenkel calls 'a striking regularity in their parallel arrangement' in the four nominal clauses that extend from 740 *φρόνημα* . . . *γαλάναϲ* to 743 *δηξίθυμον* . . . *ἄνθοϲ* that they assume that all four clauses must be parallel in sense. 'The correct feeling of translators and editors for what is expected here', writes Fraenkel, 'is all the more remarkable because they almost all accept a text which offers the addition of *δ'* or *τ'* in l. 742, with the consequence of spoiling the asyndeton.' Fraenkel himself follows suit, and accepts *τ'*, 'admittedly as a last resort, for I cannot see why in the series of four ‖ expressions the first and second should be more closely connected than one or more of the other elements with others'.

Let us try the experiment of asking what the text will mean if we follow the indication supplied by the presence of *μὲν* in l. 740 and of either *δ'* or *τ'* in l. 742; since *δ'* is the more natural particle to supply, let us first try to make sense of the passage on the assumption that *δ'* is right. This will mean that we must provisionally and for a moment clear our minds of the assumption that all four clauses are exactly parallel, that assumption which Fraenkel has called 'the correct feeling of translators and editors for what is expected here'.

If we take the particles at their face value, the 'temper of windless calm' will be one thing and the 'gentle delight of wealth' will be another; further, the 'gentle delight of wealth' will be identical with the two things that are linked with it by asyndeton, with the 'soft arrow of the eyes' and with the 'flower of love that stings the heart'. Is this possible?

What is meant by saying that there came to Troy 'a temper of windless calm'? Presumably it means that everyone felt safe and peaceful, as people do when the sky is windless and the sea still. I need hardly point out that a person can hardly be described as a *φρόνημα*, so that it is most unlikely that this expression is a periphrasis for 'Helen'.

Next, what is meant by *ἀκαςκαῖον ἄγαλμα πλούτου*? The term *ἄγαλμα* means first that in which anyone delights (*πᾶν ἐφ' ὧι τις ἀγάλλεται*, Eustathius 456. 22, etc.) and secondarily an adornment. 'The delight' (or 'the adornment') 'of wealth' is an expression whose reference is not clear if it stands by itself; it could, however, be used to describe something indicated in the context. For this reason, and because we are now considering the first of three nominal clauses linked asyndetically, it is natural to look to the two following clauses for a clue to its application here.

What is 'the soft dart of the eye'? This question is much easier to answer; from a whole number of passages in early poetry—Barrett on Eur. *Hipp*. 525–6 and 530–4 gives much of the evidence—we know that love is darted from the eyes of the beloved to those of the lover. We shall expect to find that 'the soft dart of the eyes' is Eros, Love; and if once more we look to the following clause for explanation of the one which precedes it, we shall find that expectation confirmed.

What is 'the flower of Love that stings the heart'? Clearly *ἔρωτος* is a defining genitive; we have here a periphrasis for Love. Why is Love a flower? At *PV* 7 fire is called 'the flower' of Hephaestus. The expression is not, as editors tell us, synonymous with *τὸ ςὸν . . . γέρας* at l. 38; *ἄνθος* does not mean 'the pride', 'the guerdon', 'the ‖ special distinction' of Hephaestus; nor have such expressions as *ἄνθος ςτρατιᾶς*, 'the flower of an army', which editors quote, the slightest relevance. *ἄνθος* simply means 'flower'. The most helpful parallel is a variant of *Il*. 9. 212 which is given in the scholia on the *Prometheus* passage and quoted by Plut. *Mor*. 934 B; instead of

αὐτὰρ ἐπεὶ κατὰ πῦρ ἐκάη καὶ φλὸξ ἐμαράνθη

these sources give

αὐτὰρ ἐπεὶ πυρὸϲ ἄνθοϲ ἀπέπτατο, παύϲατο δὲ φλόξ.[1]

Fire was called a flower because it is brightly coloured and because it spreads its seeds abroad; the same concept lies behind the passage of the *Odyssey* (5. 490) in which a peasant who heaps ash round the flame to keep it alive is described as

ϲπέρμα πυρὸϲ ϲώιζων, ἵνα μή ποθεν ἄλλοθεν αὔοι.

Just so in Kipling's *Jungle Book* the animals call fire 'The Red Flower'. Fire also resembles a flower, or at any rate some flowers, in having a sting, only it stings not the finger but the θυμόϲ.

We must turn back for a moment to the first of our three parallel clauses to ask why Love is called 'the delight' (or 'the adornment') of wealth'. It is a Greek commonplace that the indigent do not fall in love; it has been illustrated by Nauck on Eur. fr. 895, to whose parallels we may add Achaeus *TrGF* 20 fr. 6 and Men. *Dysk*. 341 ff. and *Heros* fr. 10. In this context it is especially natural to associate Eros with luxury; Troy was a wealthy and luxurious city, and Helen and Paris were wealthy and luxurious people.

The three nominal clauses that extend from l. 741 ἀκαϲκαῖον ... πλούτου to l. 743 δηξίθυμον ... ἄνθοϲ are seen to refer to the same thing, Love. We find here the familiar method by which something is indicated first by an obscure and riddling paraphrase, then by a paraphrase whose meaning is easier to grasp, and finally by a direct mention. We may compare the opening of the second stasimon of the *Seven against Thebes*, a passages excellently explained by Fraenkel (*Kleine Beiträge zur klassischen Philologie*, i. 399 f.): 'I shudder at her that destroys houses, the god unlike the gods, the prophet of evil who speaks all too true, her whom their father's prayer called down—the Erinys.'

πέφρικα τὰν ὠλεϲίοικον,
θεὸν οὐ θεοῖϲ ὁμοίαν,
παναληθῆ κακόμαντιν,
πατρὸϲ εὐκταίαν Ἐρινύν (720 ff.). ||

The periphrastic descriptions are at last clarified and completed by the terrible name, falling with all the more weight for having been held back, of the person they describe; just so in the corresponding lines of the antistrophe the poet holds back the name of that iron with which the curse of Oedipus has doomed his sons to divide their inheritance:

[1] [Cf. Lucr. 1. 900 *donec flammai fulserunt flore coorto*. (H.-S.). See P. Jacobsthal, *Der Blitz in der orientalischen und griechischen Kunst* (Berlin, 1906); C. Blinkenberg, *The Thunder Weapon in Religion and Folklore* (Cambridge, 1911). M. Griffith, ad loc., ought to have seen that the use of ἄνθοϲ to mean various kinds of 'excrescence' is irrelevant to *PV* 7.]

'And a stranger surveys the drawing of their lots, a Chalybian settler from Scythia, a divider of their possessions little to their liking, cruel Iron.'

ξένος δὲ κλήρους ἐπινωμᾶι
Χάλυβος Cκυθῶν ἄποικος,
κτεάνων χρηματοδαίτας
πικρός, ὠμόφρων cίδαρος (727 ff.).

For a full account of what Fraenkel (ii. 9) calls 'the way in which the riddling expression . . . is followed by the explanation that solves the conundrum', see pp. 80 f., 379, and 691 of his great work.

If the hypothesis which we provisionally adopted, that the particles *μὲν* and *δ'* are to be taken at their face value, is correct, these lines say that there came to Troy two things—a temper of windless calm, and Love. Almost all editors, acting upon what Fraenkel calls their 'correct feeling for what has been expected here', think that there came to Troy four things, each of which is in some sense a description of Helen. I do not think I need take up much space to show that the sense obtained by following our hypothesis is both a great deal more precise and a great deal more poetical than that which editors have so far extracted from the passage. The meaning is that with the arrival of Helen the Trojans acquired a feeling of blissful calm and security, of a kind that in Greek literature often precedes the moment of catastrophe. Compare, to take only one example, the description of the reception of the Trojan horse in the first stasimon of Euripides' *Troades*:

ἀνὰ δ' ἐβόαcεν λεὼc
Τρωιάδοc ἀπὸ πέτραc cταθείc·
"Ἴτ', ὦ πεπαυμένοι πόνων,
τόδ' ἱερὸν ἀνάγετε ξόανον
Ἰλιάδι Διογενεῖ κόραι."
τίc οὐκ ἔβα νεανίδων,
τίc οὐ γεραιὸc ἐκ δόμων;
κεχαρμένοι δ' ἀοιδαῖc
δόλιον ἔcχον ἄταν (522 ff.).

Also the Trojans were possessed by Eros. Eros was responsible for Helen's presence, and Paris was in a special sense his victim. The other Trojans were his victims too, not simply in that all were under Helen's ‖ spell, as the elders of Troy show themselves to be in the Teichoskopia of the third book of the *Iliad*, but in that all were the victims of Eros in its most general sense, of delusion, vain desire, infatuation. That general sense is the one in which the word is used in the speech of Diodotus in the Mytilenean debate in the third book of Thucydides (45. 5): *ἥ τ' ἐλπὶc καὶ ὁ ἔρωc ἐπὶ παντί, ὁ μὲν ἡγούμενοc, ἡ δ' ἐφεπομένη, καὶ ὁ μὲν τὴν ἐπιβουλὴν ἐκφροντίζων, ἡ δὲ τὴν εὐπορίαν τῆϲ τύχηϲ ὑποτιθεῖϲα, πλεῖϲτα βλάπτουϲι, καὶ ὄντα ἀφανῆ κρείϲϲω ἐϲτὶ τῶν ὁρωμένων δεινῶν.*[2]

What is the meaning of the words that follow at 744 f.? Fraenkel translates, 'But then she swerved from this, and accomplished a bitter end of the marriage-rites . . .'; in his view, as in that of almost all scholars, Helen is actually called 'an Erinys bringing tears to brides'. He takes *παρακλίναϲα* as intransitive, meaning 'having swerved'. This would be the only instance of an intransitive use; *Il.* 23. 424 is given as a second in LSJ, but it is there acknowledged that the word *ἵππουϲ* is easily supplied:

> *Ἀντίλοχοϲ δὲ παρατρέψαϲ ἔχε μώνυχαϲ ἵππουϲ*
> *ἐκτὸϲ ὁδοῦ, ὀλίγον δὲ παρακλίναϲ ἐδίωκεν* (423–4).

Denniston and Page take *παρακλίναϲα* to be transitive, and translate 'She turned from its course, and accomplished a bitter end of, the marriage.' Both interpretations would imply that the marriage of Paris and Helen was at first set on a course leading to happiness; later it was diverted from that course by the Erinys, who according to Denniston and Page is not identified with Helen but is named in her own person.

I agree with Denniston and Page that *Ἐρινύϲ* means 'the Erinys', and not Helen; but I doubt whether the Erinys is said to have 'turned from its course' a marriage that in reality, as distinct from appearance, was heading for disaster from the start. Rather, I believe that *παρακλίναϲα* here means 'having laid them beside each other', i.e., 'having put them to bed together'.

The use of *παρακλίνω* which I suppose to be found here is not, let it be admitted, attested early. We find it at Theocr. 2. 44:

> *εἴτε γυνὰ τήνωι παρακέκλιται εἴτε καὶ ἀνήρ.*

We find it, in the active voice and used transitively, in an epigram at *AP* 5. 2. 1–4:

> *τὴν καταφλεξίπολιν Ϲθενελαΐδα, τὴν βαρύμιϲθον,*
> *τὴν τοῖϲ βουλομένοιϲ χρυϲὸν ἀμεργομένην,*
> *γυμνήν μοι διὰ νυκτὸϲ ὅληϲ παρέκλινεν ὄνειροϲ,*
> *ἄχρι φίληϲ ἠοῦϲ προῖκα χαριζομένην.*[3] ||

The same use occurs in the narration of the Ixion myth in the A scholia on *Il.* 1. 268: *Ζεὺϲ δὲ βουλόμενοϲ αὐτὸν δοκιμάϲαι εἰ τῶι ὄντι*

[2] [S. A. Barlow, *The Imagery of Euripides* (1971), 132–3 n. 21 writes: 'It is hard to see what this interpretation adds, since Helen is still implicitly the common factor in all four phrases'; she thinks that 'it is odd, if Ll.-J. is right that *ἔρωϲ* is the referent of the last three nominal phrases, that its prominence is not matched by an earlier reference'. Words fail me.]

[3] [Cf. Euphorion, *SH* 414. 10 *τῶι ϲε παρακλίνα*[.]]

τολμᾶι ποιῆcαι τοῦτο, νεφέλην ἀπεικάcαc Ἥραι παρέκλινεν αὐτῶι: the same story is told in similar words in the epitome of Apollodorus 1. 20 (p. 181 Wagner). Hieronymus of Rhodes at Ath. 435 A quotes Theophrastus for a story of how Olympias arranged for Alexander to enjoy the company of a beautiful hetaira: *Ὀλυμπιάδοc . . . παρακλινάcηc αὐτῶι Καλλιξεῖναν τὴν Θετταλὴν ἑταίραν περικαλλεcτάτην οὖcαν . . .*

The action of the Erinys is being described, it might be argued, in terms appropriate to that of the *νυμφεύτρια* who at a Greek wedding accompanied the bride to the house of the bridegroom's parents, or to that of the bridegroom's mother, who accompanied the bridal pair to their chamber carrying a torch [cf. Eur. *Med.* 887, 1027]. That image is employed by the Virgilian Juno when she apostrophizes Lavinia before sending Allecto to Latium to provoke a war:

> sanguine Troiano et Rutulo dotabere, virgo,
> et Bellona manet te pronuba. (*Aen.* 7. 318–19.)

But the expression is best understood if we remember how on Greek vases that show wedding scenes Aphrodite, Eros, or one or other of their attendant daimones are often shown hovering in the neighbourhood of the bridal couple.[4] Just so Eros floats near Helen in the Boston skyphos showing her abduction which Fraenkel (ii. 345) so aptly recalls in connection with the passage we are discussing. Eros and Aphrodite are near the bridal couple *ὥcτε παρακλῖναι τὴν νύμφην τῶι νυμφίωι*: in this case, the daimon present for this purpose is the Erinys.

[4] See H. Metzger, *Les Représentations dans la céramique attique du IVe siècle* (1951), 15 ff.; cf. D. von Bothmer, 'New Vases by the Amasis Painter', *Antike Kunst*, 3 (1960), 73.

25(67)

Ten Notes on Aeschylus' *Agamemnon*

Coming up to Oxford in 1940 was a little like coming from Arpinum to Rome during the civil wars, when many of the greatest orators were not on view. Denys Page should have been one of my tutors while I read for Mods in 1940/1; but he was away at the war, and I met him only when I returned to read for Greats in 1946. But I was also competing for the University Scholarships, and with great kindness he found time to help me. Two terms of composition with him were as instructive as they were exciting, as he was able to make use of that exercise to teach a very great amount of Greek. But he helped me even more later, after his arrival in 1950 in Cambridge, where I had moved two years earlier. J. D. Denniston had died in 1949, leaving the draft of an edition of the *Agamemnon* which he had been meaning to revise in the light of the vast quantity of learning contained in the great commentary of Eduard Fraenkel. Fraenkel's book had appeared a year after Denniston's death, and the task of revision had devolved upon Professor Page. Once a week for two years A. J. Beattie, G. S. Kirk, and I met to work through the play under his guidance. In theory we were there to help him; but for me, at any rate, the meetings were a wonderful opportunity to learn. They were a kind of substitute for the period as a research student which I had not had; but no less valuable than the great quantity of learning which he imparted was the encouragement which he could give as no other great scholar could have done. He made one feel that there were unlimited possibilities of discovery, and that if one worked hard and used one's wits one might find out truths unknown even to one's respected elders. I shall never cease to remember these meetings with gratitude and admiration. Remembering them, I could not help deciding to offer to Professor Page a few thoughts about some of the problems of the *Agamemnon*. The drawback attaching to this choice is that it meant I could not help expressing occasional disagreement with the great scholar whom we wish to honour. But he well knows that disagreement is not necessarily

Dionysiaca: Nine Studies in Greek Poetry by Former Pupils Presented to Sir Denys Page on his Seventieth Birthday (Cambridge, 1978), 46–59.

a sign of disrespect; and whatever he may think of my attempts, he will understand the spirit in which I offer them.

I would like to thank the friends who have kindly managed to spare the time to comment on an earlier draft of these notes: Sir Kenneth Dover, Professors R. Kassel and M. L. West, and Messrs C. W. Macleod and T. C. W. Stinton.||

I. 1–4

Θεοὺϲ μὲν αἰτῶ τῶνδ' ἀπαλλαγὴν πόνων,
φρουρᾶϲ ἐτείαϲ μῆκοϲ, ἣν κοιμώμενοϲ
ϲτέγαιϲ Ἀτρειδῶν ἄγκαθεν κυνὸϲ δίκην
ἄϲτρων κάτοιδα νυκτέρων ὁμήγυριν . . .

ΣM ad v. 3 *τὸ δὲ ἄγκαθεν κατὰ ϲυγκοπὴν ἀντὶ τοῦ ἀνέκαθεν, τουτέϲτιν ἄνωθεν, ἐξ ἀρχῆϲ* cf. *ϲυναγ. λεξ. χρηϲ.*, ed. Bekker i. 337. 25 *ἄγκαθεν· κατὰ ἀποκοπὴν ἀντὶ τοῦ ἀνέκαθεν. οὕτωϲ Αἰϲχύλοϲ* Hesych. α 786, ed. Latte, i. 30 †*ἀγρίαθεν* (*ἄγκαθεν* edd.)· *ἀνέκαθεν. Αἰϲχύλοϲ Ἀγαμέμνονι.*

The ancient interpretation of *ἄγκαθεν* as equivalent to *ἄνωθεν* was accepted by Stanley, who wrote 'excubans super aedes Atridarum'; he was followed by Schütz, who at the same time refuted the scholion's absurd temporal interpretation of the word's meaning (*ἐξ ἀρχῆϲ*), which had seduced Heath and Butler. Schneidewin followed Schütz, but thought it necessary to change *ϲτέγαιϲ* to *ϲτέγηϲ*. If one could accept that *ἄγκαθεν* means the same as *ἄνωθεν*, one certainly ought to remark that in places where *ἄνωθεν* is used as the equivalent of a preposition, it seems always to be followed by a genitive and not a dative. So great a grammarian as Ahrens accepted Schneidewin's emendation. This difficulty does not make in favour of the view that *ἄγκαθεν* is equivalent to *ἄνωθεν*: and it is not the only reason for rejecting it.

Hermann pointed out that *ἄγκαθεν* could not stand for *ἀνέκαθεν*: 'mira apocope, qua vocalis necessarius abiicitur'. Pointing out that at *Eum.* 80 *ἄγκαθεν* obviously means the same as *ἐν ἀγκάλαιϲ*, Hermann argued that in our passage it must mean the same as *ἐπ' ἀγκῶνοϲ*, 'on the arms'. Hermann's interpretation has convinced most commentators since his day, including Fraenkel.

Denniston–Page offered a new way of arguing that *ἄγκαθεν* meant 'upon'. They derived it not from *ἀνέκαθεν* but from an alleged equivalent *ἀνάκαθεν*. The trouble with this suggestion is that, as Robert Renehan demonstrated at *CR*, NS 20 (1970), 125 f., no reliable instance of *ἀνάκαθεν* exists in any author. It is possible that there was such a word, all the same; but in default of proof it seems hazardous to assume it.

Must we then return to the previously accepted view? The second part of Renehan's article, in which he argues that we must, is less convincing than the first. 'Although in *Eum.* 80', Denniston–Page write (p. 66), '. . . the meaning is undoubtedly "in your arms", as 259 shows, it remains certain that ἄγκαθεν could not mean "in (or on) my arms" in *Agam.* 3: ἄγκαθεν κοιμώμενος might ‖ possibly be said of A reclining in B's arms; it could not possibly be said of a man resting on *his own* arms or elbows; moreover a man in such a position would not noticeably resemble a dog, κυνὸς δίκην, nor would it be possible to adopt a more uncomfortable position for observing the stars.'

It is no use replying, as Renehan does, that the watchman is there not to observe the stars but to watch for the beacon. He has told us that he has had ample opportunity to observe the stars, not an easy thing to do if one is so foolish as to maintain the crouching position which those who think that ἄγκαθεν is equivalent to ἄνωθεν and that κυνὸς δίκην refers to the position adopted by the watchman have supposed. West takes the watchman to be not crouching on his hands and knees, but lying stretched out propping himself up on his elbows. On that view, one would surely have to give up the notion that κυνὸς δίκην refers to the watchman's position, and indeed there is no need to believe this. Fraenkel has pointed out that the roof is a natural place to find a watchdog, and the dog is so often associated in Greek literature with the keen and watchful servant that κυνὸς δίκην need not refer to the physical attitude assumed by the watchman. A scholion on the passage says πρὸς τὸ φυλακτικὸν καὶ φιλοδέσποτον, and this is surely right. Is the watchman stretched out resting on his elbows, as West suggests? Surely not, because it is not possible for ἄγκαθεν to be equivalent to ἐπ' ἀγκῶνος, or ἐπ' ἀγκαλῶν. Renehan says that words like ἄγκαθεν and ἀγκάς by their nature normally imply a reflexive notion. But when they actually occur they always mean the same as ἐν ἀγκάλαις: they never mean the same as ἐπ' ἀγκῶνος, and I cannot believe that they can do so.

This constitutes a case for examining a new possibility. Greek poetry contains a number of passages in which the sea or sky is said to hold 'in its arms' people or things that it contains. Archilochus spoke of sailors as ψυχὰς ἔχοντες κυμάτων ἐν ἀγκάλαις (fr. 213 West, cited in a scholion on Ar. *Ran.* 704, where there is an echo of the passage). At *Cho.* 586 the πόντιαι ἀγκάλαι are said to teem with monsters; and Euripides twice speaks of offerings being committed πελαγίους ἐς ἀγκάλας (*Hel.* 1062 and 1436; cf. Nausicrates frr. 1, 3 K.–A. In the dithyrambic language of Helen's Phrygian slave Ocean is said to enfold Earth in his arms (*Or.* 1378); cf. Eur. fr. 941:

> ὁρᾶις τὸν ὑψοῦ τόνδ' ἄπειρον αἰθέρα
> γαῖαν (καὶ γῆν codd.) πέριξ ἔχονθ' ὑγραῖς ἐν ἀγκάλαις;

It is not only of sea and sky that such expressions are used. In a Hellenistic dithyramb (*PMG* 926(*a*)) the meadows are said to receive dancing maidens in their arms (*ἀγκάλαις*); and more significantly Hermes tells Prometheus (*PV* ‖ 1019) that when Zeus' thunder has split the rock *πετραία . . . ἀγκάλη σε βαστάσει*.

I suggest that l. 3 of the *Agamemnon* means 'lying in the arms of the house of the Atreidae, like a dog'; *κυνὸς δίκην* has nothing to do with the watchman's physical attitude, but alludes to the fact that dogs, keen and watchful servants of their masters, do lie on housetops. The dative *στέγαις* is a pure dative, an instance of 'der eigentliche Dativ', as Kühner–Gerth call it, of a perfectly normal kind; as one can express 'the child is lying in my arms' by saying *ἄγκαθέν μοι κοιμᾶται ὁ παῖς*, so one can express the Watchman as lying in the arms of the house by saying (*ταῖς*) *στέγαις ἄγκαθεν κοιμᾶται ὁ φύλαξ*.[1]

II. 16–18

ὅταν δ' ἀείδειν ἢ μινύρεσθαι δοκῶ,
ὕπνου τόδ' ἀντίμολπον ἐντέμνων ἄκος,
κλαίω, κτλ.

Some commentators derive the metaphor in l. 17 from one kind of cutting, some from another; not all seem to be aware, however, that several different possibilities exist. Four types of view may be distinguished; I list them in what seems to me reverse order of probability.

1. Blomfield writes 'medicamentum paro, herbas concido'. Paley takes *ἐντέμνων ἄκος* to mean 'shredding herbs into a remedy'; Verrall and Sidgwick accepts this, and LSJ follow them. Fraenkel quotes Kennedy as saying that 'this cannot be proved from any ancient author', and I can find nothing that makes against his statement.

2. Fraenkel finds here the highly technical sense of tapping a remedy like sap from a root; Rose and Thomson follow him. Parallels are adduced from Theophrastus and the elder Pliny, not from any poet. The usage is surely too technical, too late in its attestation, and too much belonging to the world of prose to be in question here.

3. The two chief methods of ancient chirurgery were cautery and the surgeon's knife; see Fraenkel on l. 849. Denniston–Page find a metaphor from surgery in this place. ' "Incising a cure" ', they write, 'is a typically condensed expression for "curing by means of incision", *τομῇι*, a common metaphor (cf. l. 849 below, *Cho.* 539 *ἄκος τομαῖον*) in fifth-century literature.'

On this view the watchman by the act of singing is applying the surgeon's knife to the disease of drowsiness. In that case the exact function of the preverb *ἐν-* is not easy to determine. 'The object of *ἐν-*',

[1] [Against, H. Neitzel, *Glotta*, 57 (1979), 191f., who does not persuade me.]

Denniston–Page continue, 'is the patient (here myself); *ἐντεμνόμενος* would have been apt, but is not specially called for.' ‖

4. The explanation which has been most popular in the past is not mentioned by either Fraenkel or Denniston–Page. Stanley took the metaphor to be from the cutting of a root to serve as antidote to the *φάρμακον* of a magician or a poisoner. This notion occurs first in the Homeric Hymn to Demeter, where the disguised goddess claims that if she is hired as nurse she will be able to protect her charge against sorcery: *οἶδα γὰρ ἀντίτομον μέγα φέρτερον ὑλοτομοῖο* (229.).

Since Stanley scholars have compared Pind. *Pyth.* 4. 221, where to protect Jason from the fire-breathing bulls Medea gives him *ἀντίτομα ϲτερεᾶν ὀδυνᾶν*: the pain of the injuries the bulls might inflict will be annulled by the antidote Medea's ointment offers. They have also cited Eur. *Alc.* 970 f., where Apollo is said to have given to the sons of Asclepius *φάρμακα πολυπόνοιϲ ἀντιτεμὼν βροτοῖϲιν*. Some have quoted Eur. *Andr.* 120 f. *εἴ τί ϲοι δυναίμαν | ἄκοϲ τῶν δυϲλύτων πόνων τεμεῖν*, where Stevens compares the two passages just cited. He is probably right to do so, but this might be regarded as an instance of the metaphor which Denniston–Page find in the *Agamemnon* passage.

Among those who have accepted Stanley's view are Schütz, Ahrens, Wecklein in 1888 and 1896, and Groeneboom. If it is correct, sleep is compared to a dangerous drug, or to the *νόϲοϲ* which it causes, singing to its antidote. This seems to me supported by the presence of the word *ἀντίμολπον*, which recalls *ἀντιτέμνειν, ἀντίτομον*, the words found in the three passages cited in the last paragraph. At l. 849 the case is different, for there Agamemnon is speaking of a radical remedy which will eliminate the infected part; *Cho.* 539 is similar. Since one cannot sleep and sing at the same time, it may be argued that by singing one is applying such a radical remedy against one's drowsiness. But it seems likelier that sleep is compared to a drug, or to the harm it causes, and singing to its antidote.

Denniston–Page (cited above) do not satisfactorily explain the *ἐν-* of *ἐντέμνων*. The problem has troubled several scholars; Ahrens dealt with it by accepting *ἐκτέμνων* from codd. F and G, Kennedy by reading *ἓν τέμνων*. Denniston–Page take the object to be 'the patient'. If I had to say what was the object, I had rather say 'the situation in general'. But in this and countless similar cases, the physical sense of the preverb is not important. The preverb is there primarily to make the verb perfective rather than imperfective, like the preverbs which Attic (*ἀπο-*) and Ionic (*κατα-*) add to the simple verb *κτείνω*. The importance of verbal aspect (*Aktionsart*) must be recognized not only in the Slavonic languages, where special verb-forms exist in order to convey it, but in many other languages, including Greek; see Wackernagel,

Vorlesungen über Syntax, i. 53 f. and ii. 181 f., and add to the literature he quotes the monograph ‖ of J. Brunel, *L'Aspect verbal et l'emploi des préverbes en grec, particulièrement en attique* (1939). In the common expression *ἐντέμνειν cφάγια*, for example, the function of the *ἐν-* is obviously that of rendering the verb perfective; Stinton compares *ἐντεθέρμανται* at Soph. *Tr.* 368, *ἐνημμένη* at Ar. *Pax*, 1092, and *ἐνάψαcθαι* at Lys. 1. 14.

III. 72–82

ἡμεῖc δ' ἀτίται cαρκὶ παλαῖαι
τῆc τότ' ἀρωγῆc ὑπολειφθέντεc
μίμνομεν ἰcχὺν
ἰcόπαιδα νέμοντεc ἐπὶ cκήπτροιc·
ὅ τε γὰρ νεαρὸc μυελὸc cτέρνων
ἐντὸc ἀνάccων
ἰcόπρεcβυc, Ἄρηc δ' οὐκ ἐνὶ χώραι·
τό θ' ὑπεργήρων φυλλάδοc ἤδη
κατακαρφομένηc τρίποδαc μὲν ὁδοὺc
cτείχει, παιδὸc δ' οὐδὲν ἀρείων
ὄναρ ἡμερόφαντον ἀλαίνει.

By no means every commentary or translation correctly indicates the train of thought. My own literal version (Prentice-Hall, Englewood Cliffs, NJ, 1970 [2nd edn., Duckworth, London, 1979]) fails in this respect, for it gives no clear indication that at 76 f. the child and the old man are declared to be equally incapable of fighting. The particles reveal the structure; the *γὰρ* in l. 76 shows that ll. 76–82 explain what precedes in two parallel clauses linked by *τε* in l. 76 and *θ'* in l. 79. The strength which the old men guide upon their staves is equal to a child's. For just as the vital force of a child is no greater than an old man's—the sense of the unique compound *ἰcόπρεcβυc* is easier to grasp after *ἰcόπαιδα* in l. 75—and he cannot fight, so extreme old age walks upon a stick and has no more valour than a child. Comparing ll. 80–2 with ll. 74–5, we see that we have returned to the same point, by a kind of ring-composition. The allusion to the riddle of the Sphinx, noticed first by Spanheim, is evident; it is made all the clearer by the stress laid by the riddle, in the form in which we have it, on the child's weakness. (See Appendix, p. 332 f.)

I have paraphrased *μυελὸc cτέρνων ἐντὸc ἀνάccων* (76–7) by 'vital force'; and that general sense seems to be strongly indicated by the context. But how can we extract that meaning from the words *μυελὸc cτέρνων ἐντὸc ἀνάccων* (or *ἀναΐccων*, if we accept Hermann's conjecture)? ‖

First of all, *μυελόc* normally means 'marrow'. Headlam thought *ἀνάccων* 'appropriate to the marrow, regent in its frame of bone and dominating the vital functions'. But the chest is not where we should expect to find the marrow; and Fraenkel (ii. 47) has shown that in the passage of Plato in which Headlam thought the marrow was said to 'rule' the word *ἀρχή* actually means 'beginning'.

In the physiology of the *Timaeus* (see 73 B f.), *μυελόc* denotes not only the spinal marrow, but the brain and semen. Galen, *de usu partium* 8. 4 (ed. Helmreich, p. 453) points out that Plato uses the word *μυελόc* of the substance that serves as the seat of sensation in crabs and other molluscs that do not, strictly speaking, have heads; it needs further specification, Galen adds, because it is used both of spinal and of cerebral marrow. Thomson in his note on the passage thinks, as many others have done, that the physiological doctrine of the *Timaeus* is of Pythagorean origin. But the use of *μυελόc* to refer to the brain is not confined to Pythagoreans, doctors, or philosophers; the fact is proved by an unambiguous instance in Soph. *Tr.* 781 *κόμηc δὲ λευκὸν μυελὸν ἐκραίνει.*

The etymology of *μυελόc* remains uncertain, but it may well be connected with *μυών*: 'sowohl das weiche Mark wie die weichen Muskeln bilden einen Gegensatz zu dem harten Knochen', says Frisk, *Gr. etym. Wörterbuch*, 264. If *μυελόc* is a general word covering spinal marrow, brain, and semen, it is not surprising, though it is unusual, that it should be used to indicate the substance that gives vital strength. It is significant that *μυελόc* is sometimes equated with *αἰών*: see *Il.* 19. 27 *ἐκ δ' αἰὼν πέφαται*, and cf. *HHMerc.* 42 and 119. R. B. Onians, *The Origins of European Thought* (2nd edn., 1954), 205–6 gives full evidence for the notion of a liquid vital substance, found in the brain, the spinal marrow, and the semen.

None of these things is located in the chest. But in early Greek thought the chest is normally regarded as the centre of the intelligence; the discovery of the brain's importance by Alcmaeon of Croton is not taken note of by the fifth-century tragedians. If, as I have argued, *μυελόc* is a general term which may be used to denote the vital fluid which is the stuff of life, I do not find it surprising that it should be said to rule within the chest.[2]

Both Fraenkel and Denniston–Page accept Hermann's emendation of *ἀνάccων* to *ἀνάιccων*. 'If *ἀνάccειν* really is the function of *μυελόc*,' Fraenkel writes (ii. 47), 'it cannot be limited to the tender period of youth; yet here, with *ὁ νεαρὸc μυελόc* preceding, we must have something characteristic of immaturity.' But why? If the poet were speaking of the vital fluid of an old man, he might well mention, say, *ὁ γεραιὸc μυελόc* as 'ruling within the chest'; in itself the expression is

[2] [See now F. Solmsen, *AJP* 100 (1979), 477–9.]

neutral and suited to old age as well as ‖ immaturity. 'Since the point is the weakness of the youthful body,' Denniston–Page write (p. 75), evidently echoing Fraenkel, 'the lordly word *ἀνάϲϲων* is out of place.' But if the youthful *μυελόϲ* is called 'equal to an old man's', the point is clearly made; a ruler, even if he is feeble, is still a ruler. *ἀνάιϲϲων* makes sense, and involves no real alteration of the paradosis. But since *ἀνάϲϲων* is perfectly satisfactory, I see no need to change it, for I find the notion of the vital fluid 'springing up' within the chest harder to parallel and harder to understand than the notion of the vital fluid being said to 'rule' there.

IV. 104–6

κύριόϲ εἰμι θροεῖν ὅδιον κράτοϲ αἴϲιον ἀνδρῶν
ἐντελέων· ἔτι γὰρ θεόθεν καταπνείει
πειθώ, μολπᾶν ἀλκάν, ϲύμφυτοϲ αἰών.

'The subject of *καταπνείει* is *πειθώ* . . .; it cannot be *αἰών*, for a man's "age" (time of life) could not be said to "*breathe down*" upon him, especially "*from the gods*"': Denniston–Page. But if Deianeira can say (at Soph. *Tr.* 34–5) 'such is the *αἰών* that was always sending my husband to and from our house', perhaps the Chorus can speak of an *αἰών* that breathes persuasive power into a man. The real sense is that Heracles' time, his life, is such that he has to go to and from his house, and that the time, the life of the Argive elders, is such that persuasion is breathed down upon them from the gods. One may compare *Suppl.* 45; the text is controversial, but it seems clear that the subject of the verb *ἐπεκραίνετο* is *μόρϲιμοϲ αἰών*. In Soph. *Aj.* 644 f. the *αἰών* of Ajax alone among the Aeacidae is said to have nurtured Ate.

If this is accepted one is free to take *πειθώ* as the object of *καταπνείει*, and to regard *μολπᾶν ἀλκάν* as standing in apposition to it in the way supposed by Fraenkel. Dover, however, is inclined to omit the commas, so as to give the sense that 'the persuasiveness of song' is the only *ἀλκά* which old age allows a man; he may be right. In either case we avoid the changes made in the text by Denniston–Page and mentioned in the OCT, which, in the words of Dodds, *Oxford Magazine*, 14 November 1957, p. 128 (in by far the most valuable review the book received) 'destroy a characteristic piece of Aeschylean poetry'.

V. 681–7

τίϲ ποτ' ὠνόμαζεν ὧδ'
ἐϲ τὸ πᾶν ἐτητύμωϲ,

μή τις ὅντιν' οὐχ ὁρῶμεν προνοί-
αιcι τοῦ πεπρωμένου
γλῶccαν ἐν τύχαι νέμων,
τὰν δορίγαμβρον ἀμφινει-
κῆ θ' Ἑλέναν;

Who gave Helen her mysteriously appropriate name? Perhaps someone whom we do not see. That is an unusual expression, carefully chosen, and the commentators do not quite see eye to eye about its meaning. 'Studiously vague,' says Verrall, 'an unknown someone', and Wecklein in his *Oresteia* of 1888 agrees with him ('doch nicht wer, am Ende jemand'); one may agree, but we are still entitled to try to guess at what the poet had in mind. Most believe that he was thinking of a supernatural being. 'Some god', says Paley; 'a supernatural being, who cannot be seen of mortals unless he chooses to reveal himself', says Rose; '*πρόνοιαι τοῦ πεπρωμένου* belong only to a god or a daimon', says Fraenkel (ii. 330). Thomson believes that the phrase points specifically to Dike, whose invisibility as she does her work is attested by Eur. fr. 255 and fr. adesp. 493; but though to say that Dike named her would be an accurate way of saying that her name expressed her nature, the words in the text hardly point unambiguously to this particular divinity.

Supernatural beings alone know the future; but they may communicate such knowledge to mortals. At Eur. *Phoen.* 636, Eteocles thus alludes to the etymology of his brother's name: *ἀληθῶς δ' ὄνομα Πολυνείκη πατὴρ / ἔθετό σοι θείαι προνοίαι νεικέων ἐπώνυμον.* The foreknowledge came, it is true, from the gods; but the name was given by the mortal father according with the foreknowledge.

We find something similar in a passage which may be said to offer an instance of the same *topos* as *Agam.* 681 f., the praise of the wisdom of the namer of a thing or person appropriately named. In the Tebtunis fragment assigned to the *Inachus* of Sophocles, the Chorus praises the astuteness of the namer of the wearer of the cap of Hades who is invisibly present on the stage. The wearer of the cap is in fact Hermes, but at first the Chorus infers from the cap's name and presumed ownership that he is Hades (see *CR*, NS 15 (1965), 241 f. [= ch. 34, pp. 397 f.]):

πολὺ πολυιδρίδαc
ὅτιc ὅδε προτέρων
ὄνομ' εὖ c' ἐθρόει
τὸν Ἀϊδοκυνέαc
cκότον ἄ⟨β⟩ροτον ὑπαί.

(P. Tebt. 692, fr. 1 i 16 f., in Carden, *The Papyrus Fragments of Sophocles* (1974), 73 = fr. 269c. 16 f. Radt.) ‖

In this passage the giving of the name is attributed to one of those belonging to the past; that is the sense of *προτέρων*, as Hunt saw when he first published the papyrus and as I demonstrated (loc. cit.) by arguments which need not be repeated. A god would have been named by gods, so that the distinction between divine prompter and human namer observed in the case of Polyneices would not exist. But it is worth noticing that the expression used is simply 'those of the past'; it might indicate that in this *topos* the reference is generally simply to whoever long ago named the child, a thing usually done by the child's parents.

Supernatural beings visit the earth 'shrouded in darkness' (see Hes. *Op.* 123, 223, *Il.* 14. 282, etc.); they may have been denoted by the unusual expression 'those whom we cannot see'. But so may the dead, the inhabitants of the house of Hades. Whether or not Felix Solmsen was right in upholding against Schulze and Wackernagel the derivation of *Ἀΐδηϲ* from alpha privative and the root of *Ϝιδεῖν*, ancient poets commonly assume its rightness (see ch. 34, p. 399). At *PV* 152 f. the words of the Chorus *τὰ πρὶν δὲ πελώρια νῦν ἀΐϲτοῖ* prompt Prometheus to wish that Zeus had cast him lower down even than Hades; and at *Agam.* 462 f. the black Erinyes consign the guilty man to darkness (*τιθεῖϲ' ἀμαυρόν*), and he has no protection when he is among those unseen (*ἐν ἀΐϲτοιϲ*). 'Those unseen' surely means the inhabitants of Hades.

Verrall's phrase 'studiously vague' well describes the choice of the expression here. But in the light of the passages cited from Euripides and Sophocles, I think in the first place of the now long-dead mortal parent, and only in the second of the god who guided him in choosing the child's name.

VI. 799–802

ϲὺ δέ μοι τότε μὲν ϲτέλλων ϲτρατιὰν
Ἑλένηϲ ἕνεκ', οὐκ ἐπικεύϲω,
κάρτ' ἀπομούϲωϲ ἦϲθα γεγραμμένοϲ
οὐδ' εὖ πραπίδων οἴακα νέμων, κτλ.

The adverb *ἀπομούϲωϲ* surely goes with *γεγραμμένοϲ*, as Stanley, Schütz, and most editors in all periods have believed. 'Tu autem', Schütz translated, 'admodum odiosis coloribus pictus ac deformatus es ... utpote qui male animum gubernasses, cum ...' This involves supposing that 'you were pictured by me very inartistically' can mean the same as 'the portrait I drew of you was ugly'. The two are not quite the same thing, but in this context one may argue that the difference is not significant.

A more serious obstacle in the way of this interpretation is the difficulty of ‖ seeing how the clause beginning with *οὐδ᾽* coheres with what goes before it. Schütz translates 'utpote qui', and that is what the sense requires; but the text as it stands contains no real equivalent to this. At the beginning of l. 802 we should expect *ὡϲ* or *ἅτε* followed by *οὐκ*, but what we find is simply *οὐδ᾽*.

This awkwardness led Elmsley to conjecture *ἀπόμουϲόϲ τ᾽ . . . οὔτ᾽* and Blomfield to conjecture *ἀπομούϲων*: it led Ahrens, Schneidewin–Hense, and Wecklein to take *ἀπομούϲωϲ* as parallel to *οὐδ᾽ εὖ* and like it modifying *οἴακα νέμων*. This will not do; not only does *ἀπομούϲωϲ* obviously belong with *γεγραμμένοϲ*, but the colourless *οὐδ᾽ εὖ* following the colourful *ἀπομούϲωϲ* and modifying the same participle falls intolerably flat.

It will not do simply to alter *οὐδ᾽* to *οὐκ*: we need an equivalent of Schütz's 'utpote qui'. I think one line, just possibly more, has been lost after l. 801. West remarks that that line could have begun with a vowel, and that it most likely began with *οὐ*.

VII. 1025–9

εἰ δὲ μὴ τεταγμένα
μοῖρα μοῖραν ἐκ θεῶν
εἶργε μὴ πλέον φέρειν,
προφθάϲαϲα καρδία
γλῶϲϲαν ἂν τάδ᾽ ἐξέχει.

Fraenkel is one of many who have taken *μοῖρα* to mean fate in general, *μοῖραν* the fate of an individual. But surely the choice of an identical word for both and the juxtaposition of the two suggests that they both refer to a *μοῖρα* which is the same or of the same kind.

Many again—Stanley seems to have been the first—have favoured an interpretation I have heard cogently advocated by Mr W. S. Barrett. Stanley translated 'nisi autem destinatum fatumque decretum deorum impediret quo minus plura praedicaret, cor antevertens linguam ista effudisset' (he should have written 'effunderet'). In fact, he believed that the reason why the heart did not anticipate the tongue and pour out the truth was simply that it was a heart and not a tongue. This gives sense and logic but, apart from the somewhat unusual sense *μοῖρα* would have to bear, the sentence might seem a little flat and unpoetical.

The first scholar, so far as I know, to advocate the view I find most attractive was Nägelsbach, whose study of early religion led him to reject the kind of view taken by Fraenkel. His rendering is as follows: 'nisi me locus is ac ‖ modicus dignitatis quem obtineo gradus prohibuisset ne profarer quae sentio de rebus Agamemnonis scelus divina lege consecuturi'. Keck reformulated this view as follows: 'wenn nicht im Menschenverkehr eine Lebensstellung die andere infolge höherer Anordnung beschränkte und sie hinderte mehr Vorteil zu haben (sich auf Kosten anderer geltend zu machen), d. h. wenn nicht meine dem König soweit untergeordnete Stellung (*Ch.* 49) mich hinderte mir ihm gegenüber etwas herauszunehmen.' The audience would have been aware, Keck added, that 'eine Lebensstellung engt ein und beschränkt die andere'. Schneidewin–Hense take much the same view, and so did Wecklein (1888 and 1896).

It may be objected that *μοῖρα* commonly denotes divine destiny, and not social status. Indeed the use of *μοῖρα* that would be in question is by no means common. But in archaic thought social status was thought of as god-given; and the use in question is well established in Aeschylus. At *Cho.* 238 Electra says that Orestes has for her four *μοῖραι*, for he is to her a father, a mother, a sister, and a brother. In that place *μοῖραι* cannot mean 'fates'; Tucker renders it by '"functions", "offices" (*partes*), sometimes "prerogatives"'. When Oceanus says to Prometheus that even apart from the claims of family *οὐκ ἔϲτιν ὅτωι / μείζονα μοῖραν νείμαιμ' ἢ ϲοί*, he means the allowance of respect due to one for whom he has such great regard; cf. Soph. *Trach.* 1238, where the dying Heracles says to Hyllus *ἀνὴρ ὅδ' ὡϲ ἔοικεν οὐ νεμεῖν ἐμοὶ / φθίνοντι μοῖραν*. At *Eum.* 172 the Erinyes complain that Apollo has destroyed their ancient privileges, *παλαιγενεῖϲ . . . μοίραϲ φθίϲαϲ*: Blass, rightly protesting against spelling *μοίραϲ* with a capital *M*, compares l. 727 *παλαιὰϲ διανομὰϲ καταφθίϲαϲ*. At *Eum.* 476 Athene gives warning that the Erinyes have a *μοῖραν οὐκ εὐπέμπελον*: 'the rights', as Sidgwick puts it, 'which the appointment of the gods have given them make it hard for me to deal harshly with them'.

Since the parodos the Chorus have from time to time hinted at dark secrets in the royal family's past which are the cause of their uneasiness. These secrets will come into the open only through Cassandra, a foreigner who is at first thought to be ignorant of Greek (cf. *CQ*, NS 12 (1962), 197–8 = ch. 21, pp. 297–8). In this place the Chorus are surely saying that the respect they owe their king prevents them from mentioning such matters.

One may add to Fraenkel's parallels for the notion of the heart outstripping the tongue (ii. 464) these two cases:

1. (Longinus), *de sublim.* 19. 1 . . . *⟨ἀϲύμ⟩πλοκα ἐκπίπτει καὶ οἱονεὶ*

προχεῖται τὰ λεγόμενα, ὀλίγου δεῖν φθάνοντα καὶ αὐτὸν τὸν λέγοντα.

2. Manilius 5. 223–4 *praecurrunt verba loquentis, / ante os est animus.*‖ W. Bühler, *Hermes*, 87 (1959), 475–6 shows against Bentley and Housman, who read *loquentes* and took it as a nominative, that *loquentis* is accusative. He cites the Pseudo-Longinus passage, but not the Aeschylus.

VIII. 1140–9

φρενομανής τις εἶ, θεοφόρητος, ἀμ-
φὶ δ' αὑτᾶς θροεῖς
νόμον ἄνομον, οἷά τις ξουθὰ
ἀκόρετος βοᾶς, φεῦ, ταλαίναις φρεσὶν
Ἴτυν Ἴτυν στένουσ' ἀμφιθαλῆ κακοῖς
ἀηδὼν βίον.
ἰὼ ἰὼ λιγείας ἀηδοῦς μόρον.
περέβαλον γάρ οἱ πτεροφόρον δέμας
θεοὶ γλυκύν τ' αἰῶνα κλαυμάτων ἄτερ·
ἐμοὶ δὲ μίμνει σχισμὸς ἀμφήκει δορί.

1145 *βίον* codd.: *μόρον* Page 1146 *ἀηδοῦς μόρον* Dobree: *ἀηδόνος μόρον* codd.: *μόρον ἀηδόνος* Hermann: *βίος ἀηδόνος* Page 1148 *αἰῶνα* M^{s}*γρ*: *ἀγῶνα* codd.

'*μόρον* in Aeschylus elsewhere always means "death", not "destiny" or "portion" ': so Denniston–Page, following Blass, *Mélanges Weil* (1898), 13. It is true that in all the other passages of Aeschylus where it occurs it *connotes* death; but does it actually *mean* that? Wilamowitz, surprised to find that in a document from Western Locri the word bears its etymological sense of 'part' or 'share', observed (*SBPAW* 1927, 15 = *Kl. Schr.* v/1. 478) that 'wir es doch im Epos, aus dem es die hohe Poesie übernimmt, nur als Endschicksal, meist geradezu Tod, kennen'. That applies exactly to its use in Aeschylus and Sophocles, as a look at the examples listed by Italie and Ellendt–Genthe will confirm. The latter (*Lexicon Sophocleum*, 460) remark that it sometimes stands for 'genus mortis'; and that applies to *Agam.* 1600 and *Cho.* 441, where it carries the epithet *ἄφερτος*, to *Pers.* 444, where the Queen will ask what *μόρος* it was that the men died by, to *Eum.* 502, where the Erinyes threaten to inflict 'every manner of death' and to fr. 288 R., where the speaker fears the foolish manner of death of a moth.

Procne when transformed into a nightingale did not die; but her metamorphosis was the end of her life as a human being, her *Endschicksal.* The gods invested her with a winged form and with a pleasant life 'without cries of pain'; but Cassandra will die painfully.

Denniston–Page find ll. 1144–5 'hardly less obscure' than they find

l. 1146. The ‖ nightingale, they argue, must bemoan the past, the death of Itys, but her life is free from any further troubles. This is hair-splitting; for anyone who has had Procne's experiences may legitimately lament over a life that is rich in evils on all sides, or more probably lament throughout such a life.

Before the Hellenistic age the exclamatory accusative in Greek is very rare indeed, and Blaydes may well have been right (see p. 98 of his edition of 1898) in suggesting that *μόρου* or *μόροϲ* should be read in l. 1146. But in l. 1145 we need *βίον*, in l. 1146 some case of the word *μόροϲ*. That is what the manuscripts offer, and transposition is unnecessary.

IX. 1440–3

ἥ τ' αἰχμάλωτοϲ ἥδε καὶ τεραϲκόποϲ
καὶ κοινόλεκτροϲ τοῦδε θεϲφατηλόγοϲ
πιϲτὴ ξύνευνοϲ, ναυτίλων δὲ ϲελμάτων
†ἱϲτοτριβήϲ†· ἄτιμα δ' οὐκ ἐπραξάτην.

Denniston–Page obelize *ἱϲτοτριβήϲ*, but still find Pauw's conjecture *ἰϲοτριβήϲ* to be 'attractive'; the OCT restores Pauw's conjecture to the text. Yet Fraenkel has conclusively shown that *ἰϲοτριβήϲ* is a compound that cannot have stood in Aeschylus (iii. 680 f.). Cf. J. Diggle, *CR*, NS 18 (1968), 2–3, who finds it necessary to emend.

Latte, *Hermes*, 66 (1931), 155 f. = *Kl. Schr.* 290 f., describes how in a number of Greek communities offenders were rendered *ἄτιμοι* by being exposed in humiliating positions. In several such cases, the penalties were imposed for adultery or fornication. In Lepreon the guilty woman had to stand in the market-place on eleven successive days without a girdle and wearing a transparent chiton. In Cyme she had to ride round the city on an ass; before and after doing so, she had to stand on a stone in the agora.

Latte relates to these facts the word *ἀκρίϲτιοϲ*, known to us only from Hesychius α 2576 (ed. Latte, i. 91), where it is thus explained: *ἡ ἐπάνω τοῦ ἱϲτοῦ καθεζομένη. ἔϲτι δὲ ἄτιμοϲ.*

The only kind of loom used by the Greeks of the classical age is the warp-weighted loom, fully described and illustrated by Grace M. Crowfoot, *BSA* 34 (1936–7), 36 f.; cf. R. J. Forbes, *Studies in Ancient Technology*, iv (1964), 204 f. If one wished to expose an offender to public view, one might very well seat him or her upon the horizontal crossbar of a loom of this type, rather than upon the stone used for this purpose at Cyme (see above, and cf. Plut. *Quaest. Gr.* 291 E *ἐπὶ λίθου τινὸϲ ἐμφανῆ καθίϲταϲαν*).

Clytemnestra is aware that Cassandra has lately made a long sea-voyage, during which she is likely to have shared with Agamemnon the comparatively ‖ comfortable accommodation provided by the platform at the bow of the ship (*cέλματα*: see J. S. Morrison and R. T. Williams, *Greek Oared Ships, 900–322 B.C.* (1968), 48 and 196–7). Even on the poop, one would have only a wooden bench to sit on; and the picture of Cassandra seated on such a bench might well recall to Clytemnestra the undignified posture of the dishonoured woman seated upon the horizontal top-beam of the loom.

Clytemnestra continues with *ἄτιμα δ' οὐκ ἐπραξάτην*, 'their fate is not without its privilege', as Denniston–Page render it. Perhaps the association of the punishment just alluded to with the notion of *ἀτιμία* has suggested the use of this particular expression.

X. 1541–2, 1551–4

τίς ὁ θάψων νιν; τίς ὁ θρηνήcων;
ἦ cὺ τόδ' ἔρξαι τλήcηι . . .;

οὐ cὲ προcήκει τὸ μέλημ' ἀλέγειν
τοῦτο· πρὸς ἡμῶν
κάππεcε κάτθανε καὶ καταθάψομεν
οὐχ ὑπὸ κλαυθμῶν τῶν ἐξ οἴκων.

In the eighth oration of Isaeus we read how after the death of Ciron his grandson and his nephew disputed his inheritance. The grandson, who was Isaeus' client, claimed that the other party had allowed him to share the expenses of the funeral, thus tacitly admitting his claim to be a legitimate grandson of the deceased. Had he not really been Ciron's grandson, he says, his adversary Diocles *ἐκείνους ἂν τοὺς λόγους ἔλεγε· "cὺ δὲ τίς εἶ; coὶ δὲ τί προcήκει θάπτειν; οὐ γιγνώcκω cε· οὐ μὴ εἴcει τὴν οἰκίαν*" (8. 24, ed. Wyse, p. 121). Similar passages are cited by Headlam on Herondas 6. 56; Dover adds Isaeus 9. 4 and other evidence given in his *Greek Popular Morality* (1974), 243 f.; cf. D. C. Kurtz and J. Boardman, *Greek Burial Customs* (1971), 142.‖

APPENDIX

The riddle is quoted by Athen. 10. 456 B; *AP* 14. 64 (cod. Paris. gr. 3058); Hyp. Soph. *OT* (ed. Dain–Mazon ii. 71); Hyp. Eur. *Phoen.* (ed. Schwartz, i. 243–4); Σ *Phoen.* 50 (ed. Schwartz, i. 256); ΣLycophr. 7 (ed. Kinkel, p. 65); Tzetz. ad Lycoph. 7 (ed. Scheer, ii. 11). A text is given by Lesky, *Mitt. Ver. kl. Phil.*, Vienna, 5 (1928) 3 f. = *Ges. Schr.*

318; like him I have recorded only the more significant variants, but I have recorded more of them, corrected a few errors and have chosen differently among them in certain places in the text.

ἔcτι δίπουν ἐπὶ γῆc καὶ τετράπον, οὐ μία μορφή,
καὶ τρίπον, ἀλλάccει δὲ φυὴν μόνον ὅcc' ἐπὶ γαῖαν
ἑρπετὰ κινεῖται καὶ ἀν' αἰθέρα καὶ κατὰ πόντον.
ἀλλ' ὁπόταν τριccοῖcιν ἐπειγόμενον ποcὶ βαίνηι,
ἔνθα τάχοc γυίοιcιν ἀφαυρότατον πέλει αὐτοῦ.

1 *μορφή* Tzetz.: *φωνή* cett. omn.   2 *φυὴν μόνον* Hyp. Soph. *OT* '*φ*', Hyp. Eur. *Phoen.* cod. A, v. l. ap. Tzetz.: *φύcιν μόνον* Athen.: *μόνον φύcιν* ΣEur.: *βοὴν μόνον* Hyp. Soph. codd. LA, *AP*, ΣLycophr., v. l. ap. Tzetz.: *φωνὴν μόνον* Hyp. Eur. cod. B   3 *κινεῖται* Hyp. Soph. codd. A '*φ*', Hyp. Eur. ed. Ald., *AP*: *γίνεται* ΣEur.: *γίνηται* Hyp. Soph., cod. L, Tzetz.: *γ(ε)ίνονται* Athen., Hyp. Eur., ΣLycophr.   *καὶ ἀν' αἰθέρα* Athen. cod. A: *ἀνά τ' αἰθέρα* cett. omn.   4 *τριccοῖcιν* ΣEur.: *πλείcτοιcιν* Hyp. Soph. (praeter cod. G), Athen. cod. A, *AP*, ΣLycophr., Tzetz.: *πλείοcιν* Athen. cod. C: *πλεόνεccιν* Hyp. Eur., Hyp. Soph. cod. G: *ἐπειγόμενον* Hyp. Soph. codd. LA, *AP*, ΣLycophr., Tzetz.: *ἐρειδόμενον* Hyp. Soph. codd. '*φ*', Athen., Hyp. Eur., ΣEur.   5 *μένοc* Hyp. Eur., Tzetz. codd. am^{2}: *τάχοc μένοc* Tzetz. cod. *γ*l.

C. Robert, *Oidipus*, i (1915), 56–7 thought this riddle might come from an early epic, the *Oedipodeia* or the *Thebais*; and when we learned from P. Oxy. 2459 that Euripides in his *Oedipus* gave a different version of the riddle of the Sphinx (see Austin, *Nova fragmenta Euripidea in papyris reperta*, fr. 83, ii. 23), I argued that this strengthened Robert's case, since this version would not have been likely to be quoted by so many writers unless it had even better authority than that of Euripides (*Gnomon*, 35 (1963) [R41], 447). I am now a good deal less certain.

Lesky in the article cited above (unknown to me in 1963) had ruled out the *Oedipodeia* on the ground that in this poem the Sphinx seems to have been merely a killer, not yet a poser of enigmas. He also points out that in early epic we should expect in l. 2 *ἀμείβειν* rather than *ἀλλάccειν*, a verb first found in Theognis 21. Lesky's suggestion that this version may come from the *Thebaid* of Antimachus does not appeal to me: Antimachus would surely have written a version of the riddle in his own manner, not in one meant to suggest archaic style. We must note that both Athenaeus and ΣEur. *Phoen.* 50 cite the riddle from the *Tragodoumena* of Asclepiades of Tragilus (*FGrH* 12 F 7; on this pupil of Isocrates, see Wilamowitz, *Analecta Euripidea* (1875), 181 n. 3 and *SBPAW* 1925, 43 = *Kl. Schr.* v/2. 56–7). This does not prove that the riddle was taken from a tragedy, but it creates a presumption that it was; hexameters were of course the regular metre for riddles even in tragedies (apart from that in Eur. *Oedipus* just mentioned, see Theodectes, *TrGF* 72 fr. 4 Snell and Sen. *Oed.* 233–8). One tragedy that may be thought to have had even greater authority than the *Oedipus* of Euripides is the *Oedipus* of Aeschylus.

But what matters for the present purpose is that something like this version of the riddle certainly existed in Aeschylus' time. Fraenkel on *Agam.* 1258 (iii. 581) mentions the famous cup in the Vatican with the Sphinx addressing Oedipus and beside her the words (κ)αὶ τρί(πον) (see Hartwig, *Meisterschalen*, pl. 73 and Beazley, *ARV* (2nd edn.), 451); this was painted in about 470 BC.

Aeschylus must have had something like this version of the riddle in his mind when he wrote *Agam.* 80–2. In my text of the riddle, l. 4, I have preferred τριϲϲοῖϲιν, offered by only one of the ‖ seven authorities that quote the text, to the more popular reading πλείϲτοιϲιν. If one reads πλείϲτοιϲιν, the slowest (or the most feeble) of the three creatures is the four-legged; if one reads τριϲϲοῖϲιν it is the three-legged. It seems likely that in the version used by Aeschylus the old man, and not the child, was the most feeble of the three.

26(13)

Some Alleged Interpolations in Aeschylus' *Choephori* and Euripides' *Electra*[1]

THE second play of the trilogy begins with the appearance before Agamemnon's tomb of the long-absent Orestes, who prays to Hermes for aid in his revenge and then dedicates upon the tomb a lock of hair cut from his own head. He is interrupted by the entrance of Electra together with the captive women who form the Chorus; in consequence of an evil dream, Clytemnestra has sent them to pour a libation to the spirit of her murdered husband. After discussion with the Chorus, Electra resolves to accompany that libation not with words of appeasement, but with a prayer for her father's help in taking revenge upon his murderers. Going to the tomb to pour the libation, she notices upon it the lock put there by Orestes (168). She notices that the hair is like her own, and at once suspects that it may be her brother's. 'How can he have dared to come here?' the Chorus ask her (179); and she replies that he must have sent the lock from his place of exile. The thought that Orestes' return is impossible plunges both the Chorus and Electra into deep sorrow, and in tears Electra broods over the lock (183 f.). It can belong only to Orestes; she wishes it could speak, so that either she could know it not to be his, or else it could at least take part with her in mourning. At this point, if we can trust the one manuscript, Electra breaks off her reflections with a prayer (201–4). 'But we call upon the gods,' she says, 'well do they know by what storms, like sailors, we are buffeted; but if we are fated to find safety, from a small seed may grow a mighty trunk.'

Next, if we are to trust what is transmitted, Electra catches sight of a new indication of her brother's presence. 'Yes, and here are tracks,' she says (205 f.), 'a second indication, the tracks of feet matching each other and resembling mine. Yes, for here are two outlines of feet, his own and those of some companion. The heels and the outlines of the tendons agree in their proportions with my prints. I am in torment, and my wits are confounded.'

[1] The text in effect of the second of my J. H. Gray Lectures, delivered at Cambridge on 1 Mar. 1961. I am greatly obliged to Professor D. L. Page for his helpful comments.

Classical Quarterly, NS 11 (1961), 171–84.

Orestes now emerges from the hiding-place where with Pylades he has been lurking ever since the entrance of Electra and the Chorus. 'Pray,' he exclaims, 'as you acknowledge to the gods your prayer's fulfilment, that in future also your fortune may be good.' During the stichomythia that follows, Electra comes by stages to accept the fact, obvious enough but hard for her to take in immediately, of her brother's identity. Orestes himself points out that her attitude is paradoxical (225 f.); 'Why, you see my very self and find me hard to recognize: yet when you had seen the hair I cut off in mourning and were scanning the traces of my footprints your thought took wings and you believed you saw me. Put the lock to the place in my hair from which I cut it, and see how your brother's head is like your own.' Orestes goes on to produce a piece of cloth woven by Electra herself, which he has brought with him as a proof of his identity.

Until the late eighteenth century no scholar questioned the authenticity of ‖ any portion of this passage. But to the keen, rationalizing intelligence of Christian Gottlob Schütz, who first edited Aeschylus in 1782, much of it seemed deplorable. 'Haec agnitionis ratio,' he wrote, 'nullo modo defendi aut excusari potest, et si quidem ab Aeschylo profecta sit, nihil aliud superest, quam ut eum hic aliquando dormitasse indignemur.' No one could possibly believe, Schütz insists, that the similarity of the footprints to her own could have made Electra think that they must be her brother's, or that the other footprints belonged to someone who had come there with him. The prayer which is Orestes' first utterance, Schütz contended, must surely have followed directly upon the prayer of Electra at ll. 201–4. He offered also certain linguistic arguments against Aeschylean authorship, which I shall come to later. He concluded that ll. 205–10 were 'a sciolo quodam confectos'; and he deleted also l. 228, thereby not only removing the only other reference to the footprints, but cutting the knot presented by the relation of l. 228 to l. 227, a problem which most editors have solved by transposition.

Schütz's drastic remedy did not win general approval; but it prompted several other scholars to suggest ways of dealing with the difficulties it was meant to solve. Butler, in his *Choephori* of 1812, agreed with Schütz in finding the recognition scene 'satis flaccidam, ne dicam ineptam' but declined to obelize a passage that he thought contained 'nonnulla . . . cothurno haud indigna'. Furthermore, was not there the best of authority for the authenticity of the lines in question, that of Euripides, who in the recognition scene of his *Electra* had notoriously held them up to ridicule? Butler's own suggestion was to transpose Electra's prayer in ll. 201–4 so that it stood between l. 210 and l. 211. This met Schütz's requirement that the prayer of Orestes should follow closely on the prayer of Electra. Henri Weil followed Butler, with one slight modification. Line 211 following on the prayer seemed to him too abrupt; he preferred to place the prayer not after l. 210, but after l. 211. But Blass in his commentary of 1906 found that the prayer followed awkwardly upon l. 211. He went back to Butler's view, though he stopped short of printing the transposition in his text. We shall return to these transpositions later.

Schütz found no follower to accept his athetesis until in 1938 R. Böhme published an article (*Hermes*, 73 (1938), 159 ff.) which he has since amplified in a book (*Bühnenbearbeitung Aeschyleischer Dramen* (2 vols., Basle, 1956–9). His treatment seems to me to require no discussion; but the same cannot be said of Schütz's other follower in modern times. In Appendix D to his great edition of the *Agamemnon*, published in 1950, Professor Eduard Fraenkel argued for the deletion of everything cut out by Schütz, and of ll. 211 and 229 into the bargain. What of the testimony of Euripides to the genuineness of the passage? Fraenkel will not admit the existence of any such testimony. Like Böhme, he holds that ll. 518–44 of the *Electra* are an interpolation made in order to ridicule a scene already interpolated in the text of Aeschylus. So startling a conclusion, reached by so eminent an authority, seems to demand a renewed examination of the problem. Following Dindorf and Wecklein, Fraenkel has also (ibid. Appendix C) discovered interpolations in the speech of Orestes over the bodies of his victims that begins at l. 973. This problem also seems to require discussion. It will be best to follow Fraenkel's example and examine first those features of the language and style of the excised passages which have seemed to the deleters to offer evidence against their authenticity; from there we shall proceed to the more important general arguments.||

In l. 206 two words for 'like' are used; this seems to Schütz, and to Fraenkel, a 'pleonasmus satis inficetus'. This may be true; but if it is, it is by no means the only such pleonasm in the works of Aeschylus. From Kumaniecki's list of pleonastic adjectives (*De elocutionis Aeschyleae natura* (Cracow, 1935), 30) I quote the following: *PV* 661 *αἰολοςτόμους*/ *χρηςμοὺς ἀςήμους δυςκρίτως τ' εἰρημένους*, *Sept.* 827 *μογεροὺς καὶ δυςδαί-μονας*, *Pers.* 289 *εὔνιδας ἠδ' ἀνάνδρους*, 665 *καινά τε κλύηις νέα τ' ἄχη*, *Agam.* 677 *χλωρόν τε καὶ βλέποντα*, *Supp.* 357 *ἐξ ἀέλπτων κἀπρομηθήτων*, 712 *ἀπροςδοκήτους τούςδε καὶ νέους λόγους*. That is not all, but that is enough. The commentators have already observed the similarity of some words in Sophocles' *Ajax* (1152–3) *τις ἐμφερὴς ἐμοὶ ὀργὴν θ' ὅμοιος*, where the qualification of *ὁμοῖος* by the limiting accusative *ὀργήν* scarcely robs the parallel of its significance.

Next, Schütz and Fraenkel make much of the difficulty of *καὶ γὰρ* in

l. 207. If *καὶ γὰρ* is right, it must mean, 'Yes, for . . .'; and they cannot see how the presence of not one, but two sets of footprints can be regarded as confirming Electra's previous statement. Against this, Conington argued that 'the condensation of the expression is as natural as the simplicity and precipitancy of the reasoning'. Electra, he held, has noticed that some of the footprints are like, but some unlike, her own. For a moment she is disconcerted by the presence of the footprints that are unlike hers; but in a moment her confidence returns as she notices that these are the prints of a second person. Conington's view is accepted by Sidgwick, Verrall, and Tucker, and it may possibly be right;[2] but it involves making out behind the words a somewhat complicated mental process. I prefer Hermann's way of resolving the difficulty, which is to suppose that something has dropped out after l. 208; in the missing part of the text, Electra will have gone on to explain the sentence beginning *καὶ γὰρ* by saying explicitly that one set of footprints is not like hers, but that the other is like them. The case for this conjecture is strengthened by the fact that l. 209 begins with an asyndeton which is in any case somewhat awkward, and might by itself have given rise to the suspicion that something was missing immediately before. Weil, Blass, Wilamowitz (in his edition of 1914), and Groeneboom all accepted Hermann's lacuna, I think rightly.

Next, Schütz objects that ll. 209–10 contain an 'inanis repetitio' of what is said in ll. 205–6. Would this be the only such such 'vain repetition' in Aeschylus? This argument, like the preceding ones, does not much help the case for believing in interpolation.

We come now to the speech of Orestes that begins at l. 225. Here are its first six lines in the order in which they stand in the manuscript:

αὐτὸν μὲν οὖν ὁρῶϲα δυϲμαθεῖϲ ἐμέ, 225
κουρὰν δ' ἰδοῦϲα τήνδε κηδείου τριχὸϲ 226
ἀνεπτερώθηϲ κἀδοκεῖϲ ὁρᾶν ἐμέ 227
ἰχνοϲκοποῦϲά τ' ἐν ϲτίβοιϲι τοῖϲ ἐμοῖϲ. 228
ϲαυτῆϲ ἀδελφοῦ ϲυμμέτρου τῶι ϲῶι κάραι 229
ϲκέψαιτο μὴ προϲθεῖϲα βόϲτρυχον τριχόϲ. 230

228 delevit Schütz, 228–9 Fraenkel 227 post 228 traiecit Robortello, 229 post 230 Bothe 229 *ϲύμμετρον* Pauw 230 *ϲκέψαι τομῆι* Hartung ‖

Editors as recent as Conington, Verrall, and Tucker have kept ll. 227–30 in the manuscript order, but only by means of perverse reasoning. First, it is strange to find a sentence consisting of a participial qualification of the subject, then the main clause, and then a

[2] Professor Page shares this view. 'The use of *καὶ γὰρ* may not be strictly logical or neatly classifiable', he writes, 'but nobody would have been much if at all bothered by it *εἰ μὴ θέϲιν διαφυλάττων*. I do not find the asyndeton at 209 at all harsh, especially by Aeschylus' standards.'

further participial qualification appended by means of *τε*. Klausen quoted *Agam.* 238 f., but a lyric parallel is not much use; no parallel from dialogue seems to have been discovered. Fraenkel points to this awkwardness as evidence in favour of his deletion of l. 228. But if the line is an interpolation, why did the interpolator put it in after, instead of before, l. 227? This interpolator must have lived during the fourth century, when tragic writing, if less vigorous, was in all likelihood more polished than during the first half of the fifth. Would not an interpolator have put it in the obvious place, before l. 227? Since the sixteenth century it has been customary to place it there; of course the possibility that the line is an interpolation and has also been misplaced by accident cannot be ruled out. But does its odd position, or indeed any intrinsic factor, offer strong grounds for its deletion?

We cannot leave the problem of ll. 227–8 till we have considered that of ll. 229–30. Conington and Tucker have both kept ll. 227–30 in the order of the manuscript, placing a comma at the end of l. 228. Tucker translates, 'So was it when thou wert questing in my tracks, that marked thy very brother, tallying in build with thee.' This involves supposing that *τῶι cῶι κάραι* is a grandiloquent periphrasis for *coι*. That is not impossible: cf. *Agam.* 905 *φίλον κάρα*, Soph. *Ant.* 1, etc. Nor is the placing of *ἀδελφοῦ* in apposition to the genitive implicit in *ἐμοῖc* in any way irregular. This interpretation is to my mind more plausible than the combination of transposition (after Bothe) and emendation (*cύμμετρον* for *cυμμέτρου* after Pauw) that has found favour with Hermann, Weil, Wecklein, Blass, Murray, and others.[3] But the chief objection to keeping ll. 228–9 as Tucker does is one I have already mentioned; the extreme difficulty of having the participial clause containing *ἰχνοcκοποῦcα* follow the main clause. This difficulty would vanish if we could suppose that a line were missing after l. 228, a line that contained the main verb of the clause linked to the preceding clause by means of the *τ'* in l. 228. By supposing this lacuna, we could account for both the problems which Murray and other editors deal with by a double transposition. Of course this solution is not certain; only it seems to me the most economical. In arguing for it I have strayed for a moment beyond my immediate purpose, which is to inquire whether the difficulties of the passage strengthen the case for deleting l. 228, with Schütz, or ll. 228 and 229, with Fraenkel. How does Fraenkel think l. 229 got in? 'After interpolating a line about the ichnoscopia', he writes (p. 819), 'he had to work his way back to 230

[3] Professor Page argues strongly for the Bothe–Pauw solution. '*In this context*', he argues, '*τῶι cῶι κάραι* equivalent to *coι* would be insufferably unstylish; and the relation of *ἀδελφοῦ* to a genitive implicit in *ἐμοῖc* seems to me grotesquely harsh here.' To illustrate the idiom supposed in the latter case I would quote Soph. *OC* 344 *τἀμὰ δυcτήνου κακά* and *Phil.* 1126 *τὰν ἐμὰν μελέου τροφάν*.

and the theme of the lock. He managed to do that, *invitissima Minerva*, by adding a perfectly superfluous explanation to τομῆι.' It is clear from Fraenkel's translation[4] that he thinks the genitive ἀδελφοῦ was meant to depend on τομῆι. This botcher of his was an exceedingly ignorant and stupid fellow. First he ‖ puts in l. 227 not after, as his purpose would seem to demand, but before l. 228; then he has the curious and wholly unnecessary notion that he must 'work back to' l. 230 by placing before it a whole line, of extreme clumsiness, designed to show from whose hair the cut had been made, as though that were in any doubt. Why should such an ignoramus be entrusted with the production of a tragedy, especially during the fourth century, an age of perhaps superficial but undeniably polished poetry? Compared to this man, the writer who finished off the *Iphigeneia in Aulis* was a genius.

Among the difficulties we have so far examined, several are real enough; but there is none for which deletion would have seemed a likely remedy, had not the footprint scene in general seemed to some critics clumsy and unnatural. Let us now consider this general argument in favour of interpolation as it is stated by Fraenkel.

> Throughout the scene, [he writes] . . . the lock is the dominating factor. Not for a moment is our attention diverted from it. Electra's fervent desire to communicate her discovery of the lock to the women of the Chorus and let them participate in her profound emotion sets the dialogue afoot. After the end of the stichomythia, the eyes of the audience as well as their thoughts are entirely concentrated on the little ringlet of hair which Electra holds in her uplifted hand. . . . After what is perhaps the most beautiful of the many aposiopeseis in Aeschylus (194), Electra's excitement reaches its culmination in the outburst φεῦ· αἴθ' εἶχε φωνήν, κτλ. At this point the actor's movements and gestures, too, must have come to a climax, when Electra, holding the dear treasure in front of her eyes, gazes at it passionately as if it were the face of her brother, until, worn out by her longing and the alternating pangs of hope and despair, she at last drops it in resignation and addresses herself to the gods, who alone know the truth. No sooner is her prayer concluded than the bringer of its fulfilment, Orestes, steps forward. His first words εὔχου τὰ λοιπά, κτλ. are a perfect echo of the beginning of her prayer (210–14) and with the following τυγχάνειν καλῶς he takes up her phrase εἰ δὲ χρὴ τυχεῖν cωτηρίας. The complete unity of purpose in this admirable scene and its concentration on a single deeply moving theme affords as strong an argument against the genuineness of 205–11 as the difficulties of detail which have caused so much trouble to the critics.

In this eloquent passage we may distinguish two arguments of substance; first, that the opening words of Orestes must have followed

[4] 'Put the lock of hair to the place from which it was cut (on the head) of your own brother, who corresponds to your head, and consider it.'

immediately upon Electra's prayer in ll. 201–4; second, that the episode of the footprints adds nothing to that of the lock, serves only to diminish its effect, and is in fact such a poor piece of work that it cannot have been written by Aeschylus.

Let us take first the former of these arguments. It may well be true that the opening words of Orestes would have more effect if they followed immediately on the prayer which they take up and which they echo. Yet can we safely insist that they *must* have followed immediately upon it? Even allowing for the pause that will have elapsed between the end of the prayer at l. 204 and the discovery of the footprints, the delivery of ll. 205–11 will not have taken long; perhaps one minute. It is hardly reasonable to deny that even if the footprint episode takes place after the prayer, the words of the prayer will still have been fresh in the minds of the audience when Orestes spoke ll. 212 f. I do not deny the ‖ possibility that the prayer *may* have immediately preceded Orestes' words; but such difficulty as there is could be met by either of the transpositions suggested by Butler and Weil without supposing an interpolation. Butler, it will be recalled, placed ll. 201–4 after l. 210; Weil placed them after l. 211. Blass pointed out that l. 201 did not follow well upon l. 211, and inclined to Butler's view. But he did not adopt it in his text—in my opinion, wisely.

We come now to Fraenkel's contention that the footprint episode is so unworthy of Aeschylus that it cannot possibly be from his pen. Most modern readers, if they are honest with themselves, are likely to feel with Fraenkel that from the artistic point of view the footprint episode would be better out. Firstly, they may well agree with Schütz and Butler, and with the author of the footprint scene in Euripides' play, that this method of recognition is in itself absurd. They may find ridiculous the notion that a sister could recognize her brother by the resemblance of his footprints to her own. Yet Fraenkel does not use this argument, and was wise not to do so. It is certainly not out of the question that in countries where people go barefoot and the ground easily preserves their prints, family resemblances of this kind can be observed; A. P. Tregenza in *Greece and Rome*, NS 2 (1955), 59 ff., gave definite proof that this is so among the Bedouin. But it is more to the point in dealing with this argument to observe that Aeschylus was not writing naturalistic drama. There is every reason to suppose that the Athenian audience of the mid-fifth century will have been tolerant of such improbabilities, particularly if they already formed part of an established traditional story, as this, in all likelihood, may have done by this date.

The modern reader will feel yet stronger sympathy with Fraenkel's main argument that the episode destroys the beauty and completeness of the scene. But before making up our minds on this point, we should bear in mind one factor that is curiously omitted from the discussion by Fraenkel, although Wilamowitz in his commentary of 1896 (pp. 170 ff.) made a particular point of stressing it. To make this factor clear, we must again briefly consider the scene from the beginning. Electra sees the lock: it looks like her own, it cannot be from Clytemnestra or Aegisthus, so it must, she concludes, belong to Orestes. How could Orestes have dared to come to Argos, the Chorus ask; he must, Electra concludes, have sent it from his place of exile. In the long speech that follows this dialogue (183 f.), Electra never expresses the hope that Orestes in person may be at hand; she wishes that either she knew the lock were not his, or else that it were his and could sympathize with her in her grief. After saying this, she utters her prayer (201–4); then—if we can believe the manuscript—she notices the footprints.

The point which Wilamowitz makes so strongly is that a lock of hair can be conveyed from abroad by a messenger; but not a footprint. Electra sees the lock, and guesses that it is her brother's; but almost at once remembers that he may have sent it. Then she sees the prints, and guesses that one of the two sets must be her brother's. The effect of that is to cause in her an extreme emotional agitation: *πάρεϲτι δ' ὠδὶϲ καὶ φρενῶν καταφθορά*. At that point Orestes finally emerges from his hiding-place. The sight of the lock had induced in Electra intense excitement and, later, prolonged reflection; the sight of the footprints, even more intense excitement than that of the lock. She must at once make the obvious inference; yet the poet, with the finest sense of psychological truth, represents her as slow to take in and to acknowledge the ‖ obvious fact that the man she sees can only be her brother. With a kind of humour Orestes himself points out the paradox. Electra sees him face to face, and will not recognize him; yet when she saw the lock she became excited and seemed to see him before her eyes; and when she saw the footprints the same thing happened. Is not this scene a trifle more effective if we recognize, with Wilamowitz, that the significance of the prints lies in the fact that they take us a stage farther in certainty than the lock has done?

What does Fraenkel think was the interpolator's motive? 'Only a general guess', he writes, 'seems possible.' But it appears, he goes on, from the sixteenth chapter of Aristotle's *Poetics* that during the fourth century the tragedians were specially interested in devising ingenious recognition scenes, and that they tried to outdo each other in inventing fresh patterns and in modifying the old ones. 'At the time', says Fraenkel, 'when so much interest was taken in the refinement of a recognition scene, some ambitious producer of the *Choephori* thought that the manner in which Aeschylus had handled the subject was really too poor. He thought he could make the speech of Electra more interesting if he brought in the footprints; and he was also glad to work in a mention of Pylades.'

The chapter of Aristotle which Fraenkel alludes to certainly offers a detailed classification of various types of recognition scene; a few examples come from fourth-century writers, but the greater part are taken from the great tragedians of the fifth century. The kind of recognition which Aristotle most approves, ἀναγνώριcιc ἡ δι' εἰκότων, is exemplified from the *Iphigeneia in Tauris* of Euripides and from the *Oedipus Tyrannus* of Sophocles. The reader of this chapter can scarcely fail to be surprised that the 'ambitious producer' of the *Choephori*, working in the fourth century, should have introduced an anagnorisis of such a primitive and unsophisticated character. We know very little about the tragedy of this period. Doubtless it was inferior to that of the fifth century; but it seems improbable that its inferiority was due to a lack of sophistication, or to crudity and over-simplicity of technique. Rather we might expect it to show an excessive facility and elaboration, contrasting unfavourably, to our taste, with the noble simplicity of the great age. Even on Fraenkel's view about its origin, the footprint scene of the *Electra* shows that before the end of the fourth century the footprint scene of the *Choephori* could be ridiculed. Are its *naïvetés* likelier to be due to an actor or producer of a sophisticated but degenerate age, or to a poet of a more primitive, though incomparably greater, period in the history of drama?

In its essence the argument of Schütz and Fraenkel boils down to this: Aeschylus was a great poet: great poets do not write bad poetry: the footprint episode is bad poetry: therefore, it was not written by Aeschylus. A reader who appreciates, as Wilamowitz did, what the footprint episode adds to the episode of the lock will not readily agree that the footprint episode is bad poetry. But here, as always, it is necessary to bear in mind, that though Aeschylus was a great poet, he lived at an early period in Greek literary history, and he was by no means a polished or a sophisticated writer. Remember the scene in which the Chorus of the *Agamemnon* solemnly debate what they are to do, while behind the stage door the murders are taking place. Strangely enough, no one has yet suggested that it is an interpolation.

If we could assume that the recognition scene in Euripides' *Electra* was free from interpolation, Fraenkel's case would of course collapse; for the author ‖ of that scene certainly knew the corresponding scene in the *Choephori* in the form in which we have it. But Fraenkel has revived the view, first advanced in 1877 by the distinguished expert on Pompeii, August Mau, that ll. 518–44 of the *Electra* are an interpolation.

That recognition scene is foreshadowed during the scene following the parodos (215 f.), the scene in which Orestes and Pylades accost Electra and without revealing their identity inform themselves about her situation and obtain a promise of hospitality. At one point (282), Orestes exclaims to Electra, 'If only Orestes were here and listening to this!' Electra answers that she would not know him if he were; the only person who could recognize him is the aged slave who was once Agamemnon's paidagogos. When the old man enters the stage, after the first stasimon (487), the recognition scene can begin. On his way to Electra's remote cottage with provisions to help her entertain her guests, he has stopped to pour a libation at the tomb of Agamemnon. To his surprise, he has found upon the tomb a sheep newly sacrificed, together with an offering of hair which cannot have been left by anyone in Argos (508–17).

At this point begins the section of the text cut out by Mau and Fraenkel (518–44). First of all, the old man suggests that Orestes may have visited the tomb in secret and have left the offerings. 'Put the lock to your own hair', he says to Electra, in language that unmistakably recalls *Cho.* 230, 'and see if the colour is the same.' This draws from Electra a severe rebuke: 'Do you think a brave man like my brother would come here in secret, for fear of Aegisthus?' The colour of the hair, says she, proves nothing; many people have hair just like each others' without being related. Then the old man suggests that Electra should approach the footprints at the tomb and see if they are like her own. 'The ground is rocky', she replies, 'so how could there be any footprints? and in any case a man's foot is bigger than a woman's, even if she is his sister.' The old man suggests, in a speech which is unfortunately incomplete, that if Orestes has returned he may have brought with him that specimen of Electra's weaving (539 *κερκίδοϲ . . . ἐξύφαϲμα ϲῆϲ*: cf. *Cho.* 231 *ὕφαϲμα . . . ϲῆϲ ἔργον χερόϲ*) which he was wearing when she saved him long before. 'Do you not realize', exclaims Electra, 'that I was then too young to weave garments? and even if I had not been, Orestes would not be wearing the same garment now, unless it had grown as he grew.' Here ends the section deleted by Mau and Fraenkel; the incomplete sentence at ll. 545–6 is regarded by them as continuing the old man's speech at ll. 508–17. We must go through the deleted section and review their arguments.

Firstly, Electra's indignant repudiation (524 f.) of the old man's suggestion that Orestes may have come to Argos in secret is regarded by Mau and Fraenkel as an argument against authenticity. In the play, this is exactly what Orestes does do; and Electra, they think, would never have suggested that it was cowardly. But would she not? The two older tragedians attached no discredit to Orestes' use of guile, which they represent as having been enjoined upon him by Apollo's oracle. Suppose the moral attitude of Euripides in this play were the same as theirs, one must acknowledge that though in the event Electra would readily understand the reasons for the use of guile, she might at first be surprised to learn of it. Her initial surprise would scarcely furnish a strong argument against the authenticity of the passage. There are, however, grounds for holding that the moral attitude towards the matricide in this play is different from that of the older tragedians. The Dioscuri in their speech from || the machine express horror at the deed, and even criticize Apollo for having commanded it. It is therefore likely enough that Euripides is adopting a disapproving attitude to the use of guile. If this is the case, we need not be surprised that Electra indignantly rejects the notion that her brother would have made a secret entry into Argos. It may be that the poet means us to reflect that such a method is in fact scarcely in keeping with the heroic code of ethics.

The main argument of Mau and Fraenkel against the rest of the scene is directed against its irrelevance to the action of the play. They allow that the lock and footprints might conceivably have been discussed at this point, for the lock has been found at the tomb, and footprints might conceivably be found there; but they insist that the following question about the *ἐξύφασμα* is in the context wholly pointless. I see little force in this argument. It is common ground that whoever wrote the scene must have written it in order to ridicule the corresponding scene in Aeschylus; the whole content makes it clear, and unmistakable echoes of Aeschylus' language reinforce the proof. Once this is allowed, and it is beyond dispute, it becomes senseless to complain that it does not expedite the action of the play. It becomes senseless to complain that at so critical a moment so pointless a discussion would not be allowed to interrupt the action; for we know that the discussion is inserted principally to ridicule the earlier poet's story. Nor is the episode worked in as unskilfully as Mau and Fraenkel suggest. As the text stands, Electra's rejection of the traditional tokens, which the audience will have been expecting her to accept, prepares the way for an unexpected climax in the old man's final recognition of Orestes by means of a scar and for Electra's acceptance of the proof. Nor have the excisors been able to point to any serious inconsistency that the scene involves. When in ll. 532 f. the old man challenges Electra to go to the tomb and compare the footprints with her own, they complain that this is impossible, as the tomb of Agamemnon is at Argos, and Electra's cottage is far away. As a matter of fact we are not told in this play that Agamemnon's tomb is at Argos. But it is hardly reasonable to suppose that the old man's suggestion of a visit to the tomb was meant seriously, and if it was, it is unlikely that the question of distance would have troubled the poet or his audience. If we hold that Euripides was incapable of writing a scene that was not entirely relevant to the dramatic action, we may find some substance in these arguments. But can we really feel quite sure that he was incapable of holding up the action of his play for a few moments in order to ridicule the primitive technique of his distinguished predecessor?[5]

Are the criticisms of Aeschylus in this scene 'unfair and unreasonable', as many scholars have insisted? Take first the recognition by means of the lock. It is true that not all people who have hair like each others' are related, and that men's and women's hair are treated differently (528–9). But it is common for relations to have hair like each other's; and it is easy to infer that a lock of hair like one's own belongs to a relation, especially if one finds it in a place where no one but a relation has any right to leave it. The footprint, in spite of Tregenza's article, seems a decidedly more unrealistic token; and what are we to say of the ἐξύφασμα? Wilamowitz (op. cit. 170) argued that it would be taken for granted, that Orestes wore clothes woven by the women of his family, and that the unlikelihood of Electra sending parcels of clothes to him in Phocis would not have troubled the original audience. But the important point with ‖ regard to the ἐξύφασμα, as also with regard to the other tokens, is surely that they served the dramatist's purpose. By the canons of the late fifth century they are certainly unrealistic; but these canons had changed since Aeschylus' time. An archaic drama like the *Choephori* demands that the audience suspend to a great extent its rationalizing tendencies. Modern drama is for the most part far more realistic than even Euripides, let alone Aeschylus; but even a modern audience finds itself unconsciously accepting the dramatist's convention in such matters.[6]

It is, in any case, not certain that Euripides' criticism of Aeschylus is quite as harsh as has been assumed. Critics seem to have forgotten that despite Electra's pungent criticism the old man turns out in the end to have been right. Ridiculous as his reasoning may have been, it leads him to a correct conclusion. When confronted a few minutes later with Orestes, the old man does recognize him by the unmistakable token of a scar from an old accident, a motive borrowed from the recognition of Odysseus by Eurycleia in the *Odyssey*. It has often been observed that Euripides' Electra is not a wholly sympathetic character; ill-tempered,

[5] [Cf. *Gnomon*, 34 (1962), 742 = ch. 12, p. 198 on *Phoen.* 751–2 as a hit at the *Seven against Thebes*.]

[6] It must be borne in mind that in all probability the tokens formed a part of the traditional story before Aeschylus. One would dearly like to know whether they figured in the *Oresteia* of Stesichorus; if so, their occurrence in Aeschylus would hardly be surprising. Not that one has any right to think Aeschylus incapable of inventing them.

arrogant, and self-pitying, she seems to modern taste, and may well be meant to seem, less attractive than the farmer to whom she is married or the aged paidagogos. I see no good reason to doubt that in this scene Euripides amused himself at the expense of what seemed to him the primitive technique of his predecessor. But I suspect that the sharpness of his criticism is somewhat mitigated by its being put into the mouth of the play's unsympathetic heroine.

To later generations of readers, who regard both poets with reverence, the idea that one might poke fun at the other, and even do so at the expense of the sacred canon of dramatic unity, may seem deeply shocking. But even a small knowledge of literary history is enough to show that great writers are by no means always respectful towards their immediate predecessors. Even when they genuinely admire them, they may be stung into pointing out their defects by the invariable tendency of the literary public to prefer what is known and established to what is new and experimental. It is natural for modern scholars to regard Plautus and Ennius with respect; but their Augustan successors, and in particular Professor Fraenkel's favourite poet, Horace, found it less easy. Horace says he does not want every copy of the *Odyssey* of Livius Andronicus to be destroyed; he merely expresses his surprise that some people should pronounce it 'faultless', 'beautiful', and 'almost perfect'.[7] This was the view which practically the whole of later antiquity took of Aeschylus; according to Quintilian, corrected versions of his plays were performed after his death. It is dangerous to talk as if the caricatures of Aeschylus and Euripides in the *Frogs* were meant for exact portraits. But if we could know what Euripides really thought of Aeschylus, it might in some ways be closer to the attitude of the Aristophanic Euripides than to that of Professor Fraenkel.

To sum up, Mau and Fraenkel have given no substantial reason for doubting the genuineness of the passage they have cut out. We can scarcely hope to prove that this, or any other passage, *must* be genuine; the possibility of a ‖ convincing interpolation is always to be reckoned with. But in default of a proof that these lines are spurious, we have no reason to doubt that the footprint episode of the *Choephori* as we know it was known also to Euripides.

Let me now turn to another difficult passage in the *Choephori* where Fraenkel has diagnosed a grave interpolation. This is part of the great speech which Orestes delivers over the bodies of Clytemnestra and Aegisthus, while he is displaying to the Chorus and the assembled people the bloodstained garment thrown over Agamemnon in his bath before his murder (973 f.). At l. 980 Orestes first draws attention to the

[7] *Epist.* 2. 1. 69 f.; see the excellent treatment by Fraenkel, *Horace*, 387 f.

garment; as far as l. 990, the subject of his speech is it alone. But at ll. 991–6, if the manuscript can be trusted, he changes to the subject of his mother. Then at l. 997 we are switched back suddenly to the subject of the garment. 'What can I call it, euphemize as I may? A trap for a wild beast, or the curtain of a bath, draped over the dead man's feet?', and so on. How can the allusive pronoun *νιν* in l. 997 follow upon the passage about Clytemnestra and yet not refer to her? It is no use arguing that Orestes somehow identifies his mother with the garment, or that since the garment has been prominently displayed and has been the main subject of Orestes' speech, the word *νιν* could easily be understood to refer to it; both these suggestions have been conclusively refuted by Fraenkel (pp. 810–12). Others have tried to solve the difficulty by transposition. Scholefield and others transposed ll. 997–1004 to follow l. 982. The conclusive objection to this was pointed out by Weil: the force of Orestes' outburst about the garment is much diminished if it is delivered *before* he has given the order for it to be spread out and displayed upon the stage. Weil himself transposed ll. 997–1004 into the next speech of Orestes, placing them after l. 1013, where it is easy to see that they are inappropriate.

Is the remedy of excision then to be preferred? It was first applied by W. Dindorf, who, in Fraenkel's words, 'applied his formidable scalpel to the speech . . . cut out twenty lines (987–1006), and killed the patient straight away'. Let us inquire how the patient stands up to the operation which Fraenkel himself, following Wecklein, has advocated. Fraenkel would amputate ll. 991–6 and also ll. 1005–6; and he has supported his diagnosis by discovering all kinds of unhealthiness in the lines in question.

'To begin with 993,' says Fraenkel, 'who or what is the subject of *φαίνει*?' Some have thought it was *τέκνον* or *τέκνων βάρος*: that seems just possible, but not likely. It can hardly be *κακόν*, as Wilamowitz in his commentary maintained; then *κακόν* would have had to have the definite article. Fraenkel does not mention another view, which was first suggested by Bothe and has been adopted by Sidgwick and by the latest commentator, Groeneboom. This is that the words mean 'as she shows', i.e. by her present condition, that is to say her fatal wounds. This seems to me the likeliest explanation, though, in default of another instance of *φαίνειν* being used in exactly the same way, it can hardly be considered certain.

'In 994 f.', writes Fraenkel, 'the construction as it stands is utterly baffling.' Most editors have followed Robortello in reading *θιγοῦϲαν* as *θιγοῦϲ' ἂν* and have supplied *δοκεῖ* from the previous sentence. In that case, Fraenkel argues, we should expect the participle *δεδηγμένον* to be negatived by *μή* and not by *οὐ*. When Fraenkel wrote this note, he cannot have seen the article 'On the negativing of Greek participles where the leading verbs are of a type to require *μή*' (*CQ* 42 (1948), 35 f.). In that article, A. C. Moorhouse makes ‖ out what seems to me a clear case for regarding the general opinion that in such a sentence the participle automatically requires *μή* as considerably over-simplified; he shows that it is the sense of each participial phrase which is the decisive factor in determining the choice of negative. The general and indefinite character of this particular sentence has already been made clear by *θιγοῦϲ' ἂν* and it is quite natural for *οὐ δεδηγμένον*, which is equivalent to *καίπερ οὐ δεδηγμένον*, to contain *οὐ* and not *μή*. This negative cannot safely be taken as a sign of late authorship. As for the content of this passage, Fraenkel 'gladly subscribes' to the verdict of Sidgwick, who wrote of 'these violent and almost grotesque words'. Certainly the words are violent and grotesque; they are in keeping with much of the language of the man who wrote *Agam.* 1232 ff.:

> *τί νιν καλοῦϲα δυϲφιλὲϲ δάκοϲ*
> *τύχοιμ' ἄν; ἀμφίϲβαιναν ἢ Ϲκύλλαν τινὰ*
> *οἰκοῦϲαν ἐν πέτραιϲι, ναυτίλων βλάβην,*
> *θύουϲαν Ἅιδου μητέρ' ἄϲπονδόν τ' Ἄρη*
> *φίλοιϲ πνέουϲαν;*

Fraenkel's attempt to show that the language of ll. 991–6 shows its alien origin is not convincing; *φαίνει* is difficult enough, but hardly suggests interpolation. In ll. 1005–6 he finds no such internal evidence of spuriousness; so we may go on to consider the grounds for his claim that 'it is the general turn of thought . . . that, more than any detail, ought to shake our belief in the genuineness of these lines'. Following Dindorf, he contends that it is wholly out of character for Orestes to devote any part of this speech to abuse of his mother. All his references to her, he argues, show the same noble restraint. But 'the noble simplicity of Aeschylus could not satisfy the coarser taste of a later generation'; and 'a clever producer, sailing in the wake of the *aura popularis*, and perhaps working in the interest of an actor who wanted a piece of rant to display the power of his own voice, thought he could improve on Aeschylus by putting in a few lines of his own'.

Let us examine what the Aeschylean Orestes says about his mother. In his long prayer to Zeus after the recognition scene, he speaks of his father as an eagle who has perished in the toils of a dread viper (247 f.). This description doubtless includes Aegisthus as well as Clytemnestra; but we must note it as by no means complimentary. The moment after Clytemnestra has caught sight of the dead body of Aegisthus, Orestes taunts her with her adulterous relationship with the dead man (894–5). *φιλεῖϲ τὸν ἄνδρα; τοίγαρ ἐν ταὐτῶι τάφωι / κείϲηι, θανόντα δ' οὔ τι μὴ* προδῶιϲ ποτε. That provokes Clytemnestra to appeal to Orestes as a son, and to display the breast from which she has given him suck. For one moment, for the only time throughout the play, Orestes hesitates; then Pylades reminds him of Apollo's command. At once Orestes resumes his earlier severity (904 ff.):

> ἕπου, πρὸϲ αὐτὸν τόνδε ϲε ϲφάξαι θέλω·
> καὶ ζῶντα γάρ νιν κρείϲϲον' ἡγήϲω πατρόϲ.
> τούτωι θανοῦϲα ϲυγκάθευδ', ἐπεὶ φιλεῖϲ
> τὸν ἄνδρα τοῦτον, ὃν δὲ χρῆν φιλεῖν ϲτυγεῖϲ.

Later in that dialogue, Orestes accuses his mother of having sold him as if he were a slave (915). She challenges him to name the price that she received,|| and he replies that he is ashamed openly to taunt her with that price. This reluctance is a mere form of rhetoric; by refusing to name it, he taunts his mother with her adultery more effectively than he could else have done. Fraenkel quotes as an example of Orestes' 'restraint' in regard to Clytemnestra his brief allusion to his father's murder in the trial scene of the *Eumenides* (458 ff.):

> ἔφθιθ' οὗτοϲ οὐ καλῶϲ, μολὼν
> εἰϲ οἶκον, ἀλλά νιν κελαινόφρων ἐμὴ
> μήτηρ κατέκτα, ποικίλοιϲ ἀγρεύμαϲιν
> κρύψαϲ' ἃ λουτρῶν ἐξεμαρτύρει φόνον.

Those words may show restraint, but the epithet κελαινόφρων is not a complimentary one. Orestes' tone towards his mother in all these passages seems to me quite fierce enough to supply a warning against the assumption that he cannot have spoken ll. 991–6. I am not at liberty to draw attention to l. 1028, where Orestes calls his mother πατροκτόνον μίαϲμα καὶ θεῶν ϲτύγοϲ: Fraenkel has thoughtfully remembered to cut it out.

We come next to Fraenkel's objections against ll. 1005–6. He thinks it 'intolerable' that the formal ending τοιοῦτον, κτλ., should be duplicated by the addition of τοιάδε at l. 1005. 'Intolerable' is perhaps too strong a word; but there is some substance in this point, and the suggestion I shall make presently will have to meet it. The thought here expressed Fraenkel finds 'not only vulgar, but revolting': 'fancy this Orestes at this point considering possibilities of marriage!' To describe Orestes as 'considering possibilities of marriage' here is not accurate. For all Greeks it was axiomatic that any man wishes to marry and to leave a son. So when Orestes wishes to condemn in the strongest possible terms his mother, his father's wife, it is natural for him to say, 'Rather than have a wife like that, I had sooner die childless.' Where is the vulgarity?

For these reasons I find Fraenkel's diagnosis of interpolation unconvincing. But how then are we to deal with the problem from which the discussion started, that of the acute difficulty presented by the reference of *νιν* in l. 997? I agree with Fraenkel that the text as it stands cannot be right; and I accept his refutation of every solution that has so far been propounded.

My own remedy would be to place ll. 997–1004 at the end of the section of Orestes' speech dealing with the garment; that is to say, after l. 990. We then get the following sequence of topics. First, Orestes orders the garment to be displayed, and describes it (980–90). What name, he asks, is he to give the horrid object? (997–1000). Such a thing might be in the possession of some murderous thief, some Sinis or Procrustes (1001–4). At this point the thought of the most appropriate possessor of such an instrument leads him back to that of the person who actually did possess it. Line 991 follows quite naturally after l. 1004: 'But as for her who devised such a hateful instrument against her husband.' 'What would be the proper name for *her*?', Orestes continues (994), just as he has asked what would be the proper name for the garment (1001–4).

The mention of the garment and the search for an apt description for it is now seen to be a prelude, and a most appropriate prelude, to the mention of its owner and the search for an apt description for her. The parallelism does not end here. After trying to find a name for the garment, Orestes goes on: 'Such an instrument might belong to a murderous thief.' After trying to find a name for Clytemnestra, he goes on: 'Such a wife may I never have; I had sooner || perish childless.' On this view, *τοιάδε* in l. 1005 will not duplicate *τοιοῦτον* in l. 1001 as Fraenkel complains it does. The transposition seems to me to have a fair degree of probability.[8]

[8] The reader may find it helpful to have the speech from l. 980 set out as it would appear with this transposition:

ἴδεcθε δ' αὖτε, τῶνδ' ἐπήκοοι κακῶν,
τὸ μηχάνημα, δεcμὸν ἀθλίωι πατρί,
πέδας τε χειροῖν καὶ ποδοῖν ξυνωρίδα,
ἐκτείνατ' αὐτὸ καὶ κύκλωι παραcταδὸν
cτέγαcτρον ἀνδρὸc δείξαθ', ὡc ἴδηι πατήρ—
οὐχ οὑμόc, ἀλλ' ὁ πάντ' ἐποπτεύων τάδε
Ἥλιοc, ἄναγνα μητρὸc ἔργα τῆc ἐμῆc,
ὡc ἂν παρῆι μοι μάρτυc ἐν δίκηι ποτέ,
ὡc τόνδ' ἐγὼ μετῆλθον ἐνδίκωc μόρον·
τὸν μητρόc. Αἰγίcθου γὰρ οὐ λέγω μόρον·
ἔχει γὰρ αἰcχυντῆροc, ὡc νόμοc, δίκην,
τί νιν προcείπω, κἂν τύχω μάλ' εὐcτομῶν; 997
ἄγρευμα θηρόc, ἢ νεκροῦ ποδένδυτον
δροίτηc καταcκήνωμα; δίκτυον μὲν οὖν,
ἄρκυν τ' ἂν εἴποιc καὶ ποδιcτῆραc πέπλουc.
τοιοῦτον ἂν κτήcαιτο φιλήτηc ἀνήρ, 1001
ξένων ἀπαιόλημα κἀργυροcτερῆ

ADDENDUM

When this article was already in proof, Mr P. W. Martin of Brasenose College told me that the transposition of *Cho.* 997–1004 which I have recommended had already been proposed by R. Proctor in his beautifully printed edition of the *Oresteia* (London, 1904). This book appeared posthumously, and in an edition limited to 225 copies; see the account of it in *CR* 18 (1904), 335. 'The text', it is there stated, 'is not a mere reprint of a vulgate or of a favoured edition, but is the result of an independent exertion of intelligence.'

βίον νομίζων, τῶιδέ τ' ἂν δολώματι
πολλούς ἀναιρῶν πολλὰ θερμαίνοι φρένα. 1004
   ἥτις δ' ἐπ' ἀνδρὶ τοῦτ' ἐμήσατο στύγος, 991
ἐξ οὗ τέκνων ἤνεγχ' ὑπὸ ζώνην βάρος,
φίλον τέως, νῦν δ' ἐχθρόν, ὡς φαίνει, κακόν—
τί σοι δοκεῖ; μύραινά γ' εἴτ' ἔχιδν' ἔφυ
σήπειν θιγοῦσ' ἂν ἄλλον οὐ δεδηγμένον 995
τόλμης ἕκατι κἀκδίκου φρονήματος; 996
τοίαδ' ἐμοὶ ξύνοικος ἐν δόμοισι μὴ 1005
γένοιτ'· ὀλοίμην πρόσθεν ἐκ θεῶν ἄπαις. 1006

I need hardly say that the disagreement with Professor Fraenkel expressed in this article is not a token of disrespect. Students of Aeschylus owe him much; those who have had, as I have had, the good fortune to be his pupils owe him even more. He will readily forgive me for defending an opinion of his teacher, Wilamowitz, against his own.

[Cf. W. Burkert, *CQ*, NS 13 (1963), 177; G. W. Bond, *Hermathena*, 118 (1974), 1 f.; Oliver Taplin, *The Stagecraft of Aeschylus* (1977), 337 f.; A. F. Garvie, *Aeschylus, Choephori* (1986), ad loc.]

27(83)

Aeschylus, *Choephori* 32–41

τορὸϲ γὰρ ὀρθόθριξ δόμων
ὀνειρόμαντιϲ ἐξ ὕπνου κότον πνέων
ἀωρόνυκτον ἀμβόα-
μα μυχόθεν ἔλακε περὶ φόβωι,
γυναικείοιϲιν ἐν
δώμαϲιν βαρὺϲ πίτνων·
κριταί τε τῶνδ᾽ ὀνειράτων
θεόθεν ἔλακον ὑπέγγυοι
μέμφεϲθαι τοὺϲ γᾶϲ νέρθεν περιθύμωϲ
τοῖϲ κτανοῦϲί τ᾽ ἐγκοτεῖν.

32 Φοῖβοϲ post ὀρθόθριξ del. Weil

THIS is Page's text (1972); Weil's deletion of Φοῖβοϲ in l. 32 was accepted by him, as by Blass (1906), Wilamowitz (1914), Mazon (1925), Groeneboom (1949), and Murray (2nd edn. 1951). Φοῖβοϲ, though kept by Wecklein (1885 and again, with Zomarides, 1910) and by Tucker (1901), will hardly do. Turnebus (1552) emended Φοῖβοϲ to φόβοϲ. Heath, accepting this, transposed and read ὀρθόθριξ φόβοϲ: he preserved strophic responsion by deleting δόμων. Schütz, who accepted his conjecture, did this by accepting Stanley's παρηΐϲ φοινίοιϲ at l. 24; Thomson (2nd edn. 1966), who also accepted Heath's conjecture, in l. 24 read παρῆιϲι (Hermann) φοινίαιϲ ἀμυγμὸϲ (Conington); A. F. Garvie (1986) follows Schütz, rightly remarking that δόμων does not look like an intrusion.

ὀρθόθριξ φόβοϲ in itself is excellent; Garvie points out that it is supported by Soph. fr. 875 ὀρθόκερωϲ φρίκη, and cf. P. Oxy. XXXIX 2881(*a*) = *TrGF* ii, adesp. fr. 659(*a*) 1–2 καὶ τότ᾽ ἐξεγείρομαι . . . ὀρθίαϲ φρίκηϲ ὕπο. Clytemnestra's fear might be called 'the dream-prophet of the house', and even be said to have uttered the cry which she has uttered, as we know from ll. 523–5 and 535. But φόβοϲ as subject of the sentence does not cohere well with περὶ φόβωι, which as Garvie says is unobjectionable; it would be absurd to emend it to παράφορον with Herwerden, and the awkward repetition is hardly protected by Soph. *Trach.* 963–7, where Meineke's ϲτάϲιϲ may well be right. Again, does

Bulletin of the American Papyrological Society, 22 (1985), 195–6.

Clytemnestra's fear breathe anger from (or after) sleep? Surely not; in Garvie's ‖ words, 'the interpretation of the dream shows that the anger is that of the powers below'. The subject of the sentence must surely be someone or something that feels anger.

This suggests that the 'dream-prophet' is the alastor, the daimon who destroys the house of Atreus. But the daimon does not utter the cry; Clytemnestra does that. For one thing, the cry is uttered περὶ φόβωι, and the daimon, though he is angry, is certainly not frightened.

The daimon does *cause* Clytemnestra to utter the cry, and if in l. 35 we alter one letter, changing a kappa to the sigma with which kappa is so easily confused, we can make him do this by writing ἔλαϲε. Fraenkel on *Agam.* 701 shows that 'ἐλαύνω means not only "drive in front of oneself" but also "spur on, set in motion"'. It seems to be used especially of the action of deities in causing events; Fraenkel mentions the instances of θεήλατοϲ at Soph. *OT* 255 and 992 (add *Ant.* 278), and quotes Dem. 9. 54 τοῦτο φοβεῖϲθαι, μή τι δαιμόνιον τὰ πράγματ' ἐλαύνηι.

The corruption may have been facilitated by the presence of ἔλακον in l. 39; for the confusion between the two words, cf. l. 939, where Pauw's alteration of ἔλακε to ἔλαϲε is generally accepted (a scholion has ἤλαϲεν). The first to conjecture ἔλαϲε was Theodor Heyse, *Die Orestie des Aeschylus* (Halle, 1884); he also conjectured περίφοβον, unnecessarily.

28(72)

Notes on P. Köln III 125 (Aeschylus, *Psychagogoi*?) [= fr. 273a Radt]

I

Ἄγε νῦν, ὦ ξεῖν', ἐπὶ ποιοφύτων
ἵςτω cηκῶν φοβερâ⟨c⟩ λίμνας
ἀπό τ' αὐχένιον λαιμὸν ἀμήcαc
τοῦδε cφαγίου πότον ἀψύχοιc
αἷμα μεθίει
δονάκων εἰc βένθοc εμ̣. υρον.

1 *νυν* ed. pr. 2 suppl. ed. pr. 3 *ἀπό τ'* Gronewald: *υποθ Π* α⟦δ⟧*μηcαc Π* 6 ⟨*ἀ*⟩*μαυρόν* ed. pr.: *ἐρεμνόν*? *ἀμαυρῶν*? *ἐρεμνῶν*?

So begins the magnificent fragment, identified with great probability by its first editor as from the *Psychagogoi* of Aeschylus, published in Kölner Papyri III (1978), pp. 11 ff. Bruno Snell, *ZPE* 40 (1980), 32 finds it strange that the dead in the underworld, who are often called *ψυχαί*, should here be called *ἄψυχοι*. But Odysseus and his men are there, in the words of Circe (*Od*. 10. 492 ff.)

ψυχῆι χρηcομένουc Θηβαίου Τειρεcίαο,
μάντιοc ἀλαοῦ, τοῦ τε φρένεc ἔμπεδοί εἰcι.
τῶι καὶ τεθνηῶτι νόον πόρε Περcεφόνεια
οἴωι πεπνῦcθαι· τοὶ δὲ cκιαὶ ἀΐccουcι.

Sophocles is echoing Homer when the Chorus of the *Electra* says of Amphiaraus (841) *καὶ νῦν ὑπὸ γαίαc . . . πάμψυχοc ἀνάccει*: that is clearly indicated by Kaibel, Jebb, and Kamerbeek, though not by J. H. Kells. On the meaning, see R. B. Onians, *The Origins of European Thought* (2nd edn. 1954), ch. 3 *passim*: 'the prophet Teiresias', he writes, 'retains his organs and intelligence. The prophet Amphiaraus, to whom he has been compared, passed beneath the ground in his body through divine intervention . . .'.

Since *ἀψύχοιc* is easily intelligible, there is no need to suggest that something is missing from the text in l. 4. Both Snell and Bärbel

Zeitschrift für Papyrologie und Epigraphik, 42 (1981), 21–2.

Kramer, the first editor, speak of a 'strophe' and 'antistrophe'; but do we usually find strophic responsion in marching anapaests? ||

II

Where are we to imagine the sacrifice as being located? Frau Kramer opts for Lake Avernus, which is where Wilamowitz (*Aischylos: Interpretationen* (1914), 246 n. 1), placed the nekyomanteion of the *Psychagogoi*; but there are other possibilities. Ar. *Ran.* 1266 quotes from that play the line

Ἑρμᾶν μὲν πρόγονοι τίομεν γένος οἱ περὶ λίμναν

(= Aeschylus fr. 273 R). This line is explained in a single manuscript, Cantabrigiensis 2626, of the early fourteenth century (see K. J. Dover, *Aristophanes, Clouds* (1968), pp. ci–cii; W. J. W. Koster, *Scholia in Aristophanem*, i. 3/2, *Schol. rec. in Nub.*, pp. v ff.) by the following scholion: *τοῦτο ἐκ τῶν Αἰςχύλου Ψυχαγωγῶν. τὸ δὲ Ἑρμᾶν μὲν τίομεν λέγουςιν οἱ Ἀρκάδες διὰ ταῦτα· ἐν τῆι Κυλλήνηι, ἥ ἐςτιν ὄρος Ἀρκαδίας, ἐτιμᾶτο ὁ Ἑρμῆς. διὰ γοῦν τὴν ἐξ ἀμνημονεύτων χρόνων τιμὴν ὡς πρόγονος τούτοις ἐδόκει. ἄιδουςι δὲ καί τινα ἱςτορίαν μυθώδη. λίμναν δὲ λέγει τὴν Cτυμφαλίδα, ἐν Ἀρκαδίαι γὰρ καὶ αὕτη.* Wilamowitz, loc. cit. dismisses this scholion as 'modern', and he may be right. But can we safely assume that the location of the scene at Lake Stymphalus is the idle conjecture of a Byzantine? Stymphalus as a location has the advantage over other places of being in a place where Hermes was the most important god; and the sentence *ἄιδουςι . . . μυθώδη* may indicate that a genealogy tracing the descent of the Arcadians from Hermes was known to the writer of the note (see L. R. Farnell, *Cults of the Greek States*, v (1909), pp. 3 f.).

From an early period, the nekyomanteion visited by Odysseus was identified with that at Ephyra, in Thesprotia.[1] I know of no special connection of Hermes with this region; but this surely constitutes a third possibility.[2]

[1] See for an account of the excavations S. Dakaris, *Das Taubenorakel von Dodona und das Totenorakel bei Ephyra* (*Antike Kunst*, Beiheft I, 1963); on Odysseus and the Thesprotians, see Ed. Schwartz, *Die Odyssee* (1924), 140 ff., 183 ff.; on Ephyra in general, see W. Burkert, *Griechische Religion der archaischen und klassischen Epoche* (1977), 186 [= *Greek Religion* (Oxford, 1985), 114–15]. I thank R. Kassel for his help with these notes.

[2] [There is a good deal of doubt about this; see C. G. Hardie, 'The Crater of Avernus', ap. R. G. Austin, *Aeneidos Liber Sextus* (1977), 279–86, esp. pp. 284 f., and J. S. Rusten, *ZPE* 45 (1982), 34 f.]

29(R110)

Review of OCT Aeschylus, ed. Page

Denys Page: *Aeschyli Septem Quae Supersunt Tragoediae.* Oxford: Clarendon Press, 1972. Pp. xxi + 335.

GILBERT MURRAY's Oxford text of 1937 (second edition, with help from Paul Maas, 1955) had long been due for supersession; and the present editor was unquestionably the best qualified person for the task. This edition is, and will long remain, the best and most serviceable modern Aeschylus.

The witnesses for the *Oresteia* and the *Suppliants* (including the minor ones) are fully utilized. To the sixteen manuscripts of the Byzantine triad collated by R. D. Dawe in his book *The Collation and Investigation of the Manuscripts of Aeschylus* (1964), Page adds eleven more. He himself has collated the two in Cambridge (Lc and Lh) personally and four others from photographs, and N. G. Wilson and the late D. C. C. Young have collated from photographs five others; for Page's explanation of his choice, he refers us to his article in *Serta Turyniana: Studies in Honor of Alexander Turyn* (1975). Page derives all the manuscripts of the triad from a single source, one full of corruptions and containing many gaps. Since it contained a number of scholia in the text itself, he thinks this cannot have been earlier than the ninth or tenth century. Like Dawe, he holds the tradition of the triad to be so full of contamination that a true reading may occur anywhere. The choice lies not between manuscripts but between variants; and it seems unlikely that many variants that stand a reasonable chance of being right are not to be discovered in the apparatus now provided.

The first signal difference between this edition and its predecessors that will strike the reader is that it greatly exceeds all of them in the number of its *cruces desperationis*, and also in the numbers of lacunae in the text indicated as certain or as possible. People who think an Oxford text should patch up a *fable convenue* whenever possible will moan over this austerity; but readers who are fully aware of the shortcomings of our evidence for the text of Aeschylus and of the uncertainty of many of the readings, especially in the two plays (*Supp.*, *Cho.*) dependent only

Classical Review, NS 26 (1976), 6–8.

on a single corrupt manuscript, will commend the editor for his honesty.

Considering that the law of diminishing returns has operated for so long in the matter of the conjectural emendation of this text, it is remarkable how many new conjectures of merit the new edition contains. I find particularly attractive the suggestions at *Pers.* 554, 978, *Supp.* 762, 817, *Cho.* 606, 620; ideas that merit careful consideration are those at *Pers.* 736, 769, *Sept.* 336, 343, 899, 903, *Supp.* 41, 51, 214, 325, *Agam.* 179, *Cho.* 406, *Eum.* 410, 845, *PV* 1031. At *Pers.* 1070 δυcβάϋκτοc may well be correct (see Dawe, op. cit. 82); but Brunck was the first to suggest it.

The lyrics are far better presented than in the old Oxford text: Murray's idiosyncratic indentation, so often recalling the Mouse's Tale in *Alice*, has been eliminated. There are many improvements in punctuation, though Page's avoid‖ance of commas often deprives the reader of guidance where he needs it. Reacting against Murray's excessive indulgence in the dash, Page eschews its employment altogether. On the whole this is for the best, but in a few places a dash would be a help.

The book contains the two lives and the catalogue of Aeschylean titles, but unlike Murray's edition omits the fragments of the plays coming from the same trilogies as those preserved. Page has taken great trouble over the attribution of conjectures, sometimes correcting even Dawe's useful appendix to Wecklein's second volume.

The difficulties of the text of Aeschylus are so great that any person who has given them serious thought will find in any new edition things with which he strongly disagrees. Most of the space available must be devoted to my own chief disagreements; but this should not obscure my admiration for a magnificent piece of work, which adds significantly to its author's tremendous contribution to the constitution of the texts of Greek poetry.

Pers. 13: the lacuna after this verse was first proposed by Mekler, who had arrived by conjecture at the reading Page now prints from nine manuscripts. 80: χρυcογόνου may be less well attested than χρυcονόμου, but it is an incomparably better reading; cf. *Soph. Ant.* 950 for the allusion to the birth of Perseus, regarded by the Greeks as the ancestor of the Persian kings. L. 114: def. Coxon, *CQ*, NS 8 (1958), 45 f.; cf. Nonn. *D.* 26. 56. L. 162: Lawson's violent conjecture hardly belongs in the text. L. 277: see Coxon, art. cit. 47 f. L. 321: the proper names can scarcely excuse the breach of Porson's Law. L. 583: τε for τὸ eliminates a normal use of the article with part of πᾶc. L. 604: Blomfield's τὰ θεῶν is preferable to Blaydes's remedy. L. 647: if one accepts ἀνήρ, why strain at Teuffel's οὖχθος? L. 860: why cannot a law be like a tower? See Coxon, art. cit. 50.

Sept. 73: Dawe's deletion of this line may be right, but has entailed the acceptance of Karsten's ἐφέςτιος at *Agam.* 851. L. 75: the emendation printed removes an instance of the infinitive for the imperative, a usage not uncommon in prayers. L. 140: see P. Maas, *Kl. Schr.* 37.

Supp. 46: why eliminate a normal use of εὐλόγως in an etymology? cf. ll. 258, 591, *Sept.* 508, fr. 6. 3. L. 81: what is wrong with the aorist στυγόντες? L. 168: I would keep οὐ and take the sentence as a question, with Friis Johansen. L. 386: Burges's δυςπαραθέλκτους deserves a mention. L. 405: μεγαίρω (Stadtmüller, and later A. Y. Campbell) is attractive. L. 491: Porson's εὑρεθέντα together with λαβεῖν seems to me to yield very strange Greek. L. 630: how can a comma directly follow καί? L. 636: I would keep ἐν ἄλλοις: for the 'sinister' sense of ἄλλος, cf. ἕτερος at *Agam.* 150 and Pind. *Pyth.* 3. 34. L. 926: why worry about the asyndeton? L. 971: is the lacuna here really necessary? L. 999: why change δὲ to cφε?

Agam. 48: 'cry for mighty war' makes, in poetry, perfectly good sense, and the conjecture μεγάλ' for μέγαν is unnecessary. L. 106: μολπᾶν ἀλκάν is not corrupt; the ἀλκά of the old men consists in song. L. 129: Aeschylus surely cannot have used πρόςθετα; see *RhM* 103 (1960), 76 f. [= ch. 23, p. 305]. L. 323: there is nothing wrong with φίλως: 'they stand apart in no friendly fashion'. L. 365: 'beyond the stars' is a strange sense, and I am attracted by Murray's device of reading ὕπερ and taking ἄςτρων βέλος as meaning ἀςτραπή. L. 423: Salzmann's conjecture hardly solves the difficulty. L. 474: the analogy of the construction used with verbs of knowing does not support ἀλοὺς ... βίον κατίδοιμι. By citing Q.S. 3. 456 βίοτον κατέδων ὀδύνῃςιν, I removed the last objection to the minimal change to κατέδοιμι (*CQ*, NS 12 (1962), 193 n. 1 [= ch. 21 n. 11]), which Page should at least have mentioned. L. 626: in deference to Dawe's objection (op. cit. 184), Page puts a crux against ‖ ἐμφανῶς. At *CR*, NS 16 (1966) [R54], 21, I pointed out that according to the *Nostoi* Menelaus quarrelled with his brother and set off alone: I should have quoted *Od.* 3. 130 f. For a similar allusion cf. *Agam.* 841–2. L. 1095: Page's removal of the mark of punctuation at the end of this line makes for a very awkward construction. Ll. 1145–6: Page's observation that the exclamatory accusative accepted by many editors in l. 1146 is open to objection is valuable. But why does he transpose the two nouns in these two lines? The nightingale, according to the Chorus, bemoans a life beset with evils all around. Then Cassandra calls out upon the μόρος, the fated end, of Procne; Procne was painlessly transformed into a bird (keep ἀγῶνα in the sense of 'ordeal'), but for Cassandra there waits a painful death [see ch. 25, p. 330]. Ll. 1271–2: the crux is not needed, provided we accept Hermann's simple change of μέτα to μέγα. L. 1423: why the comma? L. 1443: if any word in Aeschylus needs a crux, it is ἱςτοτριβής, as Fraenkel showed [but see ch. 25, p. 331]. L. 1605: Page in his commentary of 1957 regarded thirteen children of Thyestes as 'a ludicrous multitude in this context', and still daggers ἐπὶ δέκ'. But Aeschylus wrote a play about the pursuit of the fifty daughters of Danaus fleeing from their cousins and fifty sons of Aegyptus; no crux is needed. L. 1652: like Lobel and Fraenkel, Page cannot accept the asyndeton. See *RhM* 103 (1960), 79 [= ch. 23, p. 308]; in all the three places there discussed, Page prefers to emend, wrongly.

Cho. 23: προπομποῦς' is hardly a form to introduce by emendation. Ll. 98–9: Diggle's transposition seems to me possible, but not certain. Ll. 330–1: a perfectly good Aeschylean expression is violently eliminated. L. 390: another unnecessary change. L. 438: we do not need to insert ⟨ς'⟩. L. 740: how can ἔχειν be retained? L. 791: the word-order makes against the conjecture ἆραι, and θέλων here is equivalent to πρόφρων. L. 816: ἔπος λέγων is certainly puzzling, but Page's βλέπων is far from being the only possible solution. L. 834: this passage needs a crux. L. 929: Page's transfer of this line to Orestes is possible, but he should not say 'correxi'. L. 1015: the emendation is attractive, but does not deserve to be in the text. L. 1048: against Lobel's δμοιάς, see A. Sideras, *Aeschylus Homericus* (Hypomnemata 31, 1971), 139, n. 3.

Eum. 112: I would print a semicolon after κούφως. L. 132: Dawe's φόνου, like almost all the suggestions adopted from this critic, makes sense but is not necessary. L. 165: the apparently exact symmetry with l. 159 suggests that there should be a full stop after κάρα, and then ⟨τ'⟩ in l. 166 is not needed. L. 667: the lacuna posited by Dawe is not required. L. 892: Elmsley's ἕξειν is surely needed.

PV 117: E. *Hipp.* 1108 and Ar. *Lys.* 262 cast doubt on the need for Wilamowitz's transposition. L. 187: Page omits οἴω, and records that it is oddly presented in M: I would prefer to follow Bothe in keeping it but deleting Ζεύς· ἀλλ'. L. 266: at the end of this line, a semicolon would be better than a full stop. L. 328: I would place commas after ἀκριβῶς and before ὅτι. L. 354: the simplest remedy is Wellauer's deletion of ὅς. L. 425: Zakas's πρόςθ' ἕν' deserves a mention. L. 602: we need no question-mark and no accent on τινες. The sense is that the parents of those who suffer as Io suffers are unfortunate (cf. e.g. *Il.* 6. 127). L. 861: what is wrong with δαμέντων? L. 899: πόνων may be wrong, but ὕπο hardly belongs in the text. L. 902: ἔρωι is not attractive; ὄμμα means 'glance', and is an internal accusative. L. 959: the feeble conjecture τράχιστα does not belong even in the apparatus.

30(4)

Sophoclea

I. *TRACHINIAE* 94–102[1]

ὃν αἰόλα νὺξ ἐναριζομένα
τίκτει κατευνάζει τε φλογιζόμενον
Ἅλιον, Ἅλιον αἰτῶ
τοῦτο καρῦξαι, τὸν Ἀλκμή-
ναϲ πόθι μοι πόθι παῖϲ
ναίει ποτ', ὦ λαμπρᾶι ϲτεροπᾶι φλεγέθων,
ἢ ποντίαϲ αὐλῶναϲ, ἢ
διϲϲαῖϲιν ἀπείροιϲ κλιθείϲ,
εἴπ', ὦ κρατιϲτεύων κατ' ὄμμα.

All commentators so far as I know have believed that ll. 100–1 are simply a vague paraphrase for ἢ κατὰ γῆν ἢ κατὰ θάλαϲϲαν. Jebb's translation may be taken to represent the usual view: '. . . is he threading the straits of the sea, or hath he found an abode on either continent?' But this sense is not only poetically inept, but linguistically impossible.

Most editors (for instance Dindorf, Campbell, and Radermacher) explain κλιθείϲ in terms of Homer's use of κεκλιμένοϲ with the dative in the phrases λίμνηι κεκλιμένοϲ (*Il.* 5. 709), ἁλὶ κεκλιμένη (*Od.* 13. 235), etc. But LSJ (s.v. κλίνω II. 5) rightly explains this use by the words 'of Places, *lie sloping towards* the sea, etc.', and when these scholars translate the words of Sophocles by something like 'lying somewhere in the two continents', they not only give a wrong sense to the verb, but utterly fail to explain the nature of the dative. λίμνηι and ἁλί in Homer are common datives of proximity; but most interpreters of the Sophoclean passage seem to take διϲϲαῖϲιν ἀπείροιϲ as a vague locative, for which they can produce no parallel. In fact κλιθείϲ cannot mean the same as κείμενοϲ; and even if it could, διϲϲαῖϲιν ἀπείροιϲ is not the Greek for 'somewhere in the two continents'.

Jebb saw that the common view did not explain the dative. He

[1] This note formed part of a paper read to the Cambridge Philological Society in Feb. 1953. I am obliged to Professor D. L. Page, Mr R. M. Cook, and Mr D. W. Lucas for their help with it.

Classical Quarterly, NS 4 (1954), 91–5.

thought the phrase was 'suggested by' the epic use of the perfect passive; but he took the meaning to be ' "resting upon" the continents, as on a support; i.e. having found an abode on land, instead of roaming over sea'. But since the epic use is different and involves a different construction, it is hard to see how it could 'suggest' the use Jebb has in mind; and Jebb's attempted paraphrase is hopelessly inaccurate.

The only scholars who have made an honest attempt to translate *κλιθείϲ* have been Paley and Whitelaw, who thought that Heracles was imagined as standing at the Hellespont with one foot on either shore. They placed him at the Hellespont because 'the two continents' normally means 'Europe and Asia' (so Aesch. *PV* 790, Soph. fr. 881 P. = R., Hdt. 4. 118. 1, Isocr. 4. 35, 179; for an explicit statement, see Varro, *De ling. Lat.* 4). But in fact the expression could just as well refer to Europe and Africa. The usual view in antiquity was that there were two continents only. But the belief that there were three was entertained by Pindar (*Pyth.* 9. 8; cf. *Pyth.* 4. 48); and it is ‖ probably still earlier. Herodotus in two passages (2. 16; 4. 42) polemizes against the opinion of *οἱ Ἴωνεϲ* that the continents were three, and Jacoby plausibly infers that Hecataeus was the object of his attack (*RE* vii/2. 2703). And since even those who preferred to speak of two continents regarded Africa as part of Asia, there is no reason why the two continents against which Heracles is pictured as leaning should not be Europe and Africa.

This will, I suggest, have been clear to Sophocles' audience the moment it heard the word *κλιθείϲ*. This verb is commonly used of 'leaning against' a pillar; e.g. at *Il.* 11. 371 *ϲτήλῃ κεκλιμένοϲ* and *Od.* 14. 307 *κίονι κεκλιμένη*. And the character and situation of the *ϲτῆλαι* or *κίονεϲ* of Heracles were in Sophocles' time a matter of general knowledge; see Hdt. 2. 33. 3 and Pind. *Ol.* 3. 44, *Nem.* 3. 21, *Isth.* 4. 14. Surely the poet imagined Heracles as resting from his labours with an arm around each of the gigantic columns which, in the words of Diodorus (4. 18), *ϲτήλαϲ ἔθετο καθ' ἑκατέραν τῶν ἠπείρων*. Are we then to suppose that the same Heracles whom we are presently to see upon the stage is here conceived as a gigantic figure? I do not think this altogether impossible; but one may easily suppose that Sophocles was either ignorant of or permissibly indifferent to the actual size and distance apart of the Pillars of Hercules. After all, he was familiar with the name and the legend that attached to it. If this view can be accepted, the difficulties attaching to the sense of *κλιθείϲ* and the construction of *διϲϲαῖϲιν ἀπείροιϲ* disappear, and the expression acquires a poetical point which, if Jebb is right, is altogether lacking.

Now for *ποντίαϲ αὐλῶναϲ*. If my interpretation of the second clause of the disjunction is correct, it is surprising that this should bear a sense so vague as 'the straits of the sea', as Jebb translates it. And in any case, there is no ground for believing that when a Greek poet meant 'the sea', he could say 'the straits, or narrow channels, of the sea'. Which straits did Sophocles mean?

I believe the answer to this problem is contained in a scholion on the passage: ἢ πρὸϲ τὰ ἑῶια ἢ πρὸϲ τὰ δυτικά. In the Laurentianus this note appears under the rubric διϲϲαῖϲιν ἀπείροιϲ κλιθείϲ; and, taken as an explanation of this phrase, it is of course absurd. But suppose it was originally meant to paraphrase the whole disjunctive clause; then the Chorus would be inquiring of the all-seeing Sun, who every day travels from the land of the Ethiopians in the far east to beyond Gades in the far west, whether Heracles was in the east or in the west. In the course of his labours Heracles had already travelled to the far east, in search of the girdle of Hippolyte, and to the far west, in search of the cattle of Geryones and of the golden apples. The allusion to Gibraltar would certainly call to mind one or the other of these western expeditions. Can we so interpret the first half of the disjunction as to obtain the meaning 'Is Heracles in the distant east?'

This sense can be obtained if we take ποντίαϲ αὐλῶναϲ to mean the Bosporus, the Propontis, and the Hellespont. αὐλών can certainly be used of a sea strait, as it is at Aesch. *PV* 731. And πόντιοϲ is found in Aeschylus in the sense of 'Pontic', at *Sept.* 941–3:

> πικρὸϲ λυτὴρ νεικέων ὁ πόντιοϲ
> ξεῖνοϲ ἐκ πυρὸϲ ϲυθεὶϲ
> θηκτὸϲ ϲίδαροϲ.

Here Verrall and Tucker take πόντιοϲ rightly, though LSJ glosses it 'brought by sea or from beyond the sea', a meaning for which there is no evidence.‖ Πόντοϲ in the sense of 'the Black Sea' is also Aeschylean; disregarding doubtful instances like *PV* 726, 792, *Pers.* 72, see *Pers.* 876–7:

> Ἕλλαϲ ἀμφὶ πόρον πλατὺν †εὐχόμεναι μυχία τε Προποντὶϲ
> καὶ ϲτόμωμα Πόντου.

Surely it is to these three channels that the Sophoclean phrase refers.

II. *ANTIGONE* 750–5

ΚΡ. ταύτην ποτ᾽ οὐκ ἔϲθ᾽ ὡϲ ἔτι ζῶϲαν γαμεῖϲ.
ΑΙ. ἡ δ᾽ οὖν θανεῖται καὶ θανοῦϲ᾽ ὀλεῖ τινά.
ΚΡ. ἦ κἀπαπειλῶν ὧδ᾽ ἐπεξέρχηι θραϲύϲ;
ΑΙ. τίϲ δ᾽ ἔϲτ᾽ ἀπειλὴ πρὸϲ κενὰϲ γνώμαϲ λέγειν;
ΚΡ. κλαίων φρενώϲειϲ, ὢν φρενῶν αὐτὸϲ κενόϲ.

ΑΙ. εἰ μὴ πατὴρ ἦϲθ', εἶπον ἄν ϲ' οὐκ εὖ φρονεῖν.

Creon first declares that Haemon shall never marry Antigone, at least never in her lifetime (l. 750). Haemon replies (l. 751) with what the audience will have recognized as a veiled threat of suicide. But Creon takes it to be a threat of violence against himself (l. 752). 'What manner of threat is it', replies Haemon, if the manuscript text of l. 753 is right, '*to speak in reply to empty judgements*?' Now the only utterance of Haemon's that can be called a threat is that in l. 751: and we should therefore expect Haemon's self-defence in l. 753 to have reference to that utterance only. But in the words πρὸϲ κενὰϲ γνώμαϲ λέγειν we have what can only be a vague description of Haemon's conduct throughout the altercation.

I suggest that at l. 753 Sophocles wrote:

τίϲ δ' ἔϲτ' ἀπειλὴ πρόϲ ϲ' ἐμὰϲ γνώμαϲ λέγειν;

'What manner of threat is it to acquaint you with my resolutions?' Haemon is using the word γνώμαϲ to mean 'intentions, purposes, resolves' (sense III. 5 in LSJ). But Creon thinks he is using it to mean 'judgements, opinions' (sense III. 1a in LSJ). That is why he complains that Haemon is trying to lecture him: a complaint that might, it is true, have been provoked by the words the manuscripts assign to Haemon at l. 753. But it is more likely that this retort was stimulated by a sentence which Creon could misinterpret to mean 'to tell you my judgements or opinions'.

No two pairs of letters are more easily confused than *K* and *C* and *M* and *N*. And once the text had been corrupted, many readers will have thought that the echoing of κενάϲ by κενόϲ showed the excellence of the reading. But in fact the echoing of φρενώϲειϲ by φρενῶν gives l. 754 its point without. And the presence of κενόϲ in this line may well have made easier the corruption of the line before.

III. *TRACHINIAE* 187–95

ΔΗ. καὶ τοῦ τόδ' ἀϲτῶν ἢ ξένων μαθὼν λέγειϲ;
ΑΓ. ἐν βουθερεῖ[2] λειμῶνι πρὸϲ πολλοὺϲ θροεῖ
Λίχαϲ ὁ κῆρυξ ταῦτα· τοῦ δ' ἐγὼ κλυὼν[3]
ἀπῆιξ', ὅπωϲ τοι πρῶτοϲ ἀγγείλαϲ τάδε
πρὸϲ ϲοῦ τι κερδάναιμι καὶ κτώιμην χάριν.
ΔΗ. αὐτὸϲ δὲ πῶϲ ἄπεϲτιν, εἴπερ εὐτυχεῖ;
ΑΓ. οὐκ εὐμαρείαι χρώμενοϲ πολλῆι, γύναι.

[2] A proper name? so Professor Page suggests.

[3] κλύων codd.: but the tense must be aorist: cf. Ed. Fraenkel on Aesch. *Agam.* 680.

κύκλωι γὰρ αὐτὸν Μηλιεὺϲ ἅπαϲ λεὼϲ
κρίνει παραϲτάϲ, οὐδ' ἔχει βῆναι πρόϲω.

To whom does αὐτόϲ in l. 192 refer? The Messenger thought, to Lichas; and all editors, so far as I know, have thought so too. And yet this view gives rise to certain difficulties. Hermann remarked that one would have expected εὐτυχεῖ to be used, not of Lichas, but of Heracles. But if it did refer to Heracles, Hermann argued, some word like ἐκεῖνοϲ would have been inserted to show that the subject of the dependent clause was not identical with that of the main clause. He therefore concluded that αὐτόϲ must be said of Lichas, recalling that a master's good fortune may be said to belong also to his servants.

There is indeed a *gnome* to this effect; cf. Eur. *Med.* 54 f., with Professor Page's note. But there is nothing in the context here to warn us that this thought lies behind the poet's choice of verb: and to say of Lichas, not 'if there is nothing the matter with him', but 'if he is enjoying good fortune' is quite unparalleled. And yet Hermann is surely right in insisting that the subjects of both clauses must be the same. Can the subject be Heracles?

This possibility occurred to Paley, who pointed out that since all through the play so far the heaviest stress has been laid on Deianeira's state of anxious longing for her husband's return, it is odd that she does not at once ask why he is not back. And yet Paley finally took the view of Hermann and Jebb that Lichas was the subject, apparently because he was afraid to suggest that the Messenger, a character in the play, might have misunderstood Deianeira's words.

I suggest that he does, and that we have here a deliberate dramatic effect. Deianeira's mind is still on Heracles (mentioned at ll. 185–6 as about to return soon), and she asks why he is not here, since his fortune is good, and not bad as she had feared. But the Messenger takes her to refer to Lichas, whom he has spoken of in ll. 188–9. This seems to me probable, even though I cannot quote any exact parallel to this misunderstanding between characters on the stage.[3a]

IV. *OEDIPUS COLONEUS* 944–6

ᾔιδη δ' ὁθούνεκ' ἄνδρα καὶ πατροκτόνον
κἄναγνον οὐ δεξοίατ', οὐδ' ὅτωι γάμοι
ξυνόντεϲ ηὑρέθηϲαν ἀνόϲιοι τέκνων.

'ἀνόϲιοι γάμοι τέκνων', wrote Jebb, 'is said, with poetical boldness and also with a certain designed obscurity, in this sense: "a woman who has made an unholy marriage with her son".' It is hard to say whether the oddity of γάμοι or the vagueness of τέκνων is the harder to stomach.

[3a] [But see *Aj.* 339–41.]

Emendation is surely needed; but the emendations hitherto offered are not convincing. Benedict and Reiske read *τέκνωι*, a suggestion that is scarcely worth refuting; Musgrave, Hartung, and Blaydes *τοκέων*, which is both weak and, as Jebb points out, metrically objectionable.

Suppose we read:

οὐδ' ὅτωι γάμου
ξυνόντ' ἐφηυρέθηϲαν ἀνοϲίου τέκνα.

'nor yet one in whose company had been detected the children of an unholy marriage'. The corruption would be one that could easily have arisen; the ‖ substitution of *C* for *Φ* in the word *ΕΦΗΥΡΕΘΗCΑΝ*, or perhaps that of *ΓΑΜΟΙ* for *ΓΑΜΟΥ*, could easily have led to the alteration of the remaining case-endings. The sense thus given is unexceptionable; for the stigma attaching to the children of incest, cf. *OT* 1375–6, 1486 ff. There is of course no reason why a sentence, whose subject is a neuter plural noun referring to persons, should not have a plural verb; cf. Eur. *Med.* 1069 and see Wackernagel, *Vorlesungen über Syntax*, i. 101–2.

V. *ELECTRA* 1085–9

ὡϲ καὶ ϲὺ πάγκλαυτον αἰ-
ῶνα κοινὸν εἵλου,
†τὸ μὴ καλὸν καθοπλίϲα-
ϲα δύο φέρειν ⟨ἐν⟩ ἑνὶ λόγωι
ϲοφά τ' ἀρίϲτα τε παῖϲ κεκλῆϲθαι.

τὸ μὴ καλὸν καθοπλίϲαϲα gives no satisfactory sense; the Laurentian scholiast's explanation *καταπολεμήϲαϲα τὸ αἰϲχρὸν καὶ νικήϲαϲα* is generally agreed to be absurd. Most editors have taken *καθοπλίϲαϲα* to have displaced a word that means 'cast aside' or 'spurn' (*καθιππάϲαϲα* Hermann, *'πολακτίϲαϲα* J. H. H. Schmidt). But *καθοπλίϲαϲα* is unlikely to be wrong. It is an unusual word, and one not likely to have come in by a corruption; and though it does not occur elsewhere in tragedy, *ἐξοπλίζειν* is found at Aesch. *Supp.* 99, 683, 702.

I suggest that Sophocles wrote:

ἄκοϲ καλὸν καθοπλίϲαϲα.

ἄκοϲ was glossed by *τομή*, naturally enough, since the sort of 'remedy' which Electra contemplates would of course be an *ἄκοϲ τομαῖον* (Aesch. *Cho.* 539,[4] *Supp.* 268): *τομή* was then misinterpreted and wrongly

[4] Here Clytemnestra is said to send her offerings to the tomb *ἄκοϲ τομαῖον ἐλπίϲαϲα πημάτων*. *ἐλπίϲαϲα* makes sense and must not be meddled with. But if my emendation in the *Electra* is right, the possibility that Aeschylus wrote *ὁπλίϲαϲα* cannot be discounted.

divided. The construction involved is not particularly difficult. At *Supp.* 702 Aeschylus has *ἐξοπλίζειν Ἄρη*: ibid. 683 he has:

ἄχορον ἀκίθαριν
δακρυογόνον Ἄρη
βοάν τ' ἔνδημον ἐξοπλίζων.

If one can be said to 'arm Ares', and even 'to arm the rallying-cry of the people', one may surely be said to 'arm' an *ἄκος* that will be an *ἄκος τομαῖον* and therefore will require its *ὅπλον*.

[See the remarks about these passages in H. Lloyd-Jones and N. G. Wilson, *Sophoclea* (1990).]

31(7)

Notes on Sophocles' *Antigone*[1]

I. 100–9

ἀκτὶϲ ἀελίου, τὸ κάλ-
λιϲτον ἑπταπύλωι φανὲν
Θήβαι τῶν πρότερον φάοϲ,
ἐφάνθηϲ ποτ', ὦ χρυϲέαϲ
ἁμέραϲ βλέφαρον, Διρκαί-
ων ὑπὲρ ῥεέθρων μολοῦϲα,
τὸν λεύκαϲπιν Ἀπιόθεν
φῶτα βάντα πανϲαγίαι
φυγάδα πρόδρομον ὀξυτέρωι
κινήϲαϲα χαλινῶι.

106 Ahrens: Ἄργοθεν codd. 108 ΣL, R: ὀξυτόρωι λπ.

JEBB renders the last clause as follows: 'The warrior of the white shield, who came from Argos in his panoply, hath been stirred by thee to headlong flight, in swifter career.' 'In swifter career' is a discreet rendering of ὀξυτέρωι . . . χαλινῶι. 'ὀξυτέρωι', Jebb says, 'does not mean (1) "in flight swifter than their former approach" nor (2) "the reins are shaken ever faster on the horses' necks".' 'The Argives', he writes, 'began their retreat in the darkness (cf. 16): when the sun rises, the flashing steel of their bridles shows them in headlong flight.' Cf. P. Mazon, *REG* 25 (1951), 13. This view is shared by all modern scholars, except that Dain and Mazon put commas after φῶτα and πρόδρομον and translate accordingly; which I do not think is an improvement.

χαλινόϲ means 'bit', or 'bit and bridle together'. It is remarkable that the day of victory should be said to move off the Argives with sharper (or swifter) bit, or even with sharper (or swifter) bridle. Bits and bridles are not generally used to make horses go faster, but to make them go slower, or to halt them alto||gether. To make them go faster, spurs or a whip are commonly used. In practice this generalization is not always true; the bit and bridle may play a part in speeding up the

[1] Read to the Oxford Philological Society in Nov. 1955. I am very grateful to Professor D. L. Page for reading and criticizing the manuscript. I must also thank Sir John Beazley and Mr A. H. Coxon for advice on points of detail.

Classical Quarterly, NS 7 (v. 57), 17–27.

horse. But this is not their characteristic use; and in a poetic passage of this nature one would expect the characteristic use to be in question. ὀξυτέρωι could mean 'sharper'; then the sense might be that as soon as dawn broke the Argives took a firmer grip of the reins in order to accelerate. But if one wants to go faster, one lets the reins drop loose (see Jebb on *OC* 900; cf. Hes. *Sc.* 308, Soph. *El.* 713, Eur. *IA* 151). If this were meant, ὀξυτέρωι . . . χαλινῶι would be a curious way of expressing it.

The use of χαλινός seems yet more singular when it is realized that whenever the word is used in a sense wholly or partly metaphorical its mention implies restraint or compulsion, never acceleration or encouragement. Let us examine its non-literal use. Sometimes it is applied to material instruments of compulsion which are not bits or bridles; see Aesch. *PV* 562, *Agam.* 238. Pindar (*Pyth.* 4. 25) and Euripides (*IA* 1043) use it of an anchor. At Pind. *Isth.* 8. 48, παρθενίας χαλινός means the virgin zone. At Eur. fr. 463 N.[2] 1–2:

οὐ γάρ ποτ' ἄνδρα τὸν σοφὸν γυναικὶ χρὴ
δοῦναι χαλινούς, οὐδ' ἀφέντ' ἐᾶν κρατεῖν

δοῦναι χαλινούς means the same as 'give her too much rope'. At *IA* 151, Agamemnon says to the old slave who is to be his messenger: πάλιν εἴσορμα, σεῖε χαλινούς: 'shake the reins', i.e. 'let them hang loose, so that there is no check upon your speed'. Since the old man is to go on foot (139) the metaphor simply indicates great haste. χαλινός can also be used of any non-physical means of imposing constraint. Soph. fr. 869 P. = R. πολλῶν χαλινῶν ἔργον οἰάκων θ' ἅμα is rendered by Pearson 'a task requiring much restraint and much guidance'. At Soph. *El.* 1462, στόμια δέχηται τἀμά means 'submit to my control'. At Aesch. *PV* 671, Io tells how her father Inachus was forced to turn her out, against her will and his:

ἀλλ' ἐπηνάγκαζέ νιν
Διὸς χαλινὸς πρὸς βίαν πράσσειν τάδε.

'The bit of Zeus' means 'the compelling power of Zeus'. In Plato (*Rep.* 496 D) the χαλινός of Theages is the ill health that keeps him out of politics and in philosophy. For later prose instances see LSJ s.v. 1. 2; cf. s.v. χαλινόω, ἐγχαλινόω, στόμιον.

Can the word have been used here in this metaphorical sense? Lewis Campbell in his commentary wrote: 'Either literally ("riding with hastier rein") or metaphorically ("with thy hastening bridle in his lips").' 'Hastening' would be too vague a rendering of the comparative ὀξυτέρωι, which should mean 'sharper' or 'swifter'. Than what might the bridle of constraint be sharper or swifter? Nothing that I can think of.

Does this mean that Campbell's second alternative can be ruled out? I think not. ὀξυτέρωι has been put in the text by every editor since Brunck in 1786. It may have been read by the author of the note in the Laurentian scholia (οἷον παραcκευάcαca τῶν χαλινῶν μετὰ cπουδῆc ἐφάψαcθαι διὰ τὴν ὀξυτέραν φυγήν); and it is in R, the sole representative for this play of that Roman family in which Turyn hopes to find a substitute for the Paris family whose testimony he seeks ‖ to discredit (see *Studies in the Manuscript Tradition of Sophocles*, 178). But both the Laurentian family (L and its twin at Leiden) and the Parisian have ὀξυτόρωι. We cannot be sure that the Roman family, any more than we can be sure that the Paris family, is free from Byzantine interpolation; see P. Maas's review of Turyn, op. cit., in *Gnomon*, 25 (1953), 441; cf. mine of the new Budé edition, ibid. 27 (1955), 105. ὀξυτόρωι is the only reading which we can be sure has ancient authority; and though palaeographically the difference between the readings is very small, it is decidedly the *lectio difficilior*. Till ὀξυτόρωι has been shown to be impossible, we have no right to consider ὀξυτέρωι at all.

The first scholar after Brunck to defend ὀξυτόρωι was Welcker (ap. Hermann, in the preface to the third edition of his revision of Erfurdt (Leipzig, 1830), p. xvii). As both Welcker and Hermann took χαλινῶι in the usual way, Hermann could convincingly dispose of this suggestion by pointing out that sharp bits were less suitable for rapid flight than blunt ones. ὀξυτόρωι was also read by Kirchhoff, in *Die Parodos der Antigone* (Altona, 1862); as he took χαλινῶι to refer to the bridle of the sun, it is not surprising that he found no followers.

But suppose the metaphorical interpretation of χαλινῶι were correct? In the second edition of his commentary (1879), Campbell pointed out that the reading ὀξυτόρωι would accord well with his suggestion; but he took the argument no further. Later he seems to have given up the notion, for it is not mentioned in the school edition which he did with E. Abbott (1899) or in *Paralipomena Sophoclea* (1907). I wonder why. ὀξυτόροc, from τείρω, does not recur before Meleager (*AP* 4. 1. 16). But it is a perfectly good word (cf. τορόc, διάτοροc, ῥινοτόροc); Blaydes and Nauck need not have considered ὀξυτόμωι and ὀξυτόνωι. And no epithet could be better suited to describe a Greek bit. One recalls the words used by Clytemnestra to the newly enslaved Cassandra (Aesch. *Agam.* 1067):

> χαλινὸν δ' οὐκ ἐπίcταται φέρειν
> πρὶν αἱματηρὸν ἐξαφρίζεcθαι μένοc.

Ed. Fraenkel, ad loc., gives much information about Greek bits. In particular, he describes the effects of the sharp snaffle (ὁ τραχὺc χαλινόc): 'If the horse ever failed to obey, or if the rider were unsympathetic or impatient, then the spiked rolls and the disks of the bit would in a moment tear the membranes of the lower jaw and the tongue and draw blood.'

Just such a cruel bit figures in the metaphor at Soph. *Trach.* 1259 f. The dying Heracles urges his hard soul to let no cry escape his lips, applying 'a bit of iron welded with stones':

> ἄγε νυν, πρὶν τήνδ' ἀνακινῆcαι
> νόcον, ὦ ψυχὴ cκληρά, χάλυβος
> λιθοκόλλητον cτόμιον παρέχουc'
> ἀνάπαυε βοήν. . . .

Jebb took this to mean 'a curb as of steel on lips set like stone'. His view was challenged by F. W. Ball in *CR* 8 (1894), 197 f. Ball showed that in antiquity bits were sometimes set or welded with some kinds of stone. He quotes from Claudian and Nonnus several references to bits studded with gems. These instances are too late to decide the question, but Ball quotes from Lucian (*Prometheus es in verbis* 4) the description, doubtless taken from a Hellenistic source, of a black camel exhibited in Egypt by Ptolemy Lagus: *καὶ ὁ χαλινὸc ἦν λιθοκόλλητοc, Δαρείου τινοc ἢ Καμβύcου ἢ αὐτοῦ Κύρου κειμήλιον.* Here again the 'stones' in question were precious stones. But whatever sort of ‖ stones Sophocles had in mind, the use of the term λιθοκόλλητοc χαλινόc strongly suggests that in Sophocles also the reference is to stones with which the bridle was set. It makes against Hermann's suggestion, lately revised by Dain and Mazon, that the phrase involves a metaphor from the iron clamps used in architecture to hold blocks of stone together. What purpose are the 'stones' in the λιθοκόλλητοc χαλινόc likely to have served? I suspect that sharp stones were welded into the bit to form the ἐχῖνοι, the spines attached to the mouthpiece of the τραχὺc χαλινόc; see M. H. Morgan, *The Art of Horsemanship, by Xenophon* (London, 1894), 144. It is true that there is no proof that stones were used for this particular purpose at an early date. For example, the two ancient bits illustrated by Lechat, *BCH* 14 (1890), 385 f., one of which dates from the early fifth century, have ἐχῖνοι made of bronze. But it seems not unreasonable to suspect that stones, precious or, more likely, ordinary, may have been used for this purpose. In Sophocles the bit set with stones is of course quite different from a gag. Its function is not to muffle any cry, but to impose upon the agonized Heracles the painful constraint of a self-discipline that enforces silence.

The metaphor in the *Antigone* passage is very similar. The words mean 'moving him off in headlong flight by means of a sharply-piercing bit'. This is not an actual bit, but a 'bit of compulsion', like the 'bit of Zeus' in the *Prometheus* and the bit of Heracles in the *Trachiniae.*

These passages show that there is no reason why the metaphor of horse and bit should extend farther than the actual phrase ὀξυτέρωι . . . χαλινῶι. Just as in Aeschylus the bit of Zeus forces Inachus to turn his daughter out and in Sophocles the iron bit of discipline, self-imposed, is to force the tortured Heracles to keep silence, so here the day of victory is to send off the enemy in headlong flight from Thebes, constraining him to obedience by means of a sharply piercing bit. The application of the metaphor, and in particular the functioning of the instrumental, may be illustrated by a passage in which a poet conveys a not dissimilar sense, but marks off from each other in two separate clauses the two phases of the action, first the yoking of the victim with 'powerful constraint' and then his precipitation in headlong flight. This is the description of the maddening of the Proetides by Hera in Bacchyl. 11. 43–6 S.–M.:

> τὰϲ ἐξ ἐρατῶν ἐφόβηϲεν
> παγκρατὴϲ Ἥρα μελάθρων
> Προίτου, παραπλῆγι φρέναϲ
> καρτερᾶι ζεύξαϲ' ἀνάγκαι.

Just so Sophocles might have said ὀξ. χαλινὸν ἐπιθεῖϲα καὶ κινήϲαϲα. But he has chosen to adopt a manner of expression that is conciser and, to my taste, more poetical. It is perhaps a trifle harsh and obscure. But Sophocles is not always a clear or polished writer.[2] ‖

II. 593–8

> ἀρχαῖα τὰ Λαβδακιδῶν οἴκων ὁρῶμαι
> πήματα †φθιμένων ἐπὶ πήμαϲι πίπτοντ',
> οὐδ' ἀπαλλάϲϲει γενεὰν γένοϲ, ἀλλ' ἐρείπει
> θεῶν τιϲ, οὐδ' ἔχει λύϲιν.

φθιμένων is kept in the text by Pearson and by Dain. But there is little doubt that it is wrong. The syllable that seems to be resolved is in fact an anceps; and the resolution of an anceps is unparalleled in the dactylo-epitrites of tragedy or choral lyric. Campbell and Bruhn obelize the word; but Hermann's φθιτῶν stands a good chance of being

[2] One might well ask whether the Argives were thought of as being on horses at all. λεύκαϲπιν and πανϲαγίαι ('= πανοπλίαι' Jebb) certainly suggest the foot-soldier. And the line of the cyclic *Thebaid* (fr. iv Allen = 6a Davies) that describes the escape of Adrastus εἵματα λυγρὰ φέρων ϲὺν Ἀρείονι κυανοχαίτηι might be thought to suggest that its writer imagined Adrastus as escaping mounted on the back of that celebrated horse. But this is not a safe line of argument. The Argives may well have been imagined as using chariots to bring them up to or away from the scene of battle, but as doing the actual fighting on foot. This seems to be how Aeschylus conceived the battle (*Sept. passim*); and Antimachus (fr. 32 Wyss) describes Adrastus, at least at one stage, as riding in a chariot drawn by Cairus and Arion.

right, especially since the words are confused in the manuscripts at Eur. *Alc.* 100 (*φθιτῶν* LP, *φθιμένων* VB). But one should bear in mind that the text here is not certain.

Jebb paraphrases as follows: 'I see that, from olden times, the house-troubles (*οἴκων πήματα*) of the (living) Labdacidae are heaped upon the troubles of the dead.' 'The dead', he explains, 'are Laius, Oedipus and his two sons.' He thinks, then, that there are two sorts of *πήματα*, those of the dead and those of the living which the chorus sees following close upon them. Observe that to make his view clear he has had to interpolate 'living', which is not in the Greek, and he has had to strain the meaning of *ἀρχαῖα*. *ἀρχαῖα* is, as he remarks, predicative; this ought to be brought out by rendering: 'Ancient are the woes of the Labdacidae, and I see them falling, etc.' But it is going too far to treat *ἀρχαῖα* as though it were an adverb, as Jebb does. The poet has not used *ἀρχῆθεν* or *ἐξ ἀρχῆϲ*; he has used *ἀρχαῖα*, an adjective agreeing with *πήματα*, though it would suit Jebb much better if it agreed with *πήμαϲι*. The natural way of translating the words would be: 'Ancient are the house-woes of the Labdacidae, and I see them following hard upon the woes of the dead.' Jebb assumes that the woes which follow hard upon the woes of the dead must be the woes of the living. But can the woes of the living be called 'ancient'? Jebb is forced to give the word *ἀρχαῖα* a meaning which it cannot really bear.

Nauck long ago remarked that one would expect to find here a mention of only one kind of woes, following hard upon one another. He compared Aesch. *Th.* 740–1:

> *πόνοι δόμων νέοι παλαι-*
> *οῖϲι ϲυμμιγεῖϲ κακοῖϲ.*

Nauck renounced the attempt to wrest such a sense out of the transmitted text; but Meineke, Dindorf, Wecklein (ap. Jebb), and Schubert (ap. A. Y. Campbell, *Cambridge University Reporter*, 18 Feb. 1936) all more or less rewrote the passage in order to obtain this sense.

Though I deplore their methods, I sympathize with their aims; and I suggest that these could be more easily achieved by taking *φθιτῶν* with *πήματα*: 'Ancient are the woes that come from the dead in the house of the Labdacidae, and I see them falling hard upon one another.'

The notion of curses or evils belonging to or coming from the dead is not rare in tragedy, especially in the plays about the house of Pelops and the house of Laius. In this play we find it at ll. 2 (*τῶν ἀπ' Οἰδίπου κακῶν*) and 856 (*πατρῷιον δ' ἐκτίνειϲ τιν' ἆθλον*). A parallel to the sense of this stanza is at Aesch. *Cho.* 406 f.:

> *βοᾶι γὰρ λοιγὸϲ Ἐρινὺν*
> *παρὰ τῶν πρότερον φθιμένων ἄτην*
> *ἑτέραν ἐπάγουϲαν ἐπ' ἄτηι.*

||

It is not uncommon in tragedy to find two genitives depending on one noun (see Wilamowitz on Eur. *HF* 170). In this case, there would be three; *Λαβδακιδᾶν* would depend on *οἴκων* and *οἴκων* on *πήματα*. I do not see that the presence of the extra genitive dependent upon *οἴκων* makes the construction any harder to believe in. Cf. l. 795 *νικᾶι δ' ἐναργὴc βλεφάρων ἵμεροc εὐλέκτρου / νύμφαc*, where *βλεφάρων* is genitive of origin and *βλεφάρων ἵμεροc* is treated as a single notion and has *νύμφαc* depending on it. Cf. also ll. 929–30.

I have considered the possibility of taking *οἴκων* closely with *πήματα* as Jebb does ('house-woes') and *φθιμένων* with *Λαβδακιδᾶν* ('the house-woes of the dead Labdacidae'). Adjectives and participles sometimes follow their nouns at no less a distance than that which separates *φθιμένων* from *Λαβδακιδᾶν*; but here the interposition of another noun in the genitive (*οἴκων*) complicates the problem. One might quote in support of this view *Phil.* 598–9:

> τίνοc δ' Ἀτρεῖδαι τοῦδ' ἄγαν οὕτω χρόνωι
> τοcῶιδ' ἐπεcτρέφοντο πράγματοc χάριν;

But the harshness of the word-order it involves leads me to reject this way of taking the sentence in favour of the one I have suggested.

III. 599–603

The manuscripts have:

> νῦν γὰρ ἐcχάταc ὑπὲρ
> ῥίζαc ἐτέτατο φάοc ἐν Οἰδίπου δόμοιc.
> κατ' αὖ νιν φοινία
> θεῶν τῶν νερτέρων ἀμᾶι κόνιc,
> λόγου τ' ἄνοια καὶ φρενῶν ἐρινύc.

Most editors since Hermann have felt, as Jebb did, that a new sentence starting *κατ' αὖ νιν* would be 'intolerably abrupt'. They have therefore adopted one or other of the devices which Hermann suggested for linking the two sentences together; they have changed *ὑπὲρ* to *ὅπερ* or *ἐτέτατο* to *ὃ τέτατο*. If one reads *ὃ τέτατο*, one has to believe that whatever is the subject of *ἀμᾶι* cuts down the light of safety that but lately shone above the last root of the house. A root can easily be cut, but not a light; this remedy is therefore unacceptable. The same argument does not rule out Hermann's other suggestion. For if we change *ὑπέρ* to *ὅπερ*, *ἐcχάταc . . . ῥίζαc* is presumably a defining genitive: 'the light of safety consisting in the last root of the family that but lately shone in the house'. And though a light cannot normally be cut down, a light that consists in a root probably can. *ὅπερ* therefore could be right; but I prefer, with Pearson, to keep the manuscript reading. New sentences begin with a preverb *in tmesi*, *αὖ* and a pronoun at Aesch. *Agam.* 1216 and *PV* 877, q.v.

A root, or even a light of safety consisting in a root, can be cut down; but can the cutting be done by the bloody dust? Askew, Jortin, and Reiske seem independently to have hit upon the remedy of reading *κοπίς*. The change is minimal, and at Lucian, *Lucius sive asinus* 40 *κοπίδος* is a certain correction of *κόνιδος*, as Dobree points out (*Adversaria*, ed. Wagner (1883), iii. 34). At first this found general acceptance. But Hermann led a reaction, and after him Boeckh, Bergk, Campbell, Jebb (in the later editions), Tyrrell, Bruhn, Masqueray, and Dain have kept the manuscript reading. Antigone's fatal action, Hermann said, was in fact the sprinkling of a handful of dust due to the infernal gods. And is it safe to deny that Sophocles can have spoken of bloody dust cutting off a root? Hermann is surely right in saying, No; and if ‖ *κοπίς* is to be accepted, some stronger argument in its favour will have to be produced.

Such an argument arises from a consideration of the last line of the stanza. It is common to find three nouns, A, B, and C, linked in this fashion: A B *τε καὶ* C. But I know of no such instance in which A is the name of a material object (or even the name of a material object used figuratively, as *κόνις* or *κοπίς* is used here) and B and C are names of feminine abstracts readily personified, like *λόγου ἄνοια*, or names of feminine daemonic beings used in a quasi-abstract sense, like *φρενῶν ἐρινύς*. In other words, I do not believe one could say 'bloody dust' (or 'bloody chopper') 'of the infernal gods *and* folly in speech *and* the Erinys in the mind'. The two latter expressions are not, then, linked to *κόνις* or *κοπίς* by means of *τε καί*, but stand in apposition to it; the *τε* and *καί* serve to link them to each other.

What do these two expressions mean? Jebb renders *λόγου ἄνοια* by 'folly in speech', rightly. (The singular *λόγου* is a little strange; but cf. *OC* 66, *Aj.* 732.) Ate commonly causes folly in speech; and it is she whose operation the expression is probably meant to suggest. *φρενῶν ἐρινύς* means 'the Erinys in the mind'. The Erinys works on the mind to cause rash acts of folly; here, as often, Ate and Erinys are close collaborators.

Can the bloody dust scattered by Antigone be said to *be λόγου ἄνοια*? The linguistic difficulty of this is probably not unsurmountable; but the sense makes strongly against it. The sprinkling of the dust was certainly, in the view of the chorus, a foolish action; but how could they call it 'folly *in speech*'? It is just possible that the sprinkling of the dust could be called an Erinys, but very unlikely that it could be called 'an Erinys *in the mind*'. In sum, it seems all but impossible that the bloody dust, or the sprinkling of that dust, can have been called *λόγου τ' ἄνοια καὶ φρενῶν ἐρινύς*.

But can one really identify with either of these two things the chopper of the infernal gods? To start with, the very expression *θεῶν τῶν νερτέρων κοπίς* has given rise to difficulties. Even Bruhn accepted the argument of Chr. Heimreich (Gymnasialprogr. Plön (1884), 10) that *κοπίς* was too homely a word for tragedy. But this argument was destroyed by Arthur Platt (*CQ* 4 (1910), 250), who quoted fr. 894 P. = R.:

> ὀργὴ γέροντος, ὥστε μαλθακὴ κοπίς,
> †ἐν χειρὶ θήγει† ἐν τάχει δ' ἀμβλύνεται.

A more formidable objection was urged against me by a friend who read an early draft of this paper. The dust sprinkled on the body may indeed be said to belong to the infernal gods; it is their due, which Antigone has rendered them. But how can the instrument be theirs? They do not destroy Antigone; indeed, it is through paying them reverence that she is destroyed.

Now *νέρτεροι θεοί* is a vague expression that includes Hades and Persephone and also the other chthonic powers, including the Erinyes and those Arai with whom they are sometimes identified; cf. ll. 1074–5, Eur. *Med.* 1059. The infernal gods in general, and not only the Erinyes and Arai in particular, are sometimes said to be responsible for the working out of a family curse. At *Il.* 9. 456–7, the curse of Amyntor is fulfilled by Hades and Persephone; ibid. 571–2 the mother of Meleager prays to Hades and Persephone and is heard by the Erinys. The prayer of Electra at Soph. *El.* 110 f. is addressed to the house of Hades and Perse||phone, to Hermes *χθόνιος* and the Ara, and to the Erinyes; cf. Aesch. *Cho.* 124 f., 382 f., 405–6, 476–8.

The chorus is suggesting that Antigone is a victim of the family curse; and just as Orestes and Electra know that the infernal gods, who are the ministers of such curses, can strike down Aegisthus and Clytemnestra, so does the chorus know that they can strike down Antigone. And there is no reason why they should not credit the infernal gods with the possession of a bloody chopper. We know that the gods can use *μάστιγες*, *μάκελλαι*, *ἔγχη*, *flagella* as instruments of punishment. Perhaps the closest parallels to this particular instance are the sword of Thanatos at Eur. *Alc.* 74 and the iron swords of Hades at *Or.* 1398.

But, granted that the infernal gods can have a bloody chopper, can it be identified with folly in speech and the Erinys in the mind? I think it can. It is notorious that when the gods have decided to destroy a mortal, they first take away their victim's judgement, so that he will do or say some foolish thing that will involve him in disaster. This they do by sending Ate or the Erinys, or both, against their victim. The Homeric *locus classicus* is Agamemnon's speech at *Il.* 19. 78 f. Agamemnon himself was not responsible for his quarrel with Achilles:

ἐγὼ δ' οὐκ αἴτιός εἰμι,
ἀλλὰ Ζεὺς καὶ μοῖρα καὶ ἠεροφοῖτις Ἐρινύς,
οἵ τέ μοι εἰν ἀγορῆι φρεσὶν ἔμβαλον ἄγριον ἄτην.

What afflicted him was precisely λόγου ἄνοια (with which cf. εἰν ἀγορῆι) and φρενῶν ἐρινύς (with which cf. φρεσὶν ἔμβαλον). Cf. also Soph. *Trach.* 263–4, and similar passages assembled by Ed. Fraenkel on Aesch. *Agam.* 1192, esp. Trag. Adesp. 296 and 455; and see E. R. Dodds, *The Greeks and the Irrational*, 37–8, with p. 56 nn. 52–4. [Note too Aesch. *PV* 1079.]

The infernal gods are the ministers of family curses; instruments of destruction wielded by a god may be called scourges, spears, mattocks; the best-known divine instruments of destruction are folly in speech and the Erinys in the mind. With the reading κοπίς the sense is beyond exception.

IV. 611–13

The manuscripts have:

τό τ' ἔπειτα καὶ τὸ μέλλον
καὶ τὸ πρὶν ἐπαρκέσει
νόμος ὅδ' οὐδὲν ἕρπει
θνατῶν βιότωι πάμπολις ἐκτὸς ἄτας.

One type of remedy starts by accepting Boeckh's ἕρπων and takes νόμος ὅδ' as referring to the supremacy of Zeus, which is mentioned in the earlier part of the stanza. One remedy of this kind is adopted by Hermann, who read πάμπολυς and translated: 'in aeternum haec lex valebit, nulla in re mortalium vitam permulta sine malo expetens'. It is not clear how the supremacy of Zeus can be referred to as νόμος ὅδ', and the notion of this law 'coming to the life of mortals in full abundance' is a very odd one. Boeckh, Bergk, and Campbell operate with a similar construction, but keep πάμπολις. Campbell translates: 'The sovereignty of Zeus, an all-embracing law, in its eternal course fails not to bring calamity to men, when Hope deceives.' The clause 'when Hope deceives' is an unwarranted interpolation from the next stanza; the sense given to πάμπολις is scarcely possible; and the general sense is no less objectionable than that elicited by Hermann.‖

It seems reasonably certain that *νόμος ὅδ'* does not look back to what precedes, but forward to what follows. We should put a semicolon after it, and should look to the words that follow to tell us what the 'law' is. Heath's conjecture *πάμπολύ γ'* yields a meaning along these lines; and it has become almost a vulgate, having convinced Brunck, Meineke, Erfurdt, Jebb, Pearson, Masqueray, and Dain–Mazon. [Also Dawe (2nd edn. 1985).] Jebb thought it 'not only the best emendation, but (in my belief) a certain one'. But it is not even a good suggestion. The sense it is supposed to give, 'Nothing that is vast comes to the life of men without calamity', is not unsuited to the context, but is expressed with a suspicious vagueness. This should have led editors to inquire closely into the function of the *γε*, a particle for whose liberal use to fill hiatuses in emended passages the author of this conjecture was notorious ('remedium Heathianum'). *γε* has a strong tendency to come early in a sentence (see R. A. Neil, *The Knights of Aristophanes*, 186). When it is appended to an attribute, attribute and *γε* usually stand close to the noun qualified, and very commonly precede it. I can find in tragedy no other case of an attribute accompanied by *γε* trailing so near the end of a sentence when its noun stands at the beginning. And what difference does *γε* make to the sense? The word-order shows that it cannot modify the whole complex *οὐδὲν . . . πάμπολυ* but must affect the adjective only. It is hard, then, to see what the sentence can mean except this: 'Nothing comes to the life of mortals without Ate; nothing, that is, that is vast.' And this will hardly do.

Several other emendations are mentioned in Blaydes's *Spicilegium Sophocleum* and in Jebb's appendix on this passage. None has convinced, and none seems to me convincing. Are we then to conclude, with Dindorf, Linwood, and Bruhn, that corruption lies too deep for remedy? Perhaps; but I suspect the poet wrote:

οὐδὲν ἕρπει
θνατῶν βίοτος πάμπολυς ἐκτὸς ἄτας.

'to no mortal creature comes vast abundance without Ate'.

πάμπολυς for *πάμπολις* is hardly a change; and the case-ending of *βίοτος* could easily have been altered, probably by someone who assumed that *βίοτος* must mean 'life'. So it generally does in tragedy; but cf. Eur. *Andr.* 320 *βίοτον ὤγκωσας μέγαν*.

The sentiment is not an unfamiliar one; nor does this reading involve any serious grammatical difficulty. For *ἕρπειν* with a plain accusative cf. Eur. *Tro.* 92 *ἀλλ' ἕρπ' Ὄλυμπον*. This would be more helpful if Olympus were the name of a person; cf., therefore, Soph. *Phil.* 141 *σὲ δ', ὦ τέκνον, τόδ' ἐλήλυθεν / πᾶν κράτος ὠγύγιον*. Cf. also Eur. *Hipp.* 1371 and Ar. *Nub.* 30 (a paratragic passage).

Should we read οὐδὲν or οὐδέν'? *OT* 1196 βροτῶν οὐδὲν μακαρίζω provides no strong evidence in favour of the neuter, for it might mean, not 'I call no human creature happy', but 'I call happy nothing in the human lot'. But compare the words of Iocaste at *OT* 708–9:

καὶ μάθ' οὕνεκ' ἐϲτί ϲοι
βρότειον οὐδὲν μαντικῆϲ ἔχον τέχνηϲ.

These supply a definite reason for accepting Jebb's way of taking l. 1196 and for keeping οὐδὲν in the *Antigone* passage.

V. 625

πράϲϲει δ' ὀλιγοϲτὸν χρόνον ἐκτὸϲ ἄταϲ. ‖

ὀλιγοϲτόν cannot, I think, be right. Jebb points out that ὀλιγοϲτὸν χρόνον could only mean 'one in a small number of χρόνοι' and that does not suit the sense. A. C. Moorhouse, *CQ* 41 (1947), 40 accepts Jebb's argument, but argues that ὀλιγοϲτόν may still be right, since the words might mean 'And he avoids calamity during a season that comes but rarely.' But this is surely much too obscure to satisfy. Menander fr. 208. 3 Körte is no parallel; see Thierfelder's note on p. 291 of his revision of Körte's text.

Bergk conjectured ὀλίγιϲτον, which Campbell, Jebb, Bruhn, and Pearson all accept. If this were in the manuscripts, no one would quarrel with it. But the sense it yields is rather flat. We have just been told that 'evil seems good to him whom a god is leading to destruction'. After this, the remark that 'but for a short season does he fare outside of destruction' gives a feeble conclusion. Still, this may be a subjective prejudice. The Greeks had not our feelings about tautology and repetition, and the line may be there simply for the sake of a deliberate, solemn echo of the last line of the preceding strophe. It might even be held that it provides an impressive ending to the stanza, with its quiet statement of an obvious, but certain and formidable truth. And the statement of that truth finds its parallel in not dissimilar contexts at Hes. *Op.* 325 and Simonides fr. 22 Page = *PMG* 527.

But ὀλίγιϲτον is *not* in the manuscripts; and there is a way of keeping to the tradition and at the same time securing what to some people may seem a more effective conclusion. It is a way which at first sight may seem to involve grave linguistic objections; but on closer inspection I think these prove less formidable than they seem at first. This is to read: πράϲϲει δ' ὀλίγοϲ τὸν χρόνον ἐκτὸϲ ἄταϲ, which would mean 'But the small man, the poor man throughout his time fares outside of Ate.'

Both ὀλίγοϲ and τὸν χρόνον give rise to difficulties. ὀλίγοϲ is rare in tragedy, and particularly rare in Sophocles. Moorhouse has examined the tragic use of *μικρός* and *ὀλίγος* in the article quoted, a useful one in spite of its misleading omission of any reference to such words as *βαιός*, *βραχύς*, *σπάνιος*. He goes so far as to say that 'it seems probable that there are no genuine examples of *ὀλίγος* in Sophocles'. It is true that at fr. 904. 1 P. = R. Cobet's *ὀλίγοισιν* is unlikely to be right. But at fr. 646 P. = R. *ἐν γὰρ βραχεῖ καθεῖλε κὠλίγωι χρόνωι*, the text may be right in spite of Pearson's doubts. There is a tautology, but such things occur in Sophocles, and if, as I suspect, *βραχεῖ* is *neuter* the awkwardness becomes less. And even if Moorhouse were right in keeping *ὀλιγοστὸν* in the passage under discussion, this would surely itself constitute one instance of the use of *ὀλίγος*.

If *ὀλίγος* is right, it will be an instance of a rare, but well-attested use of words meaning 'small' in the sense of 'poor' or 'weak' and in contrast with *μέγας* in the sense of 'great', 'rich', 'strong'. Moorhouse, p. 37, quotes three examples:

1. Callinus fr. 1. 17 W.: *τὸν δ' ὀλίγος στενάχει καὶ μέγας ἤν τι πάθηι.*
2. *HHMerc.* 259–60: *ὑπὸ γαίηι / ἐρρήσεις ὀλίγοισι μετ' ἀνδράσιν ἡγεμονεύων.*
3. Soph. *Aj.* 158–61: *καίτοι σμικροὶ μεγάλων χωρὶς*
   σφαλερὸν πύργου ῥῦμα πέλονται.
   μετὰ γὰρ μεγάλων βαιὸς ἄριστ' ἂν
   καὶ μέγας ὀρθοῖθ' ὑπὸ μικροτέρων. ||

I do not think it impossible that Sophocles used *ὀλίγος* here. He may have done so more readily when it was a question of this rare usage, which Callinus and the author of the hymn to Hermes may have derived from early epic.

Now for *τὸν χρόνον*. How can it be used to mean what is commonly expressed by *τὸν πάντα χρόνον, τὸν ἀεὶ χρόνον*, etc.? The definite article provides a certain degree of determination, so that *ὁ χρόνος* can mean 'his time', 'the allotted time'. Now if it is agreed that the notion of a fixed span of life allotted to each mortal was a familiar one in early poetry, there is a possibility that the expression 'his time' could be understood in such a sense. There is, in fact, little doubt that the tragedians took over from the epic the notion of a fixed span of life allotted to mortals at their birth. In this play (896) Antigone complains that she is going down to Hades *πρίν μοι μοῖραν ἐξήκειν βίου*. Earlier on (461) she has declared that she counts it all to the good if she is to die before her time: *εἰ δὲ τοῦ χρόνου / πρόσθεν θανοῦμαι, κέρδος αὔτ' ἐγὼ λέγω*. Jebb translates *τοῦ χρόνου* here by 'the natural term of life'. Does this mean 'term' in the sense of a continuing space of time or 'term' in the sense of the end of a space of time? I suppose, the latter; but *τοῦ χρόνου* can hardly mean 'the end of my lifetime'. I take it to mean 'the time allotted to me', and would translate: 'if I am to die before my time dies'. As often, a person's lifetime is treated as though it were independent of that person; there is much material to illustrate this notion in Ed. Fraenkel's note on Aesch. *Agam.* 105. Here is Antigone, there is Antigone's time; Antigone will be killed, but her time will have had more years in hand. That is what is meant by 'dying before one's time dies'; and if this is right τοῦ χρόνου means not 'the term of my life', but 'my whole lifetime'. The passage has been rightly understood only by Ellendt–Genthe, *Lexicon Sophocleum*, 789, and by Ed. Fraenkel on Aesch. *Agam.* 1299, who quotes their rendering 'antequam vitae tempus natura praestitutum exierit'. There is another instance of the same usage at *El.* 961–2:

πάρεςτι δ' ἀλγεῖν εἰς τοσόνδε τοῦ χρόνου
ἄλεκτρα γηράςκουςαν ἀνυμέναιά τε.

'up to such a point in the span of life allotted me'.

This reading is therefore linguistically possible. Consider now the difference it would make to the general run of the sense. In the strophe the chorus has said, as many other tragic choruses say, that when great wealth comes to mortals it invites destruction. In the antistrophe it has described how the god first deprives his victims of the power to tell good from evil and then leads them to their doom. Could the antistrophe, and with it the ode, be rounded off by any saying better suited to the context than a reminder that the small man is not exposed to the same danger of divine envy as the great? The *Seven against Thebes*, a play which Sophocles certainly had in mind when dealing with this cognate subject, provides a parallel (766 f.):

τέλειαι γὰρ παλαιφάτων ἀρᾶν
βαρεῖαι καταλλαγαί. τὰ δ' ὀλο-
ὰ πενομένους παρέρχεται.
πρόπρυμνα δ' ἐκβολὰν φέρει
ἀνδρῶν ἀλφηςτᾶν
ὄλβος ἄγαν παχυνθείς. ‖

Compare, in relation to its whole context, *Agam.* 774 f.; see also Hes. *Op.* 5 f., Archilochus fr. 130 W., Eur. *Med.* 122, etc.

The parallelism with the strophe, made so clear to the ear by the repetition of ἐκτὸς ἄτας, may thus have been a feature of the meaning also.

VI. 966–73

The manuscripts give:

> παρὰ δὲ κυανέων πελαγέων πετρῶν διδύμας ἁλὸϲ
> ἀκταὶ Βοϲπόριαι ἠδ' ὁ Θρηικῶν
> Σαλμυδηϲϲόϲ, ἵν' ἀγχίπολιϲ Ἄρηϲ
> διϲϲοῖϲι Φινεΐδαιϲ
> εἶδεν ἀρατὸν ἕλκοϲ
> τυφλωθὲν, κτλ.

966 [πελαγέων fete could.; -ίων UZc] 967 L has ἠδ'. ἰδ' is a conjecture of Triclinius

The antistrophe indicates that l. 966 has two syllables too many; and as the presence in the text of κυανέων might easily cause someone to suppose a reference to the famous rocks of that name and to write in πετρῶν as an explanation, πετρῶν is usually deleted. But that is not all. παρὰ has to be provided with a dative, and this has been done in different ways. Jebb read πελάγει, translating: 'And by the waters of the Dark Rocks, the waters of the twofold sea, are the shores of Bosporus, and Thracian Salmydessus.' Campbell, Masqueray, and Dain agree. Pearson adapted a conjecture of Wieseler (ϲπιλάδων) and read παρὰ δὲ κυανέαιν ϲπιλάδοιν, 'by the dark-blue rocks'. He thought the manuscript reading had originated in a gloss (πελαγίων πετρῶν) explaining this rare word (*CQ* 13 (1919), 125; 22 (1928), 187).

Each of these solutions *might* be right; so might many others. But neither these nor any others that have been offered are particularly economical. No one has seen that the easiest way of supplying παρὰ with its dative is to write παρὰ δὲ κυανέων πελαγέων διδύμαϲ ἁλὸϲ / ἀκτᾶι Βοϲπορίαι, 'by the Bosporus shore of the dark waters of the double sea.'[3] The Symplegades are not the only things that are called κυάνεοϲ, or even the only things in this region that are called so. At Eur. *IT* 7 and 392, the word is used of waters; cf. *Hel.* 179 κυανοειδὲϲ ἀμφ' ὕδωρ. For the plural πελάγη used of the waters of a single sea, cf. *Aj.* 702; with the whole phrase, cf. *Od.* 5. 335 ἁλὸϲ ἐν πελάγεϲϲι. The gloss πετρῶν was not necessarily the gloss of an intelligent reader.

Line 967 is incomplete, as the antistrophe shows; the corresponding line (980) scans – – – ∪∪– ∪∪– – – ∪ –. Where does the gap lie? It has generally been assumed that ἀκται Βοϲποριαι is nominative and is linked to the following nominative by ἠδ' or ἰδ'. Most editors accept Triclinius's ἰδ' and make the line begin: ἀκταὶ Βοϲπόριαι ἰδ' ὁ Θρηικῶν. In consequence, most have assumed that a word scanning as a cretic is

[3] Hartung, Tyrrell, Platt, and Wilamowitz all read ἀκταῖϲ Βοϲπορίαιϲ. M. A. Bayfield in his school edition (Macmillan, 1901) conjectured ἀκτᾶι Βοϲπορίαι: but he accepted Jebb's πελάγει and inserted θ' after διδύμαϲ, unnecessarily.

missing at the end of the line; Boeckh's ἄξενοϲ and Meineke's ἠϊών are the favourite supplements. J. Jackson (*Marginalia Scaenica*, 22 f.) has shown that ἴδ' is very unlikely to be right. His own conjecture, however, is no more convincing than those of his predecessors, because he starts from the same assumption about where the gap must be, and this assumption is not justified. No one (except E. Herkenrath in an otherwise useless article in *BPhW* 1930, 252) has remarked that Θρηικων may perfectly well ‖ have not two, but three syllables. Cf. Aesch. *Pers.* 870, where M's Θρηΐκων should be kept. There is no reason why ὁ Θρηΐκων should not stand at the end of the line and respond with -ευτον γονάν in l. 980; the first of these four syllables is an anceps, so no freedom of responsion is involved. Since the place in the line where we can *see* that there is something wrong is after Βοϲποριαι, it is probable that that is where the gap really is.

What can have stood between Βοϲποριαι and ὁ Θρηΐκων? It is anybody's guess. Perhaps ἠδ' is a corruption of ἦν; if so, something like τόποϲ ἦν may be right.[4] I should obelize the passage, while indicating that the gap should probably come after Βοϲποριαι and that ἀκται Βοϲποριαι may very well be dative singular. No conjecture that rests on the assumption that ἀκται Βοϲποριαι must be nominative, that ἠδ' or ἴδ' must be right, or that the gap must lie at the end of l. 967 is worth putting in any future apparatus.[5]

VII. 983–6

τηλεπόροιϲ δ' ἐν ἄντροιϲ
τράφη θυέλληιϲιν ἐν πατρῴιαιϲ
Βορεὰϲ ἅμιπποϲ ὀρθόποδοϲ ὑπὲρ πάγου,
θεῶν παῖϲ.

Jebb translates: 'That child of Boreas, swift as a steed over the steep hills.' He accepts the explanation of ἅμιπποϲ as meaning ταχεῖα, ἴϲον ἵππωι δυναμένη τρέχειν that is in the Laurentian scholia. But is this unusual word really a mere synonym for 'swift'? Blaydes pointed out that one would expect it to mean 'simul equitans'; and this led him to conjecture Βορέαι and to think the words meant 'riding with Boreas'. But his conjecture is unnecessary. The Greek wind-gods have a curious connection with horses, which led Otto Gruppe to suggest that they were originally imagined in horse's shape (*Griech. Myth. und Religion-*

[4] Are the Thracians mentioned simply to distinguish this Salmydessus from that which Aesch. *PV* 725–7 [Lycophron 1285–90] placed in the Pontus? Cf. Apollodorus, *Bibl.* 1. 9. 21 τὴν τῆϲ Θράικηϲ Ϲαλμυδηϲϲόν.

[5] On Aesch. *Pers.* 870 see A. H. Coxon, *CQ* NS 8 (1958), 51. [On *Ant.* 977 see now L. P. E. Parker, *CQ* NS 16 (1966), 17.]

gesch. ii. 836 f.; cf. Preller–Robert, *Gr. Myth.* 470 f.; Robert, *Gr. Heldensage*, ii. 810 f.). At *Il.* 16. 150 f., Zephyrus takes the form of a horse to beget the horses of Peleus; at *Il.* 20. 223 f., Boreas does so to beget the horses of Erichthonius. Traces of this association seem to be found in the language of later poets. Euripides (*Ph.* 211–13) writes of Zephyrus *riding* in heaven; Pearson, ad loc., rightly observes that 'this is not merely a poetical fancy, but rests on an old superstition which made a wind-god a horse'. Similar expressions occur in the Roman poets, who doubtless derived them from Greek sources; cf. Virg. *Aen.* 2. 417: 'laetus Eois / Eurus equis', Hor. *carm.* 4. 4. 43: 'ceu flamma per taedas vel Eurus / per Siculas equitavit undas'.

This way of speaking helps to explain a passage of the *OT*, which I think Jebb misunderstood. In the first stasimon, the chorus declares that the murderer of Laius must fly with all speed (467 f.):

ὥρα νιν ἀελλάδων
ἵππων cθεναρώτερον
φυγᾶι πόδα νωμᾶν.

ἀελλάδων Hesych.: ἀελλοπόδων codd. contra metrum

What are ἀελλάδεc ἵπποι? Jebb says, 'storm-swift horses'. He compares *OC* 1081 ἀελλαία ταχύρρωcτοc πελειάc, 'a dove with swift strength as of the storm'. He might, indeed, have cited other evidence to show that, in the words of ‖ Pearson (*Fragments of S.* ii. 302), 'the constant association of ἄελλα is with the idea of speed'; cf. the use of ἀελλόποc, ἀελλοδρόμαc. But Campbell took a different view. 'May not some of the force of the patronymic be retained?' he asks, 'swift daughters of the storm'. J. T. Sheppard ('horses of storm') seems to accept this suggestion.

ἀελλάc could probably mean 'wind-swift'; Jebb cannot be proved wrong for certain. But it seems distinctly likelier that the sense is: 'It is time for him to ply in flight a foot mightier than the hoof of the horses of the storm.' Similarly at fr. 688 P. = R., Pearson is wrong to insist that ἀελλάδεc φωναί *must* mean 'swift voices' and in ruling out Campbell's suggestion that it might mean 'voices of the storm'.

If I am right about the *OT* passage, the ἀελλάδεc ἵπποι mentioned there will be the same as the θύελλαι of Boreas who were his daughter's playmates. And we have seen that ἱππεύειν in Greek and *equitare* in Latin are used of the wind-gods' motion through the air. I think ἅμιπποc probably means 'riding the air with' the other Boreads, the dative of θύελλαι being easily supplied from the preceding line. The implication would be that Cleopatra, like the rest, had wings, ἐν οὐρανῶι ἱππεύουcα. Abundant literary and artistic evidence shows that not only the wind-gods themselves, but Cleopatra's brothers, Zetes and

Calais, were imagined as having wings; see L. Stephani, 'Boreas und die Boreaden', *Mémoires de l'Académie Impériale des Sciences de St-Pétersbourg*, 7th ser., 16, 13 (1871), 17 f.; also the articles 'Boreas' in *RE* (Wernicke) and 'Boreaden' in Roscher (Rapp). [On the meaning of ἅμιπποϲ cf. Jacoby on Philochorus, *FGrH* 328 F 71.]

But what is the meaning of ὀρθόποδοϲ ὑπὲρ πάγου, 'beyond the steep hill'? (not 'hills', as Jebb says). Was there near the cave of Boreas any well-known hill or mountain? There is a fragment of Sophocles' Τυμπανιϲταί (637 P. = R.) which Pearson thinks was spoken by Cleopatra herself: ἡμεῖϲ δ' ἐν ἄντροιϲ (Bergk: ἄϲτροιϲ codd.) ἔνθα Ϲαρπηδὼν πέτρα. Pearson, ad loc., points out that the Sarpedonian rock was different from the promontory of that name. Apollonius (*Arg.* 1. 211 f.) describes how Boreas carried off Oreithyia to the neighbourhood of the Ϲαρπηδονίη πέτρη: the scholia ad loc. quote Pherecydes (*FGrH* 3 fr. 145) as saying it was near Mt Haemus. Callimachus clearly thought so too; for at *hy.* 4. 61 f., he makes Ares keep watch on the topmost peak of Haemus, leaving his horses stabled beside the ἑπτάμυχον ϲπέοϲ of Boreas. At *hy.* 3. 114, it is from Haemus that Boreas arrives. It may be that the Sarpedonian rock was identical with that topmost peak of Haemus on which Ares mounted guard. In any case, it seems likely that it was the ὀρθόπουϲ πάγοϲ of the *Antigone*.

This may have a bearing on the opening of Ion of Chios' tragedy *Omphale* (fr. 60 von Blumenthal, p. 35 = *TrGF* 19 fr. 17a),[6] which is preserved in one of the extensive fragments of a work of literary criticism published as P. Oxy. 1611. The relevant passage reads (ll. 121 f.): ὅρων μὲν [ἤ]δη Πέλοποϲ ἐξελαύ[νο]μεν, | Ἑρμῆ, βόρειον [ἵπ]πον, ἄνεταί θ' ὁδόϲ. We know that Hermes was ordered by Zeus to sell Heracles into slavery (Pherecydes *FGrH* 3 fr. 82b); it is likely, therefore, that Heracles spoke these words to Hermes. But what is a βόρειοϲ ἵπποϲ? T. W. Allen (ap. Grenfell and Hunt, P. Oxy. XIII, p. 145) thought it must have been one of the horses sired by Boreas upon the mares of Erichthonius and later taken by Heracles from Laomedon. Now if the horse on which Hermes and Heracles were riding was an ordinary horse, able to travel over land only, it is not likely to have taken them all the way from the Peloponnese to Omphale's ‖ court in Lydia. This difficulty has led A. von Blumenthal (*Ion von Chios*, 36) to suggest that ὅρων . . . Πέλοποϲ here means the boundaries between Phrygia and Lydia.

The legend of Pelops' Phrygian ancestry is well known (see K. Scherling in *RE*, Suppl. vii. 849–50). But there is nothing anywhere to indicate that 'the boundaries of Pelops' can mean anything but the

[6] D. L. Page, *GLP*, no. 19, p. 136.

boundaries of the Peloponnese. The difficulty vanishes if we suppose that the horse in question was not one of Laomedon's horses, which, however remarkable, could travel only along the ground,[7] but a wind-horse of Boreas, specially lent for the occasion. This horse would have wings, so that the party could travel from the Peloponnese to Lydia as the crow flies. The introduction of such a beast upon the stage would be one of those prodigious visual effects in which early tragedy delighted. One recalls, in particular, the winged bird of Oceanus in Aeschylus' *Prometheus* and the winged car of Triptolemus in Sophocles' *Triptolemus*. Boreas himself made a sensational appearance in Aeschylus' *Oreithyia* (see fr. 281 R.).[8]

VIII. 1214–18

ἀλλά, πρόcπολοι,
ἴτ' ἆccον ὠκεῖc, καὶ παραcτάντεc τάφωι
ἀθρήcαθ', ἁρμὸν χώματοc λιθοcπαδῆ
δύντεc πρὸc αὐτὸ cτόμιον, εἰ τὸν Αἵμονοc
φθόγγον cυνίημ' ἢ θεοῖcι κλέπτομαι.

Jebb translates l. 1216: 'pass through the gap, where the stones have been wrenched away'. This is certainly the sense the context would lead us to expect; but the trouble is that *ἁρμὸν* cannot mean 'gap'. Schaefer and Erfurdt thought it might mean 'chink'; they adduced two passages in which it is used of people putting an eye to a 'join' in a door (Plut. *Alex.* 3, Dion. Hal. *Ant. Rom.* 5. 7; cf. Josephus, *AJ* 1. 3. 2). But Hermann disposed of this notion by pointing out that the hole in question here is one big enough to admit a man. Hermann himself took the word 'de perangusto inter saxa intervallo, per quod homo penetrare possit'. But there is no evidence for thinking that *ἁρμὸν* could bear this meaning.

Campbell makes a better attempt to explain the word. 'In preparing the vault', he writes, 'an opening had been left, through which the victim was thrust into the chamber. This opening was then roughly closed with stones. Through these Haemon had passionately torn his way.' Campbell translates: 'Entering within the closure from which the stones have been torn even to the very mouth of the vault.' He

[7] [But Laomedon's horses were able to run over the water; see *Il.* 20. 227–8 (H.-S.).]

[8] [Drawing on the archaeological data provided by K. Neuser, *Anemoi: Studien zur Darstellung der Winde und Windgottheiten in der Antike* (1982), O. Zwierlein, *RhM* 130 (1987), 284 f. has shown that with the exception of the Great Altar at Pergamum, where the four winds in horse-shape draw the chariot of Zeus, the Greeks did not imagine the winds in the shape of horses, though they often spoke of them as 'riding'. But notice Nonn. *D.* 39. 377–8, where the winds, *ἀῆται*, are described as *ἐφιππεύοντεc ἀέλλαιc*. Was the *βόρειοc ἵπποc* of Ion an *ἄελλα*?]

evidently supposed that the opening left in the vault for the victim to be thrust through could be referred to as the 'join'. Bruhn expressed a doubt about the possibility of this; and certainly there is no other instance of the word *ἁρμόϲ* being used in a similar way. Campbell does not strengthen his case by quoting Euripidean passages in which the word is used of the fastening of a gate (*Paralipomena Sophoclea*, 42; cf. Blaydes, *Spicilegium Sophocleum*, ad loc.); this is something quite different. And even if *ἁρμὸν* could be used in this fashion, it is doubtful if the compound adjective *λιθοϲπαδῆ* could be used to perform the function of a participial expression or a relative clause. In other words, we should expect *ἁρμὸν . . . λιθοϲπαδῆ* to mean 'the join made by the tearing away of the stones'. But on Campbell's view it would have to mean either 'the join where the stones have been torn away (by || Haemon)' or 'the join, where you will tear away the stones'. And both these meanings, particularly the second, seem very improbable.

'The join made by the tearing away of the stones' does not make sense. But if we could substitute for 'join' a word that really could mean 'hole' or 'gap' or 'break', we should obtain good sense, the sense that Jebb and others have tried to elicit from the manuscript text, in defiance of the proper meaning of the word *ἁρμὸν*. This can be done by assuming that one letter has been corrupted, and that through a confusion of a very common kind. I think Sophocles wrote *ἀγμὸν*. This word is used in Hippocrates to mean 'fracture' and in Euripides (*IT* 263, *Bacch.* 1094) to mean 'broken cliff'; but its basic meaning is 'break', and there is no reason why it should not be used here of the 'break' made by Haemon as he tore the stones away. On the confusion between gamma and rho, see Bast, *Commentatio Palaeographica*, 732; and note that in the passage of the *IT* mentioned above, where cod. L rightly gives *ἀγμὸν*, cod. P offers the variant *ἁρμὸν* [*sic*].

[See the remarks about these passages in H. Lloyd-Jones and N. G. Wilson, *Sophoclea* (1990).]

32(22)

Sophocles, *Antigone* 1096–7

CREON's reply to the warning of the Chorus that Tiresias has always in the past proved right appears in modern texts as follows:

> ἔγνωκα καὐτὸϲ καὶ ταράϲϲομαι φρέναϲ·
> τό τ' εἰκαθεῖν γὰρ δεινόν, ἀντιϲτάντα δὲ
> ἄτηι πατάξαι θυμὸν ἐν δεινῶι πάρα.

Jackson's demonstration (*Marginalia Scaenica*, 150) that this cannot be right is wholly convincing, and need not be repeated; but his remedy is not acceptable. Accepting Brunck's *δειλόν* in l. 1096, he writes *ἄτηι 'μπαλάξαι τοὐμὸν ἐν δεινῶι κάρα*. This gives good sense at the cost of four changes. None of these is very great, and the corruptions postulated might indeed have occurred; but there are four of them, and a remedy that gives equally good sense with less alteration must be preferred.

I suggest we write:

> τό τ' εἰκαθεῖν γὰρ δεινόν, ἀντιϲτάντα δὲ
> ἄτηϲ πατάξαι θυμὸν ἐν λίνωι πάρα.

This would mean, 'For to give in would be grievous, and by offering resistance my anger may strike against the net of Ate.'

Brunck's *δειλόν* might be correct, but it is an advantage to be able to keep *δεινόν*, a word which gives a sense well suited to the speaker's character. The metaphor of a bird which flies into and so 'strikes against' a net stretched for the purpose among trees or bushes is found first at *Od*. 22. 468 f.:

> ὡϲ δ' ὅτ' ἂν ἢ κίχλαι τανυϲίπτεροι ἠὲ πέλειαι
> ἕρκει ἐνιπλήξωϲι, τὸ δ' ἑϲτήκηι ἐνὶ θάμνωι,
> αὖλιν ἐϲιέμεναι, ϲτυγερὸϲ δ' ὑπεδέξατο κοῖτοϲ,
> ὣϲ αἵ γ' ἑξείηϲ κεφαλὰϲ ἔχον . . .

Pfeiffer aptly cites this passage of Homer in commenting upon the passage where Callimachus describes how the priests in Ceos pray for the North Wind that brings the quails (fr. 75. 36–7):

Classical Review, NS 14 (1964), 129–30.

αἰτεῖcθαι τό τ' ἄημα παραὶ Διὸc ᾧι τε θαμεινοὶ
πλήccονται λινέαιc ὄρτυγεc ἐν νεφέλαιc.

The words *πλήccονται* and *λινέαιc* are both relevant to the passage of Sophocles we are considering. Both occur in a similar context in Oppian, *Hal.* 3. 117, where it is said (not of any birds, but of muraenae) that *ἐνιπλήξωcι λίνοιcι*. For *λίνον* of the fowler's net cf. Paulus Silentiarius, *AP* 9. 396. 1 *λίνου νεφοειδεϊ κόλπωι* / *ἔμπεcε* ... *κόccυφοc* and Archias, ibid. 343. 4 *πλεκτῶν* ... *λίνων* (following upon *κόccυφοc ἠερίηc κόλπον ἔδυ νεφέλης*, l. 2).‖

Homer, Callimachus, and Oppian all speak of the victims 'striking against' the net. Compare the description of Typhoeus at Nonn. *D* 18. 282 f., which appears in Keydell's text as follows:

ἐν ἠερίηι δὲ κελεύθωι
cτοιχάδαc ὑψιλόφωι νεφέλαc ἔcτηcε καρήνωι,
καὶ cκολιαῖc ὄρνιθαc ἐπιπλαγχθένταc ἐθείραιc
πολλάκι cυμμάρψαc πολυχάνδεϊ δαίνυτο λαιμῶι.

ἐπιπλαγχθένταc is the second hand's correction of L's original reading *ἐπιπλαγχένταc*. Adducing the passages of Homer and Callimachus here quoted I have proposed *ἐπιπληγένταc*, or *ἐνιπληγένταc* (*CR*, NS 11 (1961) [R 35], 23).

'The net of Ate' is a familiar notion in tragedy; see, for example, Aesch. *Pers.* 97–9 (with Groeneboom's note), *Agam.* 360, *PV* 1078. For the hunting of birds in antiquity with nets, as well as with snares and bird-lime see Fraenkel, *Aeschylus, Agamemnon*, iii. 612; the net of Ate may well have been imagined as a bird-catcher's net fixed between two trees or in a bush for the victim to run, as we say, smack into it.[1]

[1] [I omit a paragraph defending the paradosis at Soph. *Phil.* 324 against Lambinus' conjecture, now accepted by N. G. Wilson and myself in our new OCT: *θυμῶι* = *μοι ὠργιcμένωι* (so Σ) is unexceptionable (*θυμόν* has the same sense at *Ant.* 1097), but *χεῖρα πληρῶcαι* seems too difficult.]

33(49)

Notes on Sophocles' *Trachiniae*

I. 122–3

ὧν ἐπιμεμφομέναϲ ἁ-
δεῖα μέν, ἀντία δ' οἴϲω.

Kamerbeek's defence of L's *ὧν ἐπιμεμφομέναϲ* against the *ὧν ἐπιμεμφομένα ϲ'* of Laur. 32. 20 and other later manuscripts is rightly approved by Page, *Gnomon*, 32 (1960), 318. More questionable is his retention of the manuscript reading *ἁδεῖα* against Musgrave's *αἰδοῖα*, which has been accepted by Jebb, Pearson, Masqueray, and Dain–Mazon. Hermann pointed out long ago that *ἁδεῖα* as neuter plural was not Attic, and Kamerbeek's parallels are not relevant.

But Musgrave's *αἰδοῖα* is hardly more satisfactory. The expression *ἀντία οἴϲω* is equivalent to a single verb (cf. *ἀντιφερίζω*, etc.), so that *ἀντία* could hardly be so far disjoined from *οἴϲω* as to be made antithetic to a second neuter plural adjective governed by that verb. Further, one may say *ἀντία οἴϲω*, but it is far less easy to say *αἰδοῖα οἴϲω*. For the antithesis yielded by Musgrave's conjecture I see no real parallel.

Hermann and Campbell both took *ἁδεῖα* as nominative singular and supplied *εἰμί*. Ellipse of the first person is rarer than ellipse of the third, but Denniston on Eur. *El.* 37 and Fraenkel on Aesch. *Agam.* 806 give instances; Aesch. *Cho.* 412, Soph. *OC* 207, and Eur. *Cycl.* 503 are all good ones. Ellipse is particularly common with adjectives like *ἕτοιμοϲ* or *πρόθυμοϲ*, and *ἁδεῖα* would resemble these in signifying readiness to accept the other's point of view. But can the word bear this sense?

'"In a pleasant vein", Jebb objected, 'must mean "suggesting thoughts of comfort"; as in *OT* 82 "pleasant" = "bringing good news".' But since *ἀντία* expresses remonstrance against Deianeira's despair, there is no proper antithesis with *ἁδεῖα*. If *ἁδεῖα* can only ‖ mean 'pleasant' or 'comfortable', there is much force in this argument. We should expect the sense here to be, 'I am kindly disposed towards you, but I shall offer a remonstrance.' Can *ἡδύϲ* mean 'favourably

Yale Classical Studies, 22 (1972), 263–70.

disposed'? The only passage in tragedy where it seems to have this meaning is Eur. *Phoen.* 771. Eteocles wishes to consult Teiresias, but because of a past quarrel he asks Creon to approach him on his behalf. *ϲοὶ μὲν γὰρ ἡδὺϲ εἰϲ λόγουϲ ἀφίξεται*, he says to Creon, and in this context *ἡδύϲ* can only mean 'kindly', as Powell thinks. One could wish that there were more examples of the sense in question; but the *Phoenissae* passage seems to indicate that the view of Hermann and Campbell is likelier to be right than any other that has so far been put forward.

Pearson in his commentary renders it by 'gladly', comparing Dem. 23. 64 *ἡδίουϲ ἔϲεϲθε ἀκούϲαντεϲ*. But in that place the sense is not 'You will be the gladder for having heard it', but 'You will be in a sweeter temper, a better state of mind, for having heard it'; and in the *Phoenissae* passage the presence of the pronoun *ϲοί* clearly limits the operation of the word to the relation of Teiresias to Creon. LSJ gives as sense II. 2 of *ἡδύϲ* 'well-pleased, glad'. The only instances given besides the passage of Demosthenes already dealt with are two passages of Plutarch. At *Cam.* 32 *ἡδίουϲ . . . ταῖϲ ἐλπίϲιν* means 'more optimistic'; but we could extract this sense from 'sweeter in their hopes' without taking *ἡδύϲ* to mean 'glad'. At *Fab.* 5 *τὴν γνώμην ἡδίω πρὸϲ τὸ μέλλον ποιεῖν* yields no evidence for such a meaning.

II. 216–20

αἴρομαι οὐδ' ἀπώϲομαι
τὸν αὐλόν, ὦ τύραννε τῆϲ ἐμῆϲ φρενόϲ.
ἰδού μ' ἀναταράϲϲει
εὐοῖ ⟨εὐοῖ⟩ μ'
ὁ κιϲϲὸϲ ἄρτι βακχίαν
ὑποϲτρέφων ἅμιλλαν.

[216 *αἴρομαι* Lloyd-Jones, *CR*, NS 31 (1981) [R119], 170: *ἀείρομ'* codd.]

Jebb and Kamerbeek take *ὑποϲτρέφων* to mean 'whirling a little'. The word could certainly bear this meaning, but when one considers it in this context, it is hard not to agree with Pearson (*CR* 39 (1925), 2) that this view is 'not attractive'.‖

Others—for example Brunck, Hermann, Wunder, Campbell—have taken the sense to be 'turning me round'. Pearson (loc. cit.) objects that in this way the word would have to mean 'turning back something which is in process'. This objection has no force. The word may perfectly well mean that the speaker tries to go forwards but is whirled backwards as a ship is driven backwards by a contrary wind.

This is confirmed by what seems to be the most relevant parallel, which seems not yet to have been adduced. That is found in the words of Cassandra at the beginning of her final outburst of revelation (Aesch. *Agam.* 1214 ff.):

> ἰοὺ ἰού, ὢ ὢ κακά.
> ὑπ' αὖ με δεινὸϲ ὀρθομαντείαϲ πόνοϲ
> ϲτροβεῖ ταράϲϲων φροιμίοιϲ . . .

ϲτροβεῖν is 'an iterative-intensive form of ϲτρέφειν' (see Schwyzer, *Gr. Gram.* i. 720, l. 4) so that ὑπὸ . . . ϲτροβεῖ must be equivalent to ὑποϲτρέφει: Fraenkel rightly translates it 'whirls me round'.

At Ar. *Pax*, 174–6, the manuscripts read

> ὦ μηχανοποιέ, πρόϲεχε τὸν νοῦν, ὡϲ ἐμὲ
> ἤδη ϲτροφεῖ τι πνεῦμα περὶ τὸν ὀμφαλόν,
> κεἰ μὴ φυλάξειϲ, χορτάϲω τὸν κάνθαρον.

Coulon and Platnauer [also Sommerstein (1985)] both print Dindorf's ϲτρέφει, but I should prefer ϲτροβεῖ.

III. 831–7

> εἰ γάρ ϲφε Κενταύρου φονίαι νεφέλαι
> χρίει δολοποιὸϲ ἀνάγκα
> πλευρὰ προϲτακέντοϲ ἰοῦ,
> ὃν τέκετο θάνατοϲ, ἔτρεφε δ' αἰόλοϲ δράκων,
> πῶϲ ὅδ' ἂν ἀέλιον ἕτερον ἢ τανῦν ἴδοι,
> δεινοτάτωι μὲν ὕδραϲ
> προϲτετακὼϲ φάϲματι;

'Melted into, glued to the most dread shape of the Hydra' makes no sense, and in consequence innumerable emendations of φάϲματι have been attempted. A large collection will be found in ‖ Jebb's appendix on the passage; Dr A. A. Long, the author of a valuable book on *Language and Thought in Sophocles* (London, 1968), has just revived θρέμματι, one of several suggestions made by Blaydes (*GRBS* 8 (1967), 275 f.).

I am convinced by none of these suggestions; and I believe this to be one of the fairly numerous cases in which a comparatively rare word has been unjustly vexed, while a common and therefore innocent-looking word in the same passage conceals the canker. In this place the common word which I believe to be corrupt is δεινοτάτωι. Eduard Fraenkel has lately reminded us (*Horace*, 20 n. 4 and *SB der Bayer. Akad.* (1957), 48 n. 137) how often the endings -τατοϲ and -τεροϲ are confused; in this case the confusion may have been occasioned by a misunderstanding of the sense. I believe that Sophocles wrote δεινοτέρωι, and meant 'glued to a shape more deadly than the Hydra'.

The Lernaean Hydra Heracles had vanquished; he now found himself locked in combat with a more formidable antagonist, that *Ἐρινύων ὑφαντὸν ἀμφίβληςτρον* (1051–2) which is so vividly personified in the great speech that follows the *μέλος ἀπὸ ςκηνῆς*:

πλευραῖςι γὰρ προςμαχθὲν ἐκ μὲν ἐςχάτας
βέβρωκε ςάρκας, πλεύμονός τ' ἀρτηρίας
ῥοφεῖ ξυνοικοῦν· ἐκ δὲ χλωρὸν αἷμά μου
πέπωκεν ἤδη, καὶ διέφθαρμαι δέμας
τὸ πᾶν, ἀφράςτωι τῆιδε χειρωθεὶς πέδηι.

But can *φάςμα* be used in this way to signify the shirt of Nessus? One might quote l. 508, where Achelous as he comes in bull shape to fight Heracles is called *φάςμα ταύρου*, but there he may be called 'the apparition of a bull' because he looks like one. But consider Eur. *Cretes* fr. 82. 23–4 in Colin Austin's *Nova Fragmenta Euripidea in Papyris Reperta* (Berlin, 1968). Minos, says Pasiphae, did not sacrifice the sacred bull to Poseidon as he had promised:

ταῦρον γὰρ οὐκ ἔςφαξ[εν ὅν γ' ἐπηύ]ξατο
ἐλθόντα θύςειν φάςμα [πο]ντίω[ι θε]ῶι.

Page (*GLP* 74) translates 'that bull, that apparition'. The bull is so called because it is uncanny, strange, mysterious; so, if I am right about the Sophoclean passage, is the shirt of Nessus.||

IV. 1004–43

A. H. Coxon in *CR* 61 (1947), 7 f. performed a notable service to the text of Sophocles by challenging the arrangement of the *μέλος ἀπὸ ςκηνῆς* proposed by Seidler, *De versibus dochmiacis tragicorum graecorum* (1811–12), pp. 311 f., and accepted by all editors down to and including Pearson. Dain and Mazon in their Budé edition (vol. i, 1955) without knowledge of Coxon's article took the same view as he that the strophe extends from ll. 1004 to 1017 and the antistrophe from ll. 1025 to 1043, against Seidler, who had adopted the following unusual scheme:

(Heracles): str. *α′*, str. *β′*, hexameters, antistr. *α′*.
(Old Man, then Hyllus): hexameters.
(Heracles): str. *γ′*, antistr. *β′*, hexameters, antistr. *γ′*.

Dain and Mazon's arrangement is somewhat different from that of Coxon. Indeed, the text is so corrupt that there is room for some difference of opinion about its proper constitution; and though I wish to put forward a third arrangement that differs in some respects both from that of Coxon and that of the Budé editors, I hardly suppose that I shall say the final word on this difficult problem.

I shall start by giving my own version of the text and then try to justify it in a line-by-line commentary:

1004–17 = 1023–43

⟨– – – ∪ –⟩	ὦ παῖ, ποῦ ποτ᾽ εἶ;
ἐᾶτέ μ᾽, ἐᾶτέ με	τᾶιδέ με, τᾶιδέ με
δύϲμορον εὐνάϲαι	πρόϲλαβε κουφίϲαϲ
ἐᾶτέ με δύϲτανον	ἐέ, ἰὼ δαῖμον,
πᾶι ⟨πᾶι⟩ μου ψαύειϲ; ποῖ κλίνειϲ;	θρώιϲκει δ᾽ αὖ, θρώιϲκει δειλαία
ἀπολεῖϲ μ᾽, ἀπολεῖϲ,	διολοῦϲ᾽ ἡμᾶϲ
ἀνατέτροφαϲ ὅ τι καὶ μύϲηι.	ἀποτίβατοϲ ἀγρία νόϲοϲ.
(hexameters)	(hexameters ending with ὦ γλυκὺϲ Ἅιδαϲ, transposed by Seidler from l. 1041, after αὐθαίμων)
1015 οὐδ᾽ ἀπαράξαι ⟨μου⟩	1041 ὦ Διὸϲ αὐθαίμων
κρᾶτα βίου θέλει	εὔναϲον εὔναϲόν μ᾽,
⟨– ∪∪ –⟩ μολὼν τοῦ ϲτυγεροῦ; φεῦ φεῦ.	ὠκυπέται μόρωι τὸν μέλεον φθίϲαϲ. ‖

1004 ante hunc v. lacunam posuit Coxon 1005 ὕϲτατον Σ^γρ εὐνάϲαι A: εὐνᾶϲαι L 1006 εὔναϲαι (-ᾶϲαι L) post δύϲτανον del. Dain 1007 ⟨πᾶι⟩ Seidler 1009 ἀν⟨α⟩τέτροφαϲ Erfurdt

1015 ⟨μου⟩ post Blaydes add. Pearson 1017 ⟨παυϲίπονοϲ⟩? ⟨λυϲίπονοϲ⟩? 1023 ὦ παῖ: ὦ παῖ πᾶι L cett. 1025 ἰὼ rec.: ἰὼ ἰὼ L 1042 εὔναϲον εὔναϲόν μ᾽ Erfurdt: εὔναϲόν μ᾽ εὔναϲον codd.

1004–6. Coxon supposes that a dochmiac metron responding with ὦ παῖ, ποῦ ποτ᾽ εἶ (1023) has dropped out before ἐᾶτέ μ᾽. Dain and Mazon suppose that ἐᾶτέ μ᾽ responds with ὦ παῖ, ποῦ, that between ἐᾶτέ μ᾽ and ἐᾶτέ με a cretic has been lost, and that ἐᾶτέ με δύϲμορον responds with τᾶιδέ με τᾶιδέ με. So far there can be no doubt that Coxon's view is likelier to be right, since ἐᾶτέ μ᾽, ἐᾶτέ με and τᾶιδέ με τᾶιδέ με very probably respond with one another.

Coxon makes δύϲμορον εὐνάϲαι (1005) respond with πρόϲλαβε κουφίϲαϲ (1025). Dain, having used up δύϲμορον already, has to place in the text the ὕϲτατον which appears in L's margin as a variant on δύϲμορον, so that ὕϲτατον εὐνάϲαι is made to respond with πρόϲλαβε κουφίϲαϲ. Here again one must prefer Coxon's view. It is not likely that ὕϲτατον belongs in the text; it may have originated as a corruption of δύϲτανον in l. 1006, as Seidler thought. It is just worth contemplating the possibility that the stanza began with ἐὲ ἐᾶτέ με responding with ὦ παῖ, ποῦ ποτ᾽ εἶ;: in that case one would have to fill up the space by inserting a third ἐᾶτέ με or by accepting Dain's insertion of ὕϲτατον. But Coxon's way of dealing with the difficulty is to be preferred.

1006. Here Coxon follows Triclinius and several modern scholars in holding that the manuscript reading ἐᾶτέ με δύϲτανον εὐνάϲαι is 'a simple variant of *ἐᾶτέ με δύϲμορον εὐνάϲαι* which has displaced an exclamation metrically equivalent to *ἒὲ ἒὲ ἰὼ ἰὼ δαῖμον'*. But Dain obtains responsion simply by deleting *εὐνάϲαι*, so that *ἐᾶτέ με δύϲτανον* answers to *ἒὲ ἰὼ δαῖμον*. This responsion is perfectly legitimate; *εὐνάϲαι* could easily have got in from the line before; and in this place Dain's solution is to be preferred.

1016–17 ~ 1041–2. In the manuscripts the words *ὦ γλυκὺϲ Ἀίδαϲ* (thus L: some others have *Ἅιδαϲ*) appear after *ὦ Διὸϲ αὐθαίμων*. Seidler transposed them to fill a gap at the end of the ‖ hexameters, in l. 1040; he has been followed by all modern editors, and also by Coxon. Dain prefers to suppose that the end of l. 1040 has been lost, to keep *ὦ γλυκὺϲ Ἀίδαϲ* after *ὦ Διὸϲ αὐθαίμων* and to suppose that *ὦ γλυκὺϲ Ἀίδαϲ* responds with *κρᾶτα . . . θέλει* (1016). Does the actual sense of ll. 1015–17 make Dain's solution likely?

In l. 1016 almost all modern scholars have accepted Wakefield's alteration of *βίου* to *βίαι*. Kamerbeek is an exception; his fear of departing from the manuscripts occasionally leads him to prefer the right reading.[1] If it is true that strophic responsion is present here, something must be missing from this stanza; and once that is admitted, the rashness of altering *βίου* is obvious. Kamerbeek has observed that a scholion written against l. 1016 in L gives a distinct indication of what the sense of the complete stanza may have been: *οὐδεὶϲ ἐκείνων, φηϲί, βούλεται ἐλθὼν τὴν κεφαλήν μου ἀποτεμεῖν καὶ ἐλευθερῶϲαι τοῦ μοχθηροῦ βίου;*

What, then, is likely to have dropped out? Dain, who keeps *γλυκὺϲ Ἀίδαϲ* (*sic*) where it is in the manuscripts, thinks – ∪∪– ∪ × is missing in l. 1016 after *θέλει*. At the beginning of l. 1017, he thinks a choriamb is missing; so that *ὠκυπέται μόρωι τὸν μέλεον φθίϲαϲ* responds with – ∪∪– *μολὼν τοῦ ϲτυγεροῦ· φεῦ φεῦ*.

Coxon accepts Seidler's transposition of *ὦ γλυκὺϲ Ἅιδαϲ* (*sic*) to the vacant space at the end of l. 1040; so he does not suppose that – ∪∪– ∪ × is missing at the end of l. 1016. But in l. 1017 he accepts Erfurdt's transposition *τοῦ ϲτυγεροῦ μολών*, deletes *φεῦ φεῦ*, and prints the line like this: *τοῦ ϲτυγεροῦ μολὼν* – ∪∪– × .

Working backwards from l. 1047, it is clear the Coxon had no good reason to accept Erfurdt's transposition, or to cut out *φεῦ φεῦ*. For we have only to suppose a choriamb is missing at the beginning of the line to get responsion; in other words, *ὠκυπέται μόρωι τὸν μέλεον φθίϲαϲ* may well have responded with ⟨– ∪∪–⟩ *μολὼν τοῦ ϲτυγεροῦ· φεῦ φεῦ*. Can we hazard any guess at what the sense of the missing four syllables may have been? Let us look back at the text of the scholion which

[1] Bergk, it is true, kept *βίου*; but as he postulated no lacuna in this line, it is hard to see what sense he made of it.

Kamerbeek and I agree may furnish a clue to the missing part of this stanza. Kamerbeek's effort at a supplement may be found on p. 215 of his commentary. It might, I think, have been more convincing had he not accepted without ‖ question Dain's view that *ὦ γλυκὺς Ἅιδας* stands in its proper place in the manuscripts. For suppose for the sake of the argument that Dain is wrong, and that Seidler's transposition is correct. Then one could easily fill the space and bring the sense of the text into line with the paraphrase in the scholion by supplying before *μολών* a single compound adjective; something like *παυσίπονος* (Eur. *IT* 451, Ar. *Ran.* 1321), *λυσίπονος* (Pind. *Pyth.* 4. 41; fr. 131. 1), *παυσανίας* (Soph. fr. 887 P. = R.), *λυσανίας* (Ar. *Nub.* 1162), *ῥυσίπονος* (anon. *AP* 9. 525. 18). The sense is so neatly supplied by such an insertion that I feel the case for Seidler's transposition is considerably strengthened by these observations.

ADDENDUM

In the second of these notes, I ought to have explained that I take *μ'* to be the object both of *ἀναταράσσει* and of *ὑποστρέφων*. *Βακχίαν ... ἅμιλλαν* I take to be an accusative limiting the action of the verb, so that the words mean, 'whirling me around in the Bacchic race'.

By using the word 'race', I do not mean that there is any suggestion of a competition. *ἅμιλλα*, like 'race', may imply simply rapid movement, without any suggestion of 'racing' in the competitive sense. At Eur. *Or.* 456, for instance, Di Benedetto rightly says that in the phrase *γέροντι δεῦρ' ἁμιλλᾶται ποδί* the verb means not 'contendere', but 'affrettarsi'. Kannicht on Eur. *Hel.* 546 says that in such instances the basic meaning of 'certare' is still present, because the movement in question has to overcome a resistance; he quotes also Eur. *Hypsipyle* fr. 764 (p. 24 Bond).

In his article on 'Der Einzug des Chors im *Prometheus*' (*ASNP*, 2nd ser., 23 (1954), 269 f. = *Kleine Beiträge zur klassischen Philologie*, i. 389 f.), Eduard Fraenkel tried to show that the Oceanids entered not in a 'celestial omnibus', but each in her separate winged car. He proved that the separate winged car was not a prerogative of Triptolemus, as earlier scholars had supposed, and went on to argue that the words *πτερύγων θοαῖς ἁμίλλαις* (l. 129) showed that each Oceanid had her separate car, or they could not have raced each other. When his article first appeared, Sir Denys Page pointed out to me that, since *ἅμιλλα* need not imply a 'race', the proof was not complete.

[These notes are in part modified in the light of H. Lloyd-Jones and N. G. Wilson, *Sophoclis Fabulae* (OCT, 1990) and *Sophoclea* (1990).]

34(25)

A Problem in the Tebtunis *Inachus*-Fragment [= fr. 269c R.]

THE second column of P. Tebtunis 692, very probably from Sophocles' *Inachus*, is printed by Pfeiffer[1] as follows:

⟨*Xo. σατύρων*⟩ *πολὺ πολυιδρίδας*[2]
ὅτις ὅδε προτέρων
ὄνομ' εὖ σε θροεῖ,
τὸν Ἀϊδοκυνέας
σκότον ἄ⟨β⟩ροτον ὑπαί.

⟨*Xo. α'*⟩ *τὸν Διὸς μὲν οὖν ἐρώτων ἄ[γγ]ελον, μέγαν τρόχιν,*
⟨*Xo. β'*⟩ *εἰ̣[κ]άσαι πάρεστιν Ἑρμῆν π[ρὸ]ς τὰ σὰ ψοφήματα,*
⟨*Xo. γ'*⟩ *αὐτὸ̣ν ὄ̣ντα σ', αὐτόν, ὅ̣ς μ̣ο̣ι̣ δεῦρ' ἀνέστρεψεν πόδα·*
⟨*Xo. δ'*⟩ *δευτέρους πόνους ἔοικας πρὶν μύσαι κενοὺς ἐλᾶν.*

The word *προτέρων* in l. 17 has perplexed editors. Hunt in the *ed. pr.* inclined to take it to mean 'those of the past' and observed that in that case *σ' ἐθρόει* should be read in l. 18, but thought that if that were done *ὅδε* would be difficult. Körte, whom Page followed, read *προτερῶν*, a participle from *προτερέω*. Pfeiffer pointed out that *προτερέω* occurs only in prose and is not easy to accept in lyric verse, especially when it is a case of a participle governing *σε* as a direct object. Pfeiffer kept *προτέρων*, but took it in a local sense, 'in the front rows of the chorus'; this is unparalleled, and is hardly less objectionable than Körte's view.

One cannot help agreeing with Hunt that 'those of the past' is the most natural sense. Is the presence of *ὅδε* really a fatal objection to accepting it? *ὅδε* is normally deictic, but it may be applied to an absent person who is present *to the speaker's thoughts*. In Soph. *El.* 540 Clytemnestra applies it to her long-dead daughter Iphigenia:

[1] There is little to add to the bibliography of the piece given by Page, *GLP* no. 6, p. 22 [but now see Radt, *TrGF* iv. 267 f.]; the detailed discussion by Pfeiffer in *SB der Bayer. Akad.* 1938/2 is indispensable. P. Oxy. 2369 is certainly from the *Inachus*; this too has been discussed in detail by Pfeiffer (*SB der Bayer. Akad.* 1958/6).

[2] To the material assembled to illustrate the meaning of *πολυιδρίδας* by Pfeiffer, *SB der Bayer. Akad.* 1938/2. 30–1, we may now add *ποιμανδρίδαι* in a fragment of Aeschylus (P. Oxy. 2256, fr. 71 [= 451k. 8 R.]).

Classical Review, NS 15 (1965), 241–3.

πότερον ἐκείνωι (Menelaus) *παῖδες οὐκ ἦσαν διπλοῖ*
οὓς τῆςδε μᾶλλον εἰκὸς ἦν θνήισκειν;

Similarly at Eur. *Alc.* 881 Admetus applies it to his wife Alcestis; the usage occurs in the *Odyssey* (3. 352), where Nestor, speaking of the long-missing Odysseus, says *τοῦδ' ἀνδρὸς Ὀδυσσῆος*. Since the long-dead person is vividly present to the speaker's thought, *ὅδε* can be used when we should expect *ἐκεῖνος*; cf. Kühner–Gerth, i. 644; Page on Aesch. *Agam.* 57; Handley on Men. *Dysk.* 185, 234; [Newiger, *RhM* 108 (1965), 237 n. 26; Taplin, *The Stagecraft of Aeschylus* (1977), 150–1].

If this is right, we must read *c' ἐθρόει* and translate, 'Very, very knowing was he amongst those of the past who rightly called you by your name.' We are immediately reminded of the opening words of the second stasimon of Aeschylus' *Agamemnon*, where the Chorus wonders who gave Helen a name which they show by an etymology to be specially appropriate:

τίς ποτ' ὠνόμαζεν[3] *ὧδ'*
ἐς τὸ πᾶν ἐτητύμως;
μή τις ὅντιν' οὐχ ὁρῶμεν προνοί-
αισι τοῦ πεπρωμένου
γλῶσσαν ἐν τύχαι νέμων;
τὰν δορίγαμβρον ἀμφινει-
κῆ θ' Ἑλέναν;

It is hard to resist the conclusion that the chorus in the Tebtunis fragment also is offering the etymology of a name, the name of a person with whom it has identified the invisible presence on the stage in the lost portion of the text immediately before col. ii. Can we know what this name was?

The person in question is described as 'you who are beneath the uncanny darkness of the cap of Hades'. The cap of Hades is the cap that confers invisibility; see the data given by Pfeiffer, pp. 33 f. But in known myths it is never used by Hades, but by other gods, notably by Hermes, who lends it to Perseus; Hades is named as its owner only by Apollodorus, *Bibl.* 1. 2. 1 (perhaps following a Titanomachy). It has long been recognized that it had originally nothing to do with Hades but was associated with him simply because it was called 'the cap of invisibility', and for the ancients *Ἀΐδης* (*Ἅιδης*) meant 'the invisible one'.[4] Whether this etymology is false, as Schulze and Wackernagel

[3] Fraenkel and Denniston–Page prefer Tr's *ὠνόμαζεν* to F's *ὠνόμαξεν*, taking it as an instance of the special use of *ὀνομάζειν* in the imperfect (Fraenkel ii. 329). Barrett on Eur. *Hipp.* 32–3 (p. 162) warns us that this special use depends on two dubious instances; but if *ἐθρόει* is right here, the judgement of the recent editors of the *Agamemnon* acquires fresh support.

[4] See J. Lamer, *RE* xi. 2519 f. and J. Roeger, *Ἄϊδος κυνέη* (Diss. Graz, 1924) (reviewed by Kretschmer in *Glotta*, 15 (1927), 175).

thought, or Felix Solmsen was right in defending it against them makes no difference.[5] It is three times alluded to by Plato,[6] and is undoubtedly presupposed by several passages of early tragedy.[7] ‖

We can scarcely resist the inference that the person whose name is etymologized in this fashion by the chorus was in fact Hades. Since the cap of invisibility bore his name, it is not surprising that the chorus' first guess at the identity of the invisible person should be that he is Hades. True, Hades made only rare appearances in the upper world;

[5] For the literature of the subject, see Frisk, *Griechisches etymologisches Wörterbuch*, 34; the reference to Nilsson should now read '2nd edn. 455, with nn. 4 and 5'. Cf. Felix Solmsen, *Untersuchungen zur griechischen Lautlehre*, 71 f.

[6] See Dodds on *Gorgias* 493 B 4.

[7] At Aesch. *PV* 152 the words of the chorus *τὰ πρὶν δὲ πελώρια νῦν ἀϊϲτοῖ* prompt Prometheus to wish that Zeus had hidden him in Tartarus, lower even than Hades, instead of exposing him to derision in the upper air:

> *εἰ γάρ μ' ὑπὸ γῆν νέρθεν τ' Ἀΐδου*
> *τοῦ νεκροδέγμονοϲ εἰϲ ἀπέραντον*
> *Τάρταρον ἧκεν, κτλ.*

Did the word *ἀϊϲτοί*, by reason of this etymology, suggest the idea of Hades? Compare *Agam.* 462 f.:

> *κελαι-*
> *ναὶ δ' Ἐρίνυεϲ χρόνωι*
> *τυχηρὸν ὄντ' ἄνευ δίκαϲ*
> *τιθεῖϲ' ἀμαυρόν, ἐν δ' ἀΐ-*
> *ϲτοιϲ τελέθοντοϲ οὔτιϲ ἀλκά.* ‖

ἐν δ' ἀΐϲτοιϲ here means 'among the dead, in Hades'; the use of *ἀϊϲτόω* and its compounds to mean 'destroy' probably originated in this way.

With *τιθεῖϲ' ἀμαυρόν* compare *Eum.* 359, where in a similar context the Erinyes say

> *κρατερὸν ὄνθ' ὁμοίωϲ*
> *μαυροῦμεν ὑφ' αἵματοϲ νέου,*

and also Hes. *Op.* 323

> *ῥεῖά τέ μιν μαυροῦϲι θεοί.*

ἀμαυρόν is an epithet of an *εἴδωλον* at Hom. *Od.* 4. 804; compare Sappho fr. 55. 3–4 L.–P.:

> *. . . ἀλλ' ἀφάνηϲ κἀν Ἀΐδα δόμωι*
> *φοιτάϲηιϲ πεδ' ἀμαύρων νεκύων ἐκπεποταμένα.*

In this passage of Sappho *ἀφανὴϲ κἀν Ἀΐδα δόμωι* is a kind of oxymoron, and *ἀμαύρων*, as in the Aeschylean passages previously cited, has more force if it means 'obscure' than if it means 'weak'. Just so at *Cho.* 157 *κλύ', ὦ δέϲποτ', ἐξ ἀμαυρᾶϲ φρενόϲ* 'dark mind' is preferable to 'feeble mind'. The sense 'weak' is common, and the sense 'dark' could easily have derived from it, as Barrett on Eur. *Hipp.* 816 (p. 319) supposes. But the considerations just mentioned make me think it likelier that the sense 'dark' came first, as LSJ s.v. would have it. Maas in Passow–Croenert s.v. (p. 342) says 'schwach, matt, besonders von Licht und Sehen'; see also Wilamowitz on Eur. *HF* 124.

Certain epithets of Hades seem to glance at the etymology in question. At Aesch. *Sept.* 860 *πάνδοκον εἰϲ ἀφανῆ τε χέρϲον* refers to Hades; the text is right, and the conjecture which Wilamowitz printed with the word 'correxi' is one of the worst he ever made. At Soph. *OC* 1681 the *ἄϲκοποι πλάκεϲ* are those of Hades. Cf. Aesch. *PV* 1028–9:

> *εἰϲ ἀναύγητον μολεῖν*
> *Ἅιδην κνεφαῖά τ' ἀμφὶ Ταρτάρου βάθη,*

and Soph. *Aj.* 608

> *τὸν ἀπότροπον ἀΐδηλον Ἅιδαν.*

but we must remember that he not only raped his niece at Enna but fought his nephew at Pylos (Hom. *Il.* 5. 395 f.; Pind. *Ol.* 9. 33). Still, in the tetrameters that follow the lyric strophe the statement is corrected; *μὲν οὖν* bears its common sense of 'immo vero', and the speaker is saying, 'No, it is not Hades, it is Hermes.'

35(50)

Tycho von Wilamowitz-Moellendorff on the Dramatic Technique of Sophocles

No project lay nearer to the heart of Eduard Fraenkel[1] during his last years than that of promoting a reprint of the famous book *Die dramatische Technik des Sophokles*, by Tycho von Wilamowitz-Moellendorff, which was first published as vol. 22 of Philologische Untersuchungen in 1917. Tycho Wilamowitz, the son of Ulrich von Wilamowitz-Moellendorff and the grandson of Theodor Mommsen, was killed fighting against the Russians near Ivangorod on the night of 14/15 October 1914. After his death the manuscript was prepared for publication by his friend Ernst Kapp, who has explained in the foreword of the book the nature of his services.

Fraenkel's long efforts to arrange for a reprint of this important book seemed at one time to have been successful, and he began work on an introduction, which would have been of great interest to scholars. At his death he left a number of notes intended for this introduction, most of which had clearly been carefully written and would have been inserted in the text. After Fraenkel's death on 5 February 1970, I was asked by a publisher—not the original publisher—if I would write an introduction. I did so, but after about a year heard from this publisher that owing to legal difficulties he would be unable to proceed with the reprint. In July 1972 I became aware that a reprint by the original publisher, bearing the date 1969, was on the market.

Tycho's book, like the whole series of Philologische Untersuchungen, had been published by the firm of Weidmann. 'Der Fortschritt der Forschung', wrote Wilamowitz (*Erinnerungen*, 2nd edn., 1928), 196), 'vollzieht sich ganz besonders durch Einzeluntersuchungen, die das

[1] 'Dei tragici egli [il Fraenkel] aveva studiato soprattutto, negli anni della maturità, Eschilo ed Euripide; ma negli ultimi tempi, per un ulteriore bisogno di "classicità", la sua preferenza si era spostata su Sofocle, "più greco" nella sua serenità tragica. Di Sofocle si era occupato con crescente frequenza, in articoli e in seminari (nel '68 aveva letto, a Roma e ad Oxford, il Filottete), e avrebbe certo continuato ad occuparsene se fosse ancora vissuto': Sebastiano Timpanaro, *Atene e Roma*, NS 15 (1970), 92. [Rather than speak of 'classicism' or of 'tragic serenity', I would say Fraenkel was attracted by Sophocles' depiction of the hero who despite his suffering resists a hostile world.]

Classical Quarterly, NS 22 (1972) 214–28 = *Blood for the Ghosts* (1982), 219–37, where Fraenkel's notes are translated.

Maß von Zeitschriftartikeln überschreiten und als Buch eine zu geringe Zahl von Kaufern finden. Da wird einer vornehmen Buchhandlung, wie es die Weidmannsche zur Ehre Deutschlands immer war und ist, das Opfer durch Aufnahme in eine Serie etwas erleichtert.' This was not the only instance of what Wilamowitz called the 'großartige Opferwilligkeit' (ibid. 237) displayed by the firm of Weidmann under its then director, the enlightened Reimer, the friend of Mommsen and of Wilamowitz. All human institutions are subject to the law of change. Scholars will regret that Tycho von Wilamowitz's sister, Dorothea Freifrau Hiller von Gaertringen, who died on 24 March 1972 in her ninety-third year, did not see the book reprinted in the way she wished it to be.

In these circumstances it has proved impossible to use the introduction for the purpose for which it was composed. But Fraenkel's material should surely || be given to the world; and as my introduction provides it with a framework, they may as well appear together. I apologize for my inability to provide an introduction such as Fraenkel would have written. At least I know from many conversations, as well as from Fraenkel's notes themselves, that though we differed over some of the problems discussed in Tycho's book, over the main features of his work we were in complete agreement. Nearly all the examples cited in the footnotes to what follows come from Fraenkel's notes, and all the parts of the introduction which Fraenkel had completed are reproduced verbatim.

'Es mag verwunderlich scheinen,' Fraenkel's introduction would have begun, 'daß das unvollendete Werk eines jungen Forschers nach über fünfzig Jahren wieder gedruckt wird. In der Tat ist es jedoch notwendig, dies jetzt nur noch in großen Bibliotheken zu findende Buch wieder allgemein zugänglich zu machen. Die Forschung ist selbstverständlich weitergeschritten und manches in Tychos Buch ist überholt, aber entbehrlich geworden ist es noch längst nicht. Das Vorwort zu dem Neudruck versucht an einigen Beispielen zu zeigen, warum das so ist.' Few doctoral dissertations have had so great an influence as this, and both the study of recent literature and the experience of teaching and discussion in several countries have convinced me, as they did Fraenkel, that interest in it is particularly lively at the present time.

The book's influence took some time to make itself felt; as late as 1932 a reviewer of one of the first books markedly affected by Tycho's ideas[2] could say with truth that most later interpreters had preferred to

[2] W. Schadewaldt, reviewing E. Howald, *Die griechische Tragödie*, at *Gnomon*, 8 (1932), 2 = *Hellas und Hesperien*, i (1970), 238.

approach tragedy from the standpoint of *Geistesgeschichte*. But now the situation is altogether different.

Tycho was the first to apply to the Greek dramatists a particular critical attitude that was highly characteristic of the time at which he wrote. Towards the end of the nineteenth century and for long afterwards, even as late as the nineteen-thirties, students of drama, both ancient and modern, were accustomed to look upon character portrayal as one of the main elements, if not the main element, in dramatic art; and by character portrayal they usually meant a minute psychological analysis. Most of Tycho's examples of this attitude are taken from the respectable Sophoclean commentaries of Ludwig Radermacher and of Ewald Bruhn. But they abound also in the notable English edition of Sir Richard Jebb, in the most learned Sophoclean commentary of the age of Wilamowitz—the *Elektra* of Georg Kaibel—and in the works of Tycho's own illustrious father.[3] ‖

Tycho agreed with Aristotle that the characters are there for the sake of the plot, not the plot for the sake of the characters.[4] 'Für Sophokles', he wrote, 'ist die ausführliche, psychologisch genaue und konsequente Charakterzeichnung durchaus nicht, wie für die heutigen Ansprüche, die Hauptaufgabe des Dramatikers',[5] and he points to countless

[3] 'In seiner 1923 veröffentlichten Skizze "Die griechische Tragödie und ihre drei Dichter" (*Griechische Tragödien*, xiv) konnte Wilamowitz von dem Schlußteil der *Antigone* noch sagen: "Da vermißt man die liebenswürdige Ismene, die der Dichter als nicht mehr vorhanden behandelt." Das mag für uns, oder doch viele von modernen Lesern, zutreffen. Daß es aber für Sophokles und diejenigen seiner Zuschauer, die bereit waren ihm zu folgen, nicht zutrifft, das lernen wir, neben vielem noch Wesentlicheren, aus Tychos Buch.

'Tycho schreibt (S. 41): "Ismene als Mensch für sich ist Sophokles völlig gleichgültig, sie ist nichts als ein Werkzeug [weiter oben "nur Folie für Antigone"], das er benutzt, wenn er es braucht, und wegwirft, wenn er es nicht mehr braucht." So in der Hauptsache schon der feinfühlige Schneidewin (Einleitung zur *Antigone*, S. 26), mit leichten Zugeständnissen an den Zeitgeschmack: "Die Nebenpersonen insgesamt dienen wesentlich zur Entfaltung des Charakters der Antigone . . . Ismene . . . steht, wie Chrysothemis zu Elektra, zur Antigone, um den Heroismus dieser durch den Gegensatz eines an sich edeln weiblichen Charakters zu heben: ist das erreicht, ‖ so tritt sie ab, ohne daß weiter von ihr die Rede wäre." Allerdings ist "Ismene gar nicht, was wir einen Charakter nennen" (Tycho, S. 42)': Fraenkel.

Karl Reinhardt, *Tradition und Geist* (1960), 236, gives a particularly amusing instance of this tendency in Wilamowitz. Yet only three years after writing as he did about Ismene, Wilamowitz wrote, 'Über die verfehlte Ausdeutung der Charaktere, z. B. von Herakles und Elektra, war ich längst hinaus' (*DLZ* 47 (1926)), 854 = *Kl. Schr.* i. 466, in his review of Schadewaldt's *Monolog und Selbstgespräch*).

[4] *Poetics* 1450ᵃ20–22; on the text, see R. Kassel, *RhM* 109 (1966), 10.

[5] P. 40. Cf. p. 78: 'Der Dichter will keineswegs irgendwelche Charaktere in einem durch sie selbst bestimmten Kampfe zeigen, sondern er will die, wenn man will, stoffartige Wirkung der dramatischen Situation so stark wie möglich herausbringen'; p. 145: 'Immer freilich ist festzuhalten, daß Sophokles auch hier nicht, um diesen einen individuellen Charakter [der Deianeira] zeichnen zu können, etwa das ganze Stück gemacht hat, sondern, als er den Stoff behandelte, mußte er die ihm gegebene Tat Deianeiras motivieren, und gab ihr, als sie ihm in den Mittelpunkt der Handlung rückte, den Charakter für diese Tat'; p. 154: '. . . die Forderung durchgehender, einheitlicher Charakteristik dem Sophokles ganz fremd ist'.

instances where modern critics had put forward a psychological explanation of a problem posed by Sophocles' text which was in the light of reason totally untenable.[6] They took no account, he complained, of the conditions of performance or of the effect a given scene would have upon the audience, but tried to explain the actions of the play's characters as though they had been real people, often postulating hypothetical events off-stage and assigning primary importance to things which were secondary for the poet;[7] their explanations started from the supposed intentions of the characters, instead of from the legitimately inferred intentions of the author.[8]|| Sophocles, he concluded, cared nothing for consistency of character; the behaviour of his people was always determined by the situation of the moment.[9] Not

[6] P. 31: 'Es ist natürlich leicht, sich das psychologisch irgendwie zurechtzulegen, aber nicht das ist die Aufgabe...'; p. 162: 'Das ist eine Erklärung, die ganz vergißt, daß man die Schöpfung eines Dichters und nicht das Verhalten wirklicher Menschen zu erklären hat'; p. 171 n. 1: 'Diese Erklärung ist bezeichnend dafür, wie völlig die Bedingungen der Aufführung und die Rücksicht auf das Verständnis des Zuschauers vergessen werden können, wenn man nur irgendeinen Grund für das Verhalten der Personen zu finden sucht, ganz als ob man der Wirklichkeit gegenüberstände'; p. 281: '... wo man, statt die Angaben des Dichters unbefangen zu interpretieren, das uns unverständliche Verhalten seiner Personen auf eigene Faust zu rechtfertigen sucht'.

[7] P. 20: 'Es kommt ihm nicht darauf, eine einheitliche Vorgeschichte für die Handlung zu gewinnen, vielmehr scheut er sich gar nicht—an ein Versehen ist hier unmöglich zu denken—über ein hinter der Szene und außerhalb der Handlung geschehenes Ereignis in zwei verschiedenen Situationen von derselben Person einander geradezu entgegengesetzte Angaben machen zu lassen'; ibid.: 'Die dramatische Wirkung der einzelnen Szene steht ihm also höher als die einheitliche Anlage der Fabel, und er rechnet damit, daß, da das jedesmal vor Augen des Zuschauers wirklich Geschehende so zur stärksten Wirkung kommt, die auf vergangene und nebensächliche Dinge bezügliche Inkongruenz nicht bemerkt wird' ('Das Gleiche gilt für Aeschylus, wie schon Wilhelm von Humboldt, in betontem Gegensatz zu den üblichen Fragestellungen, klar erfaßt hat': Fraenkel); p. 33: 'Dies alles zeigt ... daß der Dichter mit Absicht all diese Vorgänge im Dunkel gelassen hat und nach ihrem genauen Zusammenhang ungefragt bleiben will'; p. 34: '... durch Weglassung alles Nebensächlichen die Handlung so sehr wie möglich gespannt und konzentriert wird'; p. 38: 'In diesem Zusammenhang ist das Interesse auf ganz anderes gerichtet und man denkt nicht weiter an diese nebensächliche Angabe'; cf. p. 62 (on the *Ajax*).

[8] P. 131 n. i: 'Eine derartige Konstruktion eines psychologischen Vorganges, ohne daß man sich dabei auf eine Angabe || des Dichters stützen könnte, ist aber immer von vornherein verfehlt, weil man sich damit erfindet, was man zur Erklärung braucht'; p. 193 n. 1: 'Darin zeigt sich eben, daß auf solche Konstruktionen nur der hinterher das Ganze analysierende Leser verfallen kann'; p. 194: 'Solche seltsamen Konstruktionen zeigen nur, wie sehr diese Szenen einem Beurteiler widerstreben müssen, der sie nicht vor allem als Zuschauer ansieht, und darum seine Kenntnis des Ausgangs und der wirklichen Situation niemals vergessen kann'; p. 144 (on the second report of Lichas in the *Trachiniae*): 'Die Absicht lag dem Sophokles völlig fern, darauf Rücksicht zu nehmen, daß man sich die hinter der Szene, vor Beginn der Handlung liegende Erzählung des Lichas auf der Wiese hieraus rekonstruieren könnte.' ('Im Falle des Aeschyleischen Agamemnons', Fraenkel writes, 'hätte ein Interpret, der sonst so viel aus Tychos Buch gelernt hat, sich nicht verleiten lassen dürfen, mittels psychologischen Nachgrübelns herausfinden zu wollen, warum der König den heuchlerischen Bitten Klytaimestras schließlich doch nachgibt und über das für einen Menschenfuß zu kostbare Purpurgewebe in das Haus geht, wo ihn der Mord erwartet'; see his *Agamemnon*, ii. 441–2.)

[9] P. 70 n.: 'Die verwandten Züge im Bilde des Königs Kreon in der Antigone und des Königs Oedipus sind gegeben durch die ähnliche Situation; man sieht aus ihnen gerade das, daß sich die Charakterisierung immer in gewissen Grenzen bewegt, daß bestimmte Motive, wie der Verdacht

only consistency of character, but consistency of fact were in his view disregarded. A motive could be exploited for its effect at a particular moment, only to be totally forgotten later; an oracle could be said to have demanded one thing at one time and another thing, inconsistent with the first, a little afterwards.

Again and again Tycho asks what will be the effect of the scene he is discussing upon the audience; again and again he ruthlessly flings out the accumulated rubbish of over-subtle psychologizing interpretation. He throws much light on Sophocles' dramatic methods. He was the first to appreciate the new effects made possible by experience in handling a third actor,[10] and with much acuteness he showed how in several instances the poet gained much by splitting up what might have been a unitary scene into two separate episodes.[11] His historical importance as the initiator of the whole genre of detailed studies of various aspects of dramatic technique which have done such valuable work || since is very great. But the importance of his work is far from being only historical, for his rigorous analysis still offers the best starting-point for the consideration of many Sophoclean problems.[12]

Part of Tycho's results are nowadays generally accepted. Most scholars would now agree that Sophocles showed his minor figures 'only in silhouette'; most would concur in countless places where he had rejected a psychological explanation of a speech or action; few would deny that the action, not the characters, is the central element of Sophoclean drama. But like most daring innovators he in some places

der Bestechung z. B., immer gern angewandt werden, und vor allem, daß immer die momentane Situation das Benehmen der Person bestimmt, so daß in ähnlicher Situation Ähnliches herauskommt'; p. 222: 'wie . . . Motive lediglich momentanen Zwecken des Dichters dienen, und daß die uns selbstverständlich scheinende Forderung, sie mit der wirklichen Handlung des Stückes in Verbindung zu bringen, was an sich sehr wohl möglich wäre, für ihn so wenig Geltung gehabt hat, daß man bei weiterem Nachfragen sofort allen Boden verliert'; p. 39: 'Immer kommt es Sophokles darauf am meisten an, die momentane Situation so zugestalten, daß alles, was sie an dramatischer Wirkung hergeben kann, zum vollen Eindruck kommt, und dabei hat er bewußt damit gerechnet, daß der Zuschauer, den er mit der gegenwärtigen Szene faßt, den hinter der Szene inzwischen geschehenen Dingen nicht weiter nachdenkt, als der Dichter will, und daß er sogar Inkongruenzen und Widersprüche auch bei den auf der Bühne geschehenden Dingen nicht bemerkt und einen Mangel an Motivierung nicht empfindet, eben weil er ganz unter dem Eindruck dessen steht, was lebendig vor seinen Augen geschieht.'

[10] 'Daß erst Tychos Buch unsern Blick erfolgreich auf eine von Sophokles eingeführte wesentliche Neuerung im dramatischen Aufbau gerichtet hat, darauf hat ein nicht zur Philologenzunft gehöriger Literaturhistoriker mit Recht hingewiesen: John Jones, *On Aristotle and Greek Tragedy* (London, 1962), 272 Anm. 2': Fraenkel. This important book has received less attention from classical scholars than it deserves, so perhaps I may be allowed to draw attention to my review of it (*Review of English Studies*, 15/58 (1964), 221 f. [= ch. 13]).

[11] See pp. 34, 62, 67, 143, 224.

[12] His book also contains some valuable contributions to the criticism of the text. Fraenkel singles out his treatment of *Ant.* 933 f. (p. 49 n. 1); of *Trach.* 362–3 (p. 109 n. 1; Wunder and Blaydes anticipated Tycho here); of *Phil.* 671–3 (p. 284 n. 1, in support of Jebb); of *Phil.* 850 f. (p. 293 n. 1). On *Tr.* 46–8 and 901–3, see below, pp. 410 f.

pushed his thesis too far. If character-drawing is held to imply an interest in individuals as individuals and in psychological peculiarities for their own sake, then indeed it is unimportant in Sophocles. But character-drawing in a different, but perfectly legitimate sense, is of very great importance in this author: idiosyncrasies may count for nothing, but the main qualities of the character considered as a human being count for a great deal. An anecdote makes Sophocles claim that during the third period of his career he has achieved the style which is 'the most expressive of ἦθοϲ and the best'.[13] The Greek concept of ἦθοϲ requires that the characters shall be represented as being the kind of people capable of the actions assigned them by the story, and this requirement Sophocles certainly fulfilled. To take an obvious example, no reasonable person would dispute that the contrast between heroic figures, like Ajax, Heracles, Antigone, the two Oedipuses, Electra, and Philoctetes, and ordinary human beings is a leading theme in Sophocles' work. Jebb once encountered George Eliot and, knowing her to be a close student of Sophocles, asked her how he had influenced her and was told, 'In the delineation of the great primitive emotions'.[14] George Eliot put her finger on a central truth about Sophocles, which has been better understood since Tycho disabused people of the belief that Sophocles took great trouble about minor psychological complexities.

Some of Tycho's critics fail to bear in mind that he offers not a general treatment of Sophoclean drama, but a study of the poet's dramatic technique. The great difficulty of this enterprise, especially when it is undertaken at the beginning of a scholar's career, is that dramatic technique is the servant of poetical purpose, which often requires not only taste but experience to understand.[15] Some men of taste have reproached Tycho with philistinism, and at ‖ least one arrogant philistine has claimed him as a predecessor. They are not wholly wrong, as a glance through the mentions of Tycho in the

[13] Plutarch, *De profectibus in virtute* 7, p. 79B. See C. M. Bowra, *AJP* 61 (1940), 385 f. = *Problems in Greek Poetry* (1953) 108 f. = H. Diller (ed.) *Sophokles* (Wege der Forschung 95, 1967), 120 f. (German version) and Giuliana Lanata, *Poetica preplatonica* (Florence, 1963), 146 f. [see M. R. Lefkowitz, *Hermes*, 112 (1984), 148.]

[14] See Caroline Jebb, *Life and Letters of Sir Richard Claverhouse Jebb* (Cambridge, 1907), 156; cf. Gordon Haight, *George Eliot* (1970), 173. For recent treatments of character in Sophocles, see H. Diller, *WS* 69 (1956), 70 f. = *Kl. Schr. zur antiken Literatur* (1971), 272 f. and *Antike und Abendland*, 6 (1957), 157 f. = Diller (ed.), *Sophokles*, 190 f. = *Kl. Schr.* 286 f. and B. M. W. Knox, *The Heroic Temper* (Berkeley, 1964).

[15] 'In dem von ihm gewählten Rahmen konnte Tycho nicht darstellen, daß für Sophokles die dramatische Technik durchweg nur als Mittel dient zur Gestaltung von etwas viel Tieferem, etwas, das im Wesen seiner Kunst und seines Menschentums ruht. Aber um dazu vorzudringen, muß man sich erst mit der eingeschränkteren Betrachtungsweise dieses Buchs gründlich vertraut gemacht haben': Fraenkel.

footnotes to Karl Reinhardt's *Sophokles*[16] will remind the reader. Repeatedly Tycho finds fault with the poet for departures from probability or consistency which simply go to prove what Tycho himself demonstrates, that Sophoclean technique is very different from nineteenth-century naturalism.[17] Modern readers, who are accustomed to more than one theatrical convention in which naturalism plays little or no part, find this unnecessary.

Reinhardt chose to treat of 'Sophokleischen Situationen, oder ... von Sophokleischen Verhältnis zwischen Mensch und Gott und zwischen Mensch und Mensch, und zwar wie es Szene für Szene, Stück für Stück, Stufe um Stufe sich entwickelt' (op. cit. 9 [English translation 1]). Starting from Tycho's results and using a similar method of close analysis, Reinhardt with his mature taste and judgement and his usually fine feeling for poetry was able to show that the essential quality of the poet's art lies not in psychological refinements but in the depiction of the emotions which Sophoclean situations evoke from those involved in them.

Reinhardt's method is unlike that of those who approach tragedy from the standpoint of *Geistesgeschichte*, who usually proceed by extrapolating passages supposed to reveal the poet's thought. They have neglected Tycho's work, and their self-protecting instinct can easily be understood; for in the field of *Geistesgeschichte* also it had revolutionary implications. First, the sharp reminder that above all things a Greek tragedy is a play, written to be acted in a theatre and designed to have a particular effect upon its audience, could not fail to distress those who are accustomed to treat it as though it were an ethical or metaphysical treatise. People who took it for granted that the main element of a tragedy was its religious or metaphysical content, and that the poets wrote to recommend beliefs and opinions, sometimes new and 'origi-

[16] 'Es wäre traurig, wenn nach Tychos Jugendwerk, und nicht zum wenigsten als Folge seiner aufrüttelnden Wirkung, unser Verständnis des Sophokles keine Fortschritte gemacht hätte. Die Fortschritte sind groß, vor allem dank Karl Reinhardts *Sophokles*. Das ist vielleicht das schönste Buch, das über den Dichter geschrieben ist, ein schweres Buch; seine Glut mag bisweilen einen keineswegs kaltsinnigen, jedoch auf kühleres Abwägen angewiesenen Betrachter zurückschrecken, so daß er sich dann aus einer höheren Sphäre gern einmal wieder von Tycho auf die wohlgegründete dauernde Erde zurückführen läßt': Fraenkel. Reinhardt's *Sophokles* first appeared in 1933; the latest edn. is the third, of 1947. [An English translation by Hazel and David Harvey was published in 1979.]

[17] 'Verfehlt, weil Tycho zur folgerichtigen Durchführung seiner grundsätzlichen Auffassung nicht mehr gekommen ist, sein mehrfach begegnender Tadel (besonders in den *Trachinierinnen*, aber auch in der *Electra* und sonst) des Dichters wegen flüchtiger oder nachlässiger Gestaltung einzelner Szenen oder Szenenteile, wofür oft Abhängigkeit von Motiven in früheren Dramen des Sophokles oder von Dramen des Euripides, bisweilen auch "ziemlich rasche Abfassung" verantwortlich gemacht wird. Anstatt sich an solchen Unvollkommenheiten zu stoßen, sollte der Leser versuchen, sie mit Hilfe von Tychos eigenen tieferen Erkenntnissen zu berichtigen': Fraenkel.

nal', were bound to resist an approach which was a step on the way to the revolutionary view that, at least in the case of early tragedy, traditional beliefs supplied a background against which the action of the play took place. Further, a challenge to the assumption that an ancient dramatist's aims and methods were like those of a modern dramatist was likely to strengthen the challenge to the equally prevalent assumption that the religion, ethics, and general world outlook of the ‖ ancient Greeks were either very like those now fashionable or were of interest chiefly because they were such as to 'lead up to' them.

Tycho's treatment of the *Antigone* appeared in print during his lifetime (see Ernst Kapp's Foreword, p. v). It is encumbered by a lengthy polemic against a now deservedly forgotten theory of A. B. Drachmann[18] and it suffers from an undue insistence on the improbability of Antigone escaping capture on the occasion of her first burial of Polyneices. But it is none the less an important contribution to the understanding of the play; and, perhaps because it appeared before the rest, it contains many important statements of Tycho's principles, several of which have already been cited in the footnotes to this article.

Minor improbabilities concerning Creon's proclamation (pp. 18 f.), Antigone's claim to have buried Eteocles (pp. 19 f.), and Haemon's movements (pp. 21 f.) exist, but are not particularly significant; more important is Tycho's argument that Ismene has no importance except as a foil for Antigone (pp. 23 f.)[19] More interesting still is his handling of the problem of the 'double burial' (pp. 26 f.). He makes somewhat heavy weather of the question of how Antigone avoids detection on the first occasion (pp. 26–30); surely it is easy for the spectator to infer that she does so under cover of the darkness.[20] He thinks that her return to the body is left unmotivated simply because it is inexplicable (p. 34); but is one guilty of psychologizing unreasonably if one takes it for granted that she has heard of the desecration of the body and hastens to repair the damage? But Tycho made one important contribution to the solution of this problem; he saw why Sophocles found it necessary for Antigone to make two visits to the body. First, she has to triumph by cleverly eluding the guards and carrying out the burial, and Creon's reaction to the news had to be presented; then she has to be apprehended and brought before Creon. Only at that moment is the identity of the guilty person known; Antigone's appearance as a prisoner is a tremendous *coup de théâtre*. Tycho (pp. 33–4) correctly explained this.[21]

[18] 'Zur rechten Würdigung des von Tycho für das Verständnis der *Antigone* Geleisteten dürfte man S. 7 (ungefähr) bis 17 (vor dem zweiten Abschnitte) beiseite lassen': Fraenkel.

[19] See above, n. 3.

[20] See A. T. v. S. Bradshaw, *CQ*, NS 12 (1962), 201–4.

[21] See above, n. 11.

His view that Sophocles cared nothing for consistency of character enabled him to deal summarily with the celebrated problem of the speech in which Antigone protests that she would have made such a sacrifice only for a brother (pp. 45 f.); but few scholars now would be content with this.[22]

In the *Ajax*, Tycho showed that during the first stasimon Ajax and Tecmessa must be together in the tent (pp. 55 f.). But though he may be right in thinking the *ekkyklema* was used, one may doubt whether ll. 579–85 are there, as he maintains, simply to fill in the interval needed for the machine to be employed; the dramatic value of the dialogue here is obvious.[23] Tycho's belief that Sophocles cared nothing for consistency of character allows him once more to take a short way with a celebrated problem, this time that of the 'deceptive speech' of Ajax at ll. 646 f.; in his view Ajax means to deceive his companions, although the audience is not deceived, and simply acts as though the deception were the truth, with no attempt at psychological verisimilitude on the poet's || part. For an understanding of the poetic purpose of the speech and an explanation of the pathos of its relentless but gentle irony, one must turn to Reinhardt.[24] One of Tycho's best achievements is his explanation of why Sophocles makes Teucer defend his brother's honour in two successive *agones* against Menelaus and Agamemnon (pp. 65 f.). Had there been only a single *agon*, settled in the end by the intervention of Odysseus, Teucer's unshaken resolution would not have received the emphasis the poet rightly wished to give it.[25]

Kranz in his review of Tycho's book[26] rightly observed that the treatment of the *Oedipus Tyrannus* is a good specimen of its author's methods. He shows convincingly that the justly famous plot, when closely scrutinized, reveals a number of small improbabilities and inconsistencies which a modern realistic writer would not allow himself. Creon's movements at the start are not quite consistently described (pp. 72–3). The Teiresias scene contains contradictions, but these do not justify the view of some modern writers that Teiresias is a 'sinister' figure, secretly hostile to Oedipus. The poet is not interested in Teiresias' character, and the contradictions are due simply to the requirements of the plot (pp. 73–89). Tycho truly says that the whole

[22] Reinhardt's (op. cit. 92–3 [English translation, 83–4]) is the most notable discussion; more recent treatments (see H. Friis Johansen, *Lustrum* (1962), 198–9) do not add much.

[23] See W. Kranz, *Sokrates*, NF 6 (1918), 333 = *Studien zur antiken Literatur und ihrem Nachwirken* (1967), 304.

[24] Op. cit. 31 f. [English translation, 22 f.]; again, later interpretations add little (Friis Johansen, op. cit. 177 f.).

[25] See above, n. 11.

[26] Loc. cit. (n. 23) 332 = 303.

plot depends on the old Theban shepherd's incorrect statement that Laius was killed by more than one person. He must also be right in saying that the old man says this because it is essential to the poet's purpose; but is the old man's inaccuracy as motiveless as he supposes? Lines 758–64 make it clear that the old man chose deliberately to get himself out of the city, and it is not hard to imagine why he should have done so. Tycho observes that Iocaste's attempt to prove the uselessness of prophecy (ll. 707 f.) is not really relevant, but is introduced in order to lead up to her use of the report of Polybus' death to comfort Oedipus (pp. 80–3); he shows also that in reality Oedipus would have known all as soon as he had learned that Polybus was not his real father, but that he was given to the Corinthian shepherd on Cithaeron by a slave of Laius; not that either inconsistency troubles the audience during a performance (pp. 83–5). His concluding remarks on characterization in the play and the poet's lack of interest in the psychology of his persons are altogether to the point (pp. 85–8).

The chapter on the *Trachiniae* must be judged less satisfactory. Tycho is unduly severe about this play, which he regards as a hurried and careless piece of work; he devotes excessive space to the refutation of a once influential, but now deservedly forgotten treatment by Zieliński;[27] and he is handicapped ‖ by his preoccupation with Dieterich's theory that the scene of Heracles' awakening shows the influence of the not dissimilar scene in Euripides' *Heracles*,[28] as well as by his own unverifiable guess that the character of Deianeira was derived by Sophocles from that of Penelope in his lost *Niptra*. His whole treatment of the obscure topic of the early treatment of the myth is not notably rewarding, though his conjecture that the false motive for Heracles' attack on Oechalia given by Lichas may represent a different version given in an earlier poem may quite easily be right.

Few would now agree with Tycho that the prologue 'ganz in der Weise des Euripides angelegt ist' (p. 116),[29] nor that it displays 'eine

[27] 'Überhaupt verstrickt das Buch sich oft in Polemik gegen irrige Deutungen, die uns heute, zum Teil gerade infolge der Wirkung von Tychos Buch, als gegenstandslos erscheinen. Das ist z.B. der Fall bei der zwar berechtigten, aber viel zu breit gewordenen Polemik gegen Zielinski in dem Kapitel über die *Trachinierinnen*. Überhaupt ist dieses Kapitel, das von ganz unsicheren Hypothesen ausgeht, unausgeglichen und darf nur als vorläufiger Versuch aufgefaßt werden; es gehört ja zu den Kapiteln, die, nach Ernst Kapps Zeugnis, nur in einer ersten Form in der Dissertation von 1911 vorliegen. Wer sich mit Tychos Betrachtungsweise vertraut gemacht hat, wird nicht zweifeln, daß er hier tiefgehende Änderungen vorgenommen hätte, wenn ihm Zeit für neues Durchdenken und Formulieren vergönnt gewesen wäre. Aber selbst so enthält dieses Kapitel nicht wenige Feststellungen von bleibendem Wert. Ein Leser, dem es auf ein vertieftes Verständnis der Sophokleischen Kunst ankommt, sollte sich bemühen, über die Niederungen polemischer Argumentation hinweg mit ungeminderter Aufmerksamkeit in die reichen positiven Erkenntnisse einzudringen, die sich jenseits der Polemik abheben': Fraenkel.

[28] See Johanna Heinz, *Hermes*, 72 (1937), 289 f.

[29] See Heinz, op. cit. 284 f.

Sorglosigkeit in der Motivierung und eine Mißachtung der Wahrscheinlichkeit, wie sie sich sonst nicht findet' (p. 117). In real life Hyllus would have told Deianeira where his father was long before, but Sophoclean technique obviously requires that he must do it when he does, nor will the audience be disturbed at his getting to Heracles and back so quickly. Tycho in his treatment of the difficulties raised by the different mentions of the oracle given at Dodona draws attention to some serious problems (pp. 119–33). But as early as 1921 Kranz showed that if the various references to this oracle are considered each in its own context, no really grave inconsistency is found, but it becomes clear that each time the poet lets the audience know just as much about the oracle as the present situation of the plot requires.[30] The obvious parallel is furnished by the oracle in the *Philoctetes*, which is far more freely handled.

The refutation of Zieliński's treatment of Deianeira's scene with Lichas (pp. 134–8) is followed by some good general remarks about the type of argument which Zieliński uses (pp. 138–42). Then comes an excellent explanation of Sophocles' reasons for arranging Deianeira's scenes with Lichas and the messenger as he did; the good news of the victory must come first and cause premature rejoicing, the bad news about Iole must follow after an interval, and the greatest possible dramatic effect must be obtained from both (pp. 142–5). Faced with Deianeira, whose character has charmed so many interpreters, not always to their advantage, Tycho concedes that there is hardly any other Sophoclean personage who comes so near to being a character in the modern sense (p. 145); but even here he finds inconsistencies. He rightly rejects the interpretation of her words at ll. 494–7 as a kind of Freudian slip revealing an unconscious malice (pp. 145–9); but he makes somewhat heavy weather of the alleged discrepancy between ll. 436 f. and 531 f. After all, Iole is the first among his mistresses whom Heracles has actually brought to live in the matrimonial home (note ll. 536 f.) (pp. 150–4). Surely he goes too far in arguing that Deianeira's use of the philtre is inconsistent with her words at ll. 436 f.; the poet conveys with great delicacy how a wife who has sincerely uttered these sentiments might still, not knowing the real nature of the philtre, be tempted to make use of it. Like many others, Tycho complains that the play falls into two halves (pp. 154–5); this complaint seems to issue from his assumption[31] that every play of Sophocles ‖ must have one central character about whose fortunes it revolves. For him Deianeira is the

[30] Tycho had predecessors in Dobree, *Adversaria*, iii. 36 and Wunder, *Sophoclis Tragoediae*, ii/3 (1841), 47 f.; the theory has now been revived in a somewhat different form by M. D. Reeve, *GRBS* (1970), 283 f. See Kranz, *Jahresberichte des Philologischen Vereins zu Berlin*, 47 (1921), 32 f. = *Studien* (n. 23), (see 283 f.)

[31] See John Jones, op. cit. (n. 10), ch. 1.

central figure of the first half, Heracles of the second, and this seems to him perplexing. But need either be the central figure? The play is about the events that led to the death of Heracles; it shows the fulfilment of the plan of Zeus. Tycho shows much acuteness in his handling of the problem of Hyllus' movements in the later scenes and of the authenticity of ll. 901–3; only lately has a convincing solution of the difficulties he exposes been put forward.[32]

The chapter on the *Electra* was left unfinished, though the first half had been revised (Ernst Kapp, p. vi); it is the least satisfactory of all. Here more than anywhere the reader is vexed by perpetual complaints of departures from naturalism which Tycho's own researches have shown to be typical of Sophocles' technique; and the long section on the play's chronological relation to Euripides' *Electra* (228–64), with the appendix on the *Helen* and *Iphigenia in Tauris* that follows, need not now be read.[33] The treatment of the *Electra* itself often leads to unacceptable conclusions; yet even here Tycho's arguments still deserve to be read carefully.

The lack of naturalism in the initial dialogue is hardly proof of carelessness (pp. 166–7); neither is the absence of a motive for the entry of the Chorus (p. 169); on the other hand, the observation regarding the use of the *Choephori* at ll. 77–85 is excellent. It is in no way strange that the Chorus makes no reply to Electra's long speech after the parodos (p. 171): a reply is not needed. Nor is it surprising, even after what has been said earlier, that after that speech the Coryphaeus should confirm the absence of Aegisthus and should enquire for news of Orestes. In terms of the technique of early tragedy, the repetition is not offensive, and it is important to the poet's purpose that both subjects should be recalled to the audience at this point. Equally in keeping with the poet's methods is the absence of motivation for the successive *agones* of Electra with Chrysothemis and Clytemnestra; however improbable it may seem that such scenes should take place at these particular moments, the plot obviously requires it. In his treatment of the first scene with Chrysothemis, Tycho insists that the threat to imprison Electra is a motive invented simply for the moment, and

[32] By R. P. Winnington-Ingram, *BICS* 16 (1969), 44 f.

[33] 'Auch der lange Abschnitt über das Verhältnis der Sophokleischen zu der Euripideischen *Elektra* (S. 228–268) geht von einem inzwischen widerlegten Zeitansatz des Euripideischen Dramas aus, ermüdet durch Polemik gegen jetzt Veraltetes, und, was hier vor allem hervorzuheben ist, enthält nur sehr wenig von Tychos eigentümlichsten und wertvollsten Gedankengängen. So mag der Leser, jedenfalls zunächst, diesen Abschnitt beiseite lassen, ohne daß ihm dabei etwas Wesentliches entginge. Was sodann die chronologische Frage angeht, so tritt Tycho allerdings mit Recht für die Priorität der Sophokleischen *Electra* ein, nur kann man das viel kräftiger begründen, als er es tut': Fraenkel. See now the useful dissertation of Armin Vögler, *Vergleichende Studien zur sophokleischen und euripideischen Elektra* (Heidelberg, 1967), reviewed by me at *CR*, NS 19 (1969) [R72], 36 f.; see p. 36 for the account of Tycho's treatment in his historical summary.

afterwards arbitrarily disregarded. Indeed it is disregarded later, but is there any further need for it to be mentioned? In the *agon* with Clytemnestra it is not required, and after that both Clytemnestra and, when he returns, Aegisthus have other things to think about.

It is hardly true that the first Chrysothemis scene is 'so sehr ohne Ergebnis, daß noch eine lange ruhige Unterhaltung folgen kann' (ll. 177–9); Electra's triumph in the *agon* is heightened by her success in forcing her sister to accompany the offering sent by Clytemnestra to the tomb with a prayer of Electra's || own dictation, and its effect will later be accentuated by a triumph over a more formidable antagonist in Clytemnestra herself.[34] To say that the first stasimon (p. 473 f.) 'nur die im allgemeinen durch den Stoff nahegelegten Gedanken ohne die Absicht einer besonderen Wirkung in fast konventioneller Weise ausdrückt' (p. 179) scarcely does justice to its significance: Dike and the Erinys have their importance in this play.[35] Sophoclean technique does not require that the debate between mother and daughter at ll. 516 f. must be specially motivated (pp. 179–82); Tycho rightly dismisses the psychological explanations offered by modern critics, but fails to recognize that no motivation is required. He rightly insists that the main feature of the *agon* is the triumph of Electra; but while he justly discounts its alleged psychological significance (p. 185), one may doubt whether its bearing on the moral issue is so slight as he implies.

Tycho has some good observations about the prayer of Clytemnestra at ll. 634 f. (pp. 186–7), and about the partial sincerity of her grief at the announcement of her son's death (pp. 187–8). But he can hardly be right in maintaining that an audience which has seen the Paidagogos talking with Orestes during the prologue is so carried away by his convincing fiction that it actually shares the delusion of his hearers on the stage. Yet it is true that the main function of the speech is to enable the audience to imagine its effect upon Electra, who believes it (pp. 188–93). Tycho rightly draws attention (p. 193) to the effect of contrast secured by the return of Chrysothemis, but in his insistence that everything must be seen from Electra's point of view he misses the full impact of the scene's irony. Chrysothemis, he says (p. 194), is made to appear childish; but the audience knows that she is right and Electra wrong. Is the scene really there, as Tycho thinks, simply because the motive of the offerings at the tomb had to be worked in? His treatment of Electra's proposal that the sisters shall try together to kill Aegisthus is very like his treatment of the threat to imprison Electra; the suggestion is by no means so unreal as he supposes (pp. 195–202).

[34] So Schadewaldt, *Monolog und Selbstgespräch* (1926), 58 n. 2.

[35] See Winnington-Ingram, *PCPS*, 183 (1954–5), 20 f. (German version in Diller (ed.), *Sophokles*, 400 f.).

While Electra is speaking, the audience will know that she means what she says; and Greek mythology offers several instances of actions such as she proposes. Naturally she makes no mention of her mother, for her aim is to persuade her sister; nor does Chrysothemis in her reply find it necessary to mention Clytemnestra. Considering what happens immediately afterwards, it is hardly surprising that we hear no more of this proposal.[36] ‖

The suggestion of the Coryphaeus that Electra as 'the nearest' shall introduce Orestes and his party is scarcely as surprising as Tycho finds it (pp. 202–4); the courts of the heroic age were not ceremonious in the manner of Versailles, and it would not have been unusual for the strangers to be conducted into the queen's presence by the only member of the royal family present at the moment of their arrival. Tycho conclusively refutes the view of Kaibel and others that Orestes really knows who Electra is at the moment when he orders the urn to be put into her hands (pp. 204–6); he seems to have convinced everybody. Less to the point are his complaints about the lack of verisimilitude in the actual anagnorisis (pp. 206–10). A conventional pattern is being followed, and naturalism is not in place; yet the scene is written with passion, so that the conventional element does not detract from the emotional effect. Tycho finds that the following scene (1233–1383) serves simply to present Electra's feelings; yet the warnings against indiscretion given her by Orestes and by the Paidagogos may serve to remind the audience that these feelings are shown against the background provided by a world inhabited by others. Like so many scholars, Tycho finds that Clytemnestra's murder gets very little of the poet's attention (pp. 215–16).[37] Certainly its representation occupies very few

[36] Fraenkel strongly agreed with Tycho here. 'In der *Elektra*', he wrote, 'versucht Elektra ihre Schwester Chrysothemis zu überreden, mit ihr gemeinsam Aigisth zu ermorden. Mit keinem Wort deutet Sophokles an, wie die beiden diesen Plan ausführen könnten. Die Sonderbarkeit des Verschweigens einer so wesentlichen Voraussetzung hat man schon im Altertum bemerkt und psychologisch zu erklären versucht: Elektra betone nur die Vorteile, die der Schwester aus der geplanten Tat erwachsen würden, unterdrücke aber, da sie die ängstliche Natur ihrer Schwester kenne, jeden Hinweis auf die Gefahr, die ihnen bei dem Unternehmen droht (siehe das Scholion zu 975 und den in der Ausgabe von Jahn–Michaelis dazu gestellten Passus aus der *Rhetorik* des Apsines, *Rhetores Graeci*, ed. Spengel–Hammer, I, S. 302, 2 ff.). Den wahren Grund eines so auffallenden Fortlassens macht Tycho klar, indem er (S. 197 ff.) zeigt, welche Funktion Elektras ganz unrealisierbarer Plan innerhalb des Dramas zu erfüllen hat. In diesem Zusammenhang macht er (S. 198, Anm. 1) die aufschlußreiche Beobachtung, daß in Hofmannsthals *Elektra* "Chrysothemis gegenüber die Überlegung, wie der Mord auszuführen ist, und die Möglichkeit des Gelingens noch vor der Überredung geltend ‖ gemacht, und der Plan in allen Einzelheiten anschaulich dargestellt wird". Das ist ein Musterbeispiel für den fundamentalen Unterschied zwischen dem, was ein moderner Leser oder Zuschauer für notwendig hält, und dem, was Sophokles auf der Bühne darstellen will und womit er und sein Publikum sich begnügen.'

[37] This opinion was evidently shared by Fraenkel, who in one of his notes has copied out the passage on p. 174 which contains the words 'Den gegebenen Stoff, Orests Muttermord, hat der Dichter für unser Interesse ganz in den Hintergrund geschoben und an ein in diesem Stoffe liegendes sittliches Problem kann bisher überhaupt niemand denken.' For a different view, see H.

lines of print; but in performance the impression made by this episode, and in particular by Electra's horrifying cries of encouragement to her brother, is far greater than Tycho's words imply; as usual, he is concerned to insist that the real climax of the whole action is the recognition-scene. For him Clytemnestra's murder is only an introduction and a means to the killing of Aegisthus (p. 218); yet Sophoclean practice suggests that a final dialogue scene may be little more than a pendant to a preceding lyric episode. He finds the concluding dialogue between Orestes and Aegisthus 'almost incomprehensible' (p. 217); it does not occur to him that this conversation, during which Aegisthus is allowed to get in at least one not ineffective retort (1500), may help the audience to see the phase in the history of the house of Atreus which it has seen enacted from a standpoint somewhat different from Electra's own.[38] Tycho's concluding remarks about the play's construction (pp. 219–28) suffer from the defects indicated by the preceding criticisms. They contain some just objections to the psychologizing interpretations current at the time when Tycho wrote; but Tycho's insistence that the play centres not upon Electra's character but upon the recognition-scene is not well founded. Indeed Sophocles is not concerned with the idiosyncrasies of his heroine; but he is concerned with her heroic nature and its divergence from the human norm, and he has shown the drawbacks as well as the advantages attaching to such a character. In the sense of the word 'character' that corresponds with the Greek term ἦθος, Sophocles is interested in Electra's ‖ character; and the final scenes contribute hardly less than the justly famous *anagnorisis* to its depiction.

The chapter on the *Philoctetes* was written in 1913, and was more carefully revised by the author than any other part of the work; it is the best of all, and its excellence makes one keenly regret the absence of a chapter on the work nearest in date and method to the *Philoctetes*.[39]

Tycho rightly concludes that it is not certain whether an account of how Philoctetes was persuaded to leave Lemnos for Troy figures in early epic, and briefly describes its treatment by Aeschylus and Euripides (pp. 269–75). He rightly recognizes that though in reality Neoptolemus would certainly have learned what was expected of him before reaching Lemnos, Sophoclean technique requires that he must be instructed by Odysseus during the prologue (pp. 274–5). Odysseus, he observes, outlines no definite plan; and in the prologue the important question of whether the bow alone is needed or Philoctetes with the bow is left

Friis Johansen, *Classica et Mediaevalia*, 25 (1964), 8 f.; C. P. Segal, *TAPA* 97 (1966), 473 f.; H.-J. Newiger, *Arcadia*, 4 (1969), 138 f. (an important article).

[38] See Friis Johansen, op. cit. (n. 37), 28–9. It is going too far to suggest that the justice of the revenge is doubted; but no attempt is made to minimize the horror of the matricide.

[39] Fraenkel says of the chapter supplied by Tycho's father that 'es bildet keinen organischen Teil von Tychos Buch'.

vague. That is true, although l. 112 seems to indicate that Neoptolemus assumes that Philoctetes as well as the bow must come to Troy (pp. 274–7). Tycho argues that during the parodos and the scene with Philoctetes that precedes the entry of the supposed merchant, Neoptolemus and the Chorus behave not as if they are skilfully playing a part, but as if the fiction they were enacting were the reality (pp. 278–81); the merchant scene, he thinks, makes no sense if one remembers that Neoptolemus means to betray Philoctetes (pp. 281–3). To modern taste it must seem strange that nothing indicates that Neoptolemus and the Chorus are skilfully playing a part. But Tycho exaggerates the improbabilities of the scenes in question: Neoptolemus' plan to get Philoctetes on board the ship, bow and all, and Odysseus' effort to hurry on the proceedings by sending the merchant make perfect sense in the context. It is true that Odysseus' stratagem works out unfortunately for himself in that it warns Philoctetes that Odysseus is on his track; but that is necessary for the working-out of the plot, and is not in itself improbable or inconsistent. Tycho is surely right to insist that the dialogue of Philoctetes and Neoptolemus at ll. 268–75 is intended to stress the importance of the bow, and not to illuminate the psychology of Neoptolemus (pp. 283–5). He observes that the stasimon that begins at l. 676 has every appearance of being a sincere expression of the sympathy of the Chorus for Philoctetes. His contention that the Chorus is here only the instrument of the poet (pp. 295–9) cannot be refuted by saying that the Chorus has misunderstood the real situation, and assumes that now that Philoctetes and Neoptolemus have made friends, they will easily reach an agreement:[40] the Chorus says that Philoctetes will be taken home, but nothing about his going to Troy. A more plausible answer would be that the Chorus is playing the part of deceiver all too well.[41] Up till now, Tycho has insisted, attention has been concentrated on Philoctetes; during the scene that extends from l. 730 to l. 864 it is concentrated, he points out, on Neoptolemus, of whose ambiguous position the audience is now directly reminded. He shows how necessary it is to the poet's purpose that Neoptolemus shall not steal the bow, as he did in Aeschylus, but have it entrusted to him by its owner. While Philoctetes is sleeping, the Chorus urges Neoptolemus to make off with the bow. Tycho was the first to point out ‖ the singularity of Neoptolemus' answer; instead of protesting that it would be dishonourable, Neoptolemus declares that the bow will be no use without its owner. And yet in the opening scene of the play Odysseus has by no means made this clear; and later, when Odysseus proposes to abandon Philoctetes and sail off with the bow, he

[40] Thus Gerhard Müller, in H. Diller (ed.), *Sophokles*, 213 f.

[41] See Reinhardt's comments on the words of the Chorus at ll. 391 f. and 507 f. (op. cit. 180, 283 nn. 2–3) [English translation, 170–1, 266–7 nn. 8–9].

declares that Philoctetes himself is not needed, and Neoptolemus and the Chorus do not contradict him (pp. 289–95). Before Tycho, the acute problem presented by these facts had not been faced; since his time it has been hotly debated, but the authors of the two most notable recent contributions both accept his general conclusion that Sophocles allows himself great freedom in respect of the content of the prophecy of Helenus.[42]

When Neoptolemus finally decides to make a clean breast of it and tell Philoctetes the truth, that happens, Tycho argues, without warning; he strongly denies that the audience has been prepared for it by any delineation of Neoptolemus' psychology. This is surely true; but when he argues that Neoptolemus acts not out of pity, but because his intrigue is now bound to be discovered in the near future, he omits to mention the obvious signs of shame with which the revelation is accompanied (pp. 295–7). Sophocles, in Tycho's view, is unable or unwilling to represent a double situation[43] in which a character has to keep up a pretence: he makes Neoptolemus first act exactly as though the part he was playing were genuine and then suddenly reveal the truth to Philoctetes and betray his shame, without abandoning the mission entrusted to him by Odysseus (pp. 297–8). Tycho finds it surprising, and by the canons of naturalism it certainly is, that Neoptolemus makes no attempt to explain the whole situation to Philoctetes in the hope of persuading him to come voluntarily to Troy; neither does Philoctetes demand an explanation of Neoptolemus. But the poet could have replied that Philoctetes in his enraged state would neither have demanded nor listened to an explanation (pp. 298–300). Tycho shows most skilfully that it is the merchant's communication of the prophecy of Helenus and the expedition of Odysseus that render Philoctetes, being such as he has been presented as being in his first scene with Neoptolemus, so determined to avoid being taken to Troy (pp. 300–2).

Tycho clearly points out the contradiction between Odysseus' assumption that the bow only is needed, and not its owner, and Neoptolemus' words both earlier and later in the play, and deals decisively with the attempt of Radermacher to show that Odysseus is

[42] D. B. Robinson, *CQ*, NS 19 (1969), 45–51 (against A. E. Hinds, ibid., 17 (1967), 169 f.), and O. Zwierlein, *GGA*, 222, Heft 3/4 (1970), 206 f. (in a review of W. Steidle, *Studien zum antiken Drama* (Munich, 1968)); cf. O. Taplin, *GRBS* 12 (1971), 35 n. 24.

[43] Kranz writes (in the review quoted above (n. 23), (332 = 303): 'Selbst Wilamowitz macht sich nicht völlig von dem Fehler frei, das Können des Dichters als in irgendeiner Beziehung zu gering oder behindert darzustellen; so wird man z. B. nicht sagen dürfen, daß "die Darstellung einer solchen doppelten, gebrochenen Situation ganz außerhalb des für Sophokles Möglichen liegt" (S. 278), sondern außerhalb des Erstrebten, denn "die Kunst kann immer, was sie will", und es kommt vielmehr darauf an, zu zeigen, warum sie dies oder jenes nicht gewollt hat.'

only bluffing. He shows how skilfully the exposition in the prologue is calculated to prepare for the vagueness about the prophecy necessary for the plot (pp. 302–6). If Odysseus were only bluffing, he points out (pp. 306–7), the lyric scene of Philoctetes with the Chorus (pp. 1081 f.) would lose all meaning. Tycho, who is so severe with the *Trachiniae* and the *Electra*, fully acknowledges the excellence of the scene during which Neoptolemus returns the bow (pp. 1218 f.); he acutely observes that if the volte-face had been ‖ led up to by psychological preparation, the poet could hardly have achieved such an effect of surprise (pp. 307–8). It may be doubted whether Tycho is right in finding it implausible, although dramatically effective, that before returning the bow Neoptolemus should make an unavailing attempt to persuade Philoctetes to come to Troy (pp. 308–9). Tycho thinks it would have been quite possible for Sophocles to allow Philoctetes, after the discomfiture of Odysseus, to be persuaded by Neoptolemus to abandon his quarrel with the Atreidae and go to Troy. Surely this would have been inconceivable for a Sophoclean hero, loving his friends and hating his enemies beyond the norm. He thinks Sophocles used the god from the machine to bring about the ending which the known myth required. In fact the poet's use of Heracles serves a deliberate poetic purpose; deeply distasteful as it is for Philoctetes to make his peace with the Atreidae and Odysseus, the will of Zeus requires that he must do so, and only the command of Heracles can persuade him.[44]

ADDENDUM

Here is an isolated observation of Fraenkel regarding p. 274 n. 1: 'Klar ist, daß Neoptolemos' Beschreibung [der Höhle Philoktets] darum so ausführlich ist, weil sie dem Zuschauer schildert, was er nicht sieht, wobei durch die Lage erklärt wird, daß Neoptolemos herangehend in die Höhle hineinsehen kann, obwohl dem Zuschauer das Innere unsichtbar bleibt.' Cf. A. M. Dale, *WS* 69 (1956), 104–5 = *Collected Papers* (1969), 127–9 = Diller (ed.), *Sophokles*, 249–50 (German version).

[44] On the appearance of Heracles in the Philoctetes, see A. Spira, *Untersuchungen zum Deus ex machina bei Sophokles und Euripides* (Diss. Frankfurt a. M., 1960), 12–32, and Karin Alt, *Hermes*, 89 (1961), 167–72 = Diller (ed.), *Sophokles*, 448–55.

36(R53)

Review of Euripides, *Hippolytos*, ed. Barrett

Euripides. *Hippolytos*, ed. W. S. Barrett. Oxford: Clarendon Press, 1964. Pp. xvi + 453.

THIS book has been eagerly awaited, but in no way does it fail to answer expectations; it deserves the careful attention of everyone seriously interested in Greek literature. The ninety pages of the introduction are equally divided between the treatment of the legend, including a careful discussion of the lost plays on the subject, whose fragments and testimonies are given, and the history of the text together with an assessment of the manuscripts. Then comes the text, together with a full apparatus criticus; apart from the four ancient manuscripts (among which the Paris papyrus is dealt with in an addendum on p. 438), the editor uses ten of the sixteen medieval ones that contain this play, all of which he has himself collated either from the originals or from photographs. The commentary, together with two sets of addenda, occupies some 280 pages; there are two brief appendixes and three indexes of extreme accuracy and great usefulness. The book is excellently printed and produced; I have detected only three misprints, all insignificant.

It has long been disputed how far we can go in using Seneca's *Phaedra* as a guide to the contents of the lost *Ἱππόλυτος καλυπτόμενος*. In our own time an optimistic view has been taken by W.-H. Friedrich and has been defended in detail in the assiduous dissertation of C. Zintzen, *Analytisches Hypomnema zu Senecas Phaedra* (Beiträge zur Klassischen Philologie, ed. R. Merkelbach, 1, Meisenheim am Glan, 1960). Barrett after a scrupulous weighing of the evidence concludes that it is highly dangerous to use Seneca to reconstruct the lost play; his warning on pp. 16–17 must be carefully heeded, supported as it is by reasoned argument. Bruno Snell in his Sather Lectures (*Scenes from Greek Drama* (Berkeley and Los Angeles, 1964), 23 f.) has not convinced me that Barrett is mistaken.[1] It is generally agreed that the scene in which Phaedra expresses the wish to hunt (358–403 of Seneca's play) is based on the sickbed scene of the extant *Hippolytus* (170 f.). 'If a scene taken

[1] [See above, chs. 14 and 15.]

Journal of Hellenic Studies, 85 (1965), 164–71.

from a context known to us,' Snell writes (p. 23), 'fits so badly into Seneca's play, we may assume that the main structure of those parts that are coherent has been taken over from one single Attic tragedy.' The logic of this argument does not bear close inspection. 'As we know from other dramas of Seneca,' Snell continues, 'he does not aim at a consequence of logical action or at a coherent psychology in his characters.' That reflection might well warn against readily assuming that a Senecan play must closely follow a particular model, except in cases where this can be fully demonstrated; but Snell goes on to argue that 'if we can trace back vital scenes to the first Hippolytus, it seems probable that other scenes that fit the same texture go back to the same tragedy'.

Apart from arguments that have been used before and have been dealt with by Barrett, he relies chiefly on fr. 430 Nauck^2 = fr. C Barrett, which he considers 'the most important evidence for the Euripidean origin of the plan of Seneca's *Phaedra*' (p. 26).

> *ἔχω δὲ τόλμηϲ καὶ θράϲουϲ διδάϲκαλον*
> *ἐν τοῖϲ ἀμηχάνοιϲιν εὐπορώτατον,*
> *Ἔρωτα, πάντων δυϲμαχώτατον θεῶν.*

Snell insists that the words *ἐν τοῖϲ ἀμηχάνοιϲιν* must denote a very special kind of moral helplessness. He finds the expression *τὰ ἀμήχανα* 'typical of the early lyric poets' (p. 33: it occurs at Archilochus fr. 128 W. and Sappho fr. 130 L.–P); 'helplessness (*ἀμηχανία*) begins when in the vicissitudes of the emotions one loses contact with other men and stands alone' (p. 34). The portrayal of that condition 'links Euripides with the archaic poets' (p. 39); 'in this condition', Snell continues, 'we find Phaedra at the beginning of Seneca's play; this condition is new and typical in Euripides' early tragedies.' I see not the slightest reason for attributing so specialized a sense to *ἐν τοῖϲ ἀμηχάνοιϲιν* in this fragment. Any person who finds it impossible to get what he desires can be said to be *ἐν τοῖϲ ἀμηχάνοιϲ*, and Phaedra in any version of this story must have been in that position. In this place the line forms part of a characterization of Eros of a not unfamiliar kind; compare Aesch. *PV* 59 (of Prometheus) *δεινὸϲ γὰρ εὑρεῖν κἀξ ἀμηχάνων πόρον*. Its value as evidence for Phaedra's moral condition in the first Hippolytus amounts to zero.

It is indeed likely enough that Seneca made considerable use of the lost *Hippolytus* of Euripides; Barrett's careful summary of what we know of the plot of that play agrees well enough with Seneca. But there is not enough evidence to prove that this play served Seneca as the basis of his work,[2] and Barrett's warning that for all we know the *Phaedra* of

[2] But see now O. Zwierlein, *Senecas Phaedra und ihrer Vorbilder*, Abh. der Mainzer Akademie, 1987, no. 5.]

Sophocles may have been to some extent made use of must not be disregarded.

Barrett's sketch of the history of the text in antiquity provides several valuable modifications of the classic treatment of the subject by Wilamowitz in his *Einleitung in die griechische Tragödie* of 1889 (never subsequently revised). Barrett warns against overestimating the importance of the edition of Aristophanes of Byzantium as a landmark in the history of the text; he points out that we cannot safely think of it as 'published' in any modern sense, nor can we suppose that no one copied a tragedy from any other text after its appearance. Pre-Alexandrian texts, if they are 'wild', are wild not simply because they are pre-Alexandrian (see pp. 56 f., and cf. pp. 438–9).‖ Barrett shows that Wilamowitz was too specific when he traced the choice of the ten non-alphabetical plays to a definite selection made by a particular man during the second century AD; he thinks that during the third century the plays that were generally read largely coincided with the ten that ultimately survived. Barrett in this section acknowledges a general debt to Pasquali, and with good reason; he shows that there is no reason to suppose that all common errors in the ancient manuscripts of Euripides go back to the same source, and warns that throughout the ancient, as well as the medieval, tradition we must reckon with the fact of extensive contamination. It is doubtless symptomatic of the general state of affairs that the sixth-century Berlin papyrus, which preserves some 150 lines, agrees in error now with one and now with another class of medieval manuscripts, but offers no new readings unknown to the medieval tradition.

The presence in the medieval tradition of pairs of variants known to have existed in antiquity makes it certain, in Barrett's view (p. 58), that we must reckon with transcription from more than one uncial manuscript. He considers the possibility that the two medieval classes of the selected plays derive from the independent transcription of different uncial manuscripts; but he prefers to suppose that only one uncial manuscript was transliterated, but that corrections were later imported from another. His stemma of the manuscripts that contain the *Hippolytus* is printed on p. 62. The class *Ω* consists of MBOA, the class *Λ* of HCDEL; within *Λ* there is a subordinate division between HCDE (*Δ*) and L. V is a hybrid, like its younger relative, the Hauniensis.

Barrett's careful analysis of the divergences between manuscripts points unmistakably to the presence of considerable contamination, a factor for which Turyn in his book on the manuscript tradition fails to make allowance. Barrett's concise verdict on that work will be found at p. 61 n. 1, his brief demonstration of P's derivation from L on p. 429; both seem to me correct. C and E, not previously collated, are of real value; the antiquity of the tradition represented by the fourteenth-century manuscripts CDE is confirmed by the adhesion of the tenth- or eleventh-century Jerusalem palimpsest.

The apparatus criticus is by far the fullest, and clearly the most accurate, yet offered; and any competent judge will agree that Barrett's text is greatly preferable to any of its predecessors. As any reader of his article on Bacchyl. fr. 4 in *Hermes*, 82 (1954), 421 f. or of his suggestions on Menander's *Dyscolos* must know, Barrett is a textual critic of the highest order, possessing learning, ingenuity, and judgement, each in high degree. He stands as far from the conservatism of those who cling desperately to the most absurd readings of the manuscripts as from the radicalism of the wild emenders. He is not an emender of the palaeographic school who acts as though all corruption were an affair of juggling with letters. He has no difficulty in making up his mind; indeed, he seems to me sometimes over-confident in advancing a solution to a vexed problem. This is true especially in the matter of diagnosing interpolation, where as the notes below will indicate I believe he often goes too far. No sensible person would deny that the text of Euripides had suffered interpolation. But in the nature of the case it is hard to detect interpolations with certainty; and ancient taste in the matter of moralizing reflections was clearly so different from our own that it is desperately difficult to draw the line.

We come now to the commentary. Barrett is intimately acquainted with the surviving texts of Greek poetry; his command of the secondary literature of his subject appears complete; and he has applied to the task of elucidation the entire technical apparatus of modern scholarship. His learning is all the more imposing because he uses it not to swamp the reader with bibliography but to save him from it. Erudite commentaries that abound with references to secondary authorities are of great value to professional scholars, who would be churlish to be ungrateful for help that greatly eases their own labours; but the reduction of such matter makes for greater readability. Barrett seems to me to strike a happy mean in this respect; he cites secondary literature where necessary, but not otherwise. Many commentaries on Greek plays make dull reading; the same old parallels, many already present in the oldest editions, are trotted out by editors who lack the discrimination to see what is relevant and what is not. Barrett cites enough evidence to support his argument, and no more; he has the finest and most delicate appreciation of that evidence's value. The minutest points are treated, and yet the reader is never bored, so lively and so lucid is the editor's presentation and so compelling the continuous activity of the keenest and sharpest critical intelligence. Sometimes ruthless logic is pushed too far. Euripides was a poet and not a scholar, and cannot have taken half the trouble to write the play that Barrett has taken to explain it; in some places he is credited with a degree of rationality hardly to be demanded of the most rational of logicians. But the occasional annoyance caused by what some readers may think pedantry or fussiness is a small price to pay for the privilege of contact with a mind of such remarkable acuteness.

One main feature of the commentary is its intelligence; another is its author's extraordinary familiarity with the Greek language. He knows classical Greek poetry by heart; and his knowledge is made more effective by his firm grasp of Greek grammar, both morphology and syntax. As a disciple of Eduard Fraenkel, whose important influence is acknowledged in the preface, Barrett is familiar with the work of Schulze and Wackernagel, the two great scholars who used their mastery of linguistic science to throw light on literary texts. The remarkable use to which Barrett puts this branch of knowledge clearly shows how grievously we are suffering from the lack of adequate liaison between linguists and literary scholars in most universities today. In ‖ English-speaking countries the grammars used in schools are for the most part the work of writers with an inadequate notion of linguistic science. Of course beginners in Greek should not be plunged into comparative linguistics; but there are many points at which the most elementary grammar is affected by it, and even a beginner will take no harm from being given a general idea of its methods and results.

Barrett often uses the results of comparative linguistics to explain the text; but his commentary is rich in observations on style and usage that spring simply from his own acute perception and painstaking gathering of facts. Often he corrects widely prevalent misconceptions; often he supersedes all previous treatments of a familiar topic; many of his remarks on style and language are wholly new and original. Scholars will find much to write into the margins of their Kühners, their Schwyzers, and above all their Liddell and Scotts; and schoolteachers, however busy, have a duty to make a special study of a book which can teach us so much Greek.

Interest in language and style, though dominant, is not out of proportion. The treatment of metre is throughout masterly; the handling of the lyrics compares well with that of Wilamowitz, undemonstrable theorizing is absent, and the book is rich in the detailed metrical observation that aids textual criticism. Questions of religion and mythology are well handled; so is the topography of Trozen, about which Barrett shows Euripides to have been well informed. Barrett never forgets that the work he is explaining is a play, and dramatic questions are admirably handled; the structure and mechanics of the action could scarcely be more clearly or more cogently explained. But no mortal can have everything; and this commentary will not inspire the reader with delight in the *Hippolytus* as work of art. The tendency to split hairs, combined with a habit of referring to dramatic, as well as linguistic matters, in technical or semi-colloquial language (as on p. 399 Artemis is said to have 'no axe to grind' for Phaedra) can convey disagreeable reminders of the schoolroom. It is mistaken to complain that in the parodos there is no 'high poetry', or that the fact that the Chorus learned about Phaedra's state while washing clothes at the spring is 'trivial'; the parodos is well suited to its function in the play. (Contrast with this the sympathetic treatment of the second stasimon; in the fine passage on pp. 297–8 only the word 'fairyland'—again on p. 306—jars.) More serious is a tendency to water down the tragic dilemma by lecturing Hippolytus for 'puritanism', 'priggishness', etc., at times almost giving the impression that the whole trouble might have been avoided if Hippolytus had shown more common sense. Barrett truly says (pp. 391 f.) that the downfall of Hippolytus 'springs from a defect that is the reverse side of his very virtue: his cult of purity, for all its beauty and nobility, is bound up with an intolerant rejection of an essential part of human life'. Hippolytus' utterance at ll. 1364–7 causes the editor to rebuke him for 'blindness to the defects of his narrow puritanism'; and the farewell of Artemis at ll. 1437–9 prompts him to complain that 'for all its beauty the love of Hippolytus for Artemis lacks something essential'. That is the attitude of a monotheist. The ancient Greeks were not Christians, and did not insist on having everything; and they knew there are some good things that one cannot have without sacrificing others. Hippolytus' view of life is not a comprehensive one; but the poet recognizes its splendour and does not reproach him for its limitations. In point of power to communicate a full enjoyment of the beauty of the poetry and the impact of the tragedy, this commentary, like all others known to me, falls short of Dodds's *Bacchae*.

Now to points of detail. P. 12 n. 1: Would a marriage between Phaedra and Hippolytus, Theseus being dead, have been thought of as incestuous? Hyllus in the *Trachiniae* is shocked by his father's command that he should marry Iole, who is undoubtedly in a sense his father's widow (1225–6); but he is shocked not because of this but because she has caused the death of both his parents, and in the end he agrees to marry her. P. 34 (cf. p. 315): it is in no way surprising that Theseus should rule in Trozen during the lifetime of the aged Pittheus. In the *Alcestis* the ruler is Admetus, not Pheres; in the *Bacchae* Pentheus, not Cadmus; in the *Andromache* Neoptolemus, not Peleus. P. 46: in view of the division of opinion on these matters, it should not be stated dogmatically that the ends of the *Seven against Thebes* and the *Phoenissae* are interpolated and that of the *Heraclidae* mutilated.

On l. 3: Aesch. *Sept.* 941 seems to me an undoubted instance of *πόντιος* meaning 'from the Black Sea', and Soph. *Trach.* 100 is another instance of the type of expression dealt with here; see *CQ*, NS 4 (1954), 91 f. [= ch. 30, pp. 361 f.]. Il. 10–11: 'Euripides obviously cannot make Aphrodite speak of Hippolytus as *ἁγνός* when his *ἁγνεία* consists precisely in the rejection of herself; he contrives nevertheless to imply the *ἁγνεία* by the otherwise pointless insistence on the *ἁγνεία* of his foster-father.' This is over-subtle, and the insistence that the standard epithet must have 'point' is misplaced. L. 17: a tasteless comment of Wilamowitz is very properly corrected. Ll. 32–3: Barrett demonstrates that Jortin's correction of *ὠνόμαζεν* to *ὀνομάσουσιν* must be accepted; 'many of the more deep-seated corruptions in our texts', he rightly reminds us, 'are due not to the eye or to the ear, but to the mind' (p. 161). The doubts cast by Barrett on the alleged use of the imperfect in verbs of name-giving (cf. Fraenkel on *Agam.* 681) may be thought to be diminished by a forthcoming note of mine on the Tebtunis fragment of Sophocles' *Inachus* [= ch. 34, p. 398]. L. 41: a good note on *χρή* and *δεῖ*. Ll. 58–60: Barrett shows that *ἄιδοντες* should be preferred. Ll. 79–81: a notoriously difficult passage is convincingly explained. L. 91: 'The old man is like the person we all know who says "Do you know what I did this morning?" and actually waits for us to say "No, what?" before he proceeds.' This illustration not only lowers the tone; it may prevent the reader from ‖ seeing that the utterance is one of a conventional type dictated by the nature of one–one stichomythia. L. 93 f.: the explanation of the way the word *σεμνός* is used in this dialogue is masterly, and the defence of the tradition in 99 unanswerable. L. 115: despite Barrett's suggestion (p. 181) that the old man's opening words at l. 87 furnish an explanation, I find *ὡς πρέπει δούλοις λέγειν* odd in its application to the prayer he says he will offer to the goddess. The old man is a slave, and must address even a human master in the words proper to a slave; how can he demonstrate his humility in face of the goddess, or imply that Hippolytus should copy it, by insisting that he will address her in such language? *τοὺς νέους γὰρ οὐ μιμητέον* can very well stand by itself, and I am attracted by the suggestion of H. Richards, *Aristophanes and Others*, p. 293, that *φρονοῦντας* may be corrupt for *φρενοῦντες*: cf. Soph. *Trach.* 51–2; and see Barrett on ll. 421–5 and 876. Ll. 121–2: to the list of places supposed to be fertilized by rivers flowing underground or under sea and then surfacing elsewhere I would add Paphos, fertilized by the Nile; cf. *Bacch.* 406, which Jackson, *Marginalia Scaenica*, 117, explained by citing Manilius 4. 635. Ll. 148–50: 'the dry land of the open sea' is shown to refer to the sandbar that separates the lagoon north of Trozen from the open sea. Ll. 191–7: Barrett doubts the genuineness of these lines, and they certainly surprise in this context; yet for the Nurse to dilate upon this theme would make less abrupt her expression of the wish for death at ll. 250–1. L. 219: a good note on *πρὸϲ θεῶν*. L. 221: it may be correct to write *ἔχου-* at the end of one line and *ϲ'*—at the beginning of the next, but the effect is ugly and the exactitude attained not worth the effort. L. 246: Wilamowitz quoted Hom. *Il.* 13. 279. Ll. 261–3: after Barrett's excellent remarks, let us hope no one will ever again echo Wilamowitz's disastrous assumption that the Nurse knows all through this scene that Phaedra is in love. More than any of his other writings, Wilamowitz's commentary on this play is bedevilled by his insistence, so strong in the first half of his career, on seeing Euripides in the light of Ibsen; many of Gilbert Murray's anachronistic misconceptions derive from Wilamowitz.

L. 271: Barrett puts a crux against the words *οὐκ οἶδ' ἐλέγχουϲ*. 'The answer to "we should like you to tell us *ἥτιϲ ἐϲτὶν ἡ νόϲοϲ*", he writes, 'is not "I don't know how to find out," but simply "I don't know".' In strict logic this is true; but in almost any conversation no one would complain if this answer were given to this question, especially if the giver happened to be the person most likely to be in the know. Nor is the ellipse in l. 272 which we must assume if we accept the tradition as awkward as Barrett contends. L. 277: Barrett accepts Wilamowitz's suggestion that *θανεῖν* has displaced *οὐκ οἶδ'*. Now since Phaedra is obviously not out of her mind, as *ὑπ' ἄτηϲ* would imply, the Nurse might conceivably assume that she meant to fast to death; and the fact that it makes the two parts of l. 277 tautological does not strengthen the case against the traditional reading, tautology being not uncommon in this author. Still, the emendation certainly gives a more satisfying sense. L. 287: in the note on p. 213 *δεϲπόταϲ* is a misprint for *δεϲπόταιϲ*. L. 290: the excellent note on the 'coincident' use of the aorist participle corrects a widespread misconception. Ll. 298–9: cf. Horace, *epist.* 1.6. 67–8 and Isoc., *ad Nicoclem* 39: was it a *locus communis* as early as Euripides? L. 307: the Nurse's hideously vulgar parody of an oath by a divinity is both linguistically abnormal and highly typical of her ethos. L. 328: *ϲοῦ μὴ τυχεῖν* is convincingly explained by means of the parallel at Aesch. *Supp.* 161. L. 352: cf. Christ's reply to Pilate (Matt. 27: 11; Mark 15: 2; Luke 23: 3; cf. John 18: 37). Was this a common formula for evading responsibility for a statement by another without denying its truth? Ll. 362–72: the one real parallel for the separation of this strophe from its antistrophe (669–79) is Soph. *Phil.* 391 f. ~ 507 f., which is not cited.

Ll. 377–81: Barrett rejects the view of Wilamowitz, Dodds, and Snell that Phaedra is here made the mouthpiece of a polemic against the Socratic paradox that οὐδεὶc ἑκὼν ἀδικεῖ. 'It is not Euripides who is polemizing', he writes (p. 226), 'but Phaedra, and she is polemizing not against the Socratic explaining away of moral weakness in terms of ignorance but against the much simpler view that all wrongdoing is ordinarily due to natural vice.' Snell in the Sather Lectures already quoted (pp. 30 f.) justly retorts that what Barrett calls 'the much simpler view' is not Greek at all, but Christian. Snell insists that Phaedra means that men do wrong not out of ignorance or stupidity, but out of a failure to control their passions; γνώμηc φύcιν at l. 377 and εὖ φρονεῖν at l. 378 should probably, as he argues on p. 58, be given a more intellectual sense than Barrett (p. 228) will allow. Yet Barrett is surely right in denying that we can be sure that in this place Euripides had Socrates in mind. In the passage of Plato's *Protagoras* (352) which Snell thinks contains Plato's reply to Euripides, the view expressed by Phaedra is cited as the ordinary man's opinion; and as Barrett says there is no reason why Plato should be taken to have one particular holder of it in mind. Snell tries to diminish the force of this objection by pointing out (p. 61) that according to Plato the holders of this opinion justify it by speaking of inner conflicts (men act ἡδονῆc ἡττωμένουc, etc.), and so does Phaedra. But this coincidence proves nothing; how else would Snell expect anyone who held this view to justify it?

Ll. 386–7: an excellent note on καιρόc. Ll. 407–9: the statement that this is 'the first surviving instance of a τόποc which continues through Greek into Roman literature' is not true; see Soph. *Aj*. 1192 f.[3] 428–30: a mirror, says Barrett, can be either self-revealing or other-revealing; Time's mirror is of course other-revealing, yet here Euripides elaborates the metaphor as though the mirror were self-revealing; and he finds the result 'an odd confusion'. But Time's mirror is very often self-revealing, and Euripides clearly meant it to be so in this place.

P. 238: in the last line on this page, the apostrophe ‖ of ταῦτ' has been omitted. P. 239: four lines from the end of the note on ll. 437–8, ἀπέcκηψεν is a mistake for ἀπέcκηψαν. Ll. 451–2: is it really so strange that the Nurse should speak of legends of the gods as having been written in books? There is no point in saying that no old paintings could have been found in houses of Euripides' day (p. 242); even if this were certain (as it is not), Euripides might well have imagined them in ancient palaces. Ll. 459–61: 'πὶ δεcπόταιc ἄλλοιc θεοῖcιν is rendered by 'to have other gods as masters, to be subject to other gods'. That would translate ἐπὶ τῶι ἔχειν ἄλλουc θεοὺc ὡc δεcπόταc: what is in the text should mean 'beneath the sway of other gods' (see LSJ s.v. ἐπί B.I. 1g,

[3] [The error goes back to Leo, *Plautinische Forschungen* (2nd edn. 1912), 151, who is cited by Pfeiffer on Call. fr. 110. 50.]

p. 622). Ll. 468–9: the famous crux is brilliantly and, to my mind, convincingly solved. Ll. 478–81: cf. *Med.* 407–9. 484–5: for the corruption of ψόγος to λόγος, cf. Aesch. *Supp.* 170, and for the reverse corruption cf. id., *Eum.* 989; though Barrett rightly rejects the idea that it has happened here. L. 496: Scaliger's προῆγον is rightly put in the text. Ll. 503–4: Barrett is surely right to read ἃ μή cε. Ll. 507–8: Barrett's interpretation, made clear by his punctuation, is convincing.

Ll. 530–1: Barrett renders, 'For the shafts neither of fire nor of the stars surpass that of Aphrodite launched from the hands of Eros, the son of Zeus'. οἷον he thinks 'is treated as though preceded not by ὑπέρτερον but by τοιοῦτον, a not unnatural irregularity'. To me the irregularity seems highly unnatural, and I know no parallel; nor do I easily understand why the shafts of love should be compared simply with shafts of light, which are like them in travelling fast and far, but not in their destructive power, which one would expect to find stressed here. Suppose that by a common type of corruption ὑπέρτερον has ousted ὑπέρτατον. In that case the shafts of love will have been compared to, first, 'the shaft of fire on high', i.e. the sun, which is not only bright and swift, but scorching, and, secondly, 'the shaft on high of the stars', i.e. the lightning. ἀcτεροπή (ἀcτραπή) derives from ἄcτρον (see Frisk, *Griechisches etymologisches Wörterbuch*, I. 170); and ἄcτρων βέλοc at Aesch. *Agam.* 365 could well mean 'lightning', as Gilbert Murray suggested in his edition of 1936. Other explanations of the passage require us to suppose that if Zeus in Olympus, aiming at Paris in Troy, were to overshoot the mark his bolt would travel 'beyond the stars', which seems an unlikely direction for it to take; if ἄcτρων βέλοc means 'lightning' we can write ὕπερ paroxytone and supply an object from καιροῦ.

Ll. 542–4: Barrett argues that in all other cases of διά with a genitive used with ἰέναι and an indirect object in the dative, the genitive denotes an activity of the subject and the dative the person to whom that activity is directed; he, therefore follows Nauck in adopting Dobree's conjecture ἰέντα . . . θνατοὺc. This may well be right, but the conclusion is put a little too forcibly; the existence of the phrase δι' ὄχλου εἶναί τινι, together with the not uncommon tragic use of ἰέναι to mean 'to come on' someone, often to his hurt (see Jebb on Soph. *Ant.* 234), should deter us, as T. C. W. Stinton warns me, from putting Dobree's reading in the text. L. 552: Barrett's conjecture φονίοιcι νυμφείοιc heals sense and metre, postulates a very likely corruption, and may well be right. L. 577: Barrett's explanation of coὶ μέλει πομπίμα φάτιc δωμάτων is to be preferred.

Ll. 584–8: Barrett accepts Murray's γεγωνεῖ δ' and renders, 'I hear a voice, but I have nothing sure; but it carries clear where the cry has come to you through the doors.' The sense would be more satisfactorily if one read *γεγώνει δ'* and translated, 'but tell us how the cry came to you', i.e. 'tell us what was its content'. L. 585: Barrett adopts Weil's *ἰὰν*, now confirmed by P. Oxy. 2224. At Soph. *OT* 1219, modifying the conjecture of Burges cited in Pearson's apparatus, I propose *ὡc ὀδύρομαι / περίαλλ' ἰὰν χέων* . . . L. 594: Barrett makes a good case for Barthold's *ἐκπέφηνε*.

L. 621: *BΛ* read *χαλκὸν . . . χρυcοῦ, Ων* have *χρυcὸν . . . χαλκοῦ*. Barrett insists that the former must be right, because gold is much heavier than bronze; but are the weights of the metals relevant at all? This manner of expression is simply a device to secure three cola of which the last is longer than the first two (cf. Hom. *Il.* 6. 48, cited by Barrett ad loc.); to refer to the metal X by the expression 'the weight of X' need not imply that X is specially heavy. Ll. 622–3: Barrett enquires whether *τῆc ἀξίαc* depends on *τοῦ τιμήματοc* or the other way round; if the former, the sense will be 'the sum at which its value is assessed', if the latter, it will be 'the amount appropriate to his estate'. Both are somewhat complicated, and the second alternative, which Barrett prefers, involves supposing a surprisingly prosaic sense of *τίμημα*. I am attracted by Stinton's suggestion that we read *του* indefinite and render 'buy children for a given price, each for what it is worth'. Ll. 625–6 are expelled by Barrett with some confidence; 'attempts at emendation', he says, 'have in any case failed'. But with Pierson's *ἐκπίνομεν* the lines give a wholly intelligible reference to the custom of the wedding feast; the intolerable *ἐκτίνομεν* and the incompatibility with ll. 628–9 vanish; and *κακὸν* in l. 627 takes up *κακὸν* in l. 625. The placing of ll. 625–6 between the protest against women's existence in ll. 616–24 and the proof of their badness of ll. 627 f. does not seriously 'disrupt' the argument; and the *μὲν πρῶτον* of l. 625 would not be alone in having no answering *ἔπειτα δὲ* or the like (see Denniston, *The Greek Particles*, 382 (iv)). Here as in other cases Barrett may well be right, and the doubt about the lines is certainly worth mentioning; but is it safe to pronounce that they must be corrupt? Ll. 634–7 are deleted with more reason. The speech is certainly more effective if Hippolytus says nothing about one's wife's relations and does not concede the possibility of *χρηcτὰ λέκτρα*, though even here I do not feel entirely sure that late actors, and not the poet, were responsible. L. 649: Wecklein's tentative conjecture *ἐννοοῦcιν* is worth a mention; cf. Plato, *Laws* 798 B. L. 659: D's reading *ἐκδημῆι*, which Hermann had conjectured, is rightly put in the text. L. 663 is objectionable; it gives an anticlimax and ‖ the sense is poor. Herwerden was the first to cut it out; Barrett's suggestion that it was a substitute for l. 661 is most attractive. Ll. 670–1: Barrett's solution of the problem is highly probable, and he is right in saying that the repetition (*λόγους . . . λόγου*) 'is not only blameless, but deliberate and effective'. L. 686: Barrett's interpretation of *κακύνομαι* as meaning 'I feel disgraced or degraded' is surely right, whatever may be said by people who will never believe anything without a parallel.

L. 691 is cut out: 'as if,' writes Barrett, 'at a moment like this she will turn her mind to her husband's maternal grandfather in his dotage'. Barrett's contempt for Pittheus is unjustified. It is clear from more than one passage that he was important to the poet and his audience, as he might well be considering his connections; he had brought up his grandson Hippolytus (11) and in view of his known *ἁγνεία* might have been expected to be specially shocked. In such a context *cυμφορὰc* is a natural way for Phaedra to refer to a situation which from her point of view may well be so described. The case against this line is feeble. Ll. 715–16: Barrett's conjecture is seductive; only in the phrase *εὕρημα . . . τῆcδε cυμφορᾶc*, 'something for this trouble of mine', as he renders it, the genitive is appreciably harder than in expressions like *πόροc κακῶν* or *πρόβλημά τινοc*, which Barrett suggests are parallel, and the possibility of a solution along the lines indicated by Wilamowitz (ad loc.) cannot be ruled out.

L. 734: Barrett's reading, with Maas's transposition in the antistrophe, seems the most acceptable solution. Ll. 739–49: Barthold's deletion of *πατρὸc* in the strophe and *μελάθρων* in the antistrophe is a bold measure; but as Barrett observes it heals sense and metre in both places, and it is an odd coincidence if it is wrong. Ll. 742–51: the myth is admirably handled. L. 756: the masterly note on what is misleadingly called the 'accusative in apposition to the sentence' must be carefully studied. Ll. 758–63: Barrett attacks the crux along the lines laid down by Weil, supposing that *ἀπ' ἀμφοτέρων*, looks forward to a double apposition, and that the sense is 'Phaedra's journey was ill-omened both when she left Crete and when she landed at Munichia'. In l. 757 he thinks *ἡ Κρηcίαc* has replaced something like ⟨*Μινωίδοc τ'*⟩, and in l. 760 he follows Weil in changing *δ'* to *τ'*. *Κρηcίαc*, as Barrett admits, is not normal glossographer's Greek; but as he says (p. 434) it may have come in from l. 752. Once it is acknowledged that *δύcορνιc* must be predicative and strongly emphatic, and that its force must extend over the whole sentence down to *ἔβαcαν*, it becomes hard to doubt Barrett's localizing of the corruption, and his conjecture may well be correct. L. 760: the list of tragic lengthenings before initial mute and liquid is most valuable. Aesch. *PV* 582 *πυρὶ φλέξον* is usually emended, but Stinton compares the epic name *Πυρῑφλεγέθων* and quotes *Bacch.* 1019, where as Dodds observes the tradition is metrically defensible.

Ll. 808–10: a good note on the use of the word *κλῆιθρα*. L. 811: the sensible remarks on the ἐκκύκλημα deserve everyone's attention. L. 816: I would take ἀμαυροῖ ζοάν to mean 'consigns your life to darkness'; see the note on a problem in Sophocles' *Inachus* mentioned under l. 32 [= ch. 34, p. 399]. Ll. 826–7: Barrett well shows how the text as it stands must be interpreted, but voices a suspicion that it may be corrupt. Prima facie, as he says, one would expect the two nouns each of which is accompanied by τίνα to be parallel, and Theseus can hardly enquire *what* τύχη of his wife is in question; both difficulties would be removed if instead of the second τίνα we read τάνδε . Ll. 840–1: Barrett marks τίνος κλύω with a crux, while making the necessary change of καρδίαν to κραδίαν. L. 850: Barrett rightly adopts Kirchhoff's φέγγος θ' ἁλίοιο. L. 863: Barrett says that σαίνειν 'is used of a person or a thing which attempts to rouse, or which in fact rouses, a person's favourable emotion'. At Aesch. *PV* 835 and [Eur.] *Rhes.* 55, for instance, *favourable* emotion is not in question; the verb means rather 'to claim someone's recognition', 'to impinge on the consciousness so that one recognizes it' (thus Stinton), as Jebb on Soph., *Ant.* 1234 clearly saw. L. 867: Barrett prints ἐπεισφρεῖ from an unpublished note of J. U. Powell, which neatly explains the variants; accepting Maas's suggestion that μὲν . . . βίου came in from 821, he puts a crux against ἐμοὶ . . . τυχεῖν. If Maas is right, as he very probably is, there is little hope of restoring the right text. Ll. 871–3: Barrett, after Nauck, deletes: certainly one is glad to have ancient authority for their removal. L. 875 was cut out by Wilamowitz on the ground that after Theseus has described the κακὸν as οὐ τλητὸν οὐδὲ λεκτόν, the Coryphaeus cannot at once ask him to describe it with the word λέξον. To modern taste, this is certainly anomalous; but to ancient? Expressions like ἄφατος, ἄρρητος, οὐ λεκτός, are not descriptive but emotive; and it is not safe to assume that after someone has called something 'unmentionable' or 'indescribable', it is not possible, for another person to say, 'Tell me what it is.' At l. 602 the fact that the Nurse's speech has been ἄρρητος has not prevented Hippolytus from hearing it; cf. *Ion* 782 ἄφατον ἄφατον ἀναύδατον λόγον ἐμοὶ θροεῖς.

L. 903: Barrett's conjecture στήσας ἔχεις may well be right. L. 908: Lehrs's χρόνος παλαιὸς is rightly adopted. Ll. 912–13 are deleted; this time Barrett has had no predecessor in diagnosing interpolation. He complains (i) that Hippolytus cannot, at a time like this, advance his own ubiquitous curiosity as a reason for admission into his father's confidence, (ii) that λίχνος and ἁλίσκεται show that this curiosity is described in terms of disapproval, (iii) that the lines destroy the pattern of short appeals and pauses, (iv) that they are inconsistent with l. 911, and so cannot be joined to it by γὰρ. But (i) the curiosity that makes one want to know all, in times of dire affliction, that affects one's family is a natural curiosity, (ii) 'is convicted of greed' is a pejorative expression, but it may be used here metaphorically to express the strength of the desire and not its reprehensibility, (iii) the 'pattern' might or might not be punctuated by a gnomic parenthesis of this sort; if it were, this would not be the only such ‖ case in Euripides, (iv) the run of the sense is, 'One (i.e., the afflicted) should not be silent in time of trouble; for the human heart (i.e., that of their friends) stands convicted of avidity even in such times.' L. 911, Barrett says, appeals for speech for the speaker's sake, ll. 912–13 for the hearer's; but neither line specifies for whose sake speech is desired, and the speaker might reasonably be taken to mean that it is desirable for the sake of both. L. 924: the manuscripts vary between present and aorist and between subjunctive and optative in all other combinations, but Barrett rightly restores ὑπερβάλληι, which happens not to be in any of them; cf. *Alc.* 1077. L. 946: Barrett rightly follows Wilamowitz in adopting Musgrave's ἐλήλυθα. Ll. 952–5 are well explained, and the nonsense about Hippolytus being an Orphic is well exploded. The nature of fifth-century Orphism is not necessarily relevant to this passage; the play is located in the heroic age, and the poet may have had in mind no more than may have been attributed to Orpheus, say, by Aeschylus. L. 983: Barrett adopts Herwerden's ξύντασις for ξύστασις. At Thuc. 7. 71, often adduced in defence of the manuscript reading, Barrett rightly takes ξύστασιν γνώμης not of an inner conflict in the minds of the combatants, but of their mental participation in the fighting. But if that is right, the word here might mean that the φρένες of Theseus συνίστανται τῶι Ἱππολύτωι (see LSJ S.V., B 11. 2). This reflection would restrain me from putting ξύντασις in the text; but it certainly gives an easier sense, and by a slight change. Ll. 986–7: does the poet mean Hippolytus to seem 'priggish' here? That is certainly the impression his words make on modern taste; but when one remembers the similar language of the virtuous Creon at Soph. *OT* 583 f., one wonders whether the effect can be intentional.

L. 1007: Murray's ἴτω is adopted; certainly Aesch. *Cho.* 565 f. does not protect the tradition, as Wilamowitz supposed. Ll. 1014–15: Barrett puts a crux against εἰ μή. If the lines are genuine, he would incline to read ἥκιστ', ἐπεί τοι (rejecting Wieseler's ἐπειδή, which is nearer to the *ductus litterarum*); but he thinks there is 'a strong but not absolutely conclusive case' for cutting out the whole of ll. 1012–15. If that were done we should be rid not only of what Barrett calls 'the nonsense of 1014–15', but of certain difficulties in ll.1012–13. Hippolytus, Barrett says (p. 352), 'treats the folly as self-evident, yet it is oddly uncertain where it is meant to lie'. Yet since in the first place Phaedra's husband is alive and well, and in the second the kingdom is not hers to give, the uncertainty is not really very perplexing. The first person singular ἦν is indeed rare in Euripides, but as Barrett on l. 700 shows, it cannot safely be altogether eliminated, and the shift in meaning between τὸ cῶφρον in l. 1007 and τοῖcι cώφροcιν in l. 1013 hardly amounts to a strong reason for excision. There is a case for cutting out the passage, but I should hesitate to call it 'strong'. L. 1029 is also deleted by Barrett; Valckenaer was the first to cut it out. Barrett argues that there is no point in adding an imprecation of exile to one that already implies utter destruction, and in strict logic this is true. Still, it would be possible to say, 'May I perish without name or reputation, wandering in exile over the earth'; ὀλέcθαι does not always connote actual death, and to call the line 'a manifest interpolation' is to go too far. L. 1032: Nauck's τί for εἰ is rightly adopted. L. 1042: Barrett's tentative suggestion that μὲν has ousted μοι seems very plausible. L. 1044: Barrett rightly prefers ἠξίουc: it looks as if ἄξιον in the next line ironically picks up this word. Ll. 1045–9 are left untouched, but following Nauck Barrett cuts out l. 1050. It is good that there is external evidence against it, and Barrett rightly calls it 'too similar to 1047 for comfort'; but was Euripides incapable of writing it?

L. 1070: Barrett must be right in deleting τ'. The ellipse of the verb in the first part of this line is most unusual; was αἰαῖ really *extra metrum*, and has, say, παίειc been omitted after it? Ll. 1082–3: Barrett thinks this speech is intended 'to throw subtle light on Hippolytus's psychology for its own sake, to suggest that this feeling of inferiority, of otherness, is what lies behind his urge to establish himself as a paragon of virtues that the common man cannot share'. No Greek hero needed a feeling of inferiority in order to practise virtues that lay beyond the reach of the common man, and Hippolytus was the son of Theseus and an Amazon. Hippolytus' point here is a rhetorical one; he finds his father's treatment of him so unaccountable that he affects to take refuge in the supposition, obviously not to be seriously entertained, that it is due to prejudice on account of his birth. Neither this line nor l. 1455 gives the slightest support to a modernizing psychological interpretation which we must utterly reject.

Ll. 1102–50: even Barrett is perplexed by the baffling problem of the two, perhaps three, masculine and two feminine participles used to refer to themselves by the singers of this ode.[4] If male singers sing all or part of it, who might they be? Barrett rules out the idea that they might be the companions of Hippolytus addressed at 1097 f. on the ground that these leave at l. 1098; but it would not be impossible for them to wait and sing the ode before they go. True, it is not announced that

[4] [See now G. W. Bond, *Hermathena*, 129 (1980), 59–63, whose view Diggle in vol. i of his OCT (1984) has rightly adopted.]

they will do so; but neither is it announced that the handmaids will sing their part in the exodos of Aeschylus' *Supplices*, a fact indicated by the deliberate mention of them at ll. 975 f., some sixty lines before. (Barrett surprisingly asserts that the presence of a chorus of Egyptians is 'announced' at *Supp.* 825 f. There is nothing in the text about these people; see *AC* 33 (1964), 366 ff [= ch. 19, pp. 270 f.].) Barrett tentatively suggests that at ll. 1105–6 the poet may have written *ξύνεϲιν δέ τιϲ ἐλπίδι κεύθων / λείπεται κτλ.* This is a most ingenious conjecture, and together with Musgrave's emendation of 1020, put in the text by Barrett, it would enable us to assign the ode to the regular chorus. But in view of the possibility that the companions of Hippolytus might be the singers, it is worth observing that one *could* emend away the two feminines. Alternation, as Barrett argues, is most unlikely; and on the balance I incline to agree ‖ with him that it is desirable to have the ode sung by the women.[5]

L. 1122: *Ἑλλανίαϲ . . . Ἀθάναϲ* is puzzling. [Arist.], *de mirab. auscult.* 840 A mentions an *Ἀθ. Ἑλλανία* near Metapontum; and in the Spartan *rhetra* at Plut. *Lyc.* 6. 2 Bryan's suggestion that the mysterious *Ζεὺϲ Cυλλάνιοϲ* and *Ἀθάνα Cυλλανία* are really *Ἑλλάνιοϲ* and *Ἑλλανία* might easily be right. I therefore suspect that *Ἑλλανία* is a cult-title of Athena at Trozen, otherwise unrecorded; but I cannot prove the truth of my surmise. L. 1146: it ought to be remarked that for a tragic chorus to say 'I am angry with the gods' is without parallel.

L. 1183: only two horses can be harnessed to the yoke of a chariot, so that in a team of four, two must be trace-horses. Hippolytus was driving a team of four (1212, 1229); but I do not find that the horses being called *ζυγηφόρουϲ* in this place involves an imprecision that is specially 'odd', but rather one of a kind with which poetry abounds. L. 1189 is rightly taken to mean 'fitting his feet right into the footstalls'. L. 1204: Barrett justly remarks that *νεανικόϲ* in the sense of 'lively' or 'violent' seems oddly undignified here; and rather than suppose with Björck that it is meant to characterize 'honest homeliness', I had rather supposed that the words meant 'we were alarmed, as was natural in young men'. Young men are not, of course, cowardly, only they are excitable, and might easily be startled by the uncanny sound. L. 1234: a good note on the use of *δεϲμόϲ* as *nomen actionis*, 'oddly ignored by LSJ'. L. 1292: Barrett puts a crux at the end of this line; but may not *βίοτον* be an instance of the plain accusative of respect? L. 1316: I take *παρεῖλεϲ*, like *παρέϲυραϲ* at Aesch. *PV* 1065, to be a metaphor from taking an arrow from its quiver; cf. the shooting metaphor in *ἐφῆκαϲ*, which Barrett rightly prefers at l. 1324. Ll. 1338–41: cf. Soph. *Ajax* 131–3. 1364–7: an excellent note on the avoidance of ◡◡ | ◡◡ in

[5] [I have now changed my mind.]

anapaests. 1375: B. rightly considers the metrical difficulty as warranting a crux. 1379: B. rightly adopts Weil's *τι*. Ll. 1385–6: Barrett follows Weil in changing *ἀναλγήτου* to *ἀνάλγητον* on the ground that 'the notion that the suffering itself feels no pain at the suffering is absurd'. But in Greek words like *νόϲοϲ, κακόν, πάθοϲ* were easily personified, and a word that had come to mean 'ruthless, unpitying' might in poetry easily come to be applied to one of them. Ll. 1416–22, a difficult passage, is convincingly explained. L. 1436: Barrett's interpretation is surely right: cf. Hom. *Il.* 20. 127 f.; 24. 209 f., etc. L. 1431: Barrett thinks 'son of ancient Aegeus' is unexpected when Aegeus is long dead, and thinks the expression is used because Aegeus was already old when it became known that he was Theseus's father. This seems to me misplaced subtlety. Ll. 1462–6: Barrett is reluctant to believe that Euripides himself can have placed at the end of his plays the moralizing tailpieces which in the manuscripts conclude so many of them. He argues that 'in four plays at least the lines were added later to cater for a public addicted to sententious commonplaces'. Can he so confidently acquit the fifth-century public of such a taste? The tailpieces may be later additions, but the fact would not be easy to demonstrate.[6] Barrett exaggerates the obscurity of the tailpiece of the Hippolytus itself. *μᾶλλον* if taken as Barrett shows it must be with *ἀξιοπενθεῖϲ* is easily intelligible; nor do I see any reason to doubt that Euripides could have used *τῶν μεγάλων* with reference to Hippolytus. Hippolytus was a son of Theseus and a hero (cf. *Hippolytus* I fr. 446 N^2 = fr. U Barrett), and that is enough to warrant such an appellation.

The above remarks fall very far short of an adequate description of the richness of this profoundly learned and supremely intelligent book. If some of them indicate disagreement, that should not obscure the reviewer's admiration of what seems to him a truly great achievement.

[6] [See now D. H. Roberts, *CQ*, NS 37 (1987), 51–64.]

37(8)

Euripidea[1]

I. *ION* 237–8

ΙΩΝ γενναιότης σοι, καὶ τρόπων τεκμήριον
τὸ σχῆμ' ἔχεις τόδ', ἥτις εἶ ποτ', ὦ γύναι.

Wilamowitz thought the text was sound. He quotes *Ph.* 1680:

γενναιότης σοι, μωρία δ' ἔνεστί τις

and argues that in the *Ion* passage the speaker starts as though he is going to use a similar expression and then changes his mind. But the switch from one construction to another is too violent for this writer; and it is odd that the temple slave should address the queen in so abrupt a fashion. Several conjectures have been offered; γενναῖος εἶ τις by Dobree, γενναία τις σύ by Reiske, γενναία τοι σύ by Musgrave, γενναιότητος by Boissonade, γενναιότητος τῶν τρόπων by Badham. It is plain that none of these will do [; two are not even metrical].

It is strange that no one has yet suggested that there may be something missing at the beginning of the speech. This assumption would explain both the harshness of the syntax and the abruptness of the sense, so much at variance with the ceremonious addresses to royal ladies which abound in tragedy. How much is missing or what that was is of course impossible to guess. But I suspect that not more than one line has dropped out, and that it was something like this:

ὦ χαῖρ', ἄνασσα· καὶ γὰρ οὖν μορφῆι τ' ἔνι . . .[2]

Compare Soph. *El.* 663–4. The Paedagogus asks the Chorus if he is right in identifying the lady whom he sees as Clytemnestra: πρέπει γὰρ ὡς τύραννος εἰσορᾶν. When the Chorus has confirmed his guess, he begins: ὦ χαῖρ', ἄνασσα.

[1] I am grateful to Professor D. L. Page for his criticism of these notes.

[2] [Cf. Eur. *Hypsipyle* fr. 60, 22–4.]

Classical Review, NS 7 (1957), 97–100.

II. *ORESTES* 1621–4

MEN. ὦ γαῖα Δαναῶν ἱππίου τ' Ἄργους κτίται,
οὐκ εἶ' ἐνόπλωι ποδὶ βοηδρομήcετε;
πᾶcαν γὰρ ἡμῶν ὅδε βιάζεται πόλιν
ζῆν, αἷμα μητρὸc μυcαρὸν ἐξειργαcμένοc.

1623 ἡμῶν codd. (Editors print Brunck's ὑμῶν; but why should not Menelaus for the moment count himself as an Argive?)

The second sentence is supposed to mean: 'For this man is constraining your whole city so that he may live, having on his hands the pollution of his mother's blood.'

βιάζεcθαι is often followed by an infinitive. But the analogy of every other ‖ instance suggests that these words ought properly to mean: 'This man is constraining your whole city to live'; and this is nonsense.

Nauck changed ζῆν to ζῆι δ': but there is a better remedy. The scholia offer two different explanations of the passage. The second is the one given by modern editors; but the first runs: πᾶcαν γὰρ ὑμῶν οὗτοc πόλιν βιάζεται διὰ τὸ ζῆν καὶ αὐτόν, μηδενὸc αὐτῶc cυγχωροῦντοc. No one seems to have observed that the writer of this note must have read ζῶν, which I think is the right reading. The sentence will then have meant: 'For this man is committing an act of violence against your city by his very being alive . . .', etc.

The corruption postulated is a very easy one, since βιάζεcθαι is so often followed by an infinitive that the participle may easily have been altered. The sense given is more forceful than that which we get from reading ζῆν; one point in its favour is that it makes it easier to see why πᾶcαν is there. Compare *Sept.* 1042:

αὐδῶ πόλιν cε μὴ βιάζεcθαι τόδε,

the use of νόμουc βιάζεcθαι at Soph. *Ant.* 663 and Thuc. 8. 53. 2, and Theocr. 22. 9, where Gow renders ἄcτρα βιαζόμενοι by 'wilfully disregarding the prognostics of the heavens'. In the *Orestes* passage also, 'do violence to' is a stronger equivalent of 'wilfully disregard'.[3]

III. *PHOENISSAE* 582–3

IO. δύο κακὼ cπεύδειc, τέκνον,
κείνων cτέρεcθαι τῶνδέ τ' ἐν μέcωι πεcεῖν.

Σ ὁ δὲ νοῦc· μήτε ἐκεῖνα μήτε ταῦτα ἔχειν. κἀκείνων cτερηθῆναι τούτων τε μὴ τυχεῖν. τὸ γὰρ ἐν μέcωι πεcεῖν οὕτωc φηcί· πρὸ τοῦ ἀφικέcθαι ἐπ' αὐτὰ ἐκπεcεῖν τῆc ἐλπίδοc (MTAB).

[3] [Against, see the note on this passage in Sir Charles Willink's commentary (1986); M. L. West in his edition of 1987 also keeps ζῆν.]

This view has been accepted by most editors; Pearson, for instance, writes: 'to lose Argos and to fall short of Thebes'. 'With ἐν μέϲωι, μεταξύ and the like', Pearson says, 'whether referring to space or time, only one limit is expressed in Greek'. He cites Ar. *Av.* 187:

ἀλλ' ἐν μέϲωι δήπουθεν ἀήρ ἐϲτι γῆϲ

where the meaning is 'between (the gods and) the earth is the air'.

Pearson takes κείνων and τῶνδε as neuter, so that they are equivalent to τὰ ἐν Ἄργει and τὰ ἐν Θήβαιϲ meaning 'what you have in Argos' and 'what you have in Thebes'. It follows that he has to take ἐν μέϲωι πεϲεῖν as meaning 'fall short of' in the sense of 'lose'. The trouble is that neither ἐν μέϲωι nor μεταξύ is ever coupled with πεϲεῖν or any other verb in quite this sense. They never seem to be used as 'citra' is in Ovid's 'citraque necem tua constitit ira' (*Tr.* 2. 127) or as 'this side' is used in Jonson's famous saying about Shakespeare. Grégoire and Parmentier translate 'perdre Argos et tomber avant d'atteindre Thèbes'. They seem to take the last words literally, as meaning 'be killed before reaching Thebes'. But Pearson must be right in taking both the pronouns to be neuter.

Powell offers a third way of taking the passage. 'If you are conquered', he paraphrases, 'you will lose the one advantage—the friendship of Argos; if you conquer, in the hour of success you will fail (for a victory over one's country is a loss).' Powell has not seen that this sentence must sum up, not the whole of Iocaste's address to Polyneices (568–83), but the second half of it (578–83). The first half of her address envisages the consequences of an Argive victory, the second half those of an Argive defeat. If Argos is defeated, Polyneices will ‖ incur a double penalty; not only will Thebes be lost to him, but his Argive friends will blame him for their calamity. It is this second half which 582–3 sum up; and there is no reason for thinking that τῶνδε . . . ἐν μέϲωι could mean 'in the hour of success' in this context. Pearson must be right in thinking that κείνων and τῶδε are antithetical; and if κείνων means τῶν ἐν Ἄργει, τῶνδε must mean τῶν ἐν Θήβαιϲ. Powell seems to have changed his view later; for a note in his copy of Prinz–Wecklein, which I now possess, suggests that he later came to regard l. 583 as an interpolation.

I suspect that Euripides wrote:

δύο κακὼ ϲπεύδειϲ, τέκνον,
κείνων ϲτέρεϲθαι τῶνδέ τ', ἐν μέϲωι πεϲών.

'You are bringing on yourself a double misfortune, my son, that of losing both what is yours in Argos and what is yours in Thebes, by falling between the two.'

Technically the emendation is an easy one. The text would obviously have been liable to corruption by someone who assumed that l. 583 consisted of two clauses coupled by the τε and that the second, like the first, ought to contain an infinitive.

Almost every modern language seems to have a proverb similar to our 'falling between two stools'; but I have found no exact equivalent in Greek or Latin. And yet the expression seems so natural and its meaning in this context so obvious that I believe Euripides used *ἐν μέϲωι πεϲών* in this sense.[4]

IV. *ELECTRA* 1058–9

ΗΛ. †ἆρα κλυοῦϲα, μῆτερ, εἶτ' ἔρξειϲ κακῶϲ;
ΚΛ. οὐκ ἔϲτι, τῇ δ' ἡδὺ προϲθήϲω φρενί.

If the text is right in l. 1059, the sense will be: 'Not so, but I shall give pleasure to your heart.' But can it be right? Denniston justifies *οὐκ ἔϲτι* by quoting *Or.* 1097; but here and everywhere it means 'It cannot be', and it is much too solemn and categorical a pronouncement to suit this context. Denniston quotes Soph. *Phil.* 1020 and Eur. fr. 263 N.[2] to support the use of *ἡδύ* as equivalent to *ἡδονή*. The former passage runs *ἀλλ' οὐ γὰρ οὐδὲν θεοὶ νέμουϲιν ἡδύ μοι*; the presence of *οὐδέν* destroys its value as evidence for the point in question. The fragment reads as follows:

ἔϲτι δὲ καὶ παρὰ δάκρυϲι
κείμενον ἡδὺ βροτοῖϲ, ὅταν
ἄνδρα φίλον ϲτενάχῃι τιϲ ἐν οἴκωι.

The text is clearly doubtful; and F. W. Schmidt (*Krit. Stud.* ii. 456) may be right in thinking that *κείμενον* has been transposed from the next clause. In any case the word is suspect, and the fragment cannot safely be used to support the alleged use of *ἡδύ* here.

John Jackson (*CQ* 35 (1941), 31 = *Marginalia Scaenica*, 176) conjectured:

οὔκ· εἴ τι ϲῆι τῆιδ' ἡδὺ προϲθήϲω φρενί

'No, if I will be giving pleasure to your heart in this way.' This suggestion might possibly be right; but the sentence will run smoothly and the corruption be more easily explained if we read:

οὔκ, εἴ τι τῆι ϲῆι γ' ἡδὺ προϲθήϲω φρενί.

γε is often used with conditional sentences with this limiting force; see Denniston, *The Greek Particles*, 142 for instances. For its occurrence in

[4] [E. M. Craik in her commentary of 1988 objects that 'rhetorically the change of construction is unlikely; no comment.]

a ‖ conditional clause limiting an answering negative in dialogue, compare Soph. *OT* 583:

οὔκ, εἰ διδοίης γ', ὡς ἐγώ, σαυτῶι λόγον.

It is of course very common for γε to stand after a pronoun or pronominal adjective. The sense given is well suited to the context: 'No, if I shall give pleasure to your heart in this way' (i.e. by permitting you to have your say).

38(69)

Euripides, *Medea* 1056–80

Critical scholarship has won many of its greatest triumphs by applying remorseless logic to the problems presented by the texts of ancient authors, so often disfigured by corruption and interpolation. Let anyone who doubts it compare the text of any Greek tragedy in an edition earlier than Hermann with a modern edition of the same play. But the very qualities which have been responsible for these successes can also lead to failures; scholars may easily come to take it for granted that poets should be as logical and as matter-of-fact as they themselves, and so themselves damage the texts which they are trying to protect. The eminent man whom we are honouring has long been conscious of this danger, and has several times come to the rescue of poets threatened by the too sharp pruning-shears of textual critics. That encourages me to offer him a few remarks about a problem that has aroused much controversy in recent years.

No ancient author seemed to the *fin de siècle* closer to itself than Euripides, and no ancient author suffered more from having its preoccupations read into his work. The Ibsenite Euripides of Wilamowitz and the Shavian Euripides of Gilbert Murray have even now not been fully exorcized. Euripides was assumed to be wholly preoccupied with the doctrines of a fifth-century enlightenment having a surprising amount in common with the enlightenment of modern times. He was assumed to have employed the traditional trappings of myth only because they were part of the furniture of tragedy; the mythical and the heroic were held to have no real importance in his work.[0] Some scholars treated him as though he were a sophist or philosopher, concerned to recommend certain opinions to the public through his plays; others saw him as chiefly interested in the psychology of his characters. For the former group, the chief interest of the *Medea* lies in the struggle between reason and passion in the mind of the heroine; in showing how she gives way to passion, Euripides is making

[0] [Against, see M. R. Lefkowitz, *SIFC*, 3rd ser., 5 (1987), 149–65.]

Würzburger Jahrbücher für die Altertumswissenschaft, NF 6a (Festschrift für Hartmut Erbse (1980), 51–9).

'polemic' against the excessive intellectualism of Socrates. For the others, Euripides' main interest in the *Medea* is in the heroine's psychology, not to say pathology. She is a barbarian witch, a prey to her excessively violent emotions, Jason a cold-hearted cynic; the play is a cautionary tale, showing us what is likely to happen when two such persons become involved with one another. Kitto[1] is uneasily conscious that Medea ought to be tragic, and sets out to prove her so. His Medea is 'tragic in that her passions are stronger than her reason (*θυμὸϲ δὲ κρείϲϲων τῶν ἐμῶν βουλευμάτων*, 1079), she is drawn with such vigour and directness, everything she does springs so immediately from her dominant motive that she is eminently dramatic; nevertheless she is no tragic heroine as we have hitherto un||derstood the term; she is too extreme, too simple'. Kitto admits that 'it may seem absurd' to call Medea 'a purely passive figure', but is driven to argue that it is not absurd. The truth is that his Medea cannot be tragic, because she has been denied her true heroic status, and a Medea who is not heroic cannot be tragic. Still less can Kitto's Jason, who is 'an unrelieved villain', 'a ready-made villain', have any tragic status.

The way to a new approach was paved by Kurt von Fritz[2] with his rehabilitation of Jason; effectively comparing the treatment of the subject by later authors, he showed that we have no reason to doubt Jason's claim that he is doing what he thinks best for his family and himself. Yet Jason has broken his oath; and he fails to understand what price he will have to pay for his disloyalty. The heroic and tragic Medea has been rehabilitated above all by Bernard Knox;[3] instead of the psychotic barbarian witch he delineated a great heroic figure, comparable with the Ajax or the Electra of Sophocles. Her notion of honour demands that she must reward her friends and be avenged upon her enemies; she is of such heroic nature that, like the Sophoclean Ajax, she conducts herself like one of the Olympian gods of tragedy who ruthlessly punish those who have refused them proper honour, the Apollo of the *Agamemnon* and the *Niobe* of Sophocles, the Athena of the *Ajax* or the *Ajax Locrensis*. The finest general study of the play in modern times is that of the late Eilhard Schlesinger,[4] who presents it not as a contribution to intellectual history or as a psychological study,

[1] H. D. F. Kitto, *Greek Tragedy* (2nd edn. 1961) 194; cf. 197.

[2] See 'Die Entwicklung der Iason-Medea-Sage und die Medea des Euripides', *Antike und Abendland*, 8 (1959), 33–106 = *Antike und moderne Tragödie* (1962), 322–429.

[3] See 'The Medea of Euripides', *YCS* 25 (1977), 193–225 = *Word and Action* (1979), 295–322. Anne Burnett, *CJ* 68 (1973), 1–24, rightly stressed the importance of revenge in the *Medea*.

[4] See 'Zu Euripides' Medea', *Hermes*, 94 (1966), 26–53; there is an English version [only without footnotes] in Erich Segal, *Euripides: A Collection of Critical Essays* (1968), 70–89. Almost simultaneously, the significance of the Aegeus scene was shown and its importance vindicated by Schlesinger, pp. 46 f. = 85 f. and by Hartmut Erbse, *WS* 79 (1966), 120–33; cf. U. Albini, *Studi classici e orientali* 19–20 (1970–1) = *Interpretazioni teatrali* (1972), 66–77.

but as a tragedy. What we witness during the great monologue that begins at l. 1021 is not a struggle between passion and reason, which may end in one way or another, but the agonized reaction of Medea's maternal feelings to the dreadful action to which her unswerving determination to obtain revenge has long before committed her. The last scene unmistakably brings out her likeness to the gods; embarrassingly for those who have worked so hard to reduce the tragedy to their own bourgeois level, the poet has chosen to lay the strongest emphasis on those supernatural affinities of the heroine which they have found so distressing. She and no other acts the part of the god from the machine. But these gods have no human pity, and though her apotheosis richly satisfies her longing for revenge, it cannot bring her human happiness.

During the prologue, Medea curses her children and her own family (112 f.); the Nurse, who knows her well, takes from the start the darkest view as to what action may be expected from her. When at the end of her first great speech (214 f.) she begs the Chorus not to betray her plans, she is already determined on revenge (259–63). In the speech she utters after her interview with Creon (364 f.) she ‖ expresses her intention of killing Creon, the princess and Jason (374–5). The ideal opportunity given by the arrival of Aegeus allows her to perfect her plan (764 f.); doubtless it is the preoccupation of Aegeus with his childlessness that makes her realize that the perfect revenge on Jason will be the murder of his children (791–7). The women of Corinth plead with her (812 f.), but come up against her absolute determination; when they sing the fourth stasimon (976 f.), they know that the children are already doomed.

After the song the Paidagogos enters with the children (1002 f.), bringing the news that the sentence of exile upon them has been rescinded; the princess has graciously received the gifts they brought. The old man is disappointed by Medea's reaction to news that he had hoped would give her pleasure; he attributes her sorrow to the knowledge that she will henceforth be separated from her children. He tries to console her, but Medea dismisses him, and in the presence of the children embarks upon her fourth great speech (1021 f.).

She tells the children that they have a city and a home where they will dwell for ever without their mother; the ancient commentary rightly says that she means Hades. As for her, she will go into exile without having had the satisfaction of seeing them married; nor will they, as she had hoped, be there to care for her in old age and to render her the last rites. That hope has vanished, for she will live out a sad life without them; and they will no longer look with loving eyes upon their mother, having moved away from her into another form of life; that means, of course, the life of the world of the dead.

Already in the last stanza of the third stasimon (860) the Corinthian women have asked Medea how she will be able to refrain from tears as she looks upon the children whom she is about to kill. Now (1040) she asks the children why they are looking upon her and smiling for the last time. What can she do, she asks the women; her heart has left her as she looks upon the children's smiling faces. 'I could not do it!', she exclaims, 'farewell to my earlier plans! I will take my children out of the country! Why must I bring twice as much evil on myself by using their ruin to hurt their father? No, not I; goodbye to my plans!'

At this moment Medea is checked by the thought that she will be mocked by the enemies whom she has allowed to go unpunished. Throughout the play that consideration affects her as no other does (note ll. 381 f. 404–6. 807–10. 1354–5. 1361–2). She must go through with it; it was cowardice even to listen for a moment to soft-hearted counsels. She tells the children to go into the house, declaring that she will not fail to carry out the action on which she has decided.

We now come to a passage twenty-five lines long which many scholars have excised. Theodor Bergk[5] in 1884 was the first; his view was revived by Erich Bethe in 1918 and by Günther Jachmann in 1934. Gerhard Müller trenchantly restated it in 1951; yet in 1972 Albin Lesky was able to remark that he had rightly found no followers. He spoke too soon; during that same year, Müller found a powerful || advocate in M. D. Reeve. Since then Reeve's argument for athetesis has won the support of Otto Zwierlein (1978).

At the beginning of the passage in question, Medea again momentarily relents (1056); addressing her *θυμός*, she forbids it to act and begs it to let them be, to spare the children; they shall live with her there and gladden her heart. The address to one's *θυμός* is of course Homeric; but as often in tragedy the *θυμός* is not merely one of several more or less vague terms for the seat of the intelligence but connotes pride, spirit, anger, something like what Plato means by *τὸ θυμοειδές*.

'Medea has changed her mind again,' Reeve writes (p. 52); 'she must spare the children and let them gladden her heart "there".' I detect a certain impatience in his way of referring to the second change of mind; perhaps he feels that as a rational being she should have made up her mind once for all and got it over with. At such a moment we might expect emotion to have some effect upon Medea's utterances, so that this consideration cannot weigh heavily against authenticity. Also, 'change of mind' is too strong a way of referring to what is after all a

[5] Th. Bergk, *Gr. Literaturgeschichte,* iii (1884), 512 n. 140; E. Bethe, *Ber. über die Verh. der Kl. Sächs. Ges. der Wissenschaften*, ph.-hist. Kl. 70/1 (1918), 8 f.; G. Jachmann, *NGG*, NS 1/9 (1936), 193 n. 1, 214 n. 1 [= *Textgeschichtliche Studien* (1982), 558 n. 1, 579 n. 1]; G. Müller, *SIFC* 25 (1951), 65–82; M. D. Reeve, *CQ*, NS 22 (1972), 51–61; O. Zwierlein, *Literaturwissenschaftliches Jahrbuch*, 19 (1978), 33 f.; for further bibliography see Reeve's article.

momentary hesitation; we find it natural to talk in these terms because of our long conditioning by the school of critics who find the main interest of the tragedy in the struggle between reason and passion in Medea's mind.

Reeve notes that the word ἐκεῖ (1058) is interpreted in the scholia as meaning 'in Athens'; 'but Medea', he complains, 'last mentioned Athens at 771. More vaguely "in exile?"' But a member of the audience who has heard Medea explain what her plan is, if only she can find a place where she can take refuge after its execution, who has heard Aegeus offer her that place, and has listened to the great ode (824 f.) in which the Chorus remarks on the inappropriateness of Athens, of all places, sheltering the murderess of her children is not likely to have forgotten what place Medea has it in mind to go to. Reeve seems to find 1058, like 1024, somewhat vague; but can we really doubt what ἐκεῖ means, in either place?

In the lines that follow (1059 f.) Medea declares that she will never abandon her children to the mercy of her enemies. 'By the infernal avengers in Hades!', she exclaims, 'this will never come about, that I shall leave my children to be insulted by my enemies. In any case they have to die; and since die they must, we who gave them birth will kill them.' Once more she appears to have changed her mind, with what some critics feel to be suspicious abruptness; but there is a graver difficulty. Only a moment before the alternative to killing the children seemed to be taking them to Athens; but it is now implied that if Medea spares the children they will remain in Corinth, where the Corinthians will kill them.

Many scholars have tried to remove the contradiction, some by means of emendation and others without it; let us consider the latter group first. The most popular way[6] of doing so was once to suppose that Medea is prone to lose her head and is a prey to different emotions in quick succession. But would this cause her to give a different account of the situation and its possibilities from the one she has given immediately before? Euripides does not commonly depict a rapid shift from ‖ one emotion to another by means of inconsequences of this sort. For W. Zürcher,[7] who applied to Euripides the method used by Tycho von Wilamowitz in his famous study of the dramatic technique of Sophocles, there was no problem; Medea could easily be one person one minute and another the next. Wolf Steidle[8] has argued that the

[6] See, for instance, M. Pohlenz, *Die griechische Tragödie* (2nd edn. 1954), 262; D. L. Page, *Euripides, Medea* (1938; repr. with corrections, 1952), on l. 1058; André Rivier, *Essai sur le tragique d'Euripide* (1946), 59, citing W. Schadewaldt, *Monolog und Selbstgespräch* (1926), 195.

[7] *Die Darstellung des Menschen im Drama des Euripides* (1947) 61 f.

[8] *Studien zum antiken Drama* (1968) 157 f.; cf. A. Lesky, *Die tragische Dichtung der Hellenen* (3rd edn. 1972), 311 f.

children are too young to be taken to Athens; Lesky agrees with him. Patricia Easterling[9] writes that 'she is filled with a sudden sense that she is caught in the toil of events and no longer has any choice; parents who kill their children, she tells us, often 'become convinced of a threat that clinches the feeling that they would be better dead"'. That argument will convince only those who believe that such psychological niceties are part of Euripides' stock-in-trade; even if they are, it would not be characteristic of Medea, who is not given to self-deception. Schlesinger (p. 32) says that when Medea speaks of the necessity of the children's death, she really means the necessity of her revenge; on his view she will be offering herself the alleged danger from the Corinthians as an excuse. It is true that for Medea her revenge is always necessary; but would she feel obliged to offer such an excuse to herself?

Others have emended in one way or another. Hermann changed ἐκεῖ in l. 1059 to κεἰ μή. Barthold (followed by Méridier) emended it to καὶ μή, at the same time deleting 1045 after Kvíčala. These are not very great changes, but we must note that they are made simply to remove this difficulty.

Lines 1061–2 are identical with ll. 1240–1; they were deleted by Valckenaer in the eighteenth century, and most modern editors follow him (e.g. Weil, Verrall, Méridier, Diehl, and Page). Page[10] has shown that repeated verses within the same play are unlikely to be genuine in both places; and at ll. 1240–1 the couplet is far harder to disengage from its context. G. A. Seeck[11] in 1968 put forward an ingenious but excessively complicated argument designed to show that ll. 1060–1, as well as ll. 1062–3, were interpolated; he spared ll. 1059, which as Lesky has pointed out (loc. cit.) does not go easily with l. 1064, only the second of whose two clauses is negative and so suitable to go with a phrase starting with μά.

Lines 1060–1 suspiciously resemble ll. 780–2, the first of the three passages in which the notion that the children might fall victims to the Corinthians is alluded to. This is part of the speech which Medea utters after the scene with Aegeus; she is explaining to the Corinthian women that she means to ask for the sentence of exile upon her children to be rescinded, not because she has any intention of leaving them in Corinth to become the victims of her enemies, but in order to compass the murder of the princess. In that place the allusion to the danger from the Corinthians is perfectly natural; the danger is envisaged as arising in a wholly different way from what we find at ll. 1059–61, although the words employed are notably similar. Now let us consider the

[9] 'The Infanticide in Euripides' *Medea*', *YCS* 25 (1977), 189.

[10] *Actors' Interpolations in Greek Tragedy* (1934), 103-5.

[11] *GRBS* 9 (1968), 291–307.

mention of the possibility at ll. 1240–1. The messenger has described the awful deaths of Creon and his daughter, and Medea must act swiftly and ‖ accomplish her revenge and be away before the pursuit catches up with her. 'I must not', she tells the Chorus 'by delay hand over my children to hands more inimical to slaughter.' Then follow ll. 1240–1, which are identical with ll. 1062–3. In this place there is no implication that Medea is going to kill the children simply because she cannot take them with her; there is no assertion that to take them with her would be impossible. The statement that in any case they have to die might have no reference to any danger from the Corinthians; all mortals have to die, and that may well be what Medea means to say.

I have argued that in the two other places in the play in which there is mention of possible danger to the children from the Corinthians, it is unaccompanied by any suggestion that Medea could not, if she wished, take the children with her. And this is not surprising; because the whole argument of the play demands that Medea must kill the children not because she cannot take them with her and does not want them killed by the Corinthians, but to punish Jason. We have also seen that not only are ll. 1062–3 identical with ll. 1240–1, and much easier to remove without damage to the context, but that ll. 1060–1 could very well have been suggested by ll. 780–2. I think we must reckon with the possibility that ll. 1059–63 are the work of a producer of the play, probably during the fourth century, when mummer-worship was already rife, who wanted the great speech to be as impressive a vehicle as possible for some popular performer, and could not resist working into it the powerful distich 1240–1. In order to lead up to them, he started with a striking invocation of the Erinyes (1059), perhaps taken from another play, and then made ll. 1060–1 out of ll. 780–2. This seems to me on the whole the likeliest cause of the trouble.[11a]

Suppose then that we delete ll. 1059–63; what will be the sense of l. 1064? Accompanying as it will a break in the thought and a pause in the delivery, the asyndeton will cause no trouble; Medea will be saying, 'At all events the deed is done, and she will not escape.' Reeve objects that 'recent commentators offer no parallels for this use of the perfect, and the few offered by older commentators are inadequate'. He begs the question by refusing to accept as parallels places where the perfect used of a future event as one in effect already complete stands in the apodosis of a conditional sentence; even that unreasonable stipulation does not rule out all the instances collected by Kühner–Gerth i. 384–5

[11a] [H. Erbse, *Ἀρχαιογνωσία*, 2/1 (1981), 71 ably defends ll. 1059–61. 'Sobald erkannt ist,' he writes, 'daß M. in 1058 nur mit einer Illusion spielt, ergibt sich die Folgerung, auf die allein es dem Dichter ankommt: Die Kinder sind bereits hier, in Korinth, verfallen.']

(p. 150).[12] In any case, how could any spectator, hearing Medea say 'the deed is done', fail to realize that she means that, since the gifts are already in the hands of the princess, the children's fate is now determined?

Reeve also complains that 'to understand that the princess is the subject (of *ἐκφεύξεται*) would surely have been even more difficult for the audience to understand than that *πέπρακται ταῦτα* refers to her murder'. It is true that it is some time since the princess was mentioned, and a strictly logical or a very stupid person might feel aggrieved at being expected to work out that she is the person spoken of. But the audience—an Athenian, not a Boeotian audience—that has been watching the *Medea* will be familiar with the situation; and in any case the words that follow ‖ —'already the crown is on her head'—leave little room for doubt.[13]

At ll. 1067 f. Medea expresses her intention of taking leave of the children; she implies quite clearly that she is going to kill them. At l. 1053 she has told them to go into the house; and Reeve makes much of the supposed inconsistency between this fact and her addressing them once more at ll. 1067 f., complaining that they 'show greater obtuseness than is displayed anywhere else by a character who is ordered off the stage'. Would they really have been so obtuse to hesitate, as Page and others have suggested that they do, when they notice their mother's altered demeanour as she speaks ll. 1056–8? I do not think so.

Now Medea addresses to the children words which many readers have found to be among the most moving ever written by Euripides (1069–75). Reeve agrees that *ἐκεῖ* in l. 1073 must mean in Hades, but asks what it can mean to the children. Paley, he says, thought the children are meant to understand 'in exile', and Reeve finds it strange that the children should believe this. If it were common for the tragedians to represent children in anything like a naturalistic fashion, this objection might have substance. The tragedians tend to be vague about the ages of the children whom they bring on to the stage,[14] but in

[12] e.g. Plato, *Phaedo* 80 D *ἡ δὲ ψυχὴ ἄρα, τὸ ἀϊδές, τὸ εἰς τοιοῦτον τόπον ἕτερον οἰχόμενον . . , αὕτη δὲ δὴ ἡμῖν ἡ τοιαύτη καὶ οὕτω πεφυκυῖα ἀπαλλαττομένη τοῦ σώματος εὐθὺς διαπεφύσηται καὶ ἀπόλωλεν, ὥς φασιν οἱ πολλοὶ ἄνθρωποι.*

[13] Reeve (p. 53) writes that 'a corollary of taking 1064 to refer to the murder of the princess is that 1062–3 cannot be removed, or else Medea will not have announced her intention of killing the children between the opposite announcement in 1056–8 and the parenthetic reference to killing them in 1068'. But if one imagines the speech delivered without ll. 1059–63 and with a pause before l. 1064, the audience will have no difficulty in seeing that the implication of ll. 1064 f. must be 'too late'.

[14] 'Suo nimirum iure nimis constantem in hac re rationem poeta non hoc tantum loco aspernatus est': R. Kassel, *Quomodo quibus locis apud veteres scriptores Graecos infantes atque parvuli pueri inducantur describantur commemorentur* (Diss. Würzburg, 1954), 54–5.

general they treat them as being *νήπιοι*.[15] In this particular instance, Steidle has given reasons for thinking this to be the case.[16] The children's apparent incomprehension supplies very little evidence against authenticity.

Lines 1076–7 contain, as Reeve remarks, 'a recognized crux'. But the general sense is clear, even if no single emendation stands out as the ideal solution; Medea is telling the children that she can no longer look at them, because she is overcome by her troubles (*νικῶμαι κακοῖς*). What are these *κακά*?, Reeve asks, 'scarcely', he says, 'what it means in 1078'. *κακά* is a vague expression; does it always have to refer to precisely the same thing?

Now come the lines that have caused more trouble than all the rest of the speech together. They are generally taken to mean, 'And I know what kind of evil it is that I am about to do, but my pride, my anger, my passion is stronger than any calculation—pride, that causes the greatest evils for mortals.'

Many critics, beginning apparently with Stadtmüller in 1876,[17] have objected ‖ that elsewhere in the play (at ll. 769, 772, 1044, and 1048) the word *βουλεύματα* refers to Medea's plan to kill her children. Müller strongly urged this objection. To evade it Hans Diller[18] suggested that *κρείσσων* in this place might be equivalent to *κρατῶν*, so that the words meant 'my pride is the master of, controls my plans for revenge'. This is altogether unconvincing; Rudolf Kassel[19] has pointed out that the natural way to take the word *κρείσσων* is exemplified by l. 965 of this same play, *χρυσὸς δὲ κρείσσων μυρίων λόγων βροτοῖς*. Another eminent scholar, Albrecht Dihle,[20] has tried to escape the difficulty by taking *θυμός* here not to mean 'pride' or 'spirit' but to refer to the seat of the tender emotions; Medea, he thinks, is referring to the cowardly behaviour which she knows she will be guilty of if she relents. That is a subterfuge no less unconvincing than that of Diller; as Zwierlein[21] has pointed out, it involves forcing the word to bear a sense which would be altogether unnatural in this context.

Let us consider again the word *βουλεύματα*. It is true that in four places in this play it refers to Medea's plan for revenge; in each of these the context makes it unequivocally clear that it does so. But the word in

[15] See H. Herter, *Jahrbuch für Antike und Christentum*, 4 (1961), 146 f. = *Kl. Schr.* (1975), 598 f.; note especially the words 'Nur das Eine wird nie vergessen, daß das Kind *νήπιον* ist' (151 = 605).

[16] Op. cit. (in n. 8), 158 f.

[17] *Beiträge zur Textkritik der Euripideischen Medea* (1876), 31 n. 1.

[18] *Hermes*, 94 (1966), 267–75 = *Kl. Schr.* (1971), 359–68; cf. V. Di Benedetto, *Euripide: Teatro e società* (1971), 41 f.

[19] *RhM* 116 (1973), 103 n. 21.

[20] 'Euripides' Medea', *SB Heidelberg*, ph.-hist. Kl., 1977/5, 13 f.; 'Euripides' Medea und ihre Schwestern im europäischen Drama', *Antike und Abendland* 22 (1976), 180 f.

[21] Op. cit. (in n. 5), 35 n. 24c.

itself is colourless, as one can see by looking through its numerous occurrences in Euripides. At l. 449 it is applied to Creon's decision that Jason shall marry the princess; at l. 886 to Jason's plan to leave Medea and marry Creon's daughter; fortunately in each of these places the context makes its meaning clear, so that no scholar has yet complained that it must refer to Medea's plan for revenge, because it does in the other four places. If we read this speech as most readers have read it in the past, taking each word and phrase as it is natural to take it, Medea will be saying that she knows what evil she is about to do—and *δρᾶν κακά* of course implies positive action—but that her pride, her anger is stronger than her plans, her calculations. In itself the word *βουλεύματα* is colourless; it takes its colour from its context. Here its meaning is made clear by Medea's immediately preceding statement that she knows what evil she is about to do; that knowledge would counsel her to abstain from action, but her *θυμός* is more powerful than such counsels. She has been aptly compared to Achilles in the ninth book of the *Iliad*,[22] who recognizes the truth of what Ajax has said to him, but who tells Ajax that his *θυμός* still swells with anger at the thought of how Agamemnon treated him.

Zwierlein in approving Reeve's arguments for athetesis says that 'if one were to allow the celebrated lines 1078 f. to guide one in one's assessment of the play's tragic content, one would be compelled to import into the play a conflict between passion and knowledge of what is best that otherwise plays no part in it'. No one could sympathize more strongly than I do with Zwierlein's rejection of the views of ‖ those who so much exaggerated the importance of such a conflict that they have seemed to imagine that there is a real possibility of Medea's renouncing her revenge; but I do not believe that the lines in question oblige us to take such a view. The fate of the children has long since been decided;[23] Medea has never seriously contemplated renouncing her revenge, for if she did so, she would not be Medea.[24] The poet is not

[22] Ibid. 29; K. Matthiessen, *Das griechische Drama* (1979), 117 also makes the comparison. Di Benedetto cites the Homeric passage, but introduces subtle distinctions which seem to me to lack substance (43–4).

[23] See H. Strohm, *Euripides: Interpretationen zur dramatischen Form* (1957), 101–3; his excellent exposition makes this point particularly clear.

[24] Cf. W.-H. Friedrich, 'Medeas Rache', *Nachr. der Akad. der Wiss. in Göttingen* phil.-hist. Kl. 1960/4, 110: 'Barbarei, Zauberei, Ekstase: damit reicht nachher die andere Welt gleichsam bis auf die Bühne; mit der Wegnahme der Kinder und überhaupt jeglichem Beweggrund, der Medeas Tat verständlicher, verzeihlicher machen könnte, reicht unsere Einfühlung bis auf die Bühne, bis in das Geschehen hinein. Von entgegengesetzten Seiten nähern sich die Pole einander, bis die Entladung unvermeidlich scheint. Wieviel Hölle die Stücke auch in Rede und Aktion aufbieten mögen: alle die Erfindungen, die dem Kindermord aus verschiedenen Richtungen entgegenkommen und ihn zeitigen helfen, suchen uns moralisch und ästhetisch mit ihm zu versöhnen, soweit dies irgend möglich ist. Was nicht als Aufbegehren einer gekränkten Seele, als Ausgeburt einer wenigstens im Keim noch allgemein menschlichen Leidenschaft verstanden werden kann, soll aus

showing us how Medea makes her decision; that decision has been taken, swiftly and silently, long before; rather, he is showing us the full extent of the agony which the prospect of the decision's execution causes. Medea's human instincts are for mercy, but what determines her decision is her pride; as she remarks, *θυμός* is the cause of the greatest harm to mortals.

Medeas dämonischem Wesen mit eigener Folgerichtigkeit hervorgehen; was uns nicht mehr natürlich ist, soll doch der Tochter des Fabelreiches natürlich sein. Euripides hat sein Stück anders angelegt; die Tat bleibt ganz Medeas Eigentum, nichts und niemand nimmt sie ihr ganz oder teilweise ab.' Cf. Di Benedetto, op. cit. (n. 18), 45–6.

[I regret that J. Diggle in *Euripidis Fabulae*, i (1984) has bracketed *Medea* 1056–80; against his view see D. Kovacs, *CQ*, NS 36 (1986), 343–52.]

39(R88)

Review of Euripides, *Phaethon*, ed. Diggle

Euripides, *Phaethon*, James Diggle. Cambridge: University Press, 1970. Pp. xi + 244; 6 plates.

ONE of the most exciting discoveries of the early nineteenth century was Hase and Bekker's finding of the splendid fragments of the *Phaethon* contained on two leaves of the Codex Claromontanus. Gottfried Hermann, after considerably improving the text, sent a copy to Goethe, who, 'ehrfurchtsvoll an solche köstlichen Reliquien herantretend', produced a reconstruction that contains wonderful translations of the text (in *Kunst und Altertum*: Hamburger Ausgabe, xii (4th edn. 1969), 310 f.). In 1883 Wilamowitz in an early article threw some light, but also much darkness, upon the perplexing problems presented by the fragment; in 1907 a Berlin papyrus improved our text of the exquisite lyric verses of the parodos. The task of re-editing and explaining these precious remains has supplied an ideal subject for the doctoral dissertation of a learned and ingenious young scholar. The result is a valuable and interesting book; and we must greatly regret that, like all the books in this admirable series, it is inexcusably expensive.

Dr Diggle gives an excellent account of the history of the legend; he has wisely chosen not to reassemble but to reassess the extensive material which Knaack collected, and he has done so with sound judgement. He has carefully re-examined the palimpsest, and also the fragmentary hypothesis of the play in P. Oxy. 2455, and Dr Colin Austin has re-examined the papyrus in Berlin on his behalf. Diggle prints photographs of the palimpsest and the papyrus; those of the former are no doubt the best that can be done, but I doubt if even expert palaeographers will get much out of them. The text is preceded by a reconstruction of the plot, which in several places looks forward to detailed discussions contained in the commentary.

Clymene, according to the hypothesis, told her son Phaethon after he had grown up that he was not the son of her husband Merops, but of Helios. Diggle has allowed a friend to convince him that she must

Classical Review, NS 21 (1971), 341–5.

have done this before the beginning of the action, and not, as others have supposed, during the dialogue between mother and son that followed the prologue speech. 'There is', he writes, 'no parallel in extant plays to the presence of an incident of such dramatic weight... between the end of the opening monologue and the arrival of the chorus.' What about the momentous conversation between Athene and Poseidon in the prologue of the *Troades*? And what about the alarming communication of the servant to the heroine that follows the prologue speech of the *Andromache*? It has usually been thought that Clymene's revelation was made in order to overcome her son's reluctance to marry, as he supposes, above his station, and if that is right it could have been used most effectively in the dialogue in question.

The greatest problem of the plot is that of the identity of Phaethon's prospective bride, and since the only evidence for this is contained in the lyrics of the subchorus at ll. 227–44, they must be set out here:

> Ὑμὴν Ὑμήν.
> τὰν Διὸς οὐρανίαν ἀείδομεν,
> τὰν ἐρώτων πότνιαν, τὰν παρθένοις
> γαμήλιον Ἀφροδίταν.
> πότνια, σοὶ τάδ' ἐγὼ νυμφεῖ' ἀείδω,
> Κύπρι θεῶν καλλίστα,
> τῶι τε νεόζυγι σῶι
> πώλωι, τὸν ἐν αἰθέρι κρύπτεις,
> σῶν γάμων γένναν.
> ἃ τὸν μέγαν
> τᾶσδε πόλεως βασιλῆ νυμφεύεται,
> ἀστερωποῖσιν δόμοισι χρυσέοις
> ἀρχὸν φίλον Ἀφροδίτα.
> ὦ μακάρων βασιλεὺς μείζων ἔτ' ὄλβου,
> ὃς θεὰν κηδεύσεις
> καὶ μόνος ἀθανάτων
> γαμβρὸς δι' ἀπείρονα γαῖαν
> θνατὸς ὑμνήσηι.

Diggle strongly attacks (pp. 154 f.) the theory put forward by Wilamowitz when he was thirty-five and maintained by him, with some modifications, until the last, that the bride was to have been Aphrodite. His criticism is fully justified; the theory involves unnatural interpretations and improbable conjectures. So do the suggestions of several other scholars, which Diggle deals with no less effectively.

Diggle himself revives a little-known suggestion of H. Weil that the πῶλος whom Aphrodite conceals in the aither (234) is Hymenaios. There is a story that Hymenaios was the son of Aphrodite, and there is a story that he disappeared on his wedding-day. Without going quite so far as Diggle, who thinks this guess 'unquestionably correct' (p. 158), I agree that it is much the likeliest explanation of the passage that has so far been advanced.

Diggle thinks that it is Aphrodite who 'gives in marriage' Phaethon; he emends νυμφεύεται in l. 237 to νυμφεύεαι, comparing, for the sense of the verb, Barrett on *Hipp.* 559–62. Alternatively, the subject of the verb might be Hymenaios; in that case one would keep νυμφεύεται but change ἃ in 236 to ὃϲ—after γένναν the corruption would have been easy—and Ἀφροδίτα in 239, with Hermann, to a dative.

In 240 Diggle accepts Hermann's emendation ὦ μάκαρ, ὦ βαϲιλεὺϲ μείζων ἔτ' ὄλβον; otherwise 'the sentiment', he complains, 'is unwarranted, not to say impious'. But the presence of ἔτι shows that a comparandum is needed; the words of the maidens are indeed provocative, but they are just the kind of praise which we expect to be bestowed upon a mortal soon to meet with disaster (cf. *Rhes.* 355 f., *El.* 994, *IA* 596–7).

But who was the bride? Weil suggested that she was one of Phaethon's half-sisters, the Heliades; Diggle thinks this suggestion incapable of proof, but is 'firmly convinced that it is correct' (p. 159). After his conversation with Merops, Phaethon goes to the palace of Helios to find out whether the god is really his father (61–2); Merops, Diggle thinks, believes that he has gone there to bring home the bride. 'What better piece of dramatic οἰκονομία,' he asks, 'what finer field for tragic irony, than the double motivation afforded for Phaethon's journey to Helios' palace?' But since it is clear that the palace of Helios was conceived as being not far from the palace of Merops, Phaethon could in any case have got there and back quickly enough to satisfy any critic naturalistically minded enough to worry about such matters.‖

A clue to the bride's identity might be afforded by 241, where Diggle takes ὃϲ θεὰν κηδεύϲειϲ to mean 'who will marry the goddess'. As he himself notes, 'no example of κηδεύειν + accusative in the sense "marry" is attested'; as he himself remarks, the regular way of using this verb 'might tempt us to suggest θεᾶι'. 'But in the absence of closer analogies', he continues, 'we can only acquiesce in the reading of P.' This manner of reasoning is very unlike Diggle; and I doubt if he would have adopted it if he had not been convinced that Weil was right about the bride's identity. Weil may indeed be right, but the positive evidence in favour of his theory can hardly be said to exist at all. There is, indeed, one slight indication that makes against it; for if θεᾶι in 241 is right—as well it may be, for the change needed to restore regularity is very slight, and in this corrupt manuscript by no means violent—then it would follow that the bride was some relation of Aphrodite. The goddess's connection with the Phaethon of Hes. *Theog.* 984 f. may not be quite as irrelevant to Euripides as Diggle insists; she had to our knowledge one mortal son-in-law in Cadmus. We must be grateful to Diggle for his demolition of Wilamowitz's theory; but instead of letting him convince us that Weil was right about the bride, we must insist that her identity is not known, and is not likely to be discovered without new evidence.

Despite his strong words about previous interpretations of the ode set out above (p. 149), Diggle has retained one feature of Wilamowitz's interpretation of the play which seems to me no less objectionable. When Phaethon's body is brought in, Clymene hides it from her husband, and when he discovers it she is said to be in a danger of death at his hands (281–3, where despite the textual corruption the sense is clear). Diggle seems to follow Wilamowitz in thinking that Merops is angry because he has now learned of her 'liaison' with Helios. In Greek poetry gods do not conduct 'liaisons' with mortal wives; they rape mortal virgins and then hand them on to mortal husbands, who far from being indignant at their wives' pregnancies are no less delighted than St Joseph at the privilege granted them. μετα̣[in l. 2 of the papyrus hypothesis (p. 53) seems to show that Clymene gave birth to other children *after* Phaethon, so it is likely that his birth formed no exception to the general rule. The 'adultery', the jealous husband, the love triangle are a typical construction of that early Wilamowitz who saw Euripides as a proto-Ibsen, thus encouraging Gilbert Murray to see him as a proto-Shaw.

Now for a few remarks on points of detail. P. 28: 'the women of Adria shall have manner of grief' sounds very odd as a translation of Aesch. fr. 71 R. P. 38: the parodos, Diggle tells us, is sung 'by the housemaids'. I wonder if he thinks, as many people have done, that this parodos is very beautiful poetry. P. 39: fr. 163 surely indicates that Phaethon has threatened to leave home. P. 48: 'in calculating the frequency of resolution', says Diggle, 'I follow the procedure of E. B. Ceadel rather than that of Zieliński. Ceadel's statistics are the more reliable.' A. M. Dale, no mean authority in the matter, would not have agreed; see her edition of the *Helen*, pp. xxv–xxvi. P. 49: Diggle rightly says that 'since our material is so slight, it would be foolish to suppose that we can attain anything more than a very approximate result' in dating the play. I share his impression that it is comparatively late; but how can he then go on to say that 'however slight the evidence, it is enough to disprove Wilamowitz' ascription of the play to Euripides' early years'? P. 53: in l. 16 of the hypothesis, the supplement ἐθ]ε̣ϲπιϲ[εν should not be in the text,|| even if one could feel sure that the writer would have used this verb, or that he would have used an aorist and not a historic present in this context. Text, l. 47: Diggle's supplement is not the only possibility. Ll. 65–6: Wilamowitz read ὑπὲρ δ' ἐμᾶϲ κεφαλᾶc | Πλειὰ[c ἐκλείπει cκοτία. This seems to Diggle 'feeble'; he prints Πλειά[δων πέφευγε χορόc. Diggle quotes several passages in which verbs meaning 'to flee' are used in similar context, but in every case the stars are described as fleeing; none of the verbs is a perfect, which refers to a completed action. After 'above my head', I should expect an account of something that is seen happening, not a reference to an event now over, so that I much prefer Wilamowitz's supplement to Diggle's. P. 99: the word λεπτόc, Diggle says, is rarely applied to sound. Rarely to the voice, as LSJ, from whom Diggle's parallels are taken, remark; rarely to other sounds also, so that it is worth citing Aesch. *Agam.* 892 and A.R. 3. 708. L. 70: Goethe is rebuked by Diggle for having set the 'disastrous fashion' of finding significance in the picture of the dawn song of the nightingale. 'Man hört noch die Nachtigall singen', Goethe wrote (op. cit. 313), 'wobei es höchst wichtig ist, daß ein Hochzeitsgesang mit der Klage einer Mutter um ihren Sohn beginnt.' 'Höchst wichtig' is a little strong, and later commentators state the suggestion somewhat more crudely; but even if it is only a Romantic misconception, it is one which people who enjoy poetry may get pleasure from considering.[1] L. 79 f.: Diggle is right against Wilamowitz, but should have quoted the passages which show that ἄκατοc is used of sea-going ships as well as small boats (*Tro.* 1100, *Hec.* 446). He accepts εἰρεcίαιc from the papyrus, preferring symmetry; I prefer εἰρεcίαc, the reading of the palimpsest, thinking asymmetry more Euripidean. Ll. 82–3: Diggle accepts from Austin [Ἕπου] πότνι' αὔρ[α | ἡμῖν ὑπ'] ἀκύμονι πομπᾶι. His parallels certainly make against Krantz's [Ἄγου]: but we want not 'accompany' but 'conduct'. Why not read [Ἄγ', ὦ] πότνι' αὔρα, | ἡμᾶc cὺν] ἀκύμονι πομπᾶι (l. 83 suppl. Krantz)? L. 87: the parallel at Men. *Samia* 112–13, by indicating that we have here a standard locution, may serve to confirm Diggle's view. L. 92: cf. Aesch. *Agam.* 108, Pind. *Ol.* 1. 113. 96: of the two possible interpretations of δή ποτ' which Diggle considers, the word-order seems to me to confirm that which he rejects. Ll. 113–14: Diggle does well to put cruces here. Ion fr. 64 Blumenthal [*TrGF* 19, fr. 21] uses the form ὀρτή in a tragic trimeter, so that it can hardly be dismissed as an Ionism. Did Euripides write κηρύccω δ' ὀρτὰν βαcιλήϊον? Compare ἑορτὴ κηρύccεται in a fifth-century Milesian inscription (Schwyzer 725. 12). Ll. 158–9: add to Diggle's parallels Men. fr. 333. L. 161: Diggle's πατρῷια is the best emendation so far offered. L. 163: cf. Hdt. 2. 30. 4. L. 167: 'the only plausible suggestion yet made is Musgrave's "καὶ τῆc τύχηc fort. ad sequentia pertinent"'. Yes, unless Euripides wrote φεῦ τῆc τύχηc. Ll. 168–9: the manuscript of Ps.-Longinus offers ἔλα δὲ μήτε Λιβυκὸν αἰθέρ' εἰcβαλών· | κρᾶcιν γὰρ ὑγρὰν οὐκ ἔχων ἁψῖδα cὴν | κάτω

[1] [H.-S., citing *Od.* 19. 518 f., points out that the maternal significance was always available for any poet whom it might serve.]

διειcι. Diggle rightly rejects the vulgate, Faber's κάτω διήcει: 'it is a sad reflection on the abilities of Helios' horses if they must rely on the miraculous properties of aerial humidity to keep the chariot and themselves aloft.' Accepting H. Richards's καίων, he reads διοίcει. This verb can certainly mean 'tear to pieces'; but Diggle wants it to mean 'warp' or 'distort', which is surely not quite the same thing. If I were hazarding an emendation of what is certainly a corruption, I should prefer Kock's καταιθαλώcει (*Verisimilia*, 218). L. 214: Diggle's tentative suggestion is attractive. The use of ἐν which it involves needs illustration; one passage that illustrates it is Soph. *Aj.* 971, where Lobel ‖ has pointed out to me that ἐν κενοῖc is equivalent to 'at the expense of us who are left unprotected'. L. 223: cf. also Men., *Aspis* 356–9. L. 225: on Aesch. *Supp.* 212–14 (discussed on p. 147) see Pasquali, *Riv. fil.* 2 (1924), 246 f. L. 277: a full stop, not a comma, should stand at the end of this line. P. 148: on etymologies of names in tragedy see also Pfeiffer, *SB der Bayer. Akad.*, 1938/2, 9. n. 2. P. 153: for ἀcτερωπόc see also Aesch. fr. 170 R., where Bentley's emendation is surely right. P. 161: on sacrifices to Hestia see *RhM* 103 (1960), 78 f. [=ch. 23, p. 307 f.].

Appendix A deals with the treatment of the Phaethon legend in Ovid's *Metamorphoses* and in Nonnus' *Dionysiaca.* Diggle comes down firmly in favour of Nonnus' dependence on Ovid, and indeed the investigations of the last twenty years have tended gradually to confirm the view of Rudolf Keydell, the greatest living authority on the late Greek epic poets, that they made use of Latin sources.[2] The Hellenistic source postulated by Wilamowitz cannot be proved to have existed. But can it be proved not to have existed? I am not as sure as Diggle seems to be. A second appendix deals with the treatments by Q. Sulpicius Maximus (Peek, *Gr. Vers-Inschriften,* no. 1924), Lucian, and Philostratus, and a third with the monuments; there are also a full bibliography and useful indexes.

Like most reviewers, I have used most of my space for disagreeing with my author; but I have much enjoyed and profited from his book. He holds some of his opinions very strongly, and in consequence uses strong language, which is unfortunate when the opinions in question happen to be less than certain; but these are faults which movement in time and space will cure. He knows the relevant literature admirably, yet refrains from deluging the reader with learned references; he is critically alert and always ready to question orthodox opinion; he knows Greek well, and explains the text fully but concisely in a style which though sometimes rhetorical is not inflated. His book is by far the most useful treatment of this important text, and both deserves and will repay a careful study.

[2] [See now Alan Cameron, *Claudian: Poetry and Propaganda at The Court of Honorius* (1970), ch. 1.]

Sir Hugh Lloyd-Jones: Bibliography

SCHOLARLY BOOKS, ARTICLES, REVIEWS

[Bare chapter nos. refer to this volume; *GCHLGRM* = *Greek Comedy, Hellenistic Literature, Greek Religion, and Miscellanea*; T = Tragedy, C = Comedy, L = Lyric Poetry, H = Hellenistic Poetry, E = Epic, R = Religion and Intellectual History, M = Miscellanea]

Books

1957. Appendix (containing papyrus fragments) to *Aeschylus*, ed. H. W. Smyth (Loeb Classical Library). Harvard. T

1960. *Menandri Dyscolus* (Oxford Classical Text). C

1970. Tr. and ed. Aeschylus, *Agamemnon, Libation Bearers, and Eumenides* (published separately). Prentice-Hall. Reissued 1979 as *Oresteia*. Duckworth. T

1971. *The Justice of Zeus* (Sather Classical Lectures). Second Edition (with new Epilogue), 1983. University of California. R

1975. *Females of the Species* (An edition of Semonides of Amorgos' Satire on Women). Duckworth. L

1982. *Blood for the Ghosts* (Collected Essays). Duckworth. M

1982. *Classical Survivals* (Collected Essays). Duckworth. M

1983. *Supplementum Hellenisticum*, with P. J. Parsons. De Gruyter. H

1990. *Sophoclis Fabulae*, with N. G. Wilson (Oxford Classical Text). T

1990. *Sophoclea*, with N. G. Wilson. Oxford. T

Articles

1. (= ch. 22). 1952. 'The Robes of Iphigeneia', *CR*, NS 2, 132. T
2. 1952–3. 'The Gyges Fragment: A New Possibility', *PCPS* 182, 39 ff. T
3. 1953. 'Aeschylus, *Agamemnon* 146 f.', *CQ*, NS 3, 96. T
4. (= ch. 30). 1954. 'Sophoclea', *CQ*, NS 4, 91–5. T
5. 1955. 'A New Oracle in an Inscription', *JHS* 75, 155. M
6. (= ch. 18). 1956. 'Zeus in Aeschylus', *JHS* 76, 55–67 = H. Hommel (ed.), *Wege zu Aischylos* (Wege der Forschung 87, 1974), 265–300. T
7. (= ch. 31). 1957. 'Notes on Sophocles, *Antigone*', *CQ*, NS 7, 12–27. T
8. (= ch. 37). 1957. 'Euripidea', *CR*, NS 7, 92–7. T
9. (= *GCHLGRM*, ch. 1). 1958. 'Aristophanes, *Acharnians* 393', *CR*, NS 8, 14. C
10. 1959. 'The End of the *Seven against Thebes*', *CQ*, NS 9, 80–114. T
11. 1959. 'Preliminary Notes on Menander's *Dyskolos*,' *CR*, NS 9, 183–92. C
12. (= ch. 23). 1960. 'Three Notes on Aeschylus' Agamemnon', *RhM* 103, 76–80 (cf. *CR*, NS 22 (1972), 150). T

13. (= ch. 26). 1961. 'Some Alleged Interpolations in Aeschylus' *Choephori* and Euripides' *Electra*', *CQ*, NS 11, 171–84. T
14. (= *GCHLGRM*, ch. 38). 1961. 'Dinarchus, *In Philoclem* 4. 1 f.', *CR*, NS 11, 203.
15. (= ch. 21). 1962. 'The Guilt of Agamemnon', *CQ*, NS 12, 187–99. T
16. (= *GCHLGRM*, ch. 12). 1963. 'Callimachus, fr. 30 Pf.', *Glotta*, 41, 68–70. H
17. (= *GCHLGRM*, ch. 16). 1963. 'The Seal of Posidippus', *JHS* 83, 75–99 (cf. no. 20). H
18. (= *GCHLGRM*, ch. 17). 1963. 'Un nuovo frammento papiraceo dell'elegia ellenistica' (with J. W. B. Barns), *SIFC*, NS 35, 205–27 (cf. *SH* 970). H
19. 1963. 'P. Hamburg 118*b*: A Fragment from the Prologue of Euripides, *Hypsipyle*?', ap. G. W. Bond, *Euripides, Hypsipyle* (Oxford), 157–60. T
20. (= *GCHLGRM*, ch. 16, postscript) "The Seal of Posidippus: A Postscript,' *JHS* 84, 157. (cf. no. 18). H
21. (= *GCHLGRM*, ch. 9). 1964. 'A Fragment of New Comedy: P. Antinoopolis 15', with J. W. B. Barns, *JHS* 84, 21–34 (summary in *Actes du VIIe Congrès de l'Assoc. G. Budé*, 356). C
22. (= ch. 32). 1964. 'Sophocles, *Antigone* 1096–7 (and *Philoctetes* 324), *CR*, NS 14, 129–30. T
23. (= ch. 19). 1964. 'The "Supplices" of Aeschylus: the New Date and Old Problems', *AC* 33, 356–74 = H. Hommel (ed.), *Wege zu Aischylos* (Wege der Forschung 87, 1974), 101–24 = E. Segal (ed.), *Oxford Readings in Greek Tragedy* (Oxford, 1983), 42–56. T
24. 1964. 'SEG IX. 63 Again', *PCPS*, NS 10, 72–4. M
25. (= ch. 34). 1965. 'A Problem in the Tebtunis *Inachus*-Fragment', *CR*, NS 15, 241–3. T
26. (= *GCHLGRM*, ch. 18). 1965. "λίθος πολίτης", *CR*, NS 15, 246–7. H
27. 1964. 'Theocritus 14. 1–2 Once More', *Revue belge de philologie et d'histoire*, 42, 96–7. H
28. 1965. 'Propertianum', *Philol.* 109, 305–6. *Cancelled*: the conjecture was anticipated by A. Dieterich, *Kl. Schr.* (1911), 447–8, and see P. Fedeli, *Sesto Properzio: il primo libro delle elegie* (1980), 396–8. M
29. (= ch. 17). 1966. 'Problems of Early Greek Tragedy: Pratinas, Phrynichus, the Gyges Fragment', *Estudios sobre la tragedia griega* (Cuadernos de la Fundación Pastor, 13), Madrid, 11–33. T
30. (= *GCHLGRM*, ch. 6). 1966. 'Menander's *Sikyonios*', *GRBS* 7, 175–95. C
31. (= *GCHLGRM*, ch. 7). 1966. 'Notes on the Sikyonios of Menander', *Emerita*, 34, 131–49. C
32. 1966. 'Oracles of Apollo Kareios' (with M. L. West), *Maia*, NS 18, 262–4. M
33. 1966. 'P. Oxy. 2329. 3–4', *CR*, NS 16, 275. C
34. 1967. 'P. Oxy. 10', *CR*, NS 17, 9. C
35. (= ch. 11). 1967. 'Heracles at Eleusis: P. Oxy. 2622 and PSI 3891', *Maia*, NS 19, 206–29. L

36. (= *GCHLGRM*, ch. 13). 1967. 'Callimachus, fr. 191. 62', *CR* 17: 125–7 = A. Skiadas (ed.), *Kallimachos* (Darmstadt, 1975), 308–11. H
37. 1967. 'Sappho, fr. 111', *CQ*, NS 17, 168. L
38. 1968. 'Melanippides, fr. i, 1–2 (Page, PMG 757)', *Philol.* 112, 119. L
39. 1968. 'Again Meleager's Epigram on Heraclitus', *CR*, NS 18, 21. H
40. (= *GCHLGRM*, ch. 14). 1968. 'Callimachus, Fragments 260–1' (with John Rea), *HSCP* 72, 125–46. H
41. (= ch. 4) 1968. 'The Cologne Fragment of Alcaeus', *GRBS* 9, 125–39. L
42. (= ch. 24). 1969. 'Agamemnonea', *HSCP* 73, 97–104. T
43. 1969. 'Il Prometeo Incatenato di Eschilo', *Atti del III Congresso internazionale di studi sul dramma antico*, 211–18. T
44. 1970. 'Heracles or Dionysus?', *AJA* 74, 181 (cf. no. 35). L
45. 1971. 'La giustizia di Zeus nell' "Iliade"', *QUCC* 12, 62–91 = *The Justice of Zeus*, ch. 1. R
46. (= *GCHLGRM*, ch. 3). 1971. 'Menander's *Aspis*', *GRBS* 12, 175–95. C
47. 1971. 'Qualche considerazione a proposito dell'"Aspis" di Menandro', *Dioniso*, 45, 251–60. C
48. (= *GCHLGRM*, ch. 5). 1972. 'Menander's *Samia* in the Light of the New Evidence', *YCS* 22, 119–44. C
49. (= ch. 33). 1972. 'Notes on Sophocles' *Trachiniae*', *YCS* 22, 263–70. T
50. (= ch. 35). 1972. 'Tycho von Wilamowitz-Moellendorff on the Dramatic Technique of Sophocles', *CQ*, NS 22, 214–28 = *Blood for the Ghosts*, 219–37. T
51. (= ch. 10). 1972. 'Pindar, fr. 169', *HSCP* 76, 45–56. L
52. 1972. 'La Seconda Pitica di Pindaro', *Kokalos*, 18–19, 310–25 (cf. no. 53). L
53. (= ch. 9). 1973. 'Modern Interpretation of Pindar: The Second Pythian and Seventh Nemean Odes', *JHS* 93, 109–37. L
54. 1973. 'Stasinus and the *Cypria*', *Cτacῖνoc*, 4 (1968–72), 115–22. E[1]
55. (= *GCHLGRM*, ch. 8). 1973. 'Terentian Technique in the *Adelphi* and the *Eunuchus*', *CQ* 23, 279–84 (summary in *Actes du IXme Congrès de l'Assoc. G. Budé*, 404–11). C
56. (= *GCHLGRM*, ch. 4). 1974. 'Notes on Menander's *Perikeiromene*', *ZPE* 15, 209–13. C
57. 1974. 'Simonides, PMG 351', *CR*, NS 24, 1. L
58. 1974. 'Callimachus 191. 61–3 Again', *CR*, NS 24, 5. H
59. (= *GCHLGRM*, ch. 19). 1974. 'A New Hellenistic Fragment in the Archebulean Metre', *ZPE* 13: 209–13 (cf. *SH* 992). H
60. (= ch. 6). 1975. 'More about Antileon, Tyrant of Chalcis (Solon, frag. 33 and Aristophanes, *Eq.* 1042–4)', *CP* 70, 197. L
61. 1975. 'On the Orphic Tablet from Hipponion', *PP* 30, 225–6. Cf. no. 84. R
62. 1976. 'P. Oxy. 3151, Fr. 2 (Sophocles, Aiax Locrus)', *ZPE* 22, 40. T
63. 1977. 'P. Oxy. 14. P. Mich. Inv. 4761: A Fragment of Callimachus' Aitia', *ZPE* 26, 55–6. H
64. 1977. 'Callimachus fr. 46: An Addendum to P. Sorbonne Inv. 2248', *ZPE* 26, 57–8. H

65. (= *GCHLGRM*, ch. 25). 1977. 'Aristaios in Boeotia', *ZPE* 25, 135–6. R
66. (= *GCHLGRM*, ch. 26). 1978 'Iterum de "Catabasi Orphica" ' (with P. J. Parsons), in *Kyklos: Griechisches und Byzantinisches* (Festschrift R. Keydell), Berlin, 88–100. R
67. (= ch. 25). 1978. 'Ten Notes on Aeschylus' *Agamemnon*, *Dionysiaca* (Festschrift Denys Page), Cambridge, 45–61. T
68. 1979. 'The Michigan List of Metamorphosed Persons', *ZPE* 35, 6. M
69. (= ch. 38). 1980. 'Euripides, Medea, 1056–80', *Würzburger Jahrbücher*, NF 6a (Festschrift H. Erbse), 51–9. T
70. 1980. 'Stesicoro', *Magna Grecia* 15 (genn.-feb.), 3–8. L
71. 1980. 'Homère a-t-il écrit?', *Actes du Xme Congrès de l'Association G. Budé*, Paris: 116 (summary of no. 75). E
72. (ch. 28). 1981. 'Notes on P. Köln III. 125 (Aeschylus, Psychagogoi)', *ZPE* 42, 21–2. T
73. 1981. 'A Fragment of Achaeus?', *ZPE* 41: 27–8. *Cancelled*: cf. N. J. Richardson, *ZPE* 45 (1982). 59 f. T
74. (= *GCHLGRM*, ch. 2). 1981. 'Notes on P. Turner 4 (Aristophanes, *Ποίηϲιϲ*)', *ZPE* 42, 23–5. C
75. (= ch. 1). 1981. 'Remarks on the Homeric Question', in H. Lloyd-Jones, V. Pearl, and B. Worden (eds.), *History and Imagination (Essays Presented to H. R. Trevor-Roper)*. London, 15–29 (cf. no. 71). E
76. (= *GCHLGRM*, ch. 11). 1982. 'Callimachus, H. 4. 246', *Hermes*, 110, 119–20. H
77. (= ch. 7). 1982. 'Pindar', *PBA* 68, 139–63. L
78. (= *GCHLGRM*, ch. 24). 1983. 'Artemis and Iphigeneia', *JHS* 103, 87–102. T, R
79. (= *GCHLGRM*, ch. 21). 1984. 'A Hellenistic Miscellany', *SIFC*, 3rd ser., 2: 52–72 = *Actes du VIIme Congrès de la F.I.E.C.* i (Budapest, 1983), 265–79. H
80. (= ch. 2). 1984. 'The *Meropis* (*SH* 903a)', *Atti del XVII Congresso internazionale di papirologia*, Naples, i. 141–50. H
81. (= *GCHLGRM*, ch. 23). 1985. 'Psychoanalysis and the Study of the Ancient World', in P. Horden (ed.), *Freud and the Humanities*, London, 152–80.
82. 1985. 'Sophocles, Aiax 624 f.', *Catalepton* (Festschrift B. Wyss), Berne, 16–18. T
83. (= ch. 27). 1985. 'Aeschylus, *Choephori* 32–41', *BASP* 22 (in honour of W. H. Willis), 195–6. T
84. (= ch. 8). 1985. 'Pindar and the Afterlife', *EH* 31, 245–83. LR
85. (= ch. 5). 1987. "*Λυκαιχμίαιϲ*" (with Mary R. Lefkowitz), *ZPE* 68, 9–10. L
86. (= *GCHLGRM*, ch. 22). 1987. 'Ehre und Schande in der griechischen Kultur', *Antike und Abendland*, 33, 1–28. R
87. 1987. 'A Note on Homeric Morality', *CP* 82, 307–10. R
88. 1987. 'The Structure of Menander's Comedies', *Dioniso*, 57, 313–21. C
89. 1989. 'Les Erinyes dans la tragédie grecque', *REG* 102, 1–9 (summary in *Actes du XIIme Congrès de l'Assoc. G. Budé* (Paris, 1989), 255–6) = 'Erinyes

Semnai Theai, Eumenides', in *Owls to Athens: Essays in Classical Culture presented to Sir Kenneth Dover* (Oxford, 1990), 203–11. R

Forthcoming:

90. 'Cratinus, fr. 171 K.–A. (= Austin, *CFGPR*, fr. 73), 12–15', in *Miscellanea in onore di Giusto Monaco*. C

Reviews

R1. 1949. G. B. Grundy, *Thucydides and the History of his Age*: *Cambridge Review*, 70 (29 Jan.), 328. M

R2. 1949. W. H. D. Rouse, *The March Up Country*: *Cambridge Review*, 70 (5 Feb.), 357. M

R3. 1949. A. M. Dale, *The Lyric Metres of Greek Drama*: *Cambridge Review*, 70 (21 May), 602. T

R4. 1951. D. W. Lucas, *The Greek Tragic Poets*: *Cambridge Journal*, 5, 59–64. T

R5. 1951. A. Turyn, *Pindari Carmina*: *JHS* 71, 267–8. L

R6. 1953. A. S. F. Gow, *Bucolici Graeci*: *Cambridge Review*, 74 (28 Feb.), 354. H

R7. 1953. A. S. F. Gow, *The Greek Bucolic Poets: A Translation into English Prose*: *Cambridge Review*, 74 (9 May), 480. H

R8. 1953. A. S. F. Gow and A. F. Scholfield, Nicander: *The Poems and Poetical Fragments*: *Cambridge Review*, 75 (21 Sept.), 162. H

R9. 1954. A. S. F. Gow and A. F. Scholfield, *Nicander: The Poems and Poetical Fragments*: *CR*, NS 4, 231–3. H

R10. 1954. L. H. G. Greenwood, *Aspects of Euripidean Tragedy*: *Cambridge Journal*, 7, 251–4. T

R11. 1954. E. T. Owen, *The Harmony of Aeschylus*: *JHS* 74, 195. T

R12. 1954. A. Maddalena, *Interpretazioni eschilee*: *JHS* 74, 237. T

R13. 1954. G. Baglìo, *Il 'Prometeo' di Eschilo alla luce della storia di Erodoto*: *JHS* 74, 237–8. T

R14. (= *GCHLGRM*, ch. 30). 1955. C. M. Bowra, *Problems in Greek Poetry*: *JHS* 158–9. M

R15. 1956. A. Dain and P. Mazon, *Sophocle I (Les Trachiniennes, Antigone)*: *Gnomon*, 28, 105–10. T

R16. 1956. H. D. F. Kitto, *Form and Meaning in Drama*: *Cambridge Review*, 78 (3 Nov.), 127–9. T

R17. 1956. J. C. Kamerbeek, *The Plays of Sophocles. Commentaries*, Part I. *Ajax*: *JHS* 76, 111–12. T

R18. 1957. T. B. L. Webster, *Greek Theatre Production*: *CR*, NS 7, 111–13. M

R19. (= *GCHLGRM*, ch. 28). 1957. Ernst Siegmann, *Literarische griechische Texte der Heidelberger Papyrussammlung*: *Gnomon*, 29, 425–7. M

R20. 1958. A. Turyn, *The Byzantine Manuscript Tradition of the Plays of Eruipides*: *Gnomon*, 30, 503–10. M

R21. 1958. E. Lobel, *The Oxyrhynchus Papyri*, XXIII: *CR*, NS 8, 16–22. M

R22. 1959. Wiktor Steffen, *Studia Aeschylea praecipue ad deperditarum fabularum fragmenta pertinentia*: *CR*, NS 9, 284–5. T

R23. 1959. H. J. Rose, *A Commentary on the Surviving Plays of Aeschylus* (2 vols.): *JHS* 79, 162–3. T

R24. (= *GCHLGRM*, ch. 29). 1959. V. Bartoletti, *Papiri greci e latini della Società Italiana*, XIV: *Gnomon*, 31, 109–14. M
R25. 1959. C. A. Trypanis, *Callimachus* (Loeb Classical Library): *CR*, NS 9, 244–6. H
R26. 1959. A. Dain and P. Mazon, *Sophocle II (Ajax, Œdipe Roi, Electre)*: *Gnomon* 31, 478–81. T
R27. 1960. R. Pfeiffer, *Ein neues Inachos-Fragment des Sophokles* (*SB der Bayer. Akad. der Wissenschaften*, phil.-hist. Kl. 1958, 6): *CR*, NS 10, 25–6. T
R28. 1960. A. Körte, *Menandri quae supersunt, pars altera*, rev. A. Thierfelder: *CR*, NS 10, 256. C
R29. 1961. W. Bühler, *Die Europa des Moschos* (*Hermes* Einzelschriften 13): *Gnomon*, 33, 36–8. H
R30. 1961. B. Gentili, *Anacreon*: *Gnomon*, 33, 718–19. L
R31. 1961. A. Dain and P. Mazon, *Sophocle III (Philoctète, Œdipe à Colone)*: *Gnomon*, 33, 544–6. T
R32. 1961. E. Lobel and E. G. Turner, *The Oxyrhynchus Papyri*, XXV: *CR*, NS 11, 17–21. L
R33. 1961. H. J. Mette, *Die Fragmente der Tragödien des Aischylos*: *CR*, NS 11, 15–16. T
R34. 1961. R. Pfeiffer, *Ausgewählte kleine Schriften*: *CR*, NS 11, 125–7. M
R35. 1961. R. Keydell, *Nonni Panopolitani Dionysiaca* (2 vols.): *CR*, NS 22–4. H
R36. (= ch. 12). 1962. K. von Fritz, *Antike und moderne Tragödie*: *Gnomon*, 34, 737–47. T
R37. 1962. A. Körte, *Die hellenistische Dichtung* (Zweite, vollständig neubearbeitete Auflage von Paul Händel): *CR*, NS 12, 245–6. H
R38. 1962. F. Vian, *Histoire de la tradition manuscrite de Quintus de Smyrne, Recherches sur les* Posthomerica *de Quintus de Smyrne*: *CR*, NS 12, 32–3. H
R39. 1962. K. Reinhardt, *Tradition und Geist*: *CR*, NS 12, 243–5. M
R40. 1963. F. Vian, *Apollonius de Rhodes, Argonautiques, chant III*, *CR*, NS 13, 156. H
R41. 1963. E. G. Turner, John Rea, L. Koenen, and J. M. F. Pomar, *The Oxyrhynchus Papyri*, XXVII: *Gnomon*, 35, 433–55. M
R42. 1963. U. Schindel, *Demosthenes im 18. Jahrhundert: Zehn Kapitel zum Nachleben des Demosthenes in Deutschland, Frankreich, England*: *Gnomon*, 35, 831–2. M
R43. 1964. D. L. Page, *Poetae Melici Graeci*: *CR*, NS 14, 16–19. L
R44. 1964. Hermann Fränkel, *Dichtung und Philosophie des frühen Griechentums* (2nd ed.): *CR*, NS 14, 209–10. M
R45. 1964. L. Massa Positano, *Demetrii Triclinii in Aeschyli Persas Scholia* (2nd edn.): *CR*, NS 14, 210. T
R46. 1964. H. J. Mette, *Der verlorene Aischylos*: *CR*, NS 14, 244–7. T
R47. (= ch. 13). 1964. John Jones, *On Aristotle and Greek Tragedy*: *Review of English Studies*, 15/58, 221–4 = *Classical Survivals*, 111–15. T
R48. 1964. Ernst Buschor, *Über das griechische Drama*: *Gnomon*, 36, 418–19. T
R49. 1965. Denys Page, *The Oxyrhynchus Papyri, Part XXIX*: *CR*, NS 15, 70–2. M

R50. (= ch. 3). 1964. J. Pouilloux. *et al.*, *Archiloque* (*EH* 10): *CR*, NS 15, 263–7. L
R51. 1965. F. Vian, *Les Origines de Thèbes*: *CR*, NS 15, 322–3. R
R52. 1965. R. D. Dawe, *Repertory of Conjectures on Aeschylus*: *Gnomon*, 37, 656–7. T
R53. (= ch. 36). 1965. W. S. Barrett, *Euripides, Hippolytos*: *JHS* 85, 164–71. T
R54. 1966. R. D. Dawe, *The Collation and Investigation of the Manuscripts of Aeschylus*: *CR*, NS 16, 18–21. T
R55. 1966. G. Zuntz, *An Inquiry into the Transmission of the Plays of Euripides*: *CR*, NS 16, 156–9. T
R56. (= ch. 14). 1966. Bruno Snell, *Scenes from Greek Drama*: *Gnomon*, 38, 12–17. T
R57. 1967. W. Kiefner, *Der religiöse Allbegriff des Aischylos*: *CR*, NS 17, 104–5. T
R58. (= *GCHLGRM*, ch. 31). 1967. Bruno Snell, *Gesammelte Schriften*: *CR*, NS 17, 214–17. M
R59. 1967. Edith Mora, *Sappho*: *CR*, NS 17, 269–71. L
R60. 1967. F. R. Adrados, *Ésquilo, Tragedias*: *CR*, NS 17, 271–2. T
R61. 1967. F. Vian, *Quintus de Smyrne, La suite d'Homère*, vol. ii: *CR*, NS 17, 275–6. H
R62. 1967. A. Lesky, *Gesammelte Schriften*: *CR*, NS 17, 404. M
R63. 1967. A. J. Podlecki, *The Political Background of Aeschylus*: *Phoenix* 21, 222–4. T
R64. 1968. F. Mosino, *Ibico*: *CR*, NS 18. 107. L
R65. 1968. G. Capovilla, *Callimaco*, vol. i: *Gnomon*, 40, 505–6. H
R66. 1969. G. Capovilla, *Callimaco*, vol. ii: *Gnomon*, 41, 205–6. H
R67. 1969. D. L. Page, *Lyrica Graeca Selecta*: *CR*, NS 19, 20–1. L
R68. 1969. D. A. Campbell, *Greek Lyric Poetry*: *CR*, NS 19, 22–4. L
R69. 1969. Gerhard Müller, *Sophokles, Antigone*: *CR*, NS 19, 25–30. T
R70. 1969. Hans Diller, *Sophokles* (Wege der Forschung 95): *CR*, NS 19, 30–4. T
R71. 1969. A. A. Long, *Language and Thought in Sophocles: A Study of Abstract Nouns and Poetic Techniques*: *CR*, NS 19, 34–5. T
R72. 1969. Armin Vögler, *Vergleichende Studien zur sophokleischen und euripideischen Elektra*: *CR*, NS 19, 36–8. T
R73. 1969. L. Di Gregorio, *Le scene d'annuncio nella tragedia greca*: *CR*, NS 19, 38–9. T
R74. 1969. K. Latte, *Hesychii Alexandrini Lexicon*, vol. ii: *CR*, NS 19, 50–1. M
R75. 1969. R. Führer, *Formproblem-Untersuchungen zu den Reden in der frühgriechischen Lyrik*: *CR*, NS 19, 100. L
R76. 1969. F. M. Combellack, *The War at Troy: What Homer Didn't Tell*: *CR*, NS 19, 101–2. H
R77. (= *GCHLGRM*, ch. 32). 1969.K. Latte, *Kleine Schriften*: *CR*, NS 19, 187–9. M
R78. 1969. B. Gygli-Wyss, *Das nominale Polyptoton im älteren Griechisch*: *CR*, NS 19, 243. M
R79. 1970. Rose Unterberger, *Der Gefesselte Prometheus des Aischylos: Eine Interpretation*: *CR*, NS 20, 241–2. T

R80. 1970. L. Ferrari, *I drammi perduti di Eschilo*: *CR*, NS 20, 294–6. T

R81. 1970. T. B. L. Webster, *An Introduction to Sophocles*, 2nd edn.: *CR*, NS 20, 299–300. T

R82. (= ch. 16). 1970. Jacqueline de Romilly, *Time in Greek Tragedy*: *CR*, NS 20, 302–4. T

R83. 1970. H. D. Broadhead, *Tragica: Elucidations of Passages in Greek Tragedy*: *CR*, NS 20, 304–5. T

R84. 1970. M. Fuhrmann, *Die Antike und ihre Vermittler*: *CR*, NS 20, 384–7. M

R85. 1970. L. Ferrari, *Congetture stesicoree*: *CR*, NS 20, 398. L

R86. 1970. M. J. O'Brien (ed.), *Twentieth-Century Interpretations of* Oedipus Rex: *CR*, NS 20, 398–9. T

R87. 1971. E. A. Havelock, *Prometheus, with a Translation of Prometheus Bound*: *CR*, NS 21, 288. T

R88. (= ch. 39). 1971. J. Diggle, *Euripides, Phaethon*: *CR*, NS 21, 341–5. T

R89. 1971. Willy Theiler, *Untersuchungen zur antiken Literatur*: *CR*, NS 21, 405–7. M

R90. 1971. A. M. Dale, *Collected Papers*: *CR*, NS 21, 407–9. M

R91. 1971. Paul Friedländer, *Studien zur antiken Literatur und Kunst*: *CR*, NS 21, 409–11. M

R92. 1972. W. Peek, *Lexikon zu den Dionysiaka des Nonnos*, I. Lieferung: *Gnomon*, 44, 717–18. H

R93. (= ch. 20). 1972. Anne Lebeck, *The* Oresteia: *A Study in Language and Structure*: *JHS* 92, 193–5. T

R94. 1972. T. B. L. Webster, *Sophocles, Philoctetes*: *CR*, NS 22, 102. T

R95. 1972. G. Grossmann, *Prometheus und Orestie: Attischer Geist in der attischen Tragödie*: *CR*, NS 22, 178–80. T

R96. 1972. F. Dirlmeier, *Ausgewählte Schriften zur Dichtung und Philosophie der Griechen*: *CR*, NS 22, 233–5. M

R97. 1972. K. A. Pfeiff, *Sophokles, König Ödipus*: *CR*, NS 22, 235.

R98. 1972. T. B. L. Webster, *Greek Theatre Production* (2nd edn.): *CR*, NS 22, 280. T

R99. (= ch. 15). 1973. Bruno Snell, *Szenen aus griechischen Dramen*: *CR*, NS 23, 192–4. T

R100. 1973. Hans Diller, *Kleine Schriften zur antiken Literatur*: *CR*, NS 23, 198–200. M

R101. (= *GCHLGRM*, ch. 27). 1973. J. de Romilly, *La Loi dans la pensée grecque*: *JHS* 93, 243–4. R

R102. 1974. Alexander Sideras, *Aeschylus Homericus*: *Gnomon*, 46, 507–8. T

R103. 1974. J. T. Kakridis, *Μελέτες καὶ Ἄρθρα*: *CR*, NS 24, 164–5. M

R104. 1974. F. Sbordone, *Scritti di varia filologia*: *CR*, NS 24, 150. M

R105. 1974. G. Zuntz, *Opuscula selecta: Classica, Hellenistica, Christiana*: *CR*, NS 24, 317–18. M

R106. 1975. J. H. Kells, *Sophocles, Electra*: *CR*, NS 25, 10–12. T

R107. (= *GCHLGRM*, ch. 33). 1975. P. Maas, *Kleine Schriften*: *CR*, NS 25, 138–40. M

R108. (= ch. *GCHLGRM*, ch. 36). 1976. E. J. Kenney, *The Classical Text*; *Notes and Queries*, Aug. 1976, 360–2. M

R109. 1976. V. J. Matthews, *Panyassis of Halicarnassus: Text and Commentary*: *Gnomon*, 48, 504–6. E

R110. (= ch. 29). 1976. Denys Page, *Aeschyli septem quae supersunt tragoediae*: *CR*, NS 26, 6–8. T

R111. (= *GCHLGRM*, ch. 10). 1976. Jeffrey Henderson, *The Maculate Muse: Obscene Language in Attic Comedy*: *CP* 71, 356–9. C

R112. 1978. J. C. Kamerbeek, *The Plays of Sophocles*, Part V: *Electra*: *CR*, NS 28, 221–3. T

R113. 1978. K. Wilkens, *Die Interdependenz zwischen Tragödienstruktur und Theologie bei Aischylos*: *CR*, NS 28, 213–14. T

R114. 1978. R. D. Dawe, *Studies in the Text of Sophocles*, vol. i: The Manuscripts and the Text; ii, *The Collations*. R. D. Dawe, *Sophoclis Tragoediae*, vol. i: *Ajax, Electra, Oedipus Rex*. A. Colonna, *Sophoclis Fabulae*, vol. i: Ajax, Electra: *CR*, NS 28, 214–21. T

R115. 1978. L. A. Cuenca, *Euforión de Calcis*: *CR*, NS 28, 228–9. H

R116. 1978. R. H. Beck, *Aeschylus: Playwright, Educator*: *CR*, NS 28, 342–3. T

R117. (= *GCHLGRM*, ch. 15). 1979. B. A. van Groningen, *Euphorion*: *CR*, NS 29, 14–17. H

R118. 1981. S. Radt, *Tragicorum Graecorum Fragmenta*, vol. iv: Sophocles: *CR*, NS 31, 175–8. T

R119. 1981. R. D. Dawe, *Studies in the Text of Sophocles*, vol. ii: *Women of Trachis, Antigone, Philoctetes, Oedipus at Colonus*; R. D. Dawe, *Sophoclis Tragoediae*, vol. ii: *Trachiniae, Antigone, Philoctetes, Oedipus Coloneus*; A. Colonna, *Sophoclis Fabulae*, vol. ii: *Oedipus Tyrannus, Antigone, Trachiniae*: *CR*, NS 31, 167–73. T

R120. 1981. J. C. Kamerbeek, *The Plays of Sophocles*, Part III: *Antigone*. Part VI: *Philoctetes*: *CR*, NS 31, 173–5. T

R121. 1982. E. A. Havelock, *The Greek Concept of Justice, from its Shadow in Homer to its Substance in Plato*: *JHS* 102, 258–9. R

R122. (= *GCHLGRM*, ch. 20). 1982. D. L. Page, *Further Greek Epigrams: Epigrams before* AD *50 from the Greek Anthology and other Sources, not included in 'Hellenistic Epigrams' or the 'Garland of Philip'*, rev. and prepared for publication by R. D. Dawe and J. Diggle: *CR*, NS 32, 139–44. H

R123. 1983. F. Bornmann and M. Carpitella, *Friedrich Nietzsche, Philologische Schriften* (1867–73), *Nietzsches Werke: Kritische Gesamtausgabe*, hrsg. von G. Colli und M. Montinari. Zweite Abteilung: Band i: *CR*, NS 33, 375. R

R124. 1983. G. Paduano, *Tragedie e frammenti di Sofocle*, 2 vols.: *CR*, NS 33, 311. T

R125. 1983. A. C. Moorhouse, *The Syntax of Sophocles*: *CR*, NS 33, 171–3. T

R126. 1983. M. I. Finley (ed.) *The Legacy of Greece*: *History*, 68, 92–3. M

R127. 1984. J. S. Lasso de la Vega, *Karl Reinhardt y la filología clásica en el siglo XX* (Cuadernos de la Fundación Pastor, 30): *CR*, NS 34, 365–6. M

R128. 1984. C. H. Roberts and T. C. Skeat, *The Birth of the Codex*: *Journal of Ecclesiastical History*, 35, 499. M

R129. (= *GCHLGRM*, ch. 34). 1985. W. Kraus, *Aus Allem Eines: Studien zur antiken Geistesgeschichte*. Previously unpublished. M

R130. 1986. D. E. Gerber, ed., *Greek Poetry and Philosophy: Studies in Honour of Leonard Woodbury*: *Phoenix*, 40, 275–80. M

R131. (= *GCHLGRM*, ch. 37). 1986. W. M. Calder III, H. Flashar, T. Lindken (eds.), *Wilamowitz nach 50 Jahren*: *CR*, NS 36, 295–300. M

R132. 1986. W. W. Briggs, H. W. Benario (eds.), *Basil Lanneau Gildersleeve: An American Classicist*: *CR*, NS 36, 357–8. M

R133. 1986. R. D. Dawe, *Sophoclis Tragoediae*, vol. i (2nd edn.): *Ajax, Electra, Oedipus Rex*: *CR*, NS 36, 10–12. T

R134. 1986. R. D. Dawe, *Sophoclis Tragoediae*, vol. ii (2nd edn.): *Trachiniae, Antigone, Philoctetes, Oedipus Coloneus*: *CR*, NS 36, 305. M

R135. 1987. S. Radt, *Tragicorum Graecorum Fragmenta*, vol. iii: Aeschylus: *CR*, NS 37, 142–5. T

R136. 1987. W. M. Calder III, A. Köhnken, W. Kullmann, G. Pflug (eds.), *Friedrich Gottlieb Welcker*: *Werk und Wirkung* (*Hermes*, Einzelschriften, 49): *CR*, NS 37, 294–6. M

R137. 1987. J. C. Kamerbeek, *The Plays of Sophocles*, Part VII: *Oedipus Coloneus*: *CR*, NS 37, 148–50. T

R138. (= *GCHLGRM*, ch. 35). 1987. *The Cambridge History of Classical Literature*, vol. i: *Greek Literature*, eds. P. E. Easterling and B. M. W. Knox: *CR*, NS 82, 255–65. M

R139. 1988. Roland Kany, *Mnemosyne als Programm: Geschichte, Erinnerung und die Andacht zum Unbedeutenden im Werk von Usener, Warburg und Benjamin*: *CR*, NS 38, 136–8. M

R140. 1988. K. J. Dover, *Greek and the Greeks: Collected Papers*, vol. I: *Language, Poetry, Drama*: *CR*, NS 38, 377–8. M

R141. 1989. K. J. Dover, *The Greeks and Their Legacy: Collected Papers*, vol. II: *Prose Literature, History, Society, Transmission, Influence*: *CR*, NS 39, 370–2. M

R142. 1989. A. C. Dionisotti, A. Grafton, J. Kraye (eds.), *The Uses of Greek and Latin: Historical Essays* (Warburg Institute Essays and Texts, 1–6): *CR*, NS 39, 374–6. M

Indexes

AUTHORS AND TEXTS

GREEK WORDS

MODERN AUTHORS

GENERAL